Himalayan Forests in a Changing World

S. P. Singh • Rajesh Thadani
Editors

Himalayan Forests in a Changing World

Volume I: Understanding Structure, Function and Environmental Drivers

Springer

Editors
S. P. Singh
INSA Honorary Scientist
Dehradun, Uttarakhand, India

Centre for Himalayan Studies
Graphic Era (deemed to be) University
Dehradun, Uttarakhand, India

Rajesh Thadani
Centre for Ecology Development
and Research (CEDAR)
New Delhi, India

Department of Botany
Jai Hind College
Churchgate, Mumbai, India

ISBN 978-981-95-0977-5 ISBN 978-981-95-0978-2 (eBook)
https://doi.org/10.1007/978-981-95-0978-2

This Springer imprint is published by the registered company Springer Nature Singapore Pte Ltd.
The registered company address is: 152 Beach Road, #21-01/04 Gateway East, Singapore 189721, Singapore

Dedication

Professor J. S. Singh

An eminent ecologist and visionary whose groundbreaking work laid the foundation of contemporary Himalayan Forest ecology. His ongoing inspiration continues to shape future generations and deepen our collective appreciation for the forests

Foreword

It is with pleasure that I introduce this book. It is an encyclopedic description of the forests of the Indian Himalaya. The book is largely the masterpiece of two colleagues who I have known for over 30 years—Professor S. P. Singh and Dr. Rajesh Thadani. Rajesh was my doctoral student who for his master's work and his PhD thesis studied the physiological ecology or oak-pine forests and their co-existence in the middle hills of Uttarakhand. Rajesh has continued this work with a devotion to the Himalaya's through his leadership in issues around conservation and development of the region. Professor S. P. Singh has been working on the ecology of the Himalaya for over 50 years most of which he was at the Department of Botany, Kumaun University, Nainital.

This book can perhaps be regarded as a completely new, up-to-date compilation of work which builds upon a prior book published in 1992 and an earlier general paper titled "Forest Vegetation of the Himalaya" published in *Botanical Review* in 1987, which were authored by J.S. Singh and S. P. Singh and then subsequently a multitude of research publications, some of which more recently came out as an edited text specific to the Himalayan tree line also published by Springer (2024). Together Singh and Thadani have compiled a broad but comprehensive descriptive analysis of the Himalayan forests for this book.

The book itself can be considered as a foundational text that comprises 24 chapters. It starts first with an introduction comprising two chapters describing the physical geography and climate of the Himalayas. This is followed by the core information of the book. It is arranged in 15 chapters that focus on what can be thought of as broad description of the ecology of Himalayan forests. These chapters range from subject areas as small and refined as seeds and germination of Himalayan trees and flora more generally, to the biology of tree species, and their arrangements as forest communities. It includes chapters on the physiology and phenology trees, and their productivity as stands and forests. And it includes chapters on the water relations and ecophysiology of trees and forests that scale to the hydrology of these forests as watersheds. At the end of this section of the book, the last two chapters focus on the fauna—its abundance and diversity in relation to forest type and habitat—especially for birds and mammals.

The book concludes with seven chapters. The first four cover the human impacts under topics of climate and climate change; fire and its impacts and management;

and agriculture both permanent and as a transient form of swidden. The last three chapters describe the nature of Himalayan Forest degradation and recovery, silviculture as a technique for restoration and sustainable management, and the value of Himalayan forests for their provisioning of ecosystem services. The unit of study ranges from leaves to Himalayan forests as a whole which is extraordinarily wide and interesting. The editors have made an attempt to provide a leaf-based classification of Himalayan forests which also draws attention to the coniferous forests of dry inner Himalayan ranges which generally are ignored.

The time for such a book is now given the voluminous amount of literature that has been generated by forest researchers on the trees and forests of the Himalaya. Such a book could not have been written with any comprehensive degree of understanding years ago when S. P Singh started on his journey of forest research and discovery for this region. Now it can.

This book needs to become every Indian forester's reference text for looking up information on the ecology and sustainable managed of these forests. I suspect that the text will also become a core coursebook text for university students who are studying botany or forestry. The book is both nicely written within an organizational framework that is logical and easy to understand. It includes a comprehensive set of citations available for the reader to inquire more deeply into a particular chapter and topic of interest.

In summary, I commend the authors and editors of this book. Kudos to their perseverance and tenacity and making this happen. They have done a marvelous job at organizing all this material and compiling it into something that is very readable.

Sincerely,

Mark S. Ashton
Morris K. Jesup Professor of Silviculture and Forest Ecology,
Director of Yale Forests, Senior Associate Dean for
The Forest School
Yale School of Environment
New Haven, CT, USA
May 25th 2025

Preface

The Himalayan mountain system represents one of Earth's most remarkable biogeographical features, spanning 2,500 kilometers across five nations and hosting a broad range of forest ecosystems. Himalayan forests range from evergreen rainforests in the east, to arid coniferous formations in the inner valleys of the west; from dipterocarp forests of sal (*Shorea*) in the foothills, to alpine birch-rhododendron communities near the treeline. The region is a center of pine (*Pinus*) and oak (*Quercus*), the most important forest forming genera of the mid-latitudes of the Northern Hemisphere. The exceptional ecological gradient, compressed within short geographical distances, reflects the complex interplay of elevation, aspect, precipitation, and topographical variations. Despite their ecological significance, Himalayan forests remain surprisingly understudied relative to their global importance.

Himalayan Forests in a Changing World—Volume 1: Understanding Structure, Function, and Environmental Drivers emerged from the recognition that comprehensive documentation of these ecosystems' basic ecological processes is essential before we can fully understand their responses to environmental change. The Himalayan region is experiencing climate change at rates exceeding global averages, with altered precipitation patterns, temperature regimes, and an increased frequency of extreme events. To understand these impacts, we must first establish a solid foundation of knowledge about how these forests function under current conditions.

This book is based on research carried out during the last three decades, but it draws inspiration from many past efforts, particularly the monographic book *Forests of the Himalaya* by Singh J. S. and Singh S. P. (1992). Additionally, our understanding of Himalayan forests was greatly benefitted by classical works, such as Troup (1921), Osmaston (1922), Champion and Seth (1968), Stainton (1978), and Ohsawa (1987). In addition, we acknowledge the recent volume on Nepal by Miehe and Pendry (2015). Our efforts are built on the shoulders of these giants of Himalayan ecology and forestry.

This first volume brings together current scientific understanding of the foundational aspects of Himalayan forests through multiple disciplinary lenses. Beginning with essential chapters on environmental background and climate dynamics, the volume establishes the physical framework within which these forests operate. The introduction of a novel leaf-based classification of Himalayan forests provides a fresh perspective on forest categorization, complementing traditional approaches and offering new insights into forest community organization.

Central to this volume is the comprehensive examination of forest structure and community associations across elevational gradients. Detailed coverage of forest phenology reveals the temporal dynamics that govern ecosystem functioning, while a thorough analysis of nutrient cycling illuminates the biogeochemical processes that sustain forest productivity. The integration of biomass, carbon dynamics, and productivity measurements provides quantitative understanding of ecosystem function, enhanced by cutting-edge remote sensing approaches to photosynthetic characteristics.

Special emphasis is placed on hydrological processes, recognizing water as the fundamental driver of forest distribution and function in these montane environments. The detailed examination of tree water relations in monsoonal forests addresses a critical knowledge gap, while comprehensive analysis of forest hydrological processes provides essential understanding for water resource management. The investigation of community structure and ecophysiology responses to slope aspects demonstrates how topographic variation creates the mosaic of forest type's characteristic of the Himalayas. The inclusion of tree seed ecology provides crucial understanding of regeneration processes.

Evergreen species with about a one-year leaf longevity, and simultaneous leafing and leaf shedding during the pre-monsoon (such as many oaks, chir pine) dominate over large areas. This volume gives particular prominence to ecological attributes of such forests, which are linked to the monsoon pattern of precipitation, and thus helps address a global gap in knowledge about monsoonal forest ecosystems.

What distinguishes this work is its synthesis of empirical research across multiple disciplines, unified by a systems perspective that transcends traditional academic boundaries. The contributors include leading authorities in forest ecology, hydrology, ecophysiology, and remote sensing, many with extensive Himalayan field experience. This familiarity with the system and knowledge derived from it distinguishes this volume from most general treatments of mountain forest ecology.

This volume targets forest researchers, ecosystem ecologists, graduate students, and environmental scientists seeking a comprehensive understanding of Himalayan forest structure and function. It establishes the scientific foundation necessary for the analysis of forest responses to environmental change presented in Volume 2, while providing essential baseline knowledge for informed forest management and conservation decisions. By collating scattered research and enhancing understanding of ecological processes, this volume serves as the definitive reference for understanding how Himalayan forests operate in their unique environmental context.

Dehradun, India S. P. Singh
New Delhi, India Rajesh Thadani

Acknowledgments

The conceptual framework of this book on Himalayan forests took shape over many years and was a truly joint effort. We had many discussions on its structure, objectives, possible contributors, and the thematic focus. Living in the Himalayas proved invaluable, as we could frequently interact with ecologists, conservation specialists, and local people who significantly informed our understanding of these magnificent mountain ecosystems and their human dimensions.

S. P. Singh: I owe a lot to my many research scholars and associates. The list is long, and includes Dr. G.C.S Negi, Dr. Subrat Sharma, and Dr. Ashish Tewari who are among those who have been recently associated. Ripu Daman Singh and Surabhi Gumber (my current research scholars) helped me immensely in processing the manuscript. Thanks to Rajesh Khanna for allowing me access to data from his Ph. D. thesis which helped improve the chapter on Nutrient cycling. I received valuable institutional support from Indian National Science Academy (INSA), Central Himalayan Environment Association (CHEA), Appropriate Technology India (AT India), and Graphic Era (deemed to be) University, particularly its Chair, Dr. Kamal Ghansala. I owe a lot to Dr. R.P. Singh (Director) and Dr. Hitendra Padalia of IIRS, Dr. S.P. Sati of Bharsar University, Vishal Singh of CEDAR, Rajesh Naithani (AT India), Rajiv Pandey of ICFRE, Gajendra Singh of USAC, Prof. Surekha Dangwal (VC of Doon University), Durgesh Pant (DG, UCOST, GoUK), and Prof. B.K. Joshi (ex VC of Kumaon University) for enabling me to sustain an academic life heavily oriented to Himalayas, so necessary for developing a research book on Himalayas. All of them are Dehradun based and keen students of "Himalayan Ecology." Our "mini-seminars" provided me with many insights and the benefit of their field learnings. I have always been able to depend on my co-editor Dr. Rajesh Thadani, both for his intellectual and managerial acumen and support, which enabled me to improve my contributions to this book.

Rajesh Thadani: I express my deep gratitude to the late Prof. Graeme P. Berlyn of Yale University, whose guidance helped me grasp the complexities of plant physiology and anatomy. My gratitude also to Prof. Mark Ashton who taught me a lot about trees. I extend my sincere thanks to my colleagues at CEDAR, especially Dr. Vishal Singh, for invaluable support, and to Naren Raikwal for his dedicated assistance in the field. I am indebted to my wife, Ritika, for her thoughtful help with editing, and to my son, Rishnav, for his constant encouragement. Finally, I would like to acknowledge and thank my mentor and collaborator, Professor S.P. Singh, who persisted and ensured that I never lost my connection with Himalayan forests even as I got sidetracked with many other endeavors over the years. This book is the realization of his constant perseverance.

Contents

About the Editors

S. P. Singh is easily among the top Himalayan ecologists. His research systems range from a leaf blade to inherently fragile and disaster-prone Himalayan landscapes and regions. He along with his associates have studied almost all dimensions of the Himalayan forest ecosystems, including biomass and productivity, nutrient cycling, carbon sequestration, phenology, tree water relations, forest fire adaptation, and ecosystem services. He is widely known for his research on pine and oak forest ecosystem.

Coordinating a team of researchers, he has recently added critical knowledge on the Himalayan treelines. He developed the concept of incorporating Himalayan ecosystem services in national accounting at the country level.

National academies, such as Indian National Science Academy (INSA), have acknowledged Prof. Singh's work by recognizing him their Fellow. He has published ~250 research articles, supervised 35 doctoral students, and coauthored a major book on the ecology of Himalayan forests way back in 1992. He held several major positions, such as Vice Chancellor of Garhwal University, and advisor, Planning Commission.

Rajesh Thadani works on regeneration ecology, impacts of chronic disturbance, and ecosystem services emanating from Himalayan forests. He has an MF and Ph.D. in Forest ecology and physiology from the Yale School of the Environment and is particularly interested in the oak-pine forests of the central Himalayas. He is the Founder Executive Director, and currently a Senior Fellow at the Centre for Ecology Development and Research (CEDAR), a research and advocacy organization, and Chair of Himmotthan and North East Initiative Development Agency (NEIDA), which together promote sustainable rural livelihoods in over 2,000 villages across the Indian Himalayas from Ladakh in the west to Arunachal in the east. He helped establish and is Chair of Kumaon Grameen Udyog, a Himalayan social enterprise, and is the President of Jai Hind College in Mumbai.

1 Introduction and Environmental Background

Rajesh Thadani

Abstract

The Himalayas are the youngest and highest mountain system on Earth, spanning over 2400 kilometres across five countries. These mountains harbour some of Asia's most diverse forest ecosystems with high endemism. These mountains emerged from the collision of the Indian and Eurasian plates beginning about 55 million years ago, creating a complex mosaic of parallel ranges including the Siwaliks, Lesser Himalaya, and Greater Himalaya, separated by major thrust systems. The forest ecosystems range from subtropical dipterocarp forests to temperate oak, pine, and rhododendron, and finally subalpine fir and birch forests near the treeline—compressing ecosystems that would cover 5000 kilometres of latitudinal displacement into a mere hundred kilometres. The region's geological youth manifests in active seismic activity, landslides, and erosion processes that continuously reshape forest landscapes. Historical forest development was profoundly influenced by changing elevations and rainfall patterns as these mountains rose and were influenced by multiple glaciation events. These forested mountains function as Asia's water towers and climate regulators, while supporting diverse human communities that have adapted over millennia to this region's conditions. Understanding these complex relationships is essential for forest conservation in the face of increasing anthropogenic pressures and climate change.

R. Thadani (✉)
Centre for Ecology Development and Research (CEDAR), New Delhi, India

Department of Botany, Jai Hind College, Churchgate, Mumbai, India
e-mail: thadani_rajesh@hotmail.com

S. P. Singh, R. Thadani (eds.), *Himalayan Forests in a Changing World*,
https://doi.org/10.1007/978-981-95-0978-2_1

Keywords

Tectonic uplift · Altitudinal zonation · Biogeography · Monsoon system · Lesser Himalaya · Geological thrusts · Climate regulation · Anthropogenic impact

Introduction

The Himalayas, often referred to as the 'Roof of the World', are an assemblage of majestic mountain ranges that stretch across five countries: India, Nepal, Bhutan, China, and Pakistan, and are home to some of the most complex and dynamic ecological systems on the planet. These mountains are the highest on Earth, but geologically youthful as a result of the ongoing collision of the Indian and Eurasian tectonic plates. Culturally they are ancient, Indian and Tibetan mythology and spirituality draw deeply from these mountains. Floristically the Himalayas are unique, home to a diverse array of ecosystems from wet to dry, and from tropical to alpine. Phytogeographically, they form the boundary of the Palearctic and Oriental realms, and are at the meeting point of several biogeographic zones, enhancing the diversity and richness of their flora and fauna.

The Himalayas host an unparalleled diversity of forest ecosystems, reflecting the region's dramatic altitudinal gradients, climatic variability, and rich biodiversity. Stretching from the subtropical Sal forests of the foothills, to stunted shrubs and alpine meadows above the treeline, these forests embody the interplay of natural processes and human interventions over millennia. The lower elevations are characterised by tropical and subtropical forests, transitioning into temperate broadleaf and coniferous forests at mid-elevations. Further upslope, subalpine forests dominated by hardy conifers give way to the stark beauty of alpine pastures and scrublands. Each forest type not only supports unique flora and fauna but also plays a critical role in regulating the hydrology and climate of the region, making them indispensable for the ecological and economic well-being of millions.

These forests are shaped by a myriad of factors, including altitude, precipitation, and topography, which together create a complex mosaic of habitats. The Eastern Himalayas, with their high rainfall and humidity support lush evergreen forests, while the drier forests to the west are best represented by pine and oak-dominated landscapes. Local communities, who have long depended on these forests for fuel, fodder, and timber, have also influenced their composition and spread.

The term 'Himalaya' is a combination of two Sanskrit words 'him' or snow and 'alaya' or abode. There is but one Himalayan range, and researchers such as Prof Jamuna Sharan Singh have advocated the word Himalaya as a singular. However, the plural, 'Himalayas' better recognises the totality of this immense mountain arc which consists of several geologically distinct parallel ranges, viz. the Siwalik, the Lesser Himalaya, and Greater Himalaya. While both usages find support, the prevailing convention favours the English plural 'Himalayas'.

Scale and Extent

Extending for over 2400 km from east to west, and with a mean breadth of 150 km (Zurick et al. 2005), the Himalayas are a growing mountain chain being uplifted even today as the Indian plate collides with the Eurasian plate.

The Himalayas are home to Mount Everest, the tallest peak on the planet, which stands as a symbol of human aspiration and endurance. When the Karakoram peaks are included, the Himalayan range contains all 14 of the Earth's summits over 8000 m high.

Atkinson (1882) defined the Himalayas as that 'portion of the great mountain girdle which lies between the bend of the Indus on the west' and 'the gorges of the Brahmaputra on the east', a definition that has withstood the test of time. The Himalayan arc is anchored by the summits of Nanga Parbat (8125 m) and Namcha Barwa (7756 m), respectively. These peaks help define the Western and Eastern Syntaxis bends, which are used by the two rivers—the Indus and the Tsangpo-Brahmaputra (Apollo 2017).

The northern limit of the Himalayas is marked by a 50–60 km wide tectonic valley known as the Indus-Tsangpo Suture (Valdiya 1998), which forms the boundary of these mountains, and through this two mighty rivers flow in opposite directions—the Indus flows west while the Tsangpo (Brahmaputra) flows east. Both rivers originate close to Mount Kailash (6638 m), which epitomises the imagery of the abode of snow and can be likened to the heart of the Himalayas (Box 1.1). In the south, the Himalayas are fringed by the Gangetic plains. They encompass a geographic relief from the rolling high plateaus of Tibet in the north to the outer foothills above the Gangetic plains in the South (Zurick et al. 2005).

Box 1.1: Mount Kailash: The Heart of the Himalayas

Mount Kailash, known as Kang Rimpoche in Tibet ('Precious Snow Mountain'), is situated centrally in the Himalayan arc and epitomises the cultural and political diversity of the region. Mt. Kailash is sacred to four religions, viz. Hinduism, Buddhism, Jainism, and Bon religion (the indigenous Tibetan religion). Hindus recognise Mt. Kailash as the abode of Shiva, who resides there with his consort goddess Parvati, and believe Kailash to be Mount Meru, considered the stairway to *Swarga*, or a higher realm. Jains believe this to be the site where the first Tirthankara, Rishabhadeva, attained nirvana (liberation). Tibetan Buddhists regard Mount Kailash as the navel of the world and the centre of the universe. To adherents of the ancient Bon religion, Mount Kailash is the axis mundi and spiritual centre of the world.

Overlooking Lake Mansarovar, a lake mentioned in the Hindu epics Ramayana and Mahabharata, Mt. Kailash is located north of the western trijunction of the border between China, India (Uttarakhand), and Nepal. This mountain remains shrouded in mystery and folklore and is one of the few high Himalayan peaks not scaled by humans, though local folklore puts Milarepa, a eleventh century Tibetan sage, as the only human to have climbed

(continued)

Box 1.1 (continued)
this mountain. The surroundings of Mount Kailash hold the sources of four major Himalayan rivers—the Brahmaputra (Tsangpo in Tibet), the Karnali (known as the Ghaghra in India), the Sutlej, and the Indus. The origin of these rivers around Mount Kailash is why we refer to this region as the heart of the Himalayas. These rivers, originating adjacent to Mount Kailash, not just define the far boundaries of the Himalayas but are the cultural lodestones of Himalayan countries.

The Brahmaputra, which originates near Kailash and the Mansarovar lake, is known as the Yarlung Tsangpo as it flows east along southern Tibet and along the northern boundary of the Himalayas to break through the Himalayas in great gorges and into Arunachal Pradesh. The Indus originates near Lake Manasarovar, close to Mount Kailash, and flows west through the same ITSZ, eventually entering Ladakh and then Pakistan. The Karnali River originates from the Mapchachungo Glacier near Mount Kailash and is known as the Ghaghra as it reaches downstream into India before becoming a major tributary of the Ganga, the largest river system in India. The origin of the Ganga, at Gangotri glacier, is also only about 200 km distant from Mt. Kailash. The Sutlej River originates near Lake Rakshastal and is a major river of Punjab before merging with the Indus.

These four rivers originating near Mount Kailash are major rivers of four of five countries and regions: Pakistan (the Indus), Bangladesh (the Ganga-Brahmaputra), Nepal (the Karnali), India, and Tibet (see Fig. 1.1).

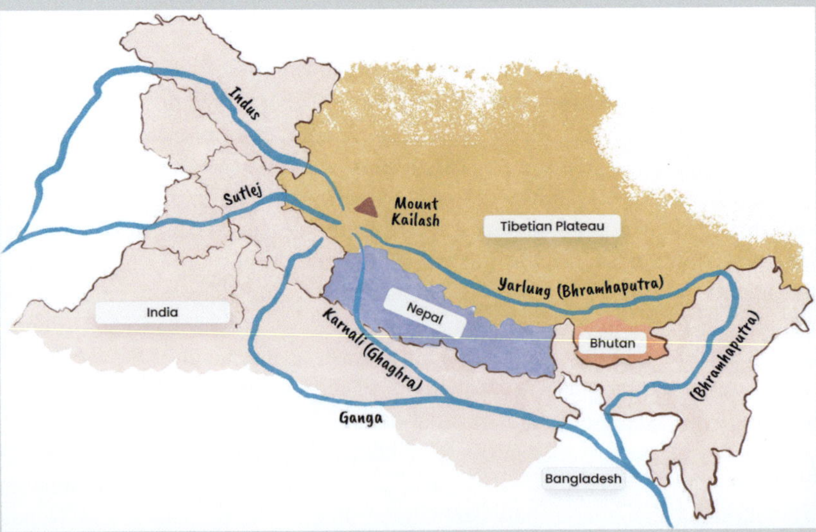

Fig. 1.1 Major Himalayan rivers emerge from the Kailash–Manasarovar region

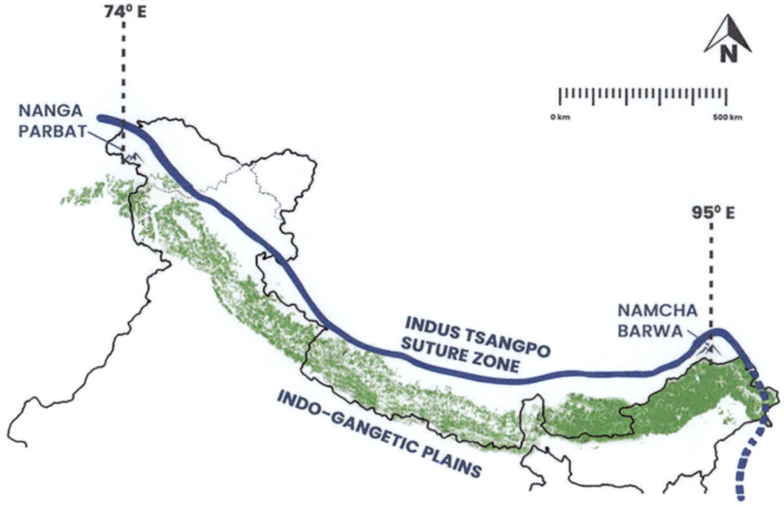

Fig. 1.2 Boundaries of the Himalayas

The Himalayas are massive. They cover ten degrees of latitudinal expanse (27° N–37°N) and over twenty degrees of longitude (74°E–95°E). On their own, they stretch in the west from Pakistan, through northern India, Nepal, Bhutan, and into part of China. This is an area of close to 600,000 sq. km (see Fig. 1.2).

A broader more inclusive definition of the Himalayas proposed by ICIMOD uses the term Hindu Kush Himalaya (HKH) which covers a vast swathe of mountain ecosystem spanning 4.2 million square kilometres over eight distinct countries (ICIMOD 2023), including not just the Himalayas and associated ranges, but also the Hindu Kush, Karakoram, and Pamir in the west, the Tibetan plateau and all the way to the Kunlun mountains in the north, the Hengduan Mountains and Shan plateau in the east, and the Arakan range in Myanmar in the southeast. This swath of land is home to 236 million people (2017 estimates Wester et al. 2019) and extends into parts of China, Afghanistan, Myanmar, Bangladesh, and encompasses the Tibet region. More importantly, over a quarter of the global population, or some 1900 million people, live in the ten river basins originating from the HKH (Sharma 2019) (Fig. 1.3).

For the purposes of this book, we do not look at the broader Hindu Kush Himalayas, but instead take a more narrow and conventional view of the Himalayas and use the political boundaries of Pakistan's Khyber Pakhtunkhwa province in the west and India's Arunachal Pradesh in the east to define the boundaries of the Himalayas. Thus, in the west the Pir Panjal, Zanskar, and associated ranges are included in our descriptions of Himalayan forests, but the Hindu-Kush and Pamirs are not (and these are areas with sparse forest cover), while in the east, our descriptions include the Mishmi hills, Naga hills, Garo-Khasi hills, and associated regions, but exclude the more eastern Indo-Burman ranges stretching into Myanmar (Fig. 1.4).

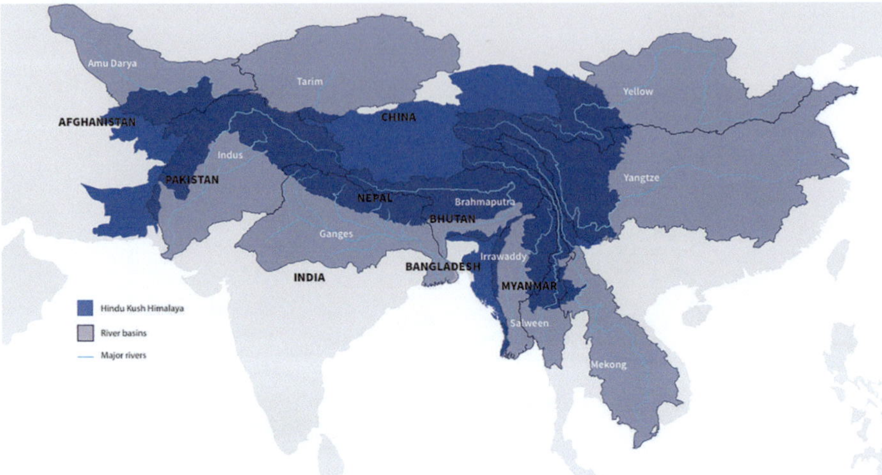

Fig. 1.3 The Hindu Kush Himalayas and its river basins. (Map by ICIMOD)

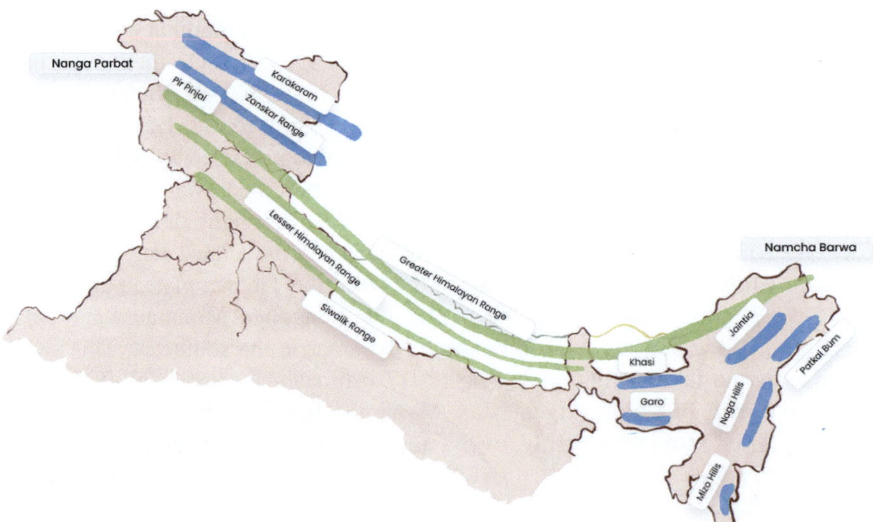

Fig. 1.4 The Himalayas and associated ranges

The Himalayas are not a single range, but in fact across much of the arc are three parallel ranges, viz. the Siwalik, Lesser Himalayas, and Greater Himalayas, with a common orogen, though structurally and lithologically distinct. The Pir Panjal and Zanskar in the west are considered part of the Himalayas, but the ranges in the east are distinct from the Himalayas, with the Naga, Mizo, and Patkai hills being part of the Indo-Burman orogen, while the Garo and Khasi hills an uplift cratonic block of the Indian plate.

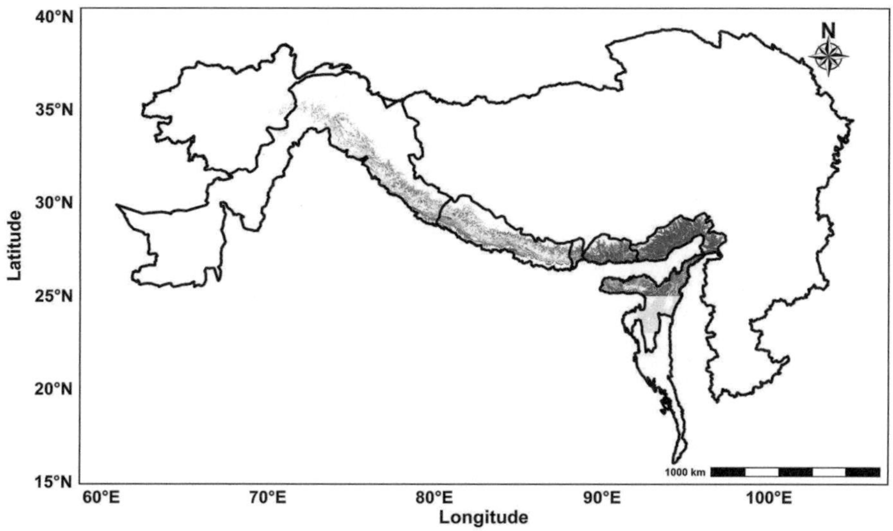

Fig. 1.5 Forests of the Himalayas (as defined in this book). (Map prepared by Zafar Reshi based on LULC map (2021) of ICIMOD)

This region makes up the river basins of three major rivers, viz. the Ganga, Indus, and Brahmaputra, and is inhabited by close to a billion people. By providing a range of ecosystem services, from climate moderation, provision of water, and the inflow of nutrient-rich soil, the Himalayas have enabled this region to support among the highest densities of human populations in the world.

We use a relatively conventional definition of the Himalayan arc (though some of the eastern Hills are additionally included) due to the forest cover of these regions. The vast expanses of the Tibetan plateau and Hindu-Kush and Pamir are both too dry and too high to support forests, and bare ground dominates these regions with grasslands and alpine pastures along the northern edges of the HKH. Consequently, these regions are of less importance for a book on forests - as forest cover is just under 20% of the HKH (Behera et al. 2018; Uddin et al. 2021). Much of the HKH forests are in the southern flanks of the Himalayas, where forests typically cover about 50% of the geographical area. Thus, Himalayan forests as described in this book are essentially the forests in the Himalayan mountains and associated montane areas of India, Nepal, Bhutan, and Pakistan (Fig. 1.5).

These are mountainous geographies that are largely conducive for forest growth up to elevations of 3500 m or higher. These are also ecosystems with considerable human influence, and for a mountainous region, this area is densely populated. About 70–75 million people[1] live in the Himalayas alone with significant additional populations in the Terai and Bhabhar foothills.

[1] India (52 million), Nepal (14 million), Pakistan (6 million), and Bhutan (>1 million).

Table 1.1 Himalayan countries and percentage of Himalayan forests within the country

Country	Area in the Himalayas (km^2)	Forest cover 2018 (km^2)	Percentage of total Himalayan forests
India	218,208	133,135	57
Nepal	111,292	54,816	23
Bhutan	38,761	30,099	13
Pakistan	125,573	17,565	7
Total	**493,834**	**235,614**	100

Data from Verma et al. (2021): drivers and mechanisms of forest change in the Himalayas

Uncertainty on a Himalayan Scale

The Himalayas transcend political boundaries. Even administrative regions within countries often overlap mountain and plains areas, leading to often varying estimates of population and land cover. Forest area estimates vary widely. Using European Space Agency data, Verma et al. (2021) estimated an area of close to half a million square kilometres of which almost 48% is under forests (Table 1.1):

However, the above numbers are indicative and appear to exclude some Terai areas. In the east, states such as Nagaland, Mizoram, and others are not considered in very strict definitions of the extent of the Himalayas. Disputed areas are often known by different names and may be double counted or not counted at all. As an example, the disputed parts of Kashmir contain about 25% of Pakistan's forest area (Qamer et al. 2016) and rich forests of fir, spruce, deodar, and pine, but are poorly documented and are often double counted as these are shown both in India and Pakistan. Definitions of forest are often confusing, as in India 'forest area' refers to the land legally designated as forest as per government records, regardless of whether it has trees, while 'forest cover' indicates the actual area covered by trees. The two numbers can be quite different.

Uncertainty is not limited to borders and definitions. There is a lack of conformity in the use of Latin names, and old and disused names often show up in common Himalayan literature. Even banj oak (*Quercus leucotrichophora*), among the most common and best known of Himalayan trees, is referred to by multiple Latin names (Thadani 2023), which is a serious issue that leads to considerable confusion and poor science. The perils of the use of poor-quality data and using generalised assumptions to build flawed models compound the uncertainty around Himalayan forests and their future trajectory (Singh and Thadani 2015).

Dividing the Himalayan Arc

Vertically, the Himalayas show a displacement of over 8 km and include a range of ecosystems from tropical and temperate to alpine and nival. A mountain range of such lofty altitudes encompasses ecosystems that correspond to a latitudinal displacement of over 5000 km. A 100 km distance can encompass ecosystems that range from

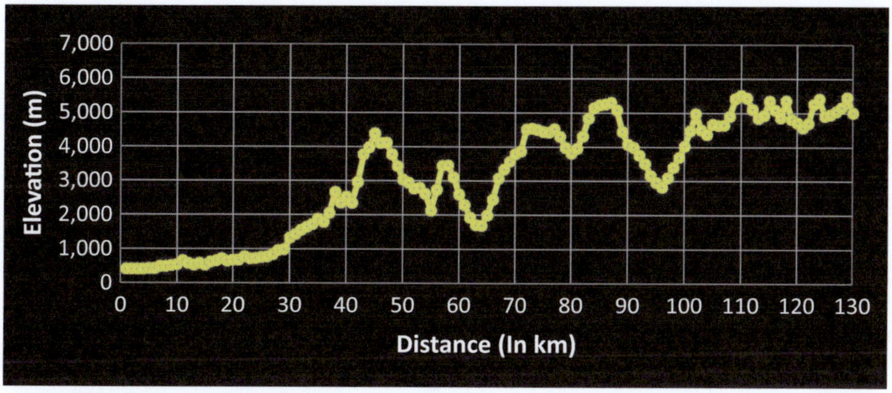

Fig. 1.6 Transect through the Himalayas near Pong dam and Dharamshala showing a 5000 m elevation rise in under 90 km

sub-tropical to alpine, and rainfall can vary manyfold from the wet monsoon-drenched outer ranges to the rain-shadow areas of the inner valleys (Fig. 1.6).

The immense length of the Himalayas leads to varying conditions. The east-to-west arc is often divided into two, three, or more divisions. A commonly used approach (Apollo 2017) is the division into three meso-physiographic regions, viz. Western, Central, and Eastern, which are as follows.

Western Himalaya

This is an arc of about 550 km long which extends from the Indus River gorge near Nanga Parbat, through Kashmir and Himachal Pradesh, and is bounded by the gorge of the Sutlej River in the east.

Central Himalaya

Extending from the Sutlej (Himachal Pradesh, India) to the Arun River (Nepal) in the east, this part encompasses the Indian state of Uttarakhand and much of Nepal (the Western and Central midlands as per Stainton 1972).

Eastern Himalaya

Includes a small part of Nepal, east of the Arun River to the eastern Syntax through which the Brahmaputra flows, this region includes Sikkim, the Darjeeling Himalayas, Bhutan, and Arunachal Pradesh. Singh (1971) included the 'Purvanchal' region[2], which is a southward extension of the Himalayas in the extreme east and runs along the Indo-Myanmar border extending from Arunachal Pradesh to Mizoram. These are lower hills than the main Himalayas and discontinuous in places, and are typically composed of sandstone. They include the Patkai

[2]See Singh (1971), p. 41 where Purvanchal is included as one of the five meso regions of the Himalayan Mountain region.

bum, the Naga hills, the Mizo hills, and other minor ranges.[3] Nonetheless, these and associated ranges in the uplands around Assam have forests contiguous with Himalayan forests, and their descriptions are included while discussing the eastern Himalayan forests.

Forests and Vegetation

In densely populated South Asia, the Himalayas are one of the few regions of good forest cover. From the moist tropical forests in the foothills of the Eastern Himalayas, to the arid zones in the west, from the tropical forests of Sal in the south, to the sub-alpine fir and birch forests at the timber-line, the variations in altitude and rainfall creates almost every conceivable climatic zone for forests. The Himalayan forests are a mosaic of ecological richness. Exposed ridges with thin soils and xeric conditions yield in just a few hundred metres to sheltered valleys with deep soils and mesic conditions. South-west aspects with strong and long sunlight hours face cooler north- and east-facing slopes. Windward slopes facing the monsoon winds get much higher precipitation than leeward slopes, and altitudes can change by several thousand metres in just a few tens of kilometres horizontal distance, bringing vegetational changes that would occur over thousands of kilometres in the plains. A subtropical forest of Sal may occur only a few kilometres from a temperate oak forest (or they may even mix as in Mangoli, Nainital), while oak and pine may alternate over a distance of a hundred metres as dry exposed ridges with shallow soils giving way to shaded valleys with deep soils and mesic conditions.

The Himalayas lie at the convergence of four floristic regions—Indian, Malay, Sino-Japanese, and Central Asian (Zurick et al. 2005), further enhancing the region's astonishing variety of flora and fauna. These forests play a crucial role in providing a range of ecosystem services that not just sustain local communities, but support downstream ecosystems and help maintain base flows in rivers that sustain livelihoods in the most densely populated part of the world.

Significance and Importance of the Himalaya

The Himalayan-Tibetan orogen, often referred to as the 'Water Towers of Asia', is the most glaciated region outside the polar realms (Haeberli et al. 1989), earning it the moniker 'The Third Pole'. While glaciers and seasonal snowmelt are well studied, less well studied are mountain aquifers. Layers of permeable rocks, as well as faults and fractures, allow for the infiltration of rain and meltwater which then collect over impermeable layers and gradually move through mountains,

[3]Purvanchal however does not include the Garo, Khasi, and Jaintia hill of Meghalaya (which is considered an extension of the main Indian peninsula) or the Mishmi and Miri hills which are an extension of the Siwalik in Arunchal Pradesh.

emerging as springs. The estimated 3–5 million springs in the Himalayan region are not just a vital source of water for Himalayan communities during the dry season, but along with snow and ice melt from glaciers, ensure that the major rivers of north India remain perennial. The continuous deposition of nutrient-rich silt and alluvial soil transported from the Himalayas is foundational to the agricultural productivity of the sub-Himalayan alluvial plains of the Ganga, Brahmaputra, and Indus. These mountains also play a crucial role in carbon sequestration, biodiversity conservation, and disaster mitigation, reinforcing their global environmental significance.

The Himalayas exert a profound influence on climate. Their rise initiated the monsoon system, upon which the entire subcontinent depends. Acting as a massive climatic barrier, they block cold northerly winds, moderating winter temperatures in South Asia and trapping the monsoon winds. The Himalayas form a formidable natural barrier impeding the free movement of flora, fauna, and people. These mountains have historically kept apart two of the world's great civilisations: India and China. While limited cultural and trade exchanges occurred through high passes like Nathu La and the Silk Route corridors, the sheer scale and harshness of the mountains discouraged large-scale invasions or migrations, allowing distinct languages, religions, and cultural systems to develop largely independently.

Through their regulation of climate, replenishment of fertile soils, and provisioning of freshwater, the Himalayas have shaped the ecological and agricultural landscape of South Asia for millennia. Supporting nearly a quarter of the global population, they are foundational to the region's high population density and agricultural productivity, underscoring their irreplaceable role in sustaining civilisations past and present.

Geological Setting

Our knowledge of the physical development of the Himalayas is tied to the theory of continental drift, which explains the world's landforms through the breakup of the primordial supercontinent, the Pangaea, and movements of the earth's lithosphere.

About 200 million years ago, Pangaea began to break up initially into two massive fragments. The northern Laurasia and the southern Gondwana. As Gondwana fragmented, Africa separated about 160–180 million years ago.[4] Western Gondwana carrying India split yet again 120–130 Ma ago, which led to the separation of Australia and Antarctica—and the Indian plate, along with Sri Lanka, Madagascar, and Seychelles started its slow but steady move north at a rate of 9–16 cm per year. By 80 Ma, India had separated from Madagascar. The speed of the Indian plate accelerated to close to 20 cm per year around 60 Ma—a speed that can be considered breathtakingly fast in geologic terms! This led to the rapid contraction of the northern Tethys Sea (Fig. 1.7).

[4]Million years is abbreviated as Ma or Mega annum.

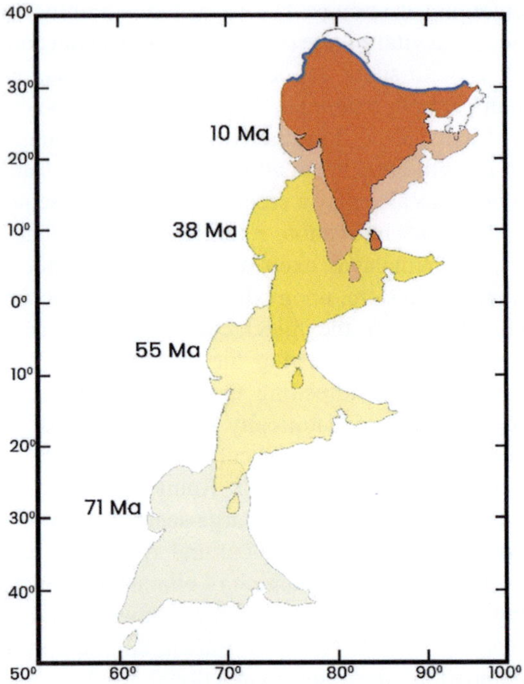

Fig. 1.7 Continental collision during the Cenozoic: The northern movement of the Indian plate, the shrinkage of the Tethys sea, and eventual formation of the Himalayas (blue line represents the ITSZ). (Adapted from Molnar and Tapponnier 1975 and other sources)

By early in the Eocene, the once vast Tethys Sea that separated Laurasia and Gondwana closed, and the Indian plate started driving into the Asian plate. This collision resulted in abruptly slowing the rapid northward flight of India to about 5 cm per year—a speed that was maintained through the convergence period (Chatterjee et al. 2013) and continues to this day (Searle 2015) (Fig. 1.8). As a result of this collision, the oceanic crustal rocks and sediments of the Tethys Sea were thrust upward along an interface known as the Indus-Tsangpo Suture Zone (ITSZ). The present-day Indus and Brahmaputra (Yarlung-Tsangpo) rivers follow this alignment in opposite directions, wrapping the Himalayas in a 'geologic embrace that stretches west to east for 2,700 kilometres' (Zurick and Pacheco 2006) and demarcates the tectonic rendering of the Himalayas.

The northern part of the Indian plate, made of thin oceanic crust, was subducted beneath Asia during the initial phases of collision (50 Ma) and now lies buried deep beneath the Tibetan Plateau. Continental lithosphere, which has a relatively thick coating of crust, is more resistant to subduction than oceanic lithosphere. This thick, buoyant crust of India resisted subduction, leading to a slowing of the Indian plate. For the Indian lithosphere to continue to be subducted, it was necessary to shave off some of its crust. These shavings are what now comprise the Himalayas (Molnar 1986).

As the lower part of the Indian plate continues to thrust under the Eurasian plate, it keeps the Tibetan plateau elevated to a high altitude. The upper part of the Indian

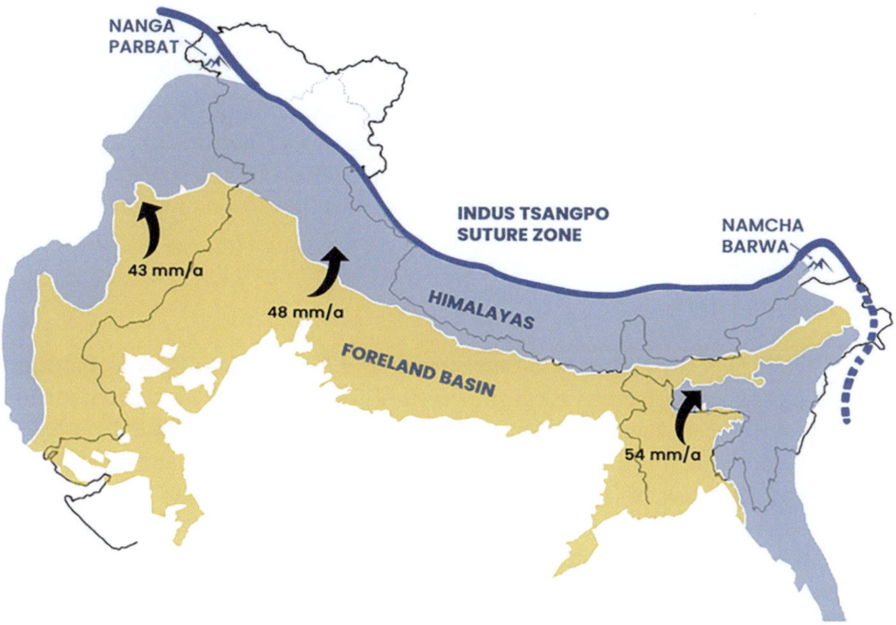

Fig. 1.8 Convergence rates of the Indian and Eurasian plates. The ITSZ (blue line) separates the two plates

crust has shortened over 500 km due to the folding and thrusting (Searle 2015). The thickening of the crust caused an increase in temperature and pressure, resulting in regional metamorphism of the greater Himalayas, localised crustal melting, and the formation of granites that form many of the highest Himalayan peaks.

By 21 Ma, elevations had risen to about 2300–2500 m. By 15 Ma, the Himalayas had begun to assume closer to their current mass with some areas elevated to at least 5000 m (Spicer et al. 2021). This period in the middle Miocene (~20–15 Ma) was the most active period of uplift (Singh and Sarkar 1990). Subsequent growth formed several peaks in excess of 8000 m with an average elevation across the range of ~6000 m. The immense weight of the rising Himalayas caused the crust beneath the Indian subcontinent to flex downward, creating the foreland basin known as the Indo-Gangetic Plain. This basin has been filled with sediments eroded from the Himalayas, forming one of the most fertile regions in the world.

The Himalayas continue to undergo rapid uplift, ranging from 4 to 8 mm/year in the Lesser Himalayas to 10–15 mm/year in the sub-Himalayas. It also continues to rapidly erode, and rivers carry down enormous silt loads to the plains below. In the sub-Himalayas, the rate of erosion is found to almost match the rate of uplift (Chatterjee et al. 2013).

Evidence of these cataclysmic events and the ancient roots of the mountains can be seen in the twisted strata of exposed rock that show the enormous pressures of buckling and folding, and in the fossilised corals, brachiopods, and skeletal fish found at the foot of glaciers several miles high. Sediments from the ancient Tethys Sea are preserved in the Tibetan Himalayas, a zone about 40 km wide at an altitude of 4000 m. Seashells have even been found near the summit of Mount Everest, demonstrating the oceanic origin of these mountains (Zurick et al. 2005). The ammonite fossils of this period—remains of the ancient Tethys sea—represent to the Hindu the Chakra of Vishnu, and to the Buddhist, the cosmic mandala.

The Rise of Tibet

The Tibetan plateau, covering an area of some 2.5 million sq. km, is the most extensive elevated surface on earth (Spicer et al. 2021) with an average altitude in excess of 4500 metres amsl. The growth of the Tibetan plateau profoundly affected Asian climate and the monsoon, thereby impacting vegetation and biodiversity across Asia. The rise of Tibet and its influences on Asian climate have occurred over a long period of time, some of which predates the collision of the Indian and Asian plates. The Tibetan plateau was not uplifted as a monolithic entity but evolved in a piecemeal fashion since early in the Mesozoic (>150 Ma) as an amalgam of Gondwanan terranes successively accreted onto the Eurasia plate, this process beginning well before the collision of the Indian plate. As with the collision with the Indian plate, each earlier accretion event resulted in the thickening of the crust and subducted ocean and continental lithosphere, with each collision leading to more surface relief (Spicer et al. 2021). The southern Tibetan plateau has a complicated geological structure, and existing plate tectonics theory struggles to fully explain its formation.

The Kailash range (Gangdese mountains) began forming some 120 million years ago. At this time, the north-central Lhasa terrain was below sea level as evidenced by marine sediments, but the rising Gangdese to the south filled these with sediment, and the area emerged from the ocean. Some 62 million years ago, the Gangdese mountains had reached an elevation of around 3500 m (Ibarra et al. 2023), and later in the Eocene, they would have been the first obstacle to moist air being drawn northwards by the low pressure forming over Siberia in the summer. Recent fossil discoveries have shown that Tibet hosts a wealth of paleontological data attesting to a past where diverse Paleogene forests (from 30 to 60 Ma) existed in subtropical intermontane lowlands in a great central valley between the Gangdese mountains (Kailash range) and Qiangtang uplands. Plant fossils of palms found in central Tibet from some 25 million years ago indicate the presence of deep valleys under 2300 m elevation, which is indicative of the terrain variability in Tibet during this period.

To the north are the Qiangtang uplands that are estimated to have reached an elevation of ~5000 m by 28 Ma (Xu et al. 2013). The increased elevation in the north and NE Tibet impacted precipitation across southeastern Asia, inducing a precipitation increase, especially in the dry (winter) season, and resulting in a transition from deciduous to evergreen broadleaf vegetation and an increase in plant diversity across southeastern Asia (Li et al. 2021).

The upliftment of the Tibetan plate was not uniform, and the region may have undergone three distinct lifting events interspersed with two lowering events over the past 40 million years (Fang and Li 2003; Zhou et al. 2007). The latest lifting event in the late Pliocene epoch and early Quaternary period (i.e. as recently as 1–2 million years ago) resulted in parts of the Tibetan plateau rising to 3500–4000 m from under 1000 m in the late Pliocene (~3–4 million years ago). However, elevations in the southern Tibetan plateau have remained largely unchanged over the past 15 million years (Coleman and Hodges 1995; Spicer et al. 2003), though elevations in North Tibet have increased significantly during this period.

Climate Change in the Miocene: The Rise of the Monsoons

The Himalayan uplift greatly impacted atmospheric circulation patterns and changed the regional climate (Zhou et al. 2007). Uplift of this region began about 50 Ma, and intensified during the Early Miocene (~23 Ma). It is likely that the onset of the Asian monsoons was triggered when the Tibetan plateau reached a height of 2–3 km (Clift et al. 2008) in the Miocene. The uplift of the Himalayas was thus among the most important geological events in the Cenozoic and changed the physiognomic and climatic character of Asia.

This story of plate tectonics is relevant to this book. Not only does it explain the formation of the Himalayas, but it also helps explain the origin of the different geological structures underlying the broad zones of the Himalayas. These in turn determine the soils which influence the vegetation and forests of the Himalayas. Most importantly, the rise of the monsoons over the past 15 million years helped mould the climatic and floral composition of the Himalayas and all of South Asia.

Formation of the Many Ranges

As India converged with Asia, the northern edge of the Indian plate was fractured, thrusted, and uplifted, resulting in the formation of the Himalayan ranges. The Greater Himalaya, Lesser Himalaya, and sub-Himalaya (Siwalik) are slices of the old Indian shield that have been stacked over each other (Chatterjee et al. 2013). These slices, or major structural divisions, as well as the major suture zone between the plates, can be mapped along almost the entire length of the east-west arc of the mountain belt, which provides an unusual simplicity (Searle and Treloar 2019). Developed progressively from north to south (Thakur et al. 2020), these are the South Tibetan Detachment (STD), the Main Central Thrust (MCT), the Main Boundary Thrust (MBT), and the Main Frontal Thrust (MFT) (also referred to as the Himalayan Frontal Thrust, HFT). While the MCT dates to the early Miocene,[5]

[5]~20 million years BP (Miocene was 23–5 Mill years BP).

the MFT, which is the most recent, was formed only in the Quaternary.[6] These roughly parallel thrusts differentiate the Himalayas into tectonically distinct and stratigraphically contrasted domains with differing histories of sedimentation, magmatic, and tectonic activities (Valdiya 1980). Given the different features of these slices and the shearing of rocks evidenced in the suture zones, a brief description of each is needed, as these underlying geologies have significant impacts on the forests above.

Indus-Tsangpo Suture Zone (ITSZ)

This is the suture zone marking the collision between the Indian and Eurasian plates. This is where the Indian plate was subducted beneath the Eurasian plate, leading to the closure of the Neo-Tethys Ocean and the uplift of the Himalayas. The ITSZ forms the northern boundary of the Himalayas, with the Indus River flowing west from the Mansarovar Lake and the Tsangpo River (Brahmaputra) flowing east. As the young Brahmaputra in the Tibetan highlands is variably referred to as the Tsangpo or Yarlung Zangbo, the suture zone is also sometimes referred to as the Indus–Yarlung Zangbo Suture Zone (IYSZ).

To its north lies the trans-Himalayan zones of Tibet. South of it lies the Tibetan Tethys zone or Tethys Himalaya, comprised of folded and thrust sedimentary rocks of the Indian plate's upper crust (Searle 2015). The youngest of these rocks are marine limestones containing foraminifera fossils characteristic of the Lower Eocene (50 Ma). The base of the Tethys Himalayan zone is the low-angle, north-dipping fault known as the Southern Tibetan Detachment System (STDS), which is the first of the faults that separate the various slices of the Himalayas (see Fig. 1.9).

The Southern Tibetan Detachment System (STDS)

The Tethys Himalaya is separated from the Greater Himalaya (or Himadri) by a series of north-dipping faults and shear zones called the STDS, which probably formed during the Miocene time. The STDS extends approximately 2500 km from Zanskar in NW India to Arunachal Pradesh in NE India.

The region is composed of once deeply buried metamorphic rocks and granites that form some of the highest peaks of the Himalayas.

Main Central Thrust (MCT)

The Main Central Thrust (MCT) divides the Lesser Himalaya from the Greater Himalaya. It marks an abrupt change in the style and orientation of structures and in the grade of metamorphism from higher metamorphic facies of the Greater or High Himalaya, to the lower amphibolite facies of the Lesser Himalaya. It is known as the Munsiari Thrust in Kumaon. It extends northwest in Himachal as the Jutogh Thrust and farther in Kashmir as the Panjal Thrust. In the Eastern Himalayas, equivalents of the Munsiari Thrust are known as the Paro Thrust and the Bomdila Thrust (Valdiya 1980).

[6]Quaternary began about 2.6 million years BP.

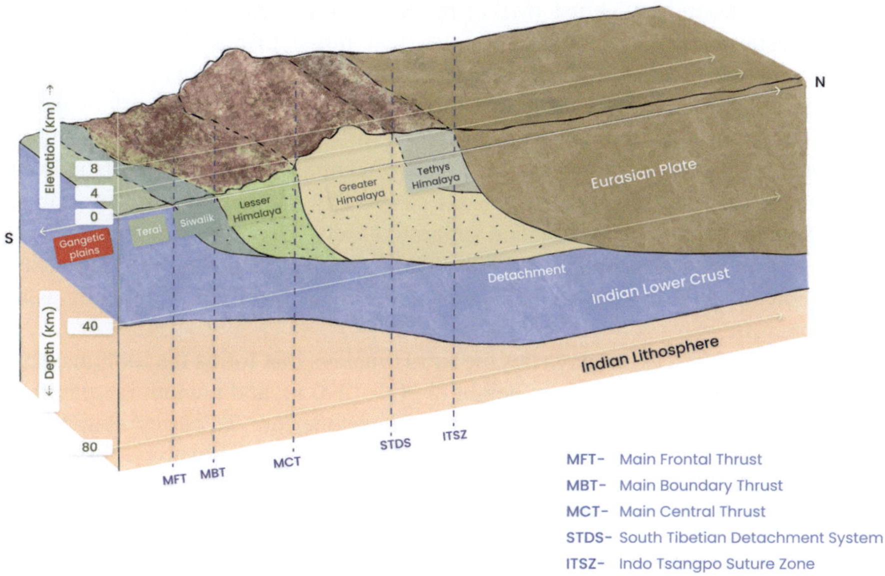

Fig. 1.9 Cross section of the Himalayas showing the main tectonic faults and the intervening zones

The formation of the MCT is an important event in the early development of the Himalayas. Put simply, it is the result of the scalping of the leading margin of the Indian plate, as the deeper parts of India slipped beneath South Tibet. The MCT is not a clearly defined plane separating two different kinds of rocks, but rather a wide zone of 10 km or more, where all of the rocks have been severely sheared and the sides of mountains appear as stacks of rock sheets (Molnar 1986). The large slats that were traditionally used for roofing in Himalayan homes are a product of the Main Central Thrust. These sheets of rock easily detach from one another and cause landslides, but this trait is seen as positive when they are mined for roofing material.

South of the Greater Himalaya, and separated from the Greater Himalaya by the Main Central Thrust, lie the Lesser or lower Himalaya, which can attain heights of up to 4000 m, and a width ranging from 32 to 80 km. This zone consists mainly of Late Proterozoic to Early Cambrian detrital sediments that were thrust over along the Main Boundary Thrust (MBT) fault, which defines their southern border. These are the Himalayan ranges of importance for this book, as much of the forests lie in this zone. The Lesser Himalaya was covered by vast forests of oak, pine, deodar, firs, and myriad other species until the nineteenth century. This is also the zone that has been heavily inhabited for millennia and has a rich cultural heritage.

Main Boundary Thrust (MBT)

The MCT ceased being an active fault about 10–20 million years ago, and a new fault, the Main Boundary Thrust (MBT), formed as India's northward motion sliced off another piece of the plate.

The Main Boundary Thrust (MBT) (or Main Boundary Fault) is not a single thrust line but rather a chain of four different thrust planes that constitute the boundary between the sub-Himalayan Cenozoic sedimentary zone, including the Siwalik and the Lesser Himalayan pre-Tertiary rock formations. At various places, this is known as the Murree thrust (in Kashmir and Himachal), the Krol thrust (In Uttarakhand), the Kathmandu thrust (east of central Nepal), and the Gondwana thrust (in the Eastern Himalaya). The MBT causes the sedimentary formations in the Lesser Himalaya to be pushed up and split into sheets. This results in repetition and inversion of lithostratigraphic units. This thrust is currently active, both tectonically and seismically (Valdiya 1980), and hence, landslides occur more frequently as the rocks are crumbly.

South of the Lesser Himalaya lies the sub-Himalaya that forms the foothills of the Himalayan range with an average height of 900–1200 m, and a width ranging from 8 to 80 km. The sediments, often rich in limestone, are in places abundant in vertebrate fossils and constitute the Murree and Siwalik formations. Sub-Himalayan rocks have been overthrust by the lesser Himalaya along the MBT that developed during the Pliocene. In turn, the sub-Himalaya is thrust along the Main Frontal Thrust (MFT) over the Holocene alluvium of the Ganga Basin deposited by the Himalayan river systems (Chatterjee et al. 2013).

In parts of the Siwaliks, wide flat valleys known as Doons (or Duns) are seen—Dehradun being the best example of this. These occur in the areas where the Main Boundary Thrust (MBT) and the Main Frontal Thrust (MFT) create spaces between the uplifted ranges. In areas where they occur, the Siwaliks can be clearly differentiated from the lesser Himalaya.

Main Frontal Thrust (MFT)

The MFT, also known as the Himalayan Frontal Thrust (HFT), is the youngest of the Himalayan faults. This frontal thrust runs along the south of the Siwalik and separates the Bhabhar, Terai, and Gangetic plains from the sub-Himalayas.

Parallel but Not Quite

While these Himalayan tectonic features extend across the Himalayan arc, it needs to be recognised that presenting them as roughly parallel is a simplified explanation, and these zones vary in width, forming undulations and ingressions. In the north-western Himalayas, for example, Ladakh lies in the trans Himalayan zone, which abuts the Greater Himalaya with no presence of a Tethys Himalayan zone in this region. In part, the more complex geological features along the collision boundaries also owe their origin to the northward translation of India and its anticlockwise rotation as it welded with Asia.

In summary, the structure of the Himalayas has been visualised as a stack of northward dipping thrust sheets, which have resulted from the progressive under-thrusting of the Indian lithosphere along the décollement (detachment plane) (Rajendran and Rajendran (2022). The limits of the Himalayan orogen are defined by the Indus-Tsangpo Suture Zone (ITSZ) to its north, the sinistral Chaman fault (CF) to its west (in current day West Pakistan and Afghanistan), the dextral Sagaing fault (SF) to its east (current day Myanmar), and the Main Frontal Thrust (MFT) to its south.

A description of these faults and shears serves to remind the reader that the Himalayas are an active mountain range. As the Indian plate continues its northward journey, these mountains are growing. Erosion rates remain high, and the steep topography and the crumbly nature of its rocks make it prone to landslides and erosion. The strong monsoon, especially in the Lesser Himalaya, washes these sediments downstream into the great Himalayan rivers, which carry to the Indo-Gangetic plains vast quantities of sediment. In the northern reaches, the two opposite-flowing rivers, the Indus and the Tsangpo-Brahmaputra, transport huge loads of sediment from the Tibetan plateau. The high denudation rates associated with these rivers are believed to have provided a positive feedback mechanism for significant localized crustal deformation (Rajendran and Rajendran 2022, p. 91).

Table 1.2 provides a brief description of the various slices or ranges of the Himalayas, each separated from the other by geology, and often by the presence of an active fault. These moving from north to south include the: trans Himalaya, Tethys Himalaya, Himadri or Great Himalaya, Lesser Himalaya, Siwalik or sub-Himalaya, the Bhabhar, and the Terai.

An additional description of the Lesser Himalaya is needed for this book. This zone is the heartland of the Himalayan forests. The climate is relatively equable and favourable for habitation, and agriculture is largely undertaken on terraces, making this zone densely populated. Most land suitable for agriculture has for long been reclaimed from forest (Mahat et al. 1986). Steeper slopes and remote areas however have good forest cover, and the richest and most diverse of Himalayan forests ranging from the sub-tropical to upper temperate and sub-alpine can be found in this zone.

The Lesser Himalaya extends 60–100 km in breadth. The region has a complex geomorphology dominated by the extreme relief provided by hills and valleys, which vary from about 600 to 4000 m. In Nepal, they are referred to as the 'Middle Hills' or the 'Mahabharat range' and form the heartland of Nepal, covering 30% of the country. Mahat et al. (1986) report that two-thirds of the population of Nepal lives in these middle hills, though this has changed in recent decades as growth rates in the Terai are much higher. Arable land is largely limited to a little above 2000 m (Mahat et al. 1986) above which forests dominate.

Himalayan Earthquakes and Natural Disasters

The youthful Himalayas are known to be highly seismically active. As plates converge Rajendran and Rajendran (2022) suggest that the stress accumulation is adequate for a magnitude 8 earthquake every decade! Given that there have been relatively few large earthquakes in the past 50 years, with the exception of the Kashmir earthquake of 2005 (Magnitude 7.6) and the Gorkha earthquake of 2015

Table 1.2 A simplified account of the various zones of the Himalayas and the tectonic features separating them

Zone	Elevation (approx.)	Forests	Geology	Examples of habitations
Terai (moist depression where Himalayan streams and rivers reappear)	100–300 m	Mostly agricultural landscapes, swampy in patches; moist Sal, *Dalbergia*	Alluvial plains from Himalayan sediment deposit	Siliguri (West Bengal); Lumbini (Nepal); Lalkuan, Pantnagar (Uttarakhand)
Bhabhar (talus land where rivers subside)	300–600 m	Dryer Sal (stunted), *Butea*, *Bombax*	Porous gravel deposits that absorb water	Kathgodam, Kotdwar, (Uttarakhand)
Main frontal thrust (MFT)				
Siwalik (known also as the outer Himalayas, or Chura range in Nepal)	600–1500 m	Sal, Shisham, *Terminalia*, *Anogeissus*, Chir, Acacia, diverse vegetation	Poorly consolidated and unconsolidated sediments of sandstone, shale, and conglomerates deposited by rivers flowing from the Miocine to Pleistocene epoch (past 20 ma)	Kalimpong (West Bengal); Jammu (J&K), Jelolikote, Mansa Devi (Uttarakhand)
Main boundary thrust (MBT)				
Lesser Himalaya (middle Himalayas or Mahabharat range in Nepal)	600–4000 m	See Chap. 4 (Forest associations). Diverse mixed conifer (pine, fir) and broadleaf (oaks, rhododendron) forest. This is the most diverse forest zone	Highly folded and faulted sedimentary rocks, in some areas metamorphic rocks	Most hill-stations and mountain towns, e.g. Darjeeling, Gangtok, Mussoorie, Nainital Shimla
Main central thrust (MCT)				
Greater Himalaya (Himadri)	3800–8848 m	Sparse or non-existent above the tree line	High-grade metamorphic rocks and granites	Nathu La (Sikkim); Badrinath (Uttarakhand)
South Tibetan detachment system (STDS)				
Tethys Himalaya	4000–7000 m	Largely absent. Some alpine meadows and sparse shrubs at	Sedimentary rocks, including limestone, shale, and sandstone,	Keylong (Lahaul—Spiti, HP); Zanskar (Ladakh),

(continued)

Table 1.2 (continued)

Zone	Elevation (approx.)	Forests	Geology	Examples of habitations
		lower elevation (precipitation <500 mm)	represent ancient marine deposits	Tawang (Arunachal)
Indus-Tsangpo suture zone (ITSZ)				
Trans-Himalaya	3500–5500 m	Sparse vegetation, cold desert flora (precipitation usually <200 mm)	Complex with a mix of sedimentary, metamorphic, and igneous rocks	Leh, Ladakh; Spiti valley, Himachal Pradesh

(Magnitude 7.8), this would indicate that the Himalayas are due for multiple future earthquakes of significance.

While the devastation of major earthquakes to humans and infrastructure is well reported, major earthquakes trigger a large number of significant landslides and lead to extensive damage to forests and vegetation. For example, the 2015 Gorkha earthquake triggered thousands of landslides in the highlands of central Nepal, causing widespread vegetation damage (Pandey et al. 2022) and triggering colonisation by pioneer species such as *Alnus nepalensis*.

Earthquakes also lead to an increase in forest use for raw materials and to rebuild livelihoods. Studies in Nepal following the Gorkha earthquake recorded a gradual increase in forest dependence, particularly on community forests, as timber was needed for reconstruction (Poudel et al. 2022). The importance of forest users and protection groups, such as the Van Panchayats of Uttarakhand and Community Forest User Groups (CFUGs) of Nepal, as civil society organisations critical during times of natural disasters has also been reported. Gentle et al. (2020) opine that CFUGs were the most effective institution to provide immediate support to disaster-affected communities post the Gorkha earthquake of 2015. Community forests are also critical in providing goods ranging from fuelwood, cattle fodder, a variety of NTFPs, including forest fruit and other edibles, critical at times of disaster. Evidence from villages impacted by the 2013 Kedarnath disaster (flooding) also indicated that impacted people often moved from roadside homes to their ancestral habitations closer to forests in the aftermath of the disaster. Given that Himalayan regions are highly prone to climatic disasters and given the imminent threat of major earthquakes, there is a need to further understand and acknowledge the importance of forests to mitigate the impacts of disasters.

In addition to earthquakes, the youthful topography of the Himalayas is prone to landslides and mass movements, especially when soils are saturated after the monsoon rains. Deforestation, unregulated construction, and roadbuilding further exacerbate and make worse these landslides. Another major natural disaster occurring with increasing frequency is Glacial Lake Flood Outbursts (GLOFs). These occur when water from glacial lakes, formed by retreating glaciers, is suddenly released. In the Himalayas, these lakes are often held back by fragile moraine dams, which can fail due to melting ice, avalanches, rockfalls, or earthquakes. Climate change accelerates glacier retreat and increases extreme weather, creating larger and more unstable lakes, raising the risk of GLOFs. These floods can devastate downstream communities, infrastructure, and ecosystems, making them a serious hazard in the region. The devastating flood of the Teesta River in Sikkim in October 2023, caused by the failure of the moraine-dammed South Lhonak Lake, is an example. A cloudburst in June 2013 near the Chorabari glacier caused a GLOF and flood that killed over 6000 people and destroyed the Kedarnath valley.

Himalayan Soil Types and Formation

Soil formation and development are dependent on several interacting factors, which in the Himalayas show great variability. In addition to the parent material, climatic conditions, vegetation cover, and human disturbance, slope and aspect are the principal factors driving soil development. The highly dissected landscapes of the Himalayas show rapid changes in bioclimatic conditions leading to a pronounced heterogeneity of soil types and their physicochemical properties over short distances (Baumler 2015).

The parallel ranges that make up the Himalayas show significant geological differences (Table 1.2). To the extreme south is the Terai, part of the Indo-Gangetic plains characterized by gravels and sands of Quaternary overlying Tertiary molasses sediments (Siwalik group). The Main Frontal Thrust (MFT) that separates the Terai from the Siwaliks consists of a succession of fluvial sediments of Neogene age and rich in fossils (Baumler 2015). To the north of the Siwalik is the Main Boundary Thrust (MBT) and between the MBT and the Main Central Thrust (MCT) lies the broad expanse of the Lesser Himalaya which is characterised by sedimentary and meta-sedimentary rocks of Precambrian to Miocene age, typically slate, limestone, phyllite, schists, quartzite, and dolomite that are poor in fossils. The MCT separates the low- to medium-grade rocks of the Lesser Himalaya from the higher-grade rocks of the high Himalayas, many of which are different types of gneiss.

Within each of these ranges, soils vary widely as parent materials are quite distinct and tectonic movements can bring very different rock types in close proximity. Consequently, soils can vary significantly at very short distances, as Himalayan farmers well know, and this perhaps is a major reason why partible inheritance traditions lead to a division of each plot of land to ensure that theirs get equal shares in both productive and unproductive lands!

Temperatures and freezing-thawing cycles vary greatly with altitude and influence soil formation. Among climatic variables, rainfall is particularly important. Parts of the Himalayas exposed to the southwestern monsoons receive well over 2000 mm of rain, which leads to strong leaching environments in the soils, in addition to eroding the topsoil in some areas. Growth of vegetation can also strongly influence soil development. Good vegetation cover, even for a few decades, can greatly enhance the quality of the young Himalayan soils, building organic matter and nutrients in the surficial layers. Vegetation cover is also the best protection against soil erosion, and it influences fundamental soil-forming processes such as aggregation and podzolisation (Baumler 2015).

Soil degradation in the Himalayas, linked to the loss of fertile topsoil, is a result of a complex interaction between physical and human factors. Physical factors relate to the steep slopes, which, when drenched with heavy monsoonal rains, are prone to high rates of erosion. Steep slopes saturated with water are prone to shallow landslides, which remove the fertile upper soil layers. These natural processes are greatly exacerbated by human factors—forest degradation, overgrazing, poor cultivation methods, and, more recently, the unplanned alignment and building of roads. These anthropogenic processes lead to the erosion of upper soil layers and the depletion of nutrients.

Soils vary significantly based on their position on slopes and the slope aspect. Soils on ridges tend to be thinner and drier than those at the base of hills, leading to differences in vegetation (Chap. 11: Slope and Aspect impacts) which in turn further influence the formation of these soils. Soils can be specific to certain geographies. For example, steeply sloping summits and ridges often have skeletal soils, which are excessively drained. This is best seen in the soils of Ladakh, which are typically calcareous, severely eroded, stony, and somewhat alkaline. In the Kashmir valley, a distinct feature found is the Karewas or 'elevated table lands', which are lacustrine deposits (i.e. deposits in lakes). These soils are deep, well-drained, fine-loamy, and good for agriculture (Sidhu and Surya 2014).

Based on several studies (in particular Singh and Singh 1992; Sharma et al. 2022), some generalisations on physico-chemical features of Himalayan soils can be made: (i) With respect to texture, soils vary from coarse loamy sand (on sandstone) to clay loam (on limestone). Sandy loam is most common at lower elevations, and clay loam at higher elevations (about 2200 m). (ii) Typically, soil textures are coarse with a high gravel content, and these are prone to displacement due to landslides. (iii) Himalayan soils are often acidic. While in the Western Himalayas acidic soils are less common, in the far eastern regions most soils are acidic (with pH below 6.5), and a large proportion of Arunachal and adjacent north eastern states have soils often with pH < 5.5. Soils at higher altitudes and under conifers tend to be acidic. (iv) Organic content in surface soil ranges from $<1\%$ to 4%. (v) Chronic human disturbance leads to lower levels of soil organic content. Nutrient concentrations are highly variable, and soils under oak forests are richer compared to those under pine forests. Human disturbance also influences nutrients. Phosphorus and nitrogen are often low and limiting to plant growth. Paucity of these nutrients may be a reason for the relative lack of deciduous forest cover in the Himalayas.

As per Champion and Seth (1968), for forest development, the depth and moisture of soils are more important than the nature of the parent rock. Given that most Himalayan soils are young without fully developed strata, it is the depth and size of particles that impact moisture supply, root aeration, and availability of nutrients. Soils with shallow depths typically support only open stunted forests as they provide neither the support nor the moisture to sustain greater growth. As for trees, many of the dominant forest-forming species, such as Chir pine, Banj oak,

Table 1.3 Brief description of some Himalayan soil types using USDA and FAO classifications[a]

USDA	IUSS WRB (FDA)	Brief description
Alfisols	Luvisols	Soils with a clay-enriched subsoil and high base saturation. Common in the Himalayas. Fertile soils
Andisols	Andosols	Soils formed in volcanic ash
Aridisols	Solonchaks, Solonetz, Calcisols, Gypsisols	Soils of arid regions, found in the Western Himalayas
Entisols	Fluvisols, Leptosols, Arenosols, Regosols	Young soils with little or no horizon development. Commonly found at high altitudes or on steep slopes
Gelisols	Cryosols	Soils with permafrost within two metres of the surface
Histosols	Histosols	Organic soils are found in high-altitude wetlands and bogs
Inceptisols	Cambisols	Young soils with minimal horizon development. Very commonly found in Himalayan forests
Mollisols	Chernozems, Kastanozems, Phaeozems	Soils with a thick, dark, organic-rich surface layer, common in grasslands
Oxisols	Ferralsols	Highly weathered tropical soils, often found in the eastern Himalayas
Spodosols	Podzols	Acid soils with a subsurface accumulation of humus and iron/aluminium often support coniferous forests
Ultisols	Acrisols, Alisols	Soils with a clay-enriched subsoil and low base saturation. Common under dense forests of the NE Himalayas
Vertisols	Vertisols	Soils with a high content of expansive clays

Compiled using Righi and Lorphelin (1987), Baumler (2015), and Sharma et al. (2022)
[a]An attempt to provide both USDA and FAO classifications is made, as both classifications are used for the Himalayas by different authors

Deodar, and Firs, are relatively tolerant of varying soil types, and altitude and rainfall are stronger determinants of vegetation.

For soil nomenclature, while the USDA classification has been traditionally used for Himalayan soils (see Table 1.3), recent authors (e.g. Baumler 2015) use the FAO classification system (International Union of Soil Sciences-Working group, or IUSS 2006). We try to provide below the USDA name, followed by the closest soil type as per FAO classification (IUSS 2006).

Inceptisols (Cambisols as per IUSS 2006) are found commonly across Himalayan forests from the sub-tropical zones all the way to the upper temperate zones (up to around 3500 m), and across the entire Himalayan arc. These are young soils with weakly developed horizons and are common in areas with moderate rainfall and some vegetation cover. They typically have a high sand and silt fraction. Studies in Nepal indicate that with increasing elevations above temperate zones, clay content decreases (Baumler 2015). Forests of Chir pine, oaks, rhododendrons, and firs are often found on Inceptisols. In the Eastern Himalayas, Inceptisols cover 44% of the land area and are the most common soil type (Sharma et al. 2022).

Alfisols (Luvisols) are commonly found in sub-tropical and lower temperate zones (~1000–2000 m), especially in the Eastern and Central Himalayas, where rainfall is abundant. These are moderately leached soils often with a subsurface horizon of clay accumulation and a high base saturation, and often rich in aluminium and iron. They may be found along with the somewhat similar lixisols. They are often fertile and support good forests and also agriculture.

Entisols (Leptosols and regosols) are often found at high altitudes where soil development is hindered by cold temperatures and steep slopes, and also in areas with active erosion or deposition. These are very young soils, often thin, with little to no horizon development. This is the most common soil type in Arunachal Pradesh as well as across much of the Western Himalayas.

Mollisols are soils with a thick, dark, organic-rich surface horizon. They are very fertile soils and often found in grassland ecosystems and in lower elevation valleys and basins where organic matter accumulates.

Spodosols (Podzols) are usually acidic soils with an organic colloid layer complexed with iron and aluminium leached from a layer above. These soils are common above 3000 m, especially in coniferous forests in the upper montane belt, all the way to the tree line. They are common in the Western Himalayas (Sharma et al. 2022).

Histosols or soils rich in organic contents are found in some of the high-altitude wetlands and bogs in the Himalayas. They are commonly referred to as peat or muck soils.

Oxisols (Ferralsols) are heavily weathered, rich in iron and aluminium oxides (sesquioxides) or kaolin, but low in silica. They have only trace nutrients due to heavy tropical rainfall and high temperatures. In the Himalayas, they are found in some tropical zones of the Eastern Himalayas.

Aridosols are soils typical of arid regions and found in parts of the dry western regions, such as Ladakh.

Ultisols (Acrisols) are found in humid regions and are characterised by a subsurface horizon with clay accumulation and significant leaching of nutrients, leading to acidic conditions and low natural fertility. They are relatively common in the Eastern Himalayas, often under dense forests that receive substantial monsoon rainfall, which contributes to the leaching and weathering processes. These soils are especially common in Meghalaya, Manipur, and Nagaland (Sharma et al. 2022).

Palaeobotany and Vegetation Development in Early Forests

Paleobotanical studies in the Himalayas use a combination of palynology (pollen and spore analysis), phytoliths (plant silica bodies) and macrofossils to reconstruct past climates and vegetation. Studies have uncovered a wealth of fossilised pollen from sites such as the Kashmir Basin and the Siwalik Hills, which provide information on ancient floras. Modern genetic analysis and phylogenetic trees also help us understand speciation and the development of diversity.

Early Vegetation

The initial period of collision between the Indian and Asian plates did not see significant uplift, but led to the establishment of a biotic corridor between India and Asia, which enabled hordes of Asian vertebrates to sweep into India in the Early and Middle Eocene. Similarly, vertebrates of Gondwanan origin migrated to the Asian plate (Chatterjee and Scotese 2010). Flora from both plates migrated along this 'intercontinental biological highway', leading to a considerable mixing of vegetation of the two landmasses (Manish and Pandit 2018). Starting with mostly immigrant flora, the newly formed Himalayas were a crucible for evolution, and today are home to some 3500–4000 endemic plant species (Manish 2019). The geophysical upheavals that disrupted ecosystem stability might be responsible for the high endemic plant diversity of the Himalayas. The changing climatic conditions, rise of mountains, and creation of various new geographic barriers as the Indian plate pressed into the Asian plate produced new geophysical niches, imposed isolation barriers, and acted as natural selection sieves that led to the formation of new species (Singh and Sarkar 1990; Manish and Pandit 2018) and has resulted in a dynamic and ever-changing ecological landscape.

Vegetation from this early period, after the collision of plates during the Tertiary, is characteristic of shallow marine to coastal conditions with a tropical climate. Pollen of genera such as *Casuarina*, *Ephedra,* and palms, as well as *Arecaceae* spp., have been identified in the Ladakh Himalayas as well as Himachal dating to the Palaeocene-Eocene (60–40 Ma). The abundance of dinoflagellate cysts further confirms this marine and aquatic environment. Further east in Meghalaya, also a coastal swamp-like environment is indicated during the Palaeocene (Singh and Sarkar 1990). In the east, the establishment of land connections with Myanmar-Malaysia occurred as late as the Oligocene (~38 Ma). While floras of the Eastern Himalayan region at this time were tropical to subtropical moist deciduous type, later enrichment of species from Sino-Japanese and Malayan regions included the entry of Rhododendrons, Tsuga, and Schima. Gymnosperm pollen is rare during this period and become common much later in Miocene-Pliocene (i.e. last 15 MA) (Kar 2000).

Significant orogeny during the Miocene epoch (~20 Ma), which continued through the Pliocene and Pleistocene, created diverse altitudinal zones, fostering distinct ecological niches and contributing to the region's rich biodiversity. Concurrently, global climate fluctuations, characterised by alternating glacial and interglacial periods, especially during the Pleistocene epoch, profoundly impacted the distribution and composition of forest types. Additionally, localised climatic changes induced by the orographic effects of the growing Himalayas further influenced precipitation patterns and temperature gradients, shaping forest dynamics. The growth of the Greater Himalaya created a rain shadow area in the trans-Himalayan region, leading to a decline in tree cover. Macrofossils of the Namling flora of southern Tibet (Spicer et al. 2003; Zhou et al. 2007) indicate the presence of a Miocene broadleaved forest flora, well before present-day rain shadow effect of the high Himalayas turned the region into a cold desert.

By the early Miocene (23–16 Ma), temperate elements such as the Pinaceae started migrating from the West Asian mainland. In the foothills and Siwaliks in the Central-Western Himalayas (Dehradun, Kalka), tropical trees similar to *Albizzia, Cassia, Dalbergia, Diospyros, Dipterocarpus, Ficus, Polyalthia,* and *Ziziphus* have been identified from plant megafossils. While the Siwaliks may have lacked gymnosperms during the Miocene, there is evidence of gymnosperm pollen that resembles *Pinus*, *Abies*, *Picea,* and *Tsuga,* which likely blew down from the surrounding higher hills (Singh and Sarkar 1990). Angiosperm families of importance today, such as the Juglandaceae and Fagaceae, are also evident. The orogeny of the middle Miocene (10–15 Ma) saw strong changes in the floristic scenario with the decline of the tropical families and the proliferation of gymnospermous elements, and also an increase in sub-tropical families such as Moraceae, Myrtaceae, Euphorbiaceae, Fabaceae, and others. Evidence indicates that the pines migrated to the Himalayas in the early Miocene and became common after the orogeny of the middle Miocene, some 10–15 million years ago (Singh and Sarkar 1990). Contemporary environmentalists who label pine as an exotic invasive would do well to examine paleobotanical evidence. It appears that ancestral pines inhabited the Himalayas at a time when the apelike ancestors of humans had not as yet diverged from chimpanzees and bonobos!

As recently as the Pliocene (<5 Ma), a continuous mesophilic south Tibetan forest fauna was still to be found, contiguous with the flora of the rising Himalayas, which lay to the south. It remains uncertain when during the Pleistocene the Himalayan uplift reduced the monsoon influence in Tibet, and mesophilic south Tibetan forests were fragmented and retreated to small refuges. Fossil evidence from species related to modern-day kharsu oak (*Quercus semecarpifolia)* in late Pliocene sediments in southern Tibet (Xu 1982) indicates that the rain shadow effect of the high Himalayas is no older than 0.5 Ma.

In the Siwaliks there was a significant expansion of C_4 plants in the late Miocene between 8 and 6 million years ago (Behrensmeyer et al. 2007). This was particularly so in the Central and Western Himalayas between Uttarakhand and Pakistan, possibly due to particular climatic factors such as decreasing influence of the westerlies, and the changing monsoon (Sanyal et al. 2024), which led to an increase in grassy vegetation.

In the Siwaliks a cool dry climate had set in by the Pliocene (<5 Ma) and species such as *Larix*, *Pinus,* and magnoliaceous trees grew luxuriantly (Singh 1982) while further east, along Nepal–Bihar, Lakhanpal and Awasthi (1983) find fossil evidence of D*albergia, Ficus, Litsea, Mangifera, Toona, Ziziphus,* and members of Fabaceae. Climatic changes during these periods led to increased continentality and semi-arid conditions, particularly in the Western Himalayas.

Late Pleistocene and the Holocene (Recent Changes Over the Past 25,000 Years)

The period of intense climate change led to several glacial periods during and since the Pleistocene. In between were warmer periods, during which the impacts of the monsoon would increase, leading to higher moisture and the expansion in range of broadleaved species. Alpine species would retreat to higher altitudes during these interglacial periods. In the glacial periods, temperatures would drop, and cold, dry conditions would prevail as a result of decreased monsoonal activity. During the Last Glacial Maximum (LGM, ~ 26,500 to 19,000 years ago), the temperatures are estimated to have been 5-8 °C lower than today. Glaciers descended up to 1500 m lower than they are today, and the Western Himalayas in particular experienced extensive glaciation. The expansion of glaciers caused a reduction in the area available for forests, as large tracts of land were covered by ice. The altitudinal zones of vegetation, especially conifers, shifted downward.

While the LGM may have been the last recognised ice age, there have been significant periods of warming and cooling since this period. Cooler periods usually coincide with a dip in the monsoon and overall drying. Peng et al. (2019), using a Beryllium-10 surface exposure dating in Bhutan, identify four glaciation events since the Last Glacial Maximum (LGM), with the most recent 'Little Ice Age' only 500–750 years BP (Peng et al. 2019). Rowan (2017) used various dating methods to determine ages of moraines and associated landforms across the Himalayas and identified two recent periods of glacial advance: around 800–900 CE and an even more recent advance of Himalayan glaciers in the Little Ice Age around 1300–1600 CE which corresponds to Peng et al. (2019).

Phadtare (2000), using pollen data from near Dokriani glacier (near Gangotri glacier), traces periods of cooling and warming through changes in the abundance of oak and pine pollen. The peat profile at the site indicated a dominance of *Quercus semecarpifolia*. When climates were cold and dry, the percentage of this evergreen oak pollen increased, indicating an increase in its dominance. This was at the cost of pine (likely *Pinus wallichiana* or blue pine), which showed a contraction of range when climates turned cold and dry. Pine pollen was abundant in periods of warmer, wetter climate. Phadtare (2000) uses the change in the ratio of oak and pine pollen to calculate the Q/P ratio (or Quercus/Pinus), which is considered a bioclimatic index for interpreting the climate trend. As per Phadtare's interpretation, cold, dry periods are identified by a high Q/P ratio, whereas warmer, wetter periods are identified by a low Q/P ratio. As can be seen in Fig. 1.10, there were several swings in climate in the past 8000 years, along from warm and wet phases to cold and dry phases where the impacts of monsoon lessened. Such climatic perturbations have occurred through the ages and influenced the distribution and spread of species.

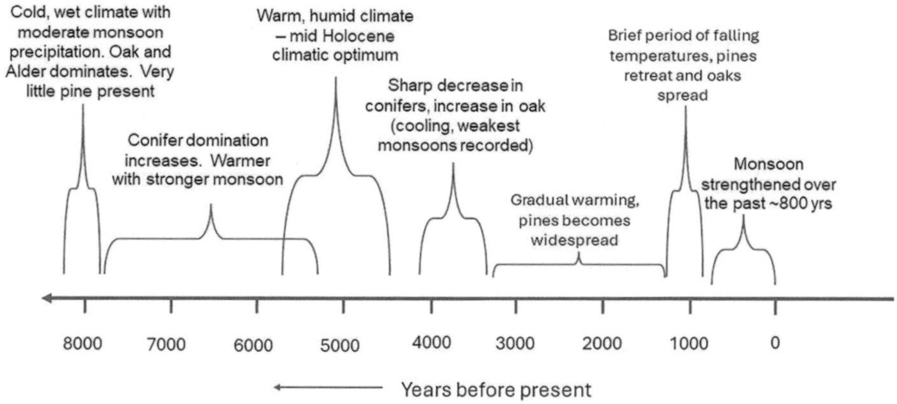

Figure made from data in Phadtare, N. R. (2000). "Sharp decrease in summer monsoon strength 4000–3500 cal yr B.P. in the central higher Himalaya of India based on pollen evidence from alpine peat." Quaternary Research, 53(1), 122-129.

Fig. 1.10 Changes in monsoonal strength over the past few millennia led to changes in the abundance of oak and pine

Himalayan lake sediments at various altitudes have been studied to provide vegetation and climate records. Most lake records show evidence of an intensification of monsoon at the beginning of the Holocene (i.e. about 10–12,000 years ago). Studies of sediments from a relict lake, Benital in the Central Himalayas (Uttarakhand), for example, suggest that the early Holocene was characterised by a wet phase with maximum rainfall about 9000 years ago (Ghosh et al. 2020). This led to stronger stream flow, erosion of bed rocks, and substantially higher discharge of sediments into the Ganga–Brahmaputra delta during this period compared to present discharge. This period of heavy rainfall, about 25% higher than present, was followed by a relatively dry period in the mid-Holocene (~6–7000 yrs. BP). This was followed by alternating cooler and drier periods over the next few thousand years (Sanyal et al. 2024). Quamar et al. (2023) studied the sedimentary profile of a palaeo lake at 1200 metres elevation in Uttarakhand and found that between 7500 and 5800 years BP, while conifers were dominant, there is a gradual decrease in broadleaved species for a period of almost a thousand years, before abundance starts increasing. Pines grow more common in this region from 7200 years BP as broadleaved trees declined, and changes in shrub and grass composition and abundance are recorded as climate changes lead to vegetational changes.

Thus, even in the past few millennia, climatic fluctuations have been significant enough to lead to significant changes in forest cover and species mix (Box 1.2).

Box 1.2: The Missing Oaks of Kashmir

Oaks are a ubiquitous Himalayan genus found from Afghanistan to Arunachal Pradesh and dominant from 1500 to 3500 m. However, they are lacking in the Kashmir valley. This observation on the lack of oaks was made as early as 1839 by Royle and has been confirmed by several authors thereafter. However, oaks existed in Kashmir in the past. As per Vishnu-Mittre (1963), pollen analysis shows that at the end of the Pliocene (1.8 Ma) the vegetation of Kashmir consisted predominantly of oak woods, with *Alnus nepalensis* as the other important constituent. With the onset of glaciations, the oaks were adversely affected, and blue pine (*Pinus wallichiana*) invaded into oak zones. In recent times, there have been relatively few or no oaks north of the Pir Panjal ranges and in the Kashmir valley. While some early authors have postulated that the uplift of the Pir Panjal created a mountain barrier that changed the climate and led to the disappearance of oaks, this has been questioned by Vishnu-Mittre (1963).

Pliocene fossils of *Quercus semecarpifolia* and other species have been found far north in Tibet at heights of 5700–5900 m on Mount Xixabangma (in Tibet, due north of Kathmandu) (Xu et al. 1973). Zhou et al. (2007) suggest an uplift of between 2100 and 3400 m for this region since the Pliocene (in the last 2–3 Ma), attesting to the rapid changes in altitudes in parts of the Himalayas.

Forests and People

Myths and Culture

References to the diversity of Himalayan flora date back to the ancient Indian scriptures. The value of medicinal plants from the Himalayas and the importance of particular trees were known to the ancient inhabitants of these lands.

The Himalayas and their forests find abundant mention in the Hindu epics, the Ramayana and Mahabharata, composed almost 2500 years ago. In the Ramayana, Sage Vishvamitra takes Rama and Lakshmana to his hermitage in the Himalayas. The rich and mystical flora of the Himalayan region is highlighted when Hanuman is sent to fetch the Sanjeevani herb to heal Lakshmana during the battle in Lanka. The herb was said to have the power to heal any wound and even bring the dead back to life. The story illustrates the value given to Himalayan plants and the appreciation for the rich diversity that existed in ancient times. Perhaps these Himalayan medicinal herbs were the first NTFPs valued in far-off lands!

These mountains are depicted as a source of River Ganga, and the majestic peaks and dense forests are described in many places during the narrative. In the *Vana Parva* section of the Mahabharata, Sage Lomasa describes the Himalayas to the Pandavas, emphasizing their sanctity, the beauty and serenity of these mountains, and the forests filled with diverse flora and fauna. It is in these forests that Arjuna undertook a severe penance to acquire celestial weapons from Lord Shiva. The description of Arjuna's penance includes vivid imagery of the Himalayan landscape, the dense forests, and the tranquil environment that aids in his meditation.

Considerable spiritual and cultural importance continues to be attached to these mountains and their forests. Badrinath and Kedarnath, high in the Garhwal Himalayas, are each visited by almost 2 million pilgrims annually (2023 data). Both of these pilgrimage centres are closely linked to Adi Shankara, who revived Vedic scholarship about 1200 years ago and are mentioned in the Mahabharata (Pattanaik 2020). A reverence for the Himalayas is not restricted to Hinduism. Mount Kailash is highly revered by Buddhists, Hindus, and Jains. The Sikhs make pilgrimages to the high-altitude lake Hem Kund sahib associated with Guru Gobind Singh ji. As per local folklore in the Swat-Kohistan region of Pakistan, there is a mountain known as Musaka Musalla, which translates to the 'prayer mat of Moses'. Local traditions hold that this mountain is considered a special place where Prophet Moses (Musa in Arabic) is expected to descend when he returns to earth. Mountain worship would undoubtedly predate the development of the established religions, and even today, mountains and forests are worshipped by many animistic traditions, especially in the Eastern Himalayas. As Reinhard (1987) writes in 'The Sacred Himalaya', 'Nowhere in the world does a mountain range figure so prominently in the religious beliefs of such a large and diverse populations'.

Early Explorations

While Himalayan trees and plants were used in Hindu rituals and Indian medicine systems since time immemorial, the systematised understanding of Himalayan flora is much more recent and begins with the European naturalists.

Early European visitors provided an indication of the vast forests and landscape of the region. Major General Thomas Hardwicke (1756–1835), an English naturalist and soldier who lived in India from 1777 to 1823, is best known for his extensive collection of paintings of natural history specimens made by unnamed Indian artists. From these paintings, many new species were described. While best known in zoological circles, his contributions to botany were immense, and he maintained an extensive correspondence with William Roxburgh and Nathaniel Wallich among others (Turner 2015). For Himalayan scholars, Hardwicke's description of his journey in Garhwal (Hardwicke 1799, 1801), published by the Asiatic Society is of great interest. His route to Sringar (Garhwal) took him through Haridwar (where the river Ganga exits the Himalayan foothills) at the time of the 1796 Kumbh mela, an enormous gathering of humanity that Hardwicke estimates 'with moderation to be computed at two and half million souls' (Hardwicke 1801), surely one of the largest aggregations of humanity to that date.

Hardwicke's travelogues give insights into the forests of the time. Even in 1796, well before any commercial forestry was envisaged, barren areas are described, likely due to the unfavourable moisture conditions of the slopes. While describing the hills near Srinagar (Garhwal), he writes:

> *Some of these ranges are covered with forests, and are always green, some are naked and stony, neither affording shelter to the birds of the air, nor the beasts of the field.*

These are among the earliest botanical descriptions, and Hardwicke described the genus level of many trees, including the common oaks, pines, and rhododendrons. Roxburgh used Hardwicke's drawings to describe several species, including *Lyonia ovalifolia* and *Prunus cerasoides*, and the description of *Rhododendron arboreum* by James Edward Smith is also entirely based on Hardwicke's account and illustration of the species.

Dr. William Roxburgh (1751–1815), a surgeon of the Madras Presidency, turned his passion for plant study into a project that created the most exhaustive listing of Indian flora and earned him the epithet of 'Founding Father of Indian Botany'. As head of the Botanical Garden in Calcutta from 1792, he founded its massive herbarium collection and made this institute into a hub of research. Other eminent botanists associated with the Calcutta Botanical Gardens included Francis Buchanan-Hamilton (1762–1829) and Nathaniel Wallich (1786–1854). Latin nomenclature was assigned to many Himalayan species during this period.

Francis Buchanan-Hamilton spent time in the Kathmandu valley (1802–03) and recorded the rich heritage of vernacular names that existed for the diverse flora. Many of these formed the basis for the approximately 800 Latin names coined by him, which have also earned him the moniker 'Father of Nepalese Botany'. Later, during his surveys of the Bengal Presidency (1807–1814), he twice sent his native collectors to Nepal to gather medicinal and other economically useful plants (Miehe et al. 2015).

Nathaniel Wallich (1786–1854) made extensive studies on the natural vegetation of India from 1813 to 1846. From his travels to Nepal in 1821, he returned with many new plants and subsequently published an important book *Tentamen Floræ Nepalensis Illustratæ* (vols I-II, 1824–26). Numerous important Himalayan trees bear his name, including *Pinus wallichiana* (Blue pine), *Taxus wallichiana* (Himalayan Yew), and *Ulmus wallichiana* (Himalayan or Kashmir Elm).

The famous English botanist and explorer Sir Joseph Dalton Hooker (1817–1911), who was a close friend and supporter of Charles Darwin, was the Director of the Botanic Garden at Kew from 1865 to 1885. Hooker carried out several expeditions in the Himalayas (1848–50) in Darjeeling, Western Nepal, and the Khasi hills of Assam, and he was one of the first European visitors to Sikkim. While Hooker is best known for publishing, along with George Bentham, *Genera Plantarum* (which describes about 97,000 species globally), his description of the

Rhododendrons of the Sikkim Himalayas (1849) and publication of the *Flora of British India* (1872–97) are significant contributions.

John Forbes Royle (1798–1858), a British botanist born in Kanpur (then Cawnpore), headed the botanical garden at Sahranpur. While his focus was on economic botany, Royle was fascinated by traditional botanical remedies used by Hindu practitioners. His reputation as a botanist rests on his 'Illustrations of the Botany and other branches of Natural History of the Himalayan Mountains and of the Flora of Cashmere' published in two volumes beginning in 1839. Victor Jacquemont, a French Botanist and Geologist, was one of the early visitors to the Western Himalayas and Ladakh in 1830–31 and provided important descriptions that helped systematise Himalayan flora.

These travellers described mountains covered with pristine and lush forests considered to be almost endless in extent. Mountain kingdoms until then had largely remained at arm's length from the great political powers of the Gangetic plains. Even the Mughals had paid little attention to the hill rajas. The need for timber and taxes was well supplied from the still abundant forests of the Gangetic plains (Tucker 1982), and there was little need to turn to the inhospitable hills. Also, the Terai—a marshy, malarial belt of dense forests, which separated the lower hills from the Gangetic plains—formed a natural barrier that was hard to traverse.

As forests of the Gangetic plains got depleted, attention turned to the Himalayas. It was only after Sir Dietrich Brandis (1824–1907) became the Inspector General of Forests in India in 1864 that the forests of the Himalayas received significant attention, though disorganised felling of trees, especially deodar, began a decade earlier. The development of silviculture and the extraction of timber from Himalayan forests are described in Chap. 10 of the second volume of this book (Silviculture), which also pays tribute to the early foresters who established sustainable forest management. However, the monumental work of Troup (1921) on the silviculture of Indian trees must be mentioned here, as it not only immensely contributed to the understanding of Himalayan trees, but it also became possibly the first major work on plant phenology, a discipline of ecology whose importance has been realised only recently.

This was also a time when the focus shifted from plant descriptions to understanding forest communities. Notable contributions were made by Kenoyer (1921) and Dudgeon and Kenoyer (1925). H.G. Champion was the first to classify and describe forests of the Himalayas (Champion 1936), and the forest classification system developed by Champion and Seth (1968) for India was used by other Himalayan nations as well and holds relevance even today. Later work done by Polunin and Stainton (1984), work in Bhutan by M. Ohsawa (1987), contributions in Nepal by JDA Stainton, and most recently by Georg Miehe are acknowledged in Chap. 4 on Forest associations.

Himalayan People

The broader HKH supports as many as 236 million people (2017 estimates Wester et al. 2019), but these include populations living in the HKH part of China, Myanmar, as well as parts of India (such as Assam) and Pakistan (Balochistan and southern Khyber Pakhtunkhwa) not normally included in the Himalayas. The Himalayas, on their own, are home to some 75 million people.

The rich diversity of the Himalayas is not restricted to its flora and fauna. The region is a cultural complex of humans of different ethnicities bound together by a strong Himalayan identity. The isolation bred by the high mountain ranges helped nurture a multiplicity of tribes with unique cultures that include languages, social structures, spiritual traditions, traditional knowledge systems, arts, and crafts (Banerji and Fareedi 2009). The Himalayas are home to 20% of all human languages and should be thought of as a linguistic and cultural 'mega centre' (Turin 2008).

Banerji and Fareedi (2009) describe cultural variation in the Himalayas as both vertical and lateral. While vertical variation predominantly flows from ecological factors, lateral variation in cultures is mediated predominantly by ethnicity and migration. Using this classification to slice the Himalayas provides a snapshot into cultural types and human diversity, no less fascinating than the diversity of forests described in later chapters of this book. It must be recognised that much of this diversity is linked to the varying climate and terrain, and just as varying altitude (temperatures) and precipitation patterns change forest types, so do they impact human culture and traditions.

Vertical Slices

As vegetation and ecosystem productivity change from tropical to temperate to alpine climes, so does human society. The foothills and the lower hills, by virtue of their rich, fertile soils brought down by the Himalayan rivers are relatively densely populated, while the Greater and Trans Himalayas are severe in terrain and climate. Populations in these higher regions are sparse, fiercely independent, and live in small communities widely dispersed across a vast, harsh terrain. Primary occupations and culture also change across these altitudinal zones:

High altitudes and inner Himalayas: The arid lands on the northern flanks of the Himalayas and at high altitudes cannot, for the most part, support cultivation. Rainfall in the inner regions is sparse, and strong winds blow wild and unchecked. These areas lack forests, though they may support a luxuriant growth of grass in the summers. The thin populations have adopted animal husbandry for their livelihoods, and follow a seasonal form of nomadism, moving with their families and herds (sheep, goats, yaks, camels) from one pasture land to the next during the growing season, and downslope during the winter. The culture of these nomadic pastoralists revolves around their herds, their produce (wool, milk, meat), and rangelands.

An important occupation of people of these altitudes was trading, an exchange of goods across high-altitude passes. However, border conflicts, especially between China and India, have largely closed these trade routes.

Mid Altitudes: These are sedentary zones with village settlements in valleys and slopes, where the soils are productive and support a wide range of crops. Terraced fields, mostly rainfed, are planted with cereals, millets, vegetables, and fruit crops. In many of these areas, access to markets allows for the cultivation of cash crops. Communities are closely tied to their forests and depend on them for a range of subsistence products including fuelwood, cattle fodder, and leaf litter to fertilise the terraced fields. These regions are densely populated and have developed elaborate social structures and cultural forms, often tied to their agricultural practices (Chap. 7 of Volume 2 of this book: Forests, fields and futures).

Foothills and low altitudes: These are irrigated lands, often fertile alluvial, planted with grain crops and supported by irrigated cultivation. Culturally, they resemble the mid altitudes, but the dependence and connection with forests is lower.

In the northeast, *jhum* or shifting cultivation may be practiced in some areas in the low to mid altitudes (see Chap. 8 of Volume 2 of this book: Shifting cultivation). Jhum communities are culturally distinct as a consequence of their livelihood strategies.

Lateral Slices

The reduction in precipitation as one moves west and north in the Himalayas has a definite impact on people and cultures. But much like the blending of different biogeographic zones which brings together flora and fauna of different affinities, varying ethnicities and migration patterns have impacted the different parts of the Himalayas. Four religions are dominant in these mountains: Buddhism, Hinduism, Islam, and Christianity, and four distinct ethnicities are commonly found. Banerji and Fareedi (2009) generalise these as:

- *Indic people* are predominantly Hindu, with Indo-Aryan languages and art forms, and settled agriculture as an occupation. Agriculture is closely integrated with animal husbandry, and there is a high dependence on cattle. Milk and milk products are economically and culturally important. These groups dominate across much of the Central Himalayas (except at higher elevations) and are found in smaller numbers at the two ends of the Himalayas. Largely of Aryan stock, they are loosely referred to as ‘Paharis’.
- *Afghan-Iranian people* following Islamic faiths dominate the Western margins of the Himalayas. Various groups follow settled agriculture or may be pastoralists. Pastoralist such as the Gujjars are found across the Western and Central Himalayas.

- *Tibetan people*: Buddhist with Tibetan and associated languages and culture. Agro-pastoralism is often their primary activity. Sheep, yaks, and goats are commonly reared. Most common in the trans Himalayan and higher Himalayan reaches and across the Himalayan belt. Pastoralist groups include the Changtang and Drokpa (Ladakh), Yolmo (central Nepal), and Sherpa (Eastern Nepal, Sikkim).
- *Burman/Southeast Asian* people with a mix of faiths. While many are of Christian faith, many have animism-influenced art forms and cultures, and some practice animistic rituals. Shifting or settled agriculture, or even pastoralism, and traditional livelihood strategies. These groups dominate the eastern parts of the Himalayas. Piggery is common, as is the rearing of Mithun (*Bos frontalis*) and occasionally yaks.

As climatic patterns become generally wetter towards the east and outer ranges of the Himalayas, livelihood strategies reflect this increasing complexity. Thus, Western Himalayan agricultural practices, while settled, reflect simpler cropping patterns, while the supposedly primitive *jhum* of the Eastern Himalayas is remarkable for the complexity of crops planted in the same plot and harvested sequentially.

The Common Cultural Fabric: A Himalayan Identity

While different cultural and ethnic groups exist, they typically co-exist and intermix peacefully. Importantly, mountain people share a common Himalayan identity and many shared traits:

- *A Connection with Nature*: Across the Himalayan arc, mountains are worshiped as the protector and life-giver. Natural resources are looked upon as living entities. Forests, trees, rivers, mountains, and lakes are worshipped. Religion and medicine are traditionally centred around nature.
- *A spiritual base*: Himalayan communities are usually deeply spiritual and strongly religious. Apart from institutionalised religious practices, tribal practices that are mostly animistic and characteristic of nature worship are prevalent in most parts of the Himalayas.
- *Rich traditional knowledge*: The Himalayas boast of a rich tapestry of traditional knowledge, spanning domains, such as architecture, water systems, medicine, and ethnobotany, that reflect the particular ecological conditions of the region.
- *Social structure*: Himalayan people have a strong sense of identity and can be quite clannish.

Thus, much like the terrain, vegetation, soils, and climate of the area, Himalayan people also display remarkable diversity while being tied together by a common Himalayan identity (Fig. 1.11).

Fig. 1.11 Mountains and forests: sub-alpine *Pinus wallichiana* in Darma valley (Uttarakhand). (Picture: Gopal Singh Rawat)

References

Apollo M (2017) The population of Himalayan regions–by the numbers: past, present and future. In: Contemporary studies in environment and tourism, pp 145–160

Atkinson ET (1882) The Himalayan districts of the North-Western provinces of India (Himalayan Gazetteer Vol I), North-Western provinces and Oudh

Banerji G, Fareedi M (2009) Protection of cultural diversity in the Himalayas. In: In A background paper for a workshop on addressing regional disparities: inclusive & culturally attuned development for the Himalayas. New Delhi

Baumler R (2015) Soils. In: Miehe G, Pendry CA, Chaudhary RP (eds) Nepal: an introduction to the natural history, ecology and human environment of the Himalayas. Royal Botanic Garden, Edinburgh, p 576

Behera MD, Murthy MS, Das P, Sharma E (2018) Modelling forest resilience in Hindu Kush Himalaya using geoinformation. J Earth Syst Sci 127:95. https://doi.org/10.1007/s12040-018-1000-x

Behrensmeyer AK, Quade J, Cerling TE, Kappelman J, Khan IA, Copeland P, Latorre C (2007) The structure and rate of late Miocene expansion of C4 plants: evidence from lateral variation in stable isotopes in paleosols of the Siwalik group, northern Pakistan. Geol Soc Am Bull 119(11–12):1486–1505

Champion HG (1936) A preliminary survey of forest types of India and Burma. Indian Forestry Record (NS) Silviculture, 1(1)

Champion HG, Seth SK (1968) A revised survey of Forest types of India. Govt. of India Press, New Delhi, p 404

Chatterjee S, Scotese CR (2010) The wandering Indian plate and its changing biogeography during the late cretaceous–early tertiary period. In: Bandopadhyay S (ed) New aspects of Mesozoic biogeography. Springer-Verlag, Berlin/Heidelberg, pp 105–126

Chatterjee S, Goswami A, Scotese CR (2013) The longest voyage: tectonic, magmatic, and paleoclimatic evolution of the Indian plate during its northward flight from Gondwana to Asia. Gondwana Res 23(1):238–267

Clift PD, Hodges KV, Heslop D, Hannigan R, Van Long H, Calves G (2008) Correlation of Himalayan exhumation rates and Asian monsoon intensity. Nat Geosci 1:875–880

Coleman M, Hodges K (1995) Evidence for Tibetan plateau uplift before 14 Myr ago from a new minimum age for east–west extension. Nature 374:49–52. https://doi.org/10.1038/374049a0

Dudgeon W, Kenoyer LA (1925) The ecology of Tehri-Garhwal. A contribution to the ecology of western Himalayas. J Indian Bot Soc 4:233–285

Fang XM, Li JJ (2003) The stages of the plateau uplift. In: Zheng D (ed) Formation, environment and development of Qinghai-Xizang (Tibetan) Plateau (in Chinese). Hebei Science & Technology Press, Shijiazhuang, pp 37–47

Gentle P, Maraseni TN, Paudel D, Dahal GR, Kanel T, Pathak B (2020) Effectiveness of community forest user groups (CFUGs) in responding to the 2015 earthquakes and COVID-19 in Nepal. Res Global 2:100025

Ghosh S, Sanyal P, Roy S, Bhushan R, Sati SP, Philippe A, Juyal N (2020) Early Holocene Indian summer monsoon and its impact on vegetation in the central Himalaya: insight from δD and δ13C values of leaf wax lipid. The Holocene 30(7):1063–1074

Haeberli W, Bosch H, Scherler K, Ostrem G, Wallen CC (1989) World Glacier inventory: status 1988. Compiled by the World Glacier Monitoring Service, IAHS-UNEP-UNESCO, Wallingford, UK

Hardwicke T (1799) Narrative of a journey to Sirinagur, Asiatick Researches, Vol. VI (1799), Calcutta, pp 309–381. https://www.biodiversitylibrary.org/item/93256#page/394/mode/1up

Hardwicke T (1801) Narrative of a journey to Sirinagur. Asiatic Annual Register 1800: 244–277

Ibarra DE, Dai J, Gao Y, Lang X, Duan P, Gao Z et al (2023) High-elevation Tibetan Plateau before India–Eurasia collision recorded by triple oxygen isotopes. Nat Geosci 16(9):810–815

ICIMOD (2023) Water, ice, society, and ecosystems in the Hindu Kush Himalaya: An outlook. (P. Wester, S. Chaudhary, N. Chettri, M. Jackson, A. Maharjan, S. Nepal, & J. F. Steiner [Eds.]). ICIMOD. https://doi.org/10.53055/ICIMOD.1028

IUSS (2006) International Union of Soil Sciences – working Group WRB (2006): World reference base for soil resources 2006 (2nd ed). World Soil Resources Report 103. FAO, Rome

Kar RK (2000) Palynostratigraphy of the tertiary sediments in north-East India with comments on the terminal Eocene events. Palaeobotanist 49:281–291

Kenoyer LA (1921) Forest formations and successions in the Sat Tal Valley, Kumaon, Himalayas. J Indian Bot Soc 2:236–258

Lakhanpal RN, Awasthi N (1983) Late Tertiary florule from near Bhikhnathoree in West Champaran District, Bihar. In Proceedings of the Symposium on Evolutionary Botany and Biostratigraphy/edited by AK Sharma, GC Mitra, Manju Banerjee; sponsored by University Grants Commission [and] Department of Botany, University of Calcutta, 1979. New Delhi: Today and Tomorrow's Printers & Publishers, 1984

Li SF, Valdes PJ, Farnsworth A, Davies-Barnard T, Tao S, Lunt DJ, Spicer RA et al (2021) Orographic evolution of northern Tibet shaped vegetation and plant diversity in eastern Asia. Sci Adv 7(5)

Mahat TBS, Griffin DM, Shepherd KR (1986) Human impact on some forests of the middle hills of Nepal 1. Forestry in the context of the traditional resources of the state. Mt Res Dev:223–232

Manish K (2019) Macroecological patterns and drivers of Himalayan plant species diversity and distribution through the ages. Front Biogeogr 11(2). https://doi.org/10.21425/F5FBG38754

Manish K, Pandit MK (2018) Geophysical upheavals and evolutionary diversification of plant species in the Himalaya. PeerJ 6:e5919

Miehe G, Gravendeel B, J Kluge…A. Subedi. (2015) Flora. In: Miehe G, Pendry CA, Chaudhary RP (eds) Nepal: an introduction to the natural history, ecology and human environment of the Himalayas. Royal Botanic Garden, Edinburgh

Molnar P (1986) The geologic history and structure of the Himalaya. Am Sci 74(2):144–154

Molnar P, Tapponnier P (1975) Cenozoic tectonics of Asia: effects of a continental collision: features of recent continental tectonics in Asia can be interpreted as results of the India-Eurasia collision. Science 189(4201):419–426

Ohsawa M (1987) Life zone ecology of the Bhutan Himalaya I. Chiba University, Chiba

Pandey HP, Gnyawali K, Dahal K, Pokhrel NP, Maraseni TK (2022) Vegetation loss and recovery analysis from the 2015 Gorkha earthquake (7.8 Mw) triggered landslides. Land Use Policy 119. https://doi.org/10.1016/j.landusepol.2022.106185

Pattanaik D (2020) Mahabharata on the Hills, Jan 26 2020. https://devdutt.com/mahabharata-on-the-hills/. Accessed Jan 2025

Peng X, Chen Y, Liu G, Liu B, Li Y, Liu Q et al (2019) Late quaternary glaciations in the Cogarbu valley, Bhutanese Himalaya. J Quat Sci 34(1):40–50. https://doi.org/10.1002/jqs.3079

Phadtare NR (2000) Sharp decrease in summer monsoon strength 4000–3500 cal yr BP in the Central Higher Himalaya of India based on pollen evidence from alpine peat. Quat Res 53(1): 122–129

Polunin O, Stainton A (1984) Flowers of the Himalaya. Oxford University Press, p 580

Poudel P, Kunwar N, Rumba YB, Paudel R (2022) The role of community forestry in catastrophic recovery: A case of reconstruction in post-earthquake in Gorkha district, Nepal. Caspian J Environ Sci 20(2):323–336

Qamer FM, Shehzad K, Abbas S, Murthy MSR, Xi C, Gilani H, Bajracharya B (2016) Mapping deforestation and forest degradation patterns in western Himalaya, Pakistan. Remote Sens 8(5): 385. https://doi.org/10.3390/rs8050385

Quamar MF, Singh AK, Joshi LM, Kotlia BS, Singh DS, Simion CA, Sava T, Prasad N (2023) Vegetation dynamics and hydro-climatic changes during the middle Holocene from the central Himalaya, India. Quaternary 6:11

Rajendran CP, Rajendran K (2022) Earthquakes of the Indian subcontinent, GeoPlanet: earth and planetary sciences. Springer Nature, Singapore. https://doi.org/10.1007/978-981-16-4748-2_7

Reinhard J (1987) The sacred Himalaya. Am Alpine J 29(61):123–132

Righi D, Lorphelin L (1987) The soils of a typical slope in the Himalayas (Nepal): their main characteristics and distribution. Catena 14(6):533–550

Rowan AV (2017) The 'little ice age' in the Himalaya: a review of glacier advance driven by northern hemisphere temperature change. Holocene 27(2):292–308. https://doi.org/10.1177/0959683616658530

Sanyal P, Adhya SP, Mandal R, Roy B, Dasgupta B, Samantaray S et al (2024) The geologic history of plants and climate in India. Annu Rev Earth Planet Sci 52

Searle M (2015) Geology. Nepal: an introduction to the natural history, ecology and human environment of the Himalayas. Royal Botanic Garden, Edinburgh

Searle MP, Treloar PJ (2019) Introduction to Himalayan tectonics: a modern synthesis. Geol Soc Lond, Spec Publ 483(1):1–7

Sharma E (2019) Introduction to the Hindu Kush Himalaya assessment. In: Wester P et al (eds) The Hindu Kush Himalaya assessment. ICIMOD and Springer

Sharma UC, Datta M, Sharma V (2022) Soils of HKH region. In: Soils in the Hindu Kush Himalayas. Geography of the physical environment. Springer, Cham. https://doi.org/10.1007/978-3-031-11458-8_6

Sidhu GS, Surya JN (2014) Soils of North-Western Himalayan eco-system and their land use, constraints, productivity potentials and future strategies. Agropedology 24(01):1–19

Singh HP (1982) Tertiary palynology of the Himalaya-a review. Palaeobotanist 30(3):268·278

Singh RL (1971) India: a regional geography. Delhi. India: National Geographical Society of India

Singh HP, Sarkar S (1990) Vegetational dynamics of tertiary Himalaya. Paleaeobotanist 38:333–344

Singh SP, Singh JS (1992) Forests of the Himalayas: structure, function and impact of man. Gyanodaya Prakashan, Nainital, p 294
Singh SP, Thadani R (2015) Complexities and controversies in Himalayan research: a call for collaboration and rigor for better data. Mt Res Dev 35(4):401–409
Spicer RA, Harris NB, Widdowson M, Herman AB, Guo S, Valdes PJ, Wolfe JA, Kelley SP (2003) Constant elevation of southern Tibet over the past 15 million years. Nature 421(6923):622–624
Spicer RA, Tao S, Valdes PJ, Farnsworth A, Wu FX, Shi G, Spicer T, Zhou Z (2021) Why 'the uplift of the Tibetan plateau' is a myth. Natl Sci Rev 8(1):nwaa091
Stainton JDA (1972) Forests of Nepal. John Murray, London
Thadani R (2023) Snow damage in a Himalayan forest maintains the dominance of evergreen oaks. Biotropica 55:737–741. https://doi.org/10.1111/btp.13236
Thakur VC, Joshi M, Jayangondaperumal R (2020) Active tectonics of Himalayan frontal fault zone in the Sub-Himalaya. In: Gupta N, Tandon SK (eds) Geodynamics of the Indian plate. Springer International Publishing, pp 439–466
Troup RS (1921) The silviculture of Indian trees. Clarendon Press
Tucker RP (1982) The forests of the Western Himalayas: the legacy of British colonial administration. Forest & conservation history. J For Hist 26(3):112–123. https://doi.org/10.2307/4004579
Turin M (2008) Living languages and troubled tongues: linguistic diversity and endangerment in the Himalayas. Language 2008:58–63
Turner IM (2015) The botanical legacy of Thomas Hardwicke's journey to Srinagar in 1796. Eur J Taxon 108
Uddin K, Matin M A, Khanal N, Maharjan S, Bajracharya B, Tenneson K, Markert A (2021) Regional land cover monitoring system for Hindu Kush Himalaya. In: Earth observation science and applications for risk reduction and enhanced resilience in Hindu Kush Himalaya Region: a decade of experience from SERVIR, pp 103–125
Valdiya KS (1980) The two intracrustal boundary thrusts of the Himalaya. Tectonophysics 66(4): 323–348
Valdiya KS (1998) Dynamic Himalaya. Universities Press, Hyderabad, pp 1–178
Verma A, Schmidt-Vogt D, De Alban JDT, Lim CL, Webb EL (2021) Drivers and mechanisms of forest change in the Himalayas. Glob Environ Chang 68:102244
Vishnu-Mittre (1963) Oaks in the Kashmir Valley with remarks on their history. Grana 4(2): 306–312
Wester P, Mishra A, Mukherji A, Shrestha AB (2019) The Hindu Kush Himalaya assessment: mountains, climate change, sustainability and people. ICIMOD. Springer Nature, p 627. https://doi.org/10.1007/978-3-319-92288-1
Xu R (1982) The discovery of Quercus semecarpifolia at Mount Xixia Bangma and its significance on Quaternary research. In: Scientific research group of Xixia Bangma area (ed) Monograph on Mt. Xixia Bangma Scientific Expedition 1964. Science Press, Beijing, pp 356–365 (in Chinese)
Xu R, Tao JR, Sun XJ (1973) On the discovery of a *Quercus semecarpifolia* bed in mount Xixabangma and its significance in botany and geology. Act Bot Sin (in Chinese) 15(1): 103–119
Xu Q, Ding L, Zhang L, Cai F, Lai Q, Yang D, Liu-Zeng J (2013) Paleogene high elevations in the Qiangtang Terrane, central Tibetan Plateau. Earth Planet Sci Lett 362:31–42
Zhou Z, Yang Q, Xia K (2007) Fossils of Quercus sect. Heterobalanus can help explain the uplift of the Himalayas. Chin Sci Bull 52(2):238–247
Zurick D, Pacheco J (2006) Illustrated atlas of the Himalaya. University Press of Kentucky, p 229
Zurick D, Pacheco J, Shrestha B, Bajracharya B (2005) Atlas of the Himalaya. ICIMOD, Kathmandu, p 96

2 Himalayan Climate and the Availability of Water

Rajesh Thadani

Abstract

The Himalayas extend over a vast longitudinal arc and are one of Earth's most climatically diverse regions. This chapter explores the complex interplay between topography, elevation gradients, and global atmospheric systems—primarily the Indian Summer Monsoon (ISM) and Western Disturbances—that create this remarkable heterogeneity. From the humid, monsoon-drenched Eastern Himalayas to the arid western reaches, climatic conditions vary dramatically over short distances, profoundly influencing forest ecosystems across the range.

The chapter examines altitudinal impacts on temperature through temperature lapse rates (TLRs), which vary seasonally, and from east to west along the Himalayan arc. It details precipitation patterns and their significant east-west and north-south gradients, explaining how orographic effects create rain shadows and localised microclimate variations. Special attention is given to mountain-valley wind systems that influence local temperatures, fog prevalence, and moisture distribution.

Climate change impacts on the Himalayas are addressed, noting that while the region is warming faster than surrounding areas, precipitation trends remain variable and inconclusive. The chapter discusses the unprecedented changes in the Himalayan cryosphere, including accelerated glacier melt and the implications for river discharge and water availability. The differential vulnerability of major river basins, viz. Indus, Ganga, and Brahmaputra, to climate change is analysed based on their varying dependence on snowmelt contribution.

R. Thadani (✉)
Centre for Ecology Development and Research (CEDAR), New Delhi, India

Department of Botany, Jai Hind College, Churchgate, Mumbai, India
e-mail: thadani_rajesh@hotmail.com

S. P. Singh, R. Thadani (eds.), *Himalayan Forests in a Changing World*,
https://doi.org/10.1007/978-981-95-0978-2_2

This exploration of Himalayan climate provides the foundation for understanding how forests in the region respond to both natural climatic variability and the accelerating pressures of global climate change, setting the stage for subsequent chapters on forest associations and ecosystem responses.

Keywords

Indian Summer Monsoon · Western Disturbances Westerlies · temperature lapse rate · orographic precipitation · rain shadow · glacier melt · river discharge · climate change · cryosphere · Karakoram anomaly · snowmelt contribution

Introduction: Climate in the Himalaya

The Himalayas, stretching across a vast longitudinal arc, represent one of the most climatically diverse regions on Earth. This diversity arises from the interplay of global atmospheric systems, viz. the Indian Summer Monsoon (ISM) and Western Disturbances, combined with the unique topography and elevation gradients of the range. Understanding the climate of the Himalayas is crucial for deciphering the spatial and temporal variations in its forest ecosystems.

The Himalayan climate exhibits remarkable heterogeneity. From the humid, monsoon-drenched Eastern Himalayas to the arid, rain-shadowed western reaches, and from the tropical lowlands to the alpine and nival zones, climatic conditions vary sharply over short distances. These variations are driven primarily by varying atmospheric circulation patterns and by vertical zonation shaped by steep elevation changes.

Climate plays a dynamic role in regulating forest phenology, productivity, and species composition, which are discussed in the subsequent chapters. Seasonal variations in snow cover and monsoonal rains influence hydrological processes. Moreover, the interactions between climate and forests form feedback loops that impact both local and regional climatic conditions. As the Himalayas experience climate change—manifested in warming temperatures, shifting precipitation patterns, and an increase in extreme weather events—the intricate balance between climate and forests is being disrupted. Understanding the region's complex and variable climatic systems is essential for understanding these impacts and guiding adaptive forest management.

This chapter explores the climatic diversity of the Himalayas and provides a foundation for understanding how forests in the region respond to both natural climatic variability and the accelerating pressures of global climate change.

Additionally, a chapter on Himalayan climate needs to mention the role the Himalayas themselves play in regulating regional climate. The Himalayas and the Hindu Kush mountains protect the Indian subcontinent from large-scale incursions of cold northern winds during the winter season and help maintain relatively moderate and humid climatic conditions in the Indian subcontinent. Additionally, the seasonal warming of the Himalayas and the Tibetan Plateau during the summer sets up a north–south thermal contrast relative to the tropical Indian Ocean, which is important for initiating the large-scale summer monsoon circulation (Krishnan et al. 2020).

Altitudinal Impacts: Temperature Lapse Rate

The primary differentiation in forests across the Himalayas is on account of altitude. As altitude increases, temperatures cool and tropical forests of dipterocarps give way to sub-tropical forests of *Castanopsis* and *Pinus*. At higher altitudes, the oaks and rhododendrons form montane temperate forests, and further up, various conifers such as firs, along with birch and rhododendrons dominate, above which it is too cold for trees to survive. The ability to tolerate frost and colder temperatures is a major differentiator between trees of the tropics and those of temperate and sub-alpine areas.

It is an accepted fact that it is colder high up on a mountain than the plains below. But it is important to understand what causes high altitudes to be cooler. When warm air rises, it expands due to the decrease in pressure. The expansion of air requires energy, which is drawn from the heat that the air is carrying with it. This is known as adiabatic cooling, and the rate of decline in temperature of an air parcel with elevation is called the temperature lapse rate (TLR).

The theoretical rate of change, or the adiabatic lapse rate for dry air is 9.8 °C/km. However, the adiabatic lapse rate for dry air alone does not explain how temperature changes in the atmosphere in relation to changing altitudes in mountain systems. Firstly, moisture in the air impacts the lapse rate. When an air parcel saturated with water vapour rises, some of the vapour will condense and release latent heat. This process causes the air to cool more slowly than it would if it were not saturated, and for moist air the adiabatic lapse rate is around 5 °C/km. Secondly, parcels of air do not simply ascend or descend unhindered across altitudinal gradients. Free-air lapse rates (vertical) are different from near-surface (slope) temperature lapse rates. In mountainous regions, the heated land surface is essentially projected into the free air, and the impacts of varying topography lead to more variable lapse rates. The lapse rate of non-rising air—commonly referred to as the 'normal' or 'environmental lapse rate' is highly variable and is affected by several factors, including radiation, convection currents, and condensation. Globally, lapse rate averages about 6.5 °C/km in the lower atmosphere, but this number changes based on terrain, prevailing temperatures, humidity, and other factors. Under certain conditions, such as inversion layers, katabatic wind, and cold air drainage, air temperature gradients in mountainous basins can even be reversed (Blandford et al. 2008), leading to 'temperature inversion' where valleys are cooler than the surrounding high grounds. In addition, seasonal and even diurnal variation in TLR also occurs (Blandford et al. 2008).

The Himalayas and Tibetan plateau have a complex terrain, which impacts the TLR, and patterns are different from those of other mountain regions. While in most mountain regions TLR is shallowest in the winter months and steepest in the summer (Kattel et al. 2015), this is not quite true in the Himalayas where a bimodal pattern of seasonal variations occurs in the TLR (Kattel et al. 2013; Joshi and Tamang 2021). Along the southern frontal Himalayan slopes, a strong TLR in the pre-monsoon is observed, followed by the post-monsoon season (Fig. 2.1). This is associated with strong dry convection corresponding to the clear weather seasons and high sensible

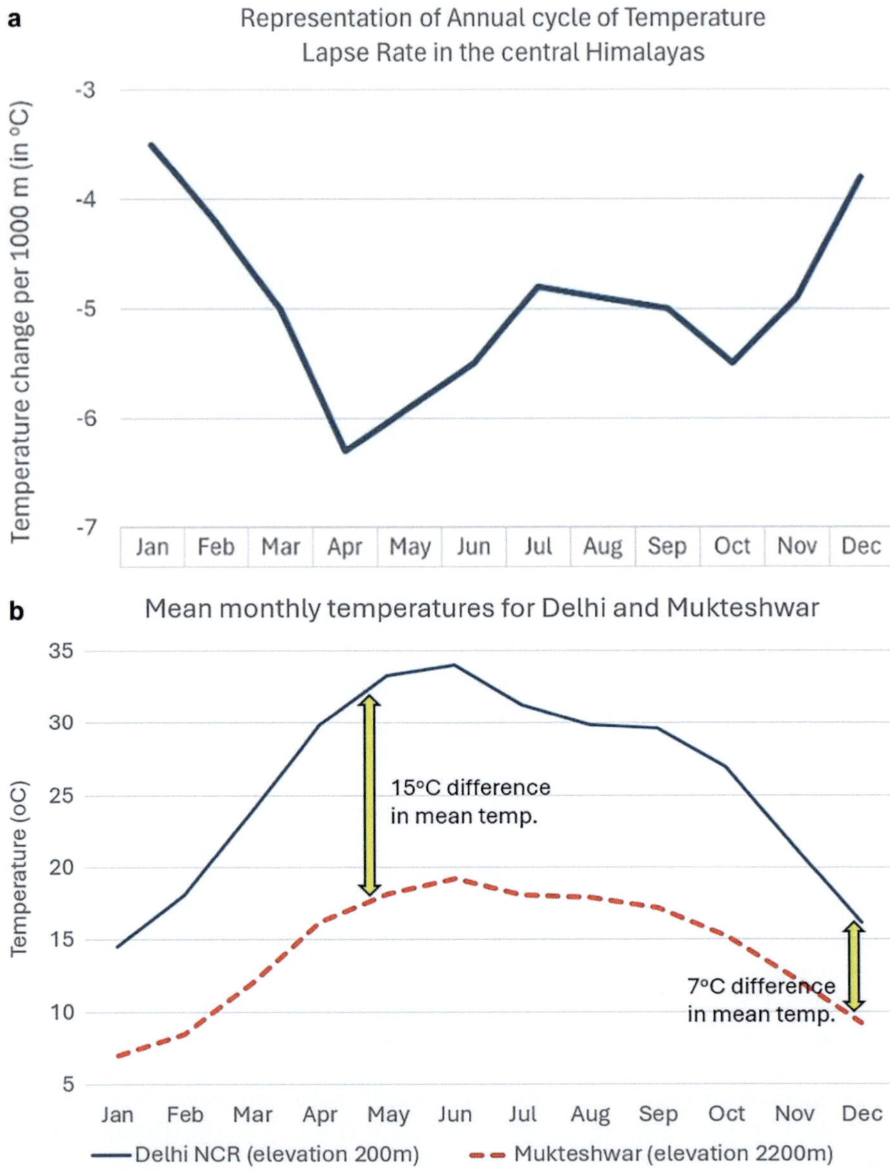

Fig. 2.1 (**a**) Representative figure to show TLR in the Central Himalayas (Developed from Kattel et al. (2013); Joshi and Tamang (2021)). (**b**) Differences in seasonal temperature between the plains and outer Himalayas due to seasonality in TLR

heat flux. Both winter and summer show low lapse rates. The lowest lapse rate occurs in the winter and is associated with strong radiative cooling and cold air flows over low elevation areas. This low TLR may also be related to winter air pollution at low elevations, which results in a cooling effect. The summer lapse rate is relatively lower due to latent heating over higher elevations and reduced solar heating over the lower elevations (Kattel et al. 2013). This helps explain why temperature differentials between the Gangetic plains (e.g. Delhi, elevation 200 m) and hill stations like Mukteshwar (elevation 2200 m) are 15 °C in the late spring, but only 7 °C in the winter (Fig. 2.1b).

In the northern reaches of the Himalayas and on the Tibetan plateau, patterns differ (Kattel et al. 2015) and lapse rates are steeper in the winter and shallower in the summer. While TLRs during the summer are comparable to the frontal (southern) ranges (~5-6 °C/km), the TLRs in Tibet during the winter are extremely high (~10-11 °C/km) as a result of intense cooling at higher elevations, corresponding to the continental dry and cold air surges, and considerable snow-temperature feedback. While this anomalous pattern is important for understanding the formation of weather systems, these are areas above the timberline and devoid of forests.

Important to note, however, are the changes in mean annual TLR as we move east to west along the Himalayan arc. As aridity increases to the west, so does the lapse rate. TLR is lowest in the humid eastern Himalayas (4.9 °C/km) and gradually increases in the western Himalayas (5.7 °C/km) with a sharper increase in the extreme west in the Pir-Panjal and Karakoram ranges (7.3 °C/km) (Joshi and Tamang 2021). This difference explains why the treelines are higher in the eastern compared to the western Himalayas.

Precipitation Along the Himalayan Arc

After temperature, it is moisture that most impacts forest type. The Himalayas are influenced by two major pluviometric regimes: The Indian Summer Monsoon (ISM) from June to September, and the Winter Westerly Disturbances (WWD) from December to March. The interplay between these systems results in two distinct climatic regimes with differing precipitation characteristics (Bookhagen and Burbank 2010), which leads to different forest types (see Chap. 4: Forest Associations).

The Indian Summer Monsoon (ISM) is the dominant source of moisture for much of the Himalayas. Originating out of the Bay of Bengal, it sweeps northward towards the Himalayas in June and abruptly declines by late September. As the monsoonal winds encounter the steep slopes of the Eastern Himalayas, the rising moist air cools and condenses to bring heavy precipitation. As a result, Cherrapunji in Meghalaya is among the wettest places on earth, with over 2000 mm of rain per month in June and July when the monsoon is most active. Most of the monsoonal winds turn to run broadly parallel to the range, with much of the moisture being directed towards the west (Burbank et al. 2012). As the currents lose moisture, there is a decrease in rainfall, especially in the foreland plains, due to the increasing distance from the source of the moisture (Singh and Kumar 1997). Much of the precipitation in the Eastern and Central Himalayas, often up to 80% of total precipitation, is a consequence of the ISM. In the western Himalayas, and in the inner ranges, the influence of the ISM wanes, but it is still associated with significant rainfall. A second monsoonal source is the East Asian monsoon (EAM) (Yihui and Chan 2005), which sweeps inland from the South China Sea and impacts inland China, Vietnam, and has minor impacts in eastern Tibet (Yao et al. 2012). While the EAM interacts with the ISM, its influence on Himalayan precipitation is minor and confined to the eastern fringes.

Winter precipitation is associated with Western disturbances, which are snow and rain-bearing storms bringing moisture from the Mediterranean and Caspian Sea (Palazzi et al. 2013). These extra-tropical upper westerlies supply smaller but significant moisture inputs to the inner ranges and are especially important in the northwestern Himalayas (Fig. 2.2).

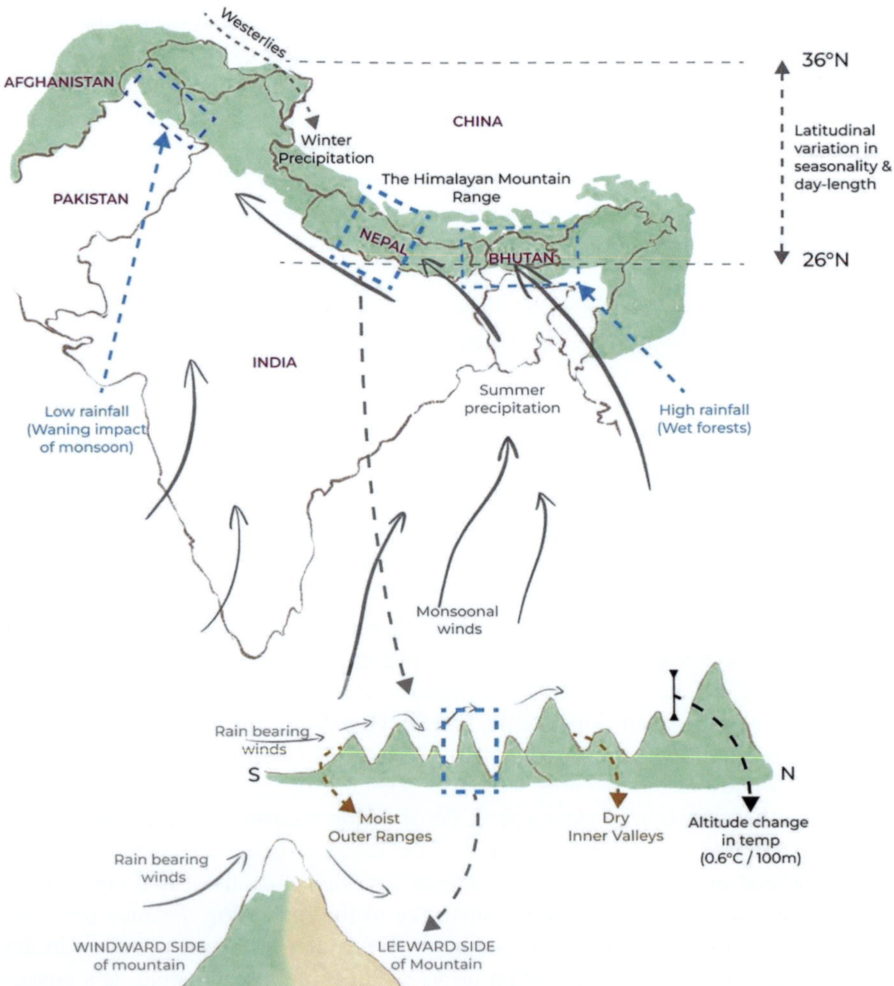

Fig. 2.2 Summary of variables impacting precipitation

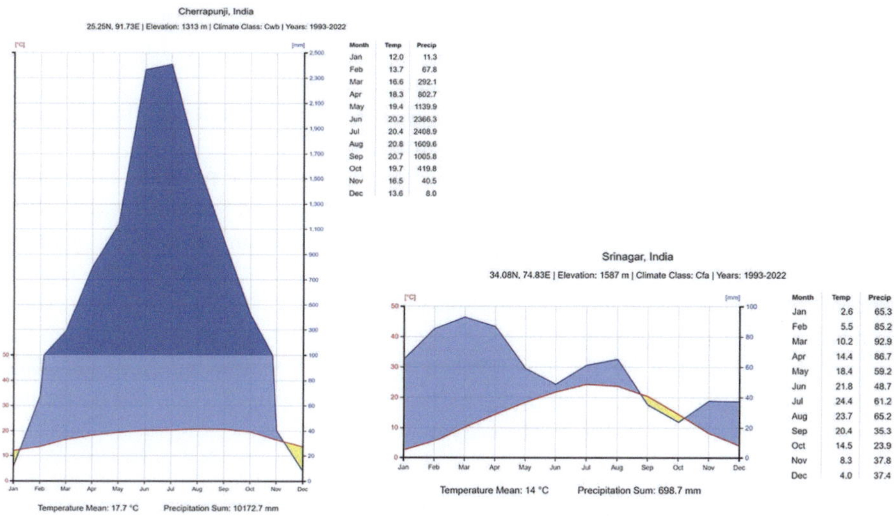

Fig. 2.3 Climate diagrams of a site in the outer Eastern (Cherrapunji) and inner Himalaya (Srinagar) demonstrate the relative importance of Monsoons and Westerlies. (Developed from Zepner et al. 2020)

The Karakoram and Western Himalayan orography induces a significant blocking effect on the westerlies, modifying circulation patterns and producing significant orographic precipitation. These lead to an increase in winter precipitation, especially snow, at higher altitudes through the western Himalayas, but they impact all the way into Nepal (Lang and Barros 2004), especially in the inner (more northern) ranges. The westerlies are more frequent and stronger in the winter season and often are the dominant source of precipitation in the western Himalayas and the Hindu Kush Karakoram region. The Kashmir Himalayas receive close to half of their precipitation during the winter months because of these winter Westerly disturbances (Cannon et al. 2015). As this system moves east, it loses moisture and precipitation decreases (Singh and Kumar 1997). Cherrapunji in the east receives only about 1% of its total rainfall in the four winter months (Nov-Feb) (see Fig. 2.3), while in Srinagar (Kashmir), February to April are the wettest months (Fig. 2.3).

The role of the Tibetan plateau needs to be emphasised as it impacts both these weather systems. In the winters, the cold Tibetan plateau acts as an elevated cold air source impacting the troposphere and causing the westerlies to split into two discrete branches. The stronger southern branch of the westerlies bring winter precipitation in the Himalayas through episodic Western Disturbances (Benn and Owen 1998; Böhner et al. 2015). The heating of the Tibetan plateau in summer raises air temperatures and enhances the pressure gradient that drives the summer monsoon (Benn and Owen 1998).

While the importance of the ISM is well understood, not just for the Himalayas but for South Asia as a whole, the role played by the westerlies is less discussed. The winter precipitation, and especially the snowfall in high elevation areas, is important for generating runoff in the warm pre-monsoon period—crucial for irrigation, for maintaining base flows of rivers during the dry period, as well as for keeping the numerous hydropower projects of the Himalayas functioning. Winter rainfall has important implications for Himalayan forests, markedly influencing phenology as well as having a considerable bearing on forest fires.

In addition, the Himalayas receive some precipitation in the pre-monsoon period (April–May). Some of this may be due to local convection and thunderstorms. While these pre-monsoonal showers occur across the Himalayan arc, in the eastern Himalayas, these are considerable enough (Stainton 1972) to influence the vegetation of the region and negate the pre-monsoon dry period evident in the western Himalayas. This pre-monsoon precipitation may be in part due to moisture from the Bay of Bengal drawn by the low-pressure systems over the warming Gangetic plains. The orographic effect forces these moist winds to ascend, cool, and condense, leading to rainfall in the Eastern Himalayas.

Orogenic Variations in Climate

Mountains create their own climate. They influence the flow of air and disturb the vertical stratification of the atmosphere by acting as physical barriers (Barros and Lettenmaier 1994). As a result of these orogenic variations, small distances of even 10 km can lead to large precipitation differences in the Himalayas (Lang and Barros 2004). Five-fold differences in precipitation can be found between major valleys and their adjacent ridges (Anders et al. 2006). Altitude can greatly impact precipitation, and the windward slope, which intercepts moisture-bearing winds, can have manifold more rainfall than the leeward slope of the same mountain (Anders et al. 2006). To sufficiently understand these differences, a dense network of rain gauges is required, which is largely lacking, especially at higher altitudes (Mal et al. 2022).

Nonetheless, there has been significant progress over the past few decades. The Tropical Rainfall Measurement Mission (TRMM)—a research satellite—provided 17 years of high-quality data on precipitation in the tropics and subtropics from 1997 to 2015 (Houze Jr. et al. 2015; Maggioni et al. 2016) and has increased our understanding of Himalayan rain and snow distribution. The installation of adequate rain gauges in a few areas, such as central Nepal (Lang and Barros 2004), has provided good short-term measurement and helped calibrate the data coming from satellites such as TRMM. Substantial gaps still remain as satellite data are flawed under certain conditions. At colder altitudes, and especially in areas with snow cover, TRMM data can be inaccurate (Shukla et al. 2019), and rain gauges remain the most accurate form of measurement. TRMM, for example, underestimates monsoonal rainfall over higher elevations. Shukla et al. (2019) found in the Upper Ganga basin that TRMM both underestimates (for heavy rainfall events) and overestimates (for low rainfall events), depending on the quantum and type of precipitation (for snowfall, the errors are significant). Further, very low precipitation rates (under 0.7 mm/hour) are not detected by TRMM radar (Anders et al. 2006). Thus, even TRMM data require correction based on field data from rain gauges, which are inadequate across the Himalayas (Shukla et al. 2019; Mal et al. 2022).

More importantly, the lack of monitoring stations, especially above 2500 m, limits our understanding of long-term trends and changes in precipitation due to climate change (Anders et al. 2006). For example, in Uttarakhand, Mal et al. (2022) identify only 19 observatories with relatively reliable long-term data. All of these were below 3000 metres, even though a third of the land area in the state is above 3000 m altitude. Further 16 of the 19 observatories, (84%) were below 2000 metres.

From the Plains to the Upper Catchments: Precipitation Changes

The central part of the Himalayas, as well as the Tibetan plateau, is largely dependent on the ISM, which provides 70–80% of the annual precipitation (Palazzi et al. 2013; Bookhagen and Burbank 2010). Above a 500 m elevation, the Central Himalayas receive a similar pattern and quantity of rainfall dominated by the ISM. However, as a result of their topographical features, both the eastern and western syntactic bends, where the Indus (West) and Brahmaputra (East) exit the Himalayas, receive considerable precipitation from winter disturbances (Bookhagen and Burbank 2010). The contribution of winter rainfall is particularly high in the west, and catchments west of the Sutlej receive a significant portion of precipitation from the winter westerlies. The impact of the western disturbances decreases as one moves east. For example, even within the state of Uttarakhand, the contribution of winter rainfall decreases from the northwestern Greater Himalayas (11% of total rainfall, >150 mm) to the lower elevations to the southeastern (6%, <90 mm) (Mal et al. 2022). At the western syntax, the Hindu Kush and Karakoram have a bimodal precipitation distribution, reflecting the significant wintertime precipitation associated with the westerlies as well as the impact of the summer monsoon.

While the east-to-west distribution of rain is influenced by the two weather systems, the sheer height and steepness of the Himalayas have a considerable impact on precipitation patterns. Along the 2500 km long Himalayan arc, there are two principal precipitation regimes as one moves perpendicular across the Himalayas (Burbank et al. 2012). The 'foreland' (the plains south of the Himalayas, generally <500 m in altitude) has a very different precipitation pattern from the main Himalayan range (500–5000 m altitude). Along the foreland plains, precipitation is greatest directly north of the Bay of Bengal and shows a strongly decreasing gradient both towards the east and west (Bookhagen and Burbank 2010).

In contrast, the mountainous regions (500–5000 m asl) receive nearly equal rainfall as one moves along the east–west direction. The integrated rainfall as one moves along the strike remains fairly similar for over 1000 km along the entire central regions of the Himalayas (Bookhagen and Burbank 2010). A significant increase in precipitation is found only in the eastern and western edges of the Himalayas.

Large precipitation variations are also found as one moves across the elevational gradient. The steepness of topography and the distance from the first ridges that rise from the plains have marked impacts on the precipitation. Windward and leeward faces of mountains can receive very different rainfall and precipitation patterns change enormously as one moves deeper into the ranges (Fig. 2.2).

A nearly continuous band of high rainfall is found along the leading edge of the Himalayas where it abuts the foreland (the initial range) as rising moist air rapidly loses the moisture it carries. Along segments of the range, a second zone of high precipitation occurs 80–100 km north of the range front. This can be typically seen when the topography rises in two significant steps (i.e. foothills to the outer range and then deeper inside the Himalayas, abruptly from the mid-elevations to the high Himalaya), such as in parts of the Central Himalayas (see Fig. 2.4b) (Shrestha et al. 2012a, b). However, when the topography rises rather steadily from the edge of the foreland up to more than 3–4 km in the Greater Himalaya, a single pronounced peak dominates monsoonal rainfall (Burbank et al. 2012) (Fig. 2.4). The height of the first ranges influences the amount of atmospheric moisture shed by the weather system. When this first range rises to about 1000–1500 m, about half the atmospheric moisture is extracted as rainfall, but if the first range approaches 3000 m, then 80% of the moisture can be extracted (Bookhagen and Burbank 2006).

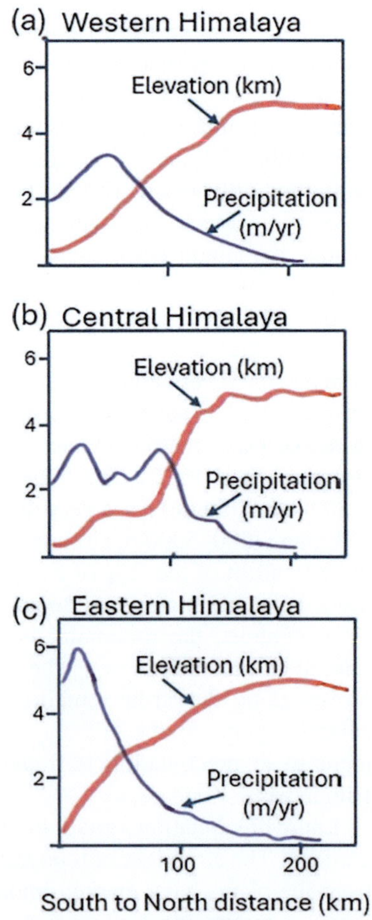

Fig. 2.4 Simplified representations of topographic change and precipitation along a south-north transect in the Western, Central, and Eastern Himalayas. (Adapted from Bookhagen and Burbank 2010)

Upper catchments that lie in the rain shadow of the Himalayas receive very little summer precipitation compared to the southern Himalayas. The monsoon winds lose most of their moisture during their passage over the plains and as they rise over the frontal mountain ranges. As these winds cross the Greater Himalayas and approach the trans-Himalayas these currents are practically dry and lead to insignificant rainfall in the trans-Himalayan regions (Singh and Kumar 1997). These catchments often receive a significant part of their precipitation from westerlies in the Western Himalayas, or the East Asian monsoon in the eastern extremities and Tibet (Burbank et al. 2012).

Consequent to this seasonality, Singh et al. (2017) call these monsoonal and non-monsoonal catchments. North of the main Himalayan ranges or in areas shielded by high mountains such as the Kashmir valley (shielded by the Pir Panjal range), the monsoon precipitation is weak, resulting in low annual precipitation and a change in seasonal distribution of rainfall. In such a non-monsoonal precipitation regime (NMPR), typically 50% or more of the annual precipitation occurs during non-monsoon periods (November to May).

In northern ranges in the central parts of the Himalayas, where neither the monsoonal nor the westerlies are strong, the rain shadow area deeper in the mountains tends to be quite dry. Studies in central Nepal (Burbank et al. 2012) show that annual precipitation decreases ten-fold between the rainfall peak (of 4000 mm/year) on the southern flank of the Himalayas and the semi-arid rain shadow on its northern flank (40–50 km farther north). Similar patterns are likely found in western Nepal and Uttarakhand as well. In Bhutan, precipitation also decreases abruptly from south to north. In the southern foothills, precipitation is often above 4000 mm, but in the rainshadow areas of the valleys of the lower midland, a mere 75 km or so north, the precipitation drops to around 600 mm (Eguchi 1991).

Monsoonal rainfall tends to be localised both in terms of intensity and quantity. Putkonen (2004) found that ridge crests can receive 10–40% more rain than adjacent valleys. In contrast, major winter snowstorms were more regionally uniform, and precipitation was a function of altitude. During winter, the valley bottoms are rather dry, but nearby ridgetops can receive 5–10 times more precipitation as snowfall (Putkonen 2004). Winter snowfall on the higher ridges can constitute as much as 40% of the annual total precipitation. The magnitude of the contrast between ridge-crest and valley-bottom stations is strongly seasonal. Modest contrasts in rainfall between ridges versus valleys during the monsoon are replaced by strong altitude-dependent precipitation contrasts in the winter (Burbank et al. 2012).

However, since air cools adiabatically at higher altitudes and moisture is condensed out (Shrestha et al. 2012b), relatively little snow accumulates above ~6200 m (max accumulation is at 4–6 km) and the highest Himalayan summits are cold and semi-arid often with little accumulation of snow or flux of snow and ice (Harper and Humphrey 2003).

Impact of Mountains on Localised Wind Systems

Characteristic of mountain systems such as the Himalayas is the development of localised mountain-valley wind systems triggered by the differential heating of slopes. These thermally driven diurnal mountain winds are produced by heating of atmospheric layers during daytime and cooling during nighttime (Zardi and Whiteman 2013). Diurnal winds are a key feature of the climatology of mountainous regions, particularly in areas where background winds are weak and skies are clear (Zardi and Whiteman 2013), and are better developed in the summer than in winter due to the stronger day–night heating contrasts. Diurnal mountain wind systems have major impacts on forest vegetation as they influence local temperatures, the prevalence of fog, frost damage, moisture, and the spread of forest fires.

There are several types of topographically influenced local wind systems. Slope wind systems are diurnal wind systems that blow up or down slopes of a valley. During the day, warmer surface temperatures on a mountain slope lead to a heated air layer that rises up the slope while continuing to gain heat from the underlying surface. These upslope moving winds are called anabatic winds. During nighttime, air in the cooled boundary layer flows down the slope while continuing to lose heat to the underlying surface (Katabatic winds). In the Himalayas, daytime anabatic circulation is much more vigorous as compared to nocturnal katabatic winds, which tend to be relatively gentle (Egger et al. 2000; Böhner et al. 2015). Diurnal valley winds are thermally driven winds that blow along the axis of a valley, with up-valley flows during daytime and down-valley flows during nighttime.

These diurnal circulation patterns cause local differences in temperature and humidity and can significantly influence vegetation. Daytime circulation models are most pronounced in the transverse valleys of the Himalayas, where they alter both thermal and moisture regimes and also alter the normal diurnal variations of temperature (Böhner et al. 2015).

Impacts of Climate Change

Long-Term Precipitation Trends in the Himalaya

The Himalayas are warming faster than surrounding areas (Shrestha et al. 2012a; Krishnan et al. 2019), and while there is evidence to support elevation-dependant warming, precipitation trends are variable and inconclusive (Krishnan et al. 2019).

Understanding how the monsoon (ISM), the main source of precipitation for the Himalayas, will change in the coming years is a challenge for climate science (Turner and Annamalai 2012), and the processes driving the monsoon are still not fully understood. The monsoon is believed to be strongly sensitive to global warming, and it is believed that rainfall magnitude may increase over the core monsoon zone in the future (Sharmila et al. 2015), with a projected increase of 4–25% monsoon precipitation in the long term (Krishnan et al. 2019). There are also strong indications of a transition towards more episodic and intense monsoonal precipitation. This would mean an increase in wet extremes and heavy precipitation events, but a possible reduction in the number of wet days. An increase in the length of dry periods in the decades to come (Palazzi et al. 2013; Sharmila et al. 2015) is likely. Such a pattern could reduce water recharge into aquifers and greatly increase both runoff and the erosivity of rainfall. While forests would be adversely impacted by these changes, they can also help mitigate some of the impacts of this increased variability in rainfall.

Winter precipitation driven by the westerlies is projected to increase by 7–15% in the Karakoram but decline slightly in the Central Himalayas (Krishnan et al. 2019). The East Asian monsoon, which in any case plays a minor role in the Himalayas, may be less susceptible to climate change in terms of its intensity, though it may change position (Li et al. 2010). The interactions of the changing ISM with the westerlies and the East Asian Monsoon are poorly understood, and there are considerable differences between various data sets on total rainfall trends across the Himalayas.

Furthermore, the extreme topographical heterogeneity of the Himalayas will lead to differential impacts of climate change. Studies that look at the effects of climate change in various parts of the Himalayas (e.g. Mal et al. 2022) give ambiguous results, with considerable regional disparity and contrasting rainfall trends. Particular stations show highly localised and peculiar trends indicative of high local scale variability (Mal et al. 2022) and hence often contradictory regional trends for precipitation. For example, long-term (1901–2020) data of northeastern India indicate that while some Himalayan states (Arunachal, Assam, Nagaland) show an overall negative trend in annual precipitation, adjacent regions (Meghalaya, Mizoram) showed a net increase in rainfall during this period (Bora et al. 2022). There were no discernible trends in pre-monsoon rainfall in this part of the Himalayas. Winter rainfall was also variable. While in Arunachal, an increase in winter rainfall was noted, most other states showed a decreasing trend.

Similarly, in the Hindu Kush Karakoram, various data sets looking at the time period between 1950 and 2010 indicate no trends of changes in winter precipitation (Palazzi et al. 2013) though earlier work has indicated an increase in winter precipitation in the HKK (Archer and Fowler 2004) which had been used to help explain what is referred to as the Karakoram anomaly—the relative stability or anomalous growth of glaciers in the Karakoram observed since the 1990s, which is in sharp contrast with glacier retreat in most parts of the world including in the adjacent Himalayas (Hewitt 2005). Recent evidence also indicates a possible end to the Karakoram anomaly (ICIMOD 2023).

Bhutiyani et al. (2009) examine long-term data for the Northwest Himalayas (Jammu, Kashmir, Ladakh, and Himachal). Some stations have data available since 1866, though many have only 30–40 years of data. While temperature increase across the region was significant, precipitation trends were variable. They found a significant decrease in annual precipitation across the region (though some stations such as Srinagar showed increasing trends). Monsoon precipitation appeared to have significantly decreased, while winter precipitation showed a non-significant increase.

Long-term data from rain gauges are perhaps the best predictor of future trends, but these data are sparse, and given local variability in trends, meaningful estimates are hard to make. It has been suggested that other proxy studies, such as dendroclimatic analysis to reconstruct past climate, may help in enhancing our understanding. Studies of tree ring records from deodar trees in Uttarakhand (Ram 2022) found that higher winter temperatures (Nov-Jan) were positively correlated with the subsequent season's growth, possibly as higher winter temperatures trigger earlier snowmelt, which increases soil moisture availability. Looking at tree cores dating back to 1740, Ram (2022) found wider growth rings of deodar since about 1950, possibly a result of increasing winter temperatures. A more detailed account of climate change impacts on Himalayan ecosystems and dendrochronological findings is provided in Chap. 3 of volume 2 of this book (Climate Change and Forests).

Strong trends of elevation-dependent warming are reported. In the Nepal Himalayas, the rate of increase of daytime temperature was found almost twice as high at ~5000 m elevation (~0.16 °C/year) compared to lower elevations (~2000 m, daytime mean annual surface temp change ~0.08 °C/year) (Zhao et al. 2019). Shekhar et al. (2010) used data from the Snow and Avalanche Study Establishment (SASE) to study temperatures in the Western Himalayas (Kashmir–Himachal region) and found a greater increase in maximum temperatures (~2.8 °C) in the period 1984–85 to 2007–08 and a smaller increase in minimum temperatures (~1 °C). A regional increase in the number of warm nights across the Hindu Kush Himalaya(HKH) and a strong decline in extreme cold nights is well established (Krishnan et al. 2019).

The Himalayan arc does exhibit considerable variability. Shekhar et al. (2010) found in the Western Himalayas, while three of the four main ranges, viz. the Greater Himalayas, Pir Panjal, and Shamshawari show increasing temperature trends during the period 1988–2008, a drop in temperature was noted in the Karakoram. Snowfall was found to be lower in all four of these Western Himalayan ranges, though the decreasing trend in snowfall was minimal for the Karakoram (Shekhar et al. 2010).

While Global Climate change is considered responsible for these variations, it should be noted that increasing temperatures and glacial melting in the Himalayas can also be attributed in part to local phenomenon, such as increased fossil fuel burning, black carbon, deforestation, and urbanisation (Shekhar et al. 2010; Zhao et al. 2019).

Changes in Himalayan Glaciers: Climate Impacts on the Cryosphere

The Hindu Kush Himalaya (HKH) contains the largest volume of ice on Earth outside the polar regions, and the unprecedented and largely irreversible changes in the Hindu Kush Himalaya (HKH) cryosphere are a matter of global concern (ICIMOD 2023). Climate change, elevation-dependant warming, and the deposition of black carbon from pollutants in the densely populated regions surrounding the Hindu Kush Himalaya (HKH) have led to accelerated rates of glacier melt. Worryingly, glacier mass loss has accelerated in the past decade and increased by 65%, from an average of −0.17 metres water equivalent (m w.e.) per year for the period

2000–2009 to −0.28 m w.e. per year for 2010–2019 (Jackson et al. 2023). Assuming global warming levels of 1.5 °C to 2 °C, the Hindu Kush Himalaya (HKH) glaciers are expected to lose 30–50% of their volume by 2100, and losses will be significantly higher if warming crosses 2 °C (ICIMOD 2023), as appears likely.

As with the impacts of climate change, glacier melt is variable, and the reduction in size and mass varies across the Himalayan arc. In general, a rapid decline of Himalayan glaciers has been reported since the 1920s. In their study of glaciers in the Tibetan plateau, Yao et al. (2012) find that glaciers in the central regions show the most extreme shrinkage among Himalayan glaciers—both in terms of glacier length and area, as well as the most negative mass balances. However, an increase in precipitation in the eastern Pamir region over the past half century (Yao et al. 2012) has led to an anomalous behaviour of glaciers in this region. In the Pamir region, north of the Hindu Kush Himalaya, glacier contraction was minimal and slow, and some glaciers even showed a positive mass balance. The *Karakoram anomaly* (Hewitt 2005) refers to this apparent expansion of many central Karakoram glaciers, which, after decades of decline, were observed to have started expanding in the late 1990s. This was in contrast to the contraction observed in glaciers across the rest of the Hindu Kush Himalaya (HKH).

Dimri (2021) suggested mechanisms to explain this Karakoram anomaly and found that geographical and elevation positioning of the Karakoram make its environmental conditions conducive for glacier stability. An increase in the frequency of western disturbances and consequently increased winter precipitation over this western extreme of the Himalayas help explain the stability of Karakoram glaciers (Dimri 2021; Ridley et al. 2013). Recent detailed inventories of the Karakoram (Xie et al. 2023) find approximately 10,500 glaciers in the Karakoram mountains alone, covering over 22,500 sq. km (see table in Fig. 2.5). The total glacial cover in the Karakoram has remained largely stable with a small increase between 1990 and 2020 (Xie et al. 2023) in the central and western Karakoram. Recent data (Hugonnet et al. 2021; Jackson et al. 2023) indicate evidence from the late 2010s suggesting an apparent end to the Karakoram anomaly of mass gain, and subsequent thinning of these glaciers in the future.

Climate Impacts on River Flow and Discharge

The hydrological budget of Himalayan rivers is dominated by monsoonal rainfall and snowmelt, but their relative impact differs across various basins (Bookhagen and Burbank 2010) with snowmelt being critical for sustaining flows in the Indus basin, important in the Brahmaputra basin, but having a limited role in the other important Himalayan watersheds such as the Ganges, Yangtze, and Yellow rivers (Immerzeel et al. 2010).

In the far western ranges, the contribution of snowmelt to river discharge is as high as 66% for the Indus as it emerges from the Himalayas and 57% for the Sutlej. In the east, 34% of the discharge for the Brahmaputra (Tsangpo) basin is contributed by snowmelt. The central Gangetic river catchments (Fig. 2.6), such as those of the Ganga, Yamuna, or Sharda, have less than 25%, and often under 20% of their annual

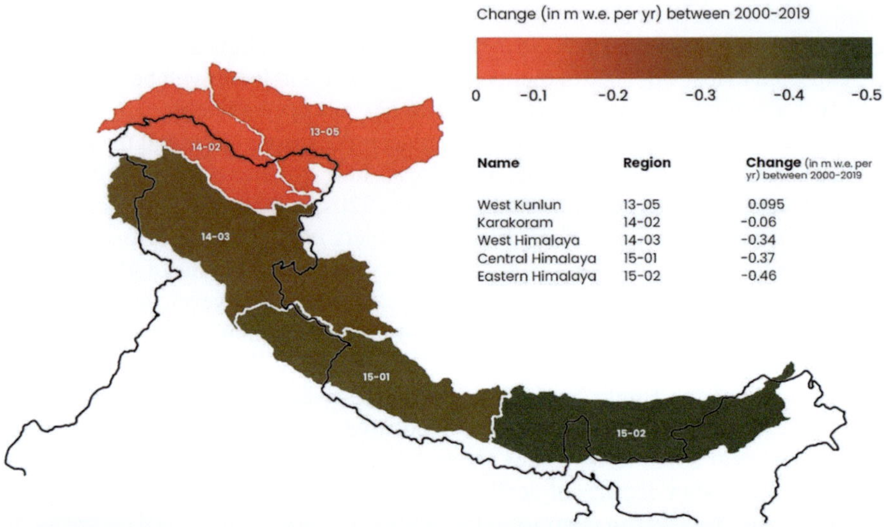

Glacier Mass Balances from different regions in the Himalaya
(adapted from Jackson et al. 2023)

Region	RGI number		Glacier Area (km²)	Change (in m w.e. per yr) 1975-99ᵃ	2000-09	2010-19
West Kunlun	13-05	NW Tibetan plateau	8,141	0.02 ± 0.14	0.20 ± 0.05	- 0.01 ± 0.04
Karakoram	14-02	Ladakh (China / Pak / Ind border areas)	22,881	- 0.07 ± 0.05	-0.03 ± 0.05	- 0.09 ± 0.04
West Himalaya	14-03	Himachal (Spiti Lahaul)	7,776	-0.12 ± 0.09	-0.32 ± 0.05	-0.36 ± 0.06
Central Himalaya	15-01	North UK, NW Nepal	5,450	-0.25 ± 0.10	-0.34 ± 0.05	-0.40 ± 0.06
Eastern Himalaya	15-02	E. Nepal, Bhutan, Arunachal	4,904	-0.25 ± 0.08	-0.41 ± 0.06	-0.51 ± 0.07

1975-99[a] Spatial coverage of glaciers studied was low in this period leading to less accurate number
RGI no. is the Randolph Glacier Inventory numbering

m w.e per yr - metres water equivalent per year
(as density of ice is 0.9 times the density of water, so a loss of 1.0 m w.e. per yr is equivalent to a loss of about 1.1 meters of ice)

Fig. 2.5 Changes reported in glaciers across some Himalayan regions. (Figure adapted from data in Jackson et al. 2023 (ICIMOD))
1975-99[a] Spatial coverage of glaciers studied was low in this period leading to less accurate number
RGI no. is the Randolph Glacier Inventory numbering
m w.e per yr - metres water equivalent per year
(as density of ice is 0.9 times the density of water, a loss of 1.0 m w.e. per yr is equivalent to ~ 1.1 meters of ice)

discharge from snowmelt (Bookhagen and Burbank 2010; Burbank et al. 2012). A global warming scenario may lead to higher levels of water in the spring season as glaciers melt, but by around 2050–2060 glacier run-offs will peak and thereafter decline (Nepal et al. 2023).

Climate change will differentially impact these river basins as the impact of monsoons and western disturbances, and the altitudinal distribution of their catchments is quite different. The Brahmaputra basin has 68% of its area at elevations above 2000 m, the Indus basin has 40%, but for the Ganga basin only

Himalayan watershed	Indus	Ganga	Brahmaputra (Tsangpo)
Downstream population	300 million	600 million	150 million
Precipitation system	Monsoon + Westerlies	Monsoon	Monsoon + some winter ppt
May-October rainfall contribution to discharge[(1)]	25-65% for the various rivers	>70%	>70%
Major rivers catchments [(1)]	Indus - 205,500 sq km Sutlej - 52,800 sq km Jhelum - 26,500 sq km	Ganga – 21,700 sq km Karnali – 45,500 sq km Arun - 33,600 sq km	Tsangpo –256,000 sq km Subansiri – 26,000 sq km Manas – 21,000 sq km
Contribution of snowmelt (primary river)[(1)(2)]	Very high (~2/3rd)	Low (~1/4th)	Moderate (~1/3rd)
Area above 2000 m elevation[(3)]	40%	14%	68%
Avg. winter ppt[(4)]	115 mm	72mm	152mm
Avg. monsoon ppt[(4)]	315 mm	1050 mm	1385 mm
Annual ppt[(4)]	610 mm	1310 mm	2220 mm
Winter ppt (as % of total)	18.9%	5.5%	6.9%

Fig. 2.6 Major river basins situated within the Himalayas
The map shows the part of the Hindu Kush Himalaya (HKH) occupied by the three major river basins of the region. Parts of these river basins, situated in the plains outside the Hindu Kush Himalaya (HKH), are not shown in this visualisation
1. Bookhagen and Burbank (2010)
2. Nepal et al. (2023)
3. Immerzeel et al. (2010)
4. Jackson et al. (2023)
5. Figure adapted from Naegeli et al. (2022)

14% of the total area is above 2000 m (Immerzeel et al. 2010). Thus, even though the reduction in snow and ice predicted by 2050 as a result of climate change is strongest in the Ganga basin, with a predicted decrease in mean upstream water supply of 17.6%, the change in flow is not very significant in part because of the low dependence on snow and glaciers and in part as any reduction is compensated by increased mean upstream rainfall (a predicted 8% increase in the case of the Ganga) (Immerzeel et al. 2010). For the major Himalayan basins, Immerzeel et al. (2010) suggest that increased upstream rainfall will, to an extent, negate the decrease due to changes in glaciers, thereby leading to a lower impact in river flow than earlier predicted.

The contribution of snowmelt is also dependent on elevation. Whereas little snow accumulates below 2000 m, snow accumulations steadily increase at higher altitudes of up to 4500–5000 m asl, before levelling off and then dramatically decreasing above 6000 m elevation (Burbank et al. 2012). Of Himalayan catchments, the Indus is most heavily dependent on snow and glaciers. The major contributors to the flow of the Upper Indus are zones of heavy snowfall and the glacierised basins of the Karakoram (Hewitt et al. 1989). The impacts of climate change were predicted to be lower than expected for this river due to the relative stability of these Karakoram glaciers and their low rates of loss (Hewitt 2005). While recent evidence indicates an increased rate of glacier melting (Jackson et al. 2023), it is as yet thought to be much lower than in other parts of the Himalayas.

Whereas monsoonal precipitation dominates the overall moisture flux, its contribution to runoff exhibits significant spatial variability (Bookhagen and Burbank 2010). In the Central Himalayas, the ISM contributes about 80% of the annual discharge from rivers, whereas in the Western Himalayas, where the westerlies are more active, the contribution to annual discharge varies from 25% to 65% (Burbank et al. 2012). Seasonality greatly influences the relative importance of snowmelt and rainfall. Snowmelt is particularly important during the late spring and summer seasons (from April to June). Changes in timing or amount of snowmelt due to rising temperatures or warmer spring can have important societal consequences (Bookhagen and Burbank 2010) (see also Box 2.1).

Box 2.1: Seasonal impact of Solar Insolation

The direction a slope faces, or its aspect, significantly influences the amount of solar insolation (sunlight) received and thereby shapes local climates and forest types. South-facing slopes receive more direct sunlight than north-facing slopes throughout the year in the Himalayas, which are located north of the Tropic of Cancer (23.5° N)—the northernmost latitude where the sun can be directly overhead at noon. The difference in solar insolation is particularly stark during the winter, and north-facing slopes have cooler and more moist conditions (see also Chap. 11: Response to Slope and Aspect).

This leads to more mesophyllic vegetation on north-facing slopes, which are shaded and retain moisture for longer, while southern slopes are more xerophytic. In Uttarakhand, for example, oaks such as *Quercus leucotrichophora* commonly dominate north slopes, while south-facing slopes have a higher component of chir pine (*Pinus roxburghii*).

Calculating the Angle of the Mid-Day Sun in the Winter and Summer

In the northern hemisphere, to calculate the sun's angle during the winter and summer solstice, one needs to add (for winter solstice) or subtract (for summer solstice) 23.5° to the latitude and subtract the value from 90°. This gives the angle of elevation from the horizon of the sun during midday of the solstice and defines the limit of the sun's angle during midsummer and midwinter.

For example, for Mussoorie in the Central Himalayas (located 30.5° N), the values for solar angle during mid-day would be:

(continued)

Box 2.1 (continued)
Summer solstice: 90°—(30.5–23.5) = 83°.
Winter solstice: 90°—(30.5 + 23.5) = 36 °.

This helps us understand the impact of seasonality on steep slopes. In the summer, when the sun can reach almost overhead (83° from the horizon), the difference in sunlight between north and south slopes is minimal, but in the winter, seasonality is very pronounced, and steep north-facing slopes may be permanently shaded (see Fig. 2.7 below). In January, the sunlit southern slopes of mountain crests can receive about four times more solar insolation than north-facing slopes (Böhner et al. 2015).

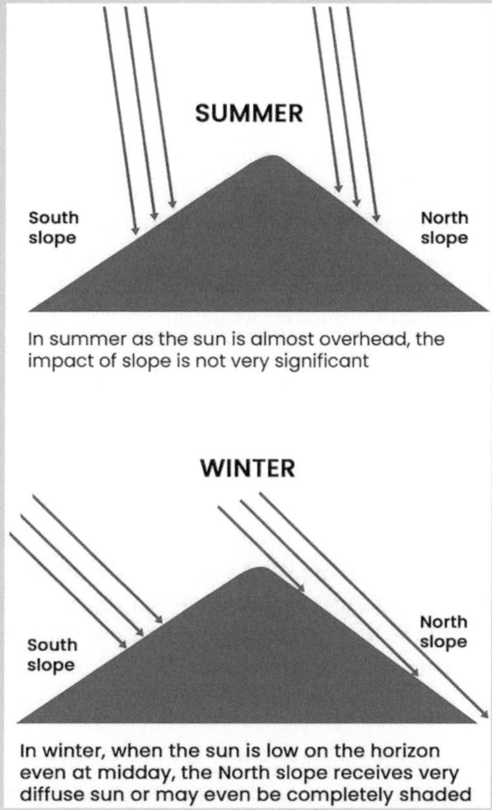

Fig. 2.7 In the summer, the difference in solar radiation received by north- and south-facing slopes across the Himalayas is relatively minor, but in winter, as the angle of the sun is low over the horizon, north slopes receive only a fraction of the solar radiation of south-facing slopes

Summary and Conclusions

The vast east-to-west expanse of the Himalayas, changing precipitation patterns, and the steep topography lead to a patchwork of climatic conditions across the region. Nonetheless, some patterns are apparent and some generalisations can be made:

- Moving east to west along the foreland plains abutting the Himalayan arc, there is a six-fold east-to-west rainfall decline along the Ganga foreland (< 500 m). Rainfall is maximum in the forelands towards the eastern edge, directly north of the Bay of Bengal, and declines to a sixth towards the western edge of the northern plains. However, in the mountains above 500 m, this gradient is not apparent, and rainfall remains nearly unchanged for about 1000 km along the east-to-west arc.
- Moving across the width of the range, in an approximately south-to-north direction, an almost ten-fold decline in precipitation can be seen in the central part of the Himalayan ranges, from 4000 mm in parts of the frontal ranges to under 500 mm as one crosses into the Tibetan Plateau.
- While the southern flanks of the Himalayas, abutting the Indian plains, and the entire central portion of the Himalayas are dependent on the ISM, both syntaxes receive significant rainfall outside the monsoon. The western parts in particular receive a large proportion (often 40%) of the total precipitation as winter rain and snow. These more interior areas at either end of the Himalayas tend to have a non-monsoonal precipitation regime, while the bulk of the Himalayas has monsoonal precipitation patterns.
- The Tibetan plateau plays an important role in regulating the precipitation systems of the Himalaya.
- Precipitation tends to be highest at the leading edge of the Himalayan ranges where the windward slopes cause the rising moist air to rapidly shed moisture. About 50–80% of the moisture can be extracted by the first ranges alone, leading to a sudden drop in precipitation as one moves inland across the Himalayan range.
- Winter storms tend to be more uniform across a region, and precipitation differences are often dependent on altitude. Ridges can sometimes receive 5–10 times more precipitation in the form of snow compared to adjacent valleys. Monsoonal storms tend to be more localised in intensity and quantity of rain, and ridge to valley differences are smaller, but the variation across small distances is greater.
- Small distances of a few kilometres can lead to large differences in precipitation due to the topography, leading to a heterogeneous mosaic of precipitation patterns.
- Changes in precipitation patterns because of climate change are inconsistent across the Himalayas and differ regionally.
- Climate change is leading to warming across the Himalayas. Warming is strongest at higher altitudes (elevation-dependant warming).
- River basins have widely differing dependence on snow and glaciers. While glacier melting is strongest in the central sections of the Himalayas, rivers flowing

in this region have the lowest dependence on snow and ice. The Indus, in contrast, which is heavily dependent on snow and ice for its discharge, also has a significant part of its catchment in the Karakoram–Pamir region, which is least impacted by glacial melt.
- Glaciers across the Himalayas are melting. Though rates of melting differ along the east-to-west arc, it appears to have accelerated in recent years due to global warming.
- The lack of an adequate network of rain gauges and the presence of very few weather stations with long-term data hinder our understanding of Himalayan precipitation and changes occurring due to global warming. Data are particularly sparse at higher altitudes.
- Vegetation composition and productivity are greatly impacted by these changing climatic conditions. Differing temperature lapse rates lead to differences in vegetation and tree line altitudes between the east and west Himalayas.

These regional changes in precipitation across the Himalayan arc, and temperature and rainfall changes due to varying topography influence productivity and diversity, lead to the dominance of particular plant families and species in different parts of the Himalayas, which are discussed in more detail in the following two chapters.

References

Anders AM, Roe GH, Hallet B, Montgomery DR, Finnegan NJ, Putkonen J (2006) Spatial patterns of precipitation and topography in the Himalaya. Special Papers-Geological Society of America, vol 398, p 39

Archer DR, Fowler HJ (2004) Spatial and temporal variations in precipitation in the upper Indus Basin, global teleconnections and hydrological implications. Hydrol Earth Syst Sci 8(1):47–61

Barros AP, Lettenmaier DP (1994) Dynamic modelling of orographically induced precipitation. Rev Geophys 32(3):265–284

Benn DI, Owen LA (1998) The role of the Indian summer monsoon and the mid-latitude westerlies in Himalayan glaciation: review and speculative discussion. J Geol Soc 155(2):353–363

Bhutiyani MR, Kale VS, Pawar NJ (2009) Climate change and the precipitation variations in the northwestern Himalaya: 1866–2006. Int J Climatol 30(4):535–548

Blandford TR, Humes KS, Harshburger BJ, Moore BC, Walden VP, Ye H (2008) Seasonal and synoptic variations in near-surface air temperature lapse rates in a mountainous basin. J Appl Meteorol Climatol 47(1):249–261

Böhner J, Miehe G, Miehe S, Nagy L (2015) Climate and weather. In: Nepal: an introduction to the natural history, ecology and human environment in the Himalayas, pp 23–90

Bookhagen B, Burbank DW (2006) Topography, relief, and TRMM-derived rainfall variations along the Himalaya. Geophys Res Lett 33(8)

Bookhagen B, Burbank DW (2010) Toward a complete Himalayan hydrological budget: spatiotemporal distribution of snowmelt and rainfall and their impact on river discharge. J Geophys Res Earth 115(F3)

Bora SL, Bhuyan K, Hazarika PJ, Gogoi J, Goswami K (2022) Analysis of rainfall trend using non-parametric methods and innovative trend analysis during 1901–2020 in seven states of North East India. Curr Sci 122(7):801

Burbank DW, Bookhagen B, Gabet EJ, Putkonen J (2012) Modern climate and erosion in the Himalaya. Compt Rendus Geosci 344(11–12):610–626

Cannon F, Carvalho LM, Jones C, Bookhagen B (2015) Multi-annual variations in winter westerly disturbance activity affecting the Himalaya. Clim Dyn 44:441–455

Dimri AP (2021) Decoding the Karakoram anomaly. Sci Total Environ 788

Egger J, Bajrachaya S, Egger U, Heinrich R, Reuder J, Shayka P, Wendt H, Wirth V (2000) Diurnal winds in the Himalayan Kali Gandaki valley. Part I: observations. Mon Weather Rev 128(4): 1106–1122

Eguchi T (1991) Regional and seasonal change of precipitation in Bhutan – analysis of daily precipitation in 1988. In: Ohsawa M (ed) Life zone ecology of the Bhutan Himalaya II. Chiba University, pp 1–20

Harper JT, Humphrey NF (2003) High altitude Himalayan climate inferred from glacial ice flux. Geophys Res Lett 30(14)

Hewitt K (2005) The Karakoram anomaly? Glacier expansion and the 'elevation effect,' Karakoram Himalaya. Mt Res Dev 25(4):332–340

Hewitt K, Wake CP, Young GJ, David C (1989) Hydrological investigations at Biafo glacier, Karakoram range, Himalaya; an important source of water for the Indus River. Ann Glaciol 13: 103–108

Houze RA Jr, Rasmussen KL, Zuluaga MD, Brodzik SR (2015) The variable nature of convection in the tropics and subtropics: a legacy of 16 years of the Tropical Rainfall Measuring Mission satellite. Rev Geophys 53(3):994–1021

Hugonnet R, McNabb R, Berthier E, Menounos B, Nuth C, Girod L, Farinotti D, Huss M, Dussaillant I, Brun F, Kääb A (2021) Accelerated global glacier mass loss in the early twenty-first century. Nature 592(7856):726–731

ICIMOD (2023) Water, ice, society, and ecosystems in the Hindu Kush Himalaya: an outlook. (P Wester, S Chaudhary, N Chettri, M Jackson, A Maharjan, S Nepal, & J F Steiner [Eds.]). ICIMOD. https://doi.org/10.53055/ICIMOD.1028

Immerzeel WW, Van Beek LP, Bierkens MF (2010) Climate change will affect the Asian water towers. Science 328(5984):1382–1385

Jackson M, Azam MF, Baral P, Benestad R, Brun F, Muhammad S, Pradhananga S, Shrestha F, Steiner JF, Thapa A (2023) Consequences of climate change for the cryosphere in the Hindu Kush Himalaya. In: Wester P, Chaudhary S, Chettri N, Jackson M, Maharjan A, Nepal S, Steiner JF (eds) Water, ice, society, and ecosystems in the Hindu Kush Himalaya: an outlook. ICIMOD, pp 17–71

Joshi R, Tamang ND (2021) Chapter 7: temperature lapse rate in the Himalaya. In: Singh SP, Singh RD, Gumber S (eds) Interpreting mountain treelines in a changing world. Central Himalayan Environment Association and International Centre for Integrated Mountain Development. https://doi.org/10.53055/ICIMOD.787

Kattel DB, Yao T, Yang K, Tian L, Yang G, Joswiak D (2013) Temperature lapse rate in complex mountain terrain on the southern slope of the Central Himalayas. Theor Appl Climatol 113:671–682

Kattel DB, Yao T, Yang W, Gao Y, Tian L (2015) Comparison of temperature lapse rates from the northern to the southern slopes of the Himalayas. Int J Climatol 35(15):4431–4443

Krishnan R, Shrestha AB, Ren G, Rajbhandari R, Saeed S, Sanjay J, Syed MA, Vellore R, Xu Y, You Q, Ren Y (2019) Unravelling climate change in the Hindu Kush Himalaya: rapid warming in the mountains and increasing extremes. In: Wester P et al (eds) The Hindu Kush Himalaya assessment: mountains, climate change, sustainability and people. ICIMOD/Springer, pp 57–97. https://doi.org/10.1007/978-3-319-92288-1_3

Krishnan R, Gnanaseelan C, Sanjay J, Swapna P, Dhara C, Sabin TP et al (2020) Introduction to climate change over the Indian region. Assessment of climate change over the Indian Region: a report of the Ministry of Earth Sciences (MoES), Government of India, pp 1–20

Lang TJ, Barros AP (2004) Winter storms in the Central Himalayas. J Meteorol Soc Jpn 82(3): 829–844

Li J, Wu Z, Jiang Z, He J (2010) Can global warming strengthen the East Asian summer monsoon? J Clim 23(24):6696–6705
Maggioni V, Meyers PC, Robinson MD (2016) A review of merged high-resolution satellite precipitation product accuracy during the Tropical Rainfall Measuring Mission (TRMM) era. J Hydrometeorol 17(4):1101–1117
Mal S, Arora M, Banerjee A, Singh RB, Scott CA, Allen SK, Karki R (2022) Spatial variations and long-term trends (1901–2013) of rainfall across Uttarakhand Himalaya, India. In: Mountain landscapes in transition. Springer, Cham, pp 163–183
Naegeli K, Franke J, Neuhaus C, Rietze N, Stengel M, Wu X, Wunderle S (2022) Revealing four decades of snow cover dynamics in the Hindu Kush Himalaya. Sci Rep 12(1):13443
Nepal S, Steiner JF, Allen S, Azam MF, Bhuchar S, Biemans H, Dhakal M, Khanal S, Li D, Lutz A, Pradhananga S, Ritzema R, Stoffel M, Stuart-Smith R (2023) Consequences of cryospheric change for water resources and hazards in the Hindu Kush Himalaya. In: Wester P, Chaudhary S, Chettri N, Jackson M, Maharjan A, Nepal S, Steiner JF (eds) Water, ice, society, and ecosystems in the Hindu Kush Himalaya: an outlook. ICIMOD, pp 73–121. https://doi.org/10.53055/ICIMOD.1031
Palazzi E, von Hardenberg J, Provenzale A (2013) Precipitation in the Hindu-Kush Karakoram Himalaya: observations and future scenarios. J Geophys Res Atmos 118(1):85–100
Putkonen JK (2004) Continuous snow and rain data at 500 to 4400 m altitude near Annapurna, Nepal, 1999–2001. Arct Antarct Alp Res 36(2):244–248
Ram S (2022) Dendroclimatic analysis of Western Himalayan tree growth in India. Curr Sci 122(7): 769–771
Ridley J, Wiltshire A, Mathison C (2013) More frequent occurrence of westerly disturbances in Karakoram up to 2100. Sci Total Environ 468:S31–S35
Sharmila S, Joseph S, Sahai AK, Abhilash S, Chattopadhyay R (2015) Future projection of Indian summer monsoon variability under climate change scenario: an assessment from CMIP5 climate models. Glob Planet Chang 124:62–78
Shekhar MS, Chand H, Kumar S, Srinivasan K, Ganju A (2010) Climate-change studies in the western Himalaya. Ann Glaciol 51(54):105–112
Shrestha UB, Gautam S, Bawa KS (2012a) Widespread climate change in the Himalayas and associated changes in local ecosystems. PLoS One 7(5):e36741. https://doi.org/10.1371/journal.pone.0036741
Shrestha D, Singh P, Nakamura K (2012b) Spatiotemporal variation of rainfall over the central Himalayan region revealed by TRMM precipitation radar. J Geophys Res Atmos 117(D22)
Shukla AK, Ojha CSP, Singh RP, Pal L, Fu D (2019) Evaluation of TRMM precipitation dataset over Himalayan catchment: the upper ganga basin, India. Water 11(3):613
Singh P, Kumar N (1997) Effect of orography on precipitation in the western Himalayan region. J Hydrol 199(1–2):183–206
Singh SP, Singh RD, Gumber SU, Bhatt SP (2017) Two principal precipitation regimes in Himalayas and their influence on tree distribution. Trop Ecol 58(4):679–691
Stainton JDA (1972) Forests of Nepal. John Murray, London
Turner AG, Annamalai H (2012) Climate change and the South Asian summer monsoon. Nat Clim Chang 2(8):587–595
Xie F, Liu S, Gao Y, Zhu Y, Bolch T, Kääb A, Duan S, Miao W, Kang J, Zhang Y, Pan X, Qin C, Wu K, Qi M, Zhang X, Yi Y, Han F, Yao X, Liu Q, Wang X, Jiang Z, Shangguan D, Zhang Y, Grünwald R, Adnan M, Karki J, Saifullah M (2023) Interdecadal glacier inventories in the Karakoram since the 1990s. Earth Syst Sci Data 15:847–867. https://doi.org/10.5194/essd-15-847-2023
Yao T, Thompson L, Yang W, Yu W, Gao Y, Guo X, Yang X, Duan K, Zhao H, Xu B, Pu J (2012) Different glacier status with atmospheric circulations in Tibetan Plateau and surroundings. Nat Clim Chang 2(9):663–667
Yihui D, Chan JC (2005) The East Asian summer monsoon: an overview. Meteorog Atmos Phys 89(1):117–142

Zardi D, Whiteman CD (2013) Diurnal mountain wind systems. In: Chow F, De Wekker S, Snyder B (eds) Mountain weather research and forecasting. Springer atmospheric sciences. Springer, Dordrecht. https://doi.org/10.1007/978-94-007-4098-3_2

Zepner L, Karrasch P, Wiemann F, Bernard L (2020) ClimateCharts.net – an interactive climate analysis web platform. Int J Digital Earth

Zhao W, He J, Wu Y, Xiong D, Wen F, Li A (2019) An analysis of land surface temperature trends in the central Himalayan region based on MODIS products. Remote Sens 11(8):900

Himalayan Forests and a Leaf-Based Classification

3

Rajesh Thadani and S. P. Singh

Abstract

This chapter provides an overview of Himalayan forest classification and summarises a novel leaf-based classification system. The Himalayas, with their vast climatic and topographical diversity, host a complex mosaic of forest ecosystems primarily determined by elevation-dependent temperature gradients and precipitation patterns. Traditional classification systems have categorised Himalayan forests based on climatic factors, but these systems often struggle to capture the ecological nuances of mountainous vegetation. We propose a classification based on leaf traits—including leaf life span, phenology, form and sclerophylly—which offers greater ecological relevance. Himalayan forests are divided into five main types: evergreen with about one-year leaf lifespan (e1), tropical evergreen with mixed leaf lifespan (e2), coniferous with multi-year leaves (e3), winter deciduous (dw) and pre-monsoon deciduous (dp). Particularly distinctive are the e1 forests, characterised by simultaneous leaf drop and flush during pre-monsoon periods, which represent a uniquely Himalayan adaptation to monsoon-dominated climate patterns. A leaf-based classification system enables more effective remote sensing applications and provides insights into forest responses to climate change. The chapter also highlights how biogeographic

R. Thadani (✉)
Centre for Ecology Development and Research (CEDAR), New Delhi, India

Department of Botany, Jai Hind College, Churchgate, Mumbai, India
e-mail: thadani_rajesh@hotmail.com

S. P. Singh
INSA Honorary Scientist, Dehradun, Uttarakhand, India

Centre for Himalayan Studies, Graphic Era (deemed to be) University, Dehradun, Uttarakhand, India
e-mail: spsecology@gmail.com

S. P. Singh, R. Thadani (eds.), *Himalayan Forests in a Changing World*,
https://doi.org/10.1007/978-981-95-0978-2_3

factors create a fundamental division between the diverse Indo-Malayan forests of the eastern Himalayas and the simpler, often monodominant Palearctic-influenced forests of the western region.

Keywords

Forest classification · Biogeographic patterns · Leaf traits · Pre-monsoon drought · Evergreen adaptations · Deciduous

Introduction

The Himalayas are vast and climatologically diverse, and describing the vegetation of these mountains is a complex task. Himalayan forests are much more than the trees that visually dominate these ecosystems. Diverse angiosperm lifeforms are found in the understory as shrubs and herbs, or as epiphytes on trees. Groups such as pteridophytes (the ferns), bryophytes (mosses), various fungi (including mycorrhizal associations), and lichens can form vital associations. Together, these influence how trees regenerate and grow, modify nutrient cycling, protect the forest soils from erosion, and influence ecosystem function. While the appearance of a forest is best described by the dominant trees, dense thickets of shrubs, understories of bamboo, or thick growths of epiphytic ferns, mosses and orchids can greatly modify the appearance of forests.

This chapter focuses on forest trees - the dominant emergent lifeform of a forest. To classify Himalayan forests, it is useful to focus on trees. Herbs and shrubs form patterns that are influenced by overstory forests. Furthermore, the effect of human disturbances (through activities such as grazing, litter removal, fodder extraction, or surface fires) can, in short time scales, very significantly impact herb and shrub flora. Given the high impact of humans across the Himalayan landscape, it becomes more difficult to use these lifeforms for forest classification. Within trees the focus is also on overstory trees.

We provide an overview of the main types of forests found in the Himalayas and describe patterns discernible across the landscapes. In mountain ecosystems, elevation-dependent temperature and precipitation are the two primary biophysical determinants that influence tree species composition. Additionally, we live in the Anthropocene, and humans greatly influence forests. While acknowledging the major role played by humans across the landscape, this chapter focuses on natural (i.e. non-human) patterns of forest differentiation. Human impacts are dealt with in later chapters of this book, and the reader is encouraged to refer in particular to the last five chapters of Volume 2 of this book, in particular the chapters on Degradation (Chap. 9) and Silviculture (Chap. 10).

In addition to engaging with established systems of forest classification, we provide an alternate classification based on leaf type, which we believe to be more ecologically relevant and appropriately characterises the Himalayan forests.

Classification Systems Developed for Himalayan Forests

Many attempts have been made to study vegetation and successional patterns in Himalayan forests. Well known are the efforts of Kenoyer (1921) and Dudgeon and Kenoyer (1925) in Nainital, Champion's (1936) forest classification for India and Burma, Schweinfurth's (1957) detailed map on Himalayan vegetation, and Singh and Singh's (1987, 1992) reviews of Himalayan vegetation. For Nepal, Stainton (1972) provides an excellent account of forest associations in his 'Forests of Nepal'. More recently, Miehe et al. (2015) enumerated some 55 vegetation units for Nepal, including over 40 types dominated by trees. For Bhutan, several works by Ohsawa and in particular 'Life zone ecology of the Bhutan Himalaya' (Ohsawa 1987) provide an excellent account of forest formations.

The most influential work, however, remains 'Forest Types of India' by Champion and Seth (1968). In this revision of an earlier publication by Champion (1936), the authors divide the forests across India into six major groups based on climatic factors. While the bulk of India's forests were part of the first two groups, viz. (I) Moist Tropical Forests and (II) Dry Tropical forests, the remaining four groups were devoted to forests that occurred in mountainous locations and especially the Himalayas. These were groups III (Montane Subtropical forests), IV (Montane Temperate forests), V (Sub-Alpine forests) and VI (Alpine forests). These six major groups were further subdivided into 16 type groups based on moisture regimes and physiognomy, and here again, nine of these types pertain to montane environments (Table 3.1). The preponderance of categories of montane forests, which make up the majority of forest divisions of India, reflected the variability of Himalayan forests when such climatic factors were used for classification.

Table 3.1 Forest types of India and their importance in the Himalayan landscape

Group	Forest Type[1]	Area[2] (sq km) in Indian Himalayas	% Himalayan forests	Geographies where found										
				Ladakh	J&K	Himachal	Uttarakhand	Sikkim	Meghalaya	Arunachal	Nagaland	Manipur	Mizoram	Tripura
1	Moist Tropical Forest	5,078	2%						■	■	■			
2	Tropical Semi-evergreen forests	39,096	18%						■	■	■	■	■	■
3	Tropical Moist deciduous forest	34,577	16%			■	■	■	■	■	■	■	■	■
4	Littoral and Swamp forests	136	0%				■					■		
5	Tropical Dry Deciduous Forests	6,580	3%		■	■	■							
8	Subtropical broadleaved hill forests	30,201	14%					■	■	■	■	■	■	
9	Sub tropical Pine forests	18,886	9%		■	■	■		■	■	■	■	■	
10	Subtropical Dry evergreen forests	170	0%		■									
11	Montane Wet Temperate forests	19,627	9%					■		■	■	■		
12	Himalayan Moist temperate forests	30,760	14%	■	■	■	■	■		■	■	■		
13	Himalayan Dry temperate forests	4,360	2%	■	■	■	■			■				
14	Sub-alpine forests	14,469	7%	■	■	■	■	■		■				
15	Moist alpine scrub	14,438	7%	■	■	■	■	■		■				
16	Dry alpine scrub	2,749	1%	■	■	■	■			■				

[1]Classification as per Champion and Seth (1968)
[2]Area data developed from FSI (2021)

Champion and Seth's classification was excellent for the country as a whole, and its popularity is such that it has been adopted by several neighbouring countries, including Pakistan and Nepal (Stainton 1972) to describe Himalayan forests. While the classification is well accepted for lowland tropical forests, there are some drawbacks with regard to montane forests. The disproportionately large number of types and subtypes allocated to montane forests is perhaps a reflection of the issues Champion and Seth faced in trying to fit diverse mountainous forests into a classification system based on temperature and moisture regimes (Singh and Thadani 2024). Nonetheless, this classification system has merit, and given its long history of use, particularly by forest departments of Himalayan countries, it remains relevant. An attempt was made to update this iconic work by ICFRE (2013), published as the 'Forest Types of India: Revisited'. However, this work did not gain popularity for various reasons, including uneven data collection. Later authors have attempted different ways of classifying forests. Singh and Singh (1987) use the basic patterns of Champion and Seth (1968) and divide Himalayan forests into 11 formations based on elevation, precipitation and leaf characters (viz. leaf drop pattern, leaf size and texture).

While temperature and rainfall are key to determining forest types, the varied topography of the Himalayas creates a tapestry of dramatically different microsites, which play a major role in forest vegetation. Schweinfurth (1957, 1984), among others, engaged with topographical heterogeneity, and we also attempt to build this into our conceptual framework, using in part the classification proposed by Singh and Thadani (2024), to enhance an understanding of Himalayan Forest systems in a global context.

Major Determinants of Forest Type

Elevational gradients have the most pronounced impacts on vegetation. Temperatures decrease with an increase in elevation. On average, the drop in mean annual temperature in the Himalayas ranges between 0.50 and 0.67 °C per 100 m increase in altitude, though there are considerable spatial and seasonal variations (Kattel et al. 2013, Joshi and Tamang 2021, see Chap. 2). The impact of this temperature decline can be seen in various ways—such as an increase in frost and freezing, and a decrease in growing period—which inhibits several species. In a South to North transect of merely 100 km in the Himalayas, moving from the foothills to the Greater Himalayas, forest ecosystems can transition from Tropical to Subtropical, Montane temperate, Sub-Alpine and Alpine.

The quantum and seasonality of *precipitation* play an equally vital role in determining forest vegetation. The east-west arc of the Himalayas covers over 20 degrees of longitude (74°E–95°E) and encompasses differing precipitation regimes. The Eastern Himalayas (EH), especially at lower altitudes, receive the full force of the monsoons and have much higher precipitation than the northwestern sections as the monsoons have lost much of their moisture by the time they hit Kashmir, Pakistan and Afghanistan. The climate also changes from marine type to

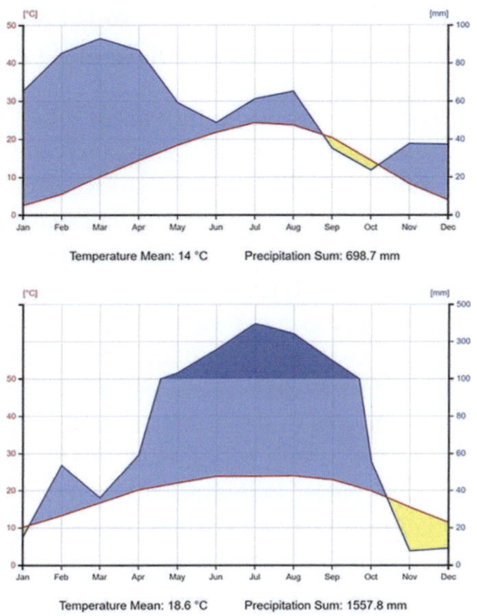

Srinagar, India
34N, 75E, 1600m

Month	Temp	Precip
Jan	2.6	65.3
Feb	5.5	85.2
Mar	10.2	92.9
Apr	14.4	86.7
May	18.4	59.2
Jun	21.8	48.7
Jul	24.4	61.2
Aug	23.7	65.2
Sep	20.4	35.3
Oct	14.5	23.9
Nov	8.3	37.8
Dec	4.0	37.4

Station	% Monsoon rainfall (May-Sept)
Srinagar	<40%
Kathmandu	~85%

Kathmandu, Nepal
28N, 85E, 1350m

Month	Temp	Precip
Jan	10.2	15.0
Feb	13.3	53.7
Mar	16.7	36.1
Apr	20.2	59.1
May	22.0	132.5
Jun	23.8	256.4
Jul	23.8	396.6
Aug	23.9	340.4
Sep	22.9	195.6
Oct	19.9	55.5
Nov	15.5	7.7
Dec	11.4	9.2

Fig. 3.1 Differing precipitation regimes of the NW Himalayas (westerlies dominated) with the Central and Eastern Himalayas (monsoon-dominated). (Source: Zepner et al. 2020. climatecharts.net/)

more continental as one moves west along the arc (Sakai and Malla 1981; Singh et al. 2021). Consequently, it is in the Eastern Himalayas where most of the wet forest types can be found (Champion and Seth 1968).

While the quantum of rainfall is the primary determinant, the *periodicity and seasonal distribution* of precipitation also has a significant impact. While the monsoon is the main bearer of rainfall for much of the Himalayas, the Westerlies often dominate in the northwestern Himalayas and bring in precipitation, often snow, in the winter and early spring (Fig. 3.1). For example, Srinagar in the Kashmir Himalayas (northwestern Himalaya) and Thimpu in Bhutan (Eastern Himalaya) receive somewhat similar rainfall (600–700 mm); however, Srinagar has a strong influence of the Westerlies and receives almost 60% of its precipitation in the winter and spring (Nov–April), while Thimpu gets much of its precipitation through monsoon rains (>80%) with negligible precipitation (~3%) in the winter and early spring (Nov–Feb).

In addition, north–south transects also show significant differences in precipitation. The Outer Himalayas surround the northern Gangetic plains and receive the full force of the monsoonal winds. As the monsoons ascend over these outer ranges, a considerable amount of moisture is lost, leading to heavy rainfall on these slopes, often in excess of 2000 mm annually (Chap. 2: Himalayan climate). Forest types such as *Quercus lanata, Quercus leucotrichophora* and *Pinus roxburghii* flourish in the outer (southern) ranges most impacted by the monsoon. North of the Himadri or Greater Himalayas lie the rain shadow regions of the Tibetan plateau and inner

Himalayas with low precipitation from the monsoon system (Ohsawa 1987; Bookhagen and Burbank 2010) and occasionally under 100 mm of precipitation. In between are the intermontane basins and river valleys, known as the 'Inner Himalayas' with a somewhat transitional character between the moist Outer Himalayas and xeric Tibetan plateau but with their own distinctive patterns (Schweinfurth 1984). These Inner Himalayan ranges include the Valley of Kashmir and the upper valley of the Sutlej and Chenab in the Western Himalayas. In the east, the drier northern part of Sikkim (Upper Sikkim), large parts of Bhutan, as well as parts of Arunachal Pradesh, such as the Ziro valley, have a typically inner Himalayan character. Here, the oaks generally do not occur. These inner valley forests may receive a significant part of their precipitation from the westerlies (largely snowfall during winters) (Miehe 2015) and forests of deodar (*Cedrus deodara*), blue pine (*Pinus wallichiana*) and spruce (*Picea smithiana*) dominate along with some broad-leaf forests of *Aesculus* and *Acer* (Champion and Seth 1968; Miehe et al. 2015).

While the Himalayas are primarily an east–west range, latitudinal variations are also considerable. Consequently, seasonal variations in day length and temperature are far more in Gilgit (35.92°N) than in Gangtok (23.36°N) as an example, with the former having days about an hour longer in the summer, but an hour shorter in the winter. The effect of this difference is apparent in vegetation. For example, the proportion of deciduous tree species declines from West to East along the Himalayan arc from Jammu & Kashmir (71.4% deciduous angiosperm trees) and Himachal (69.6%) to Sikkim (46.6%) and Arunachal Pradesh (43.4%) (Bhatt et al. 2020). It should be mentioned that while deciduous tree species are common, it is evergreens that occur in abundance, widely dominating Himalayan landscapes. Deciduous species, while diverse, form limited stands mostly confined to patches or particular habitats in a few altitudinal belts. Deciduous trees can be dominant along streams, at lower altitudes in the Western Himalayas (drier subtropical climes may have summer deciduous forests), and in patches along the tree line where winter deciduous species such as birch (*Betula utilis*) form forests. The majority of Himalayan forests, and the most abundant tree species, are evergreen.

The more mesic Eastern Himalayas have higher species richness across most taxonomic groups, including the number of tree species (Fig. 3.2). While the richness of tree species is three- to fivefold higher in the Bhutan (Eastern) Himalayas, the difference in diversity of shrubs and herbs was lower, with the Bhutan Himalayas having two-to-three times the species richness compared to the Kashmir Himalayas (Rana et al. 2019). Peak richness of all these groups typically occurs between 1000 and 2000 m elevation (Bhatt et al. 2020; Rana et al. 2019).

Rana et al. (2019) compiled a list of 8765 angiosperm species from 1912 genera and 210 families across the Himalayas combining data from 31 different floras and after removing duplicate entries (synonyms) and alien species. While these were predominantly herbs and shrubs, they identified 145 tree species in the northwestern (Jammu and Kashmir) Himalayas and 783 in the Eastern (Bhutan) Himalayas. Mean temperature and annual precipitation together explain ~60% of the variation in species richness.

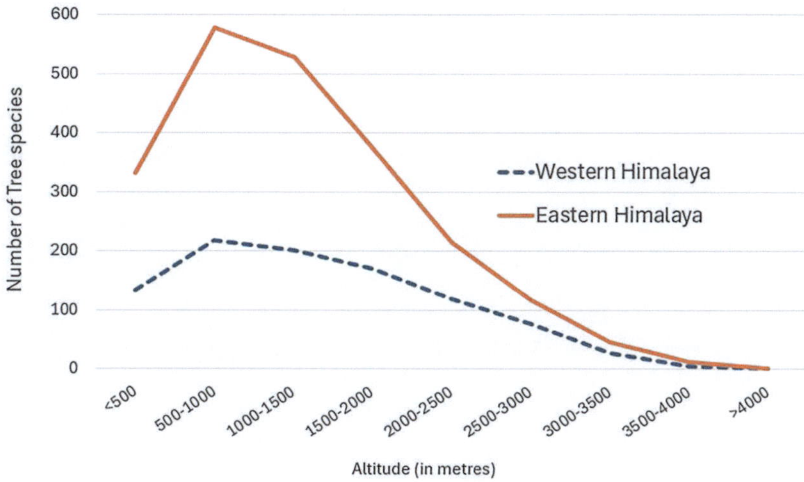

Fig. 3.2 Tree diversity at different altitudinal zones in the Himalayas. (Source: Figure developed from data published in Bhatt et al. 2020)

As temperatures decline, evapotranspiration losses decrease. Thus, for the same amount of precipitation, higher elevation sites are more mesic than lower (warmer) elevations. Topographical heterogeneity and varied landforms also influence species occurrence. Varied rock strata, folding patterns and dip and scarp slopes create myriad combinations of soil depths and soil types, which favour different species. The numerous parallel ranges, such as the Siwalik[1] and the Lesser Himalayas, reflect the differences in orogeny and lead to differences in soil composition, depth and drainage, which influence vegetation. Slope exposure and position play an important role in determining forest vegetation (Tyagi et al. 2023) (see Fig. 3.3).

Superimposed on these natural variations are human impacts. The Himalayas have been populated and used by humans for millennia, and the impact of humans can be seen in forests in every part of the Himalayas (Singh and Singh 1992; Schmidt-Vogt and Miehe 2015). Human impacts extends to the treeline, and most Himalayan treelines are anthropogenically depressed (Schickhoff et al. 2015). The population density in the Himalayas is higher than in most other mountains of the planet and grew by 250%, from under 20 million to almost 55 million, between 1961 and 2011 (Apollo 2017). Even as remote a region as the Tibetan plateau shows human habitations over 30,000 years old (Zhang et al. 2018). Subsistence agricultural activities still dominate the Himalayas and are dependent on biomass inputs from forests (see Chaps. 7–9 in Volume 2 of this book). Transhumance and pastoralism have been an integral part of the Himalayas since time immemorial, while shifting cultivation continues to be practiced in the Eastern Himalayas. Burning of forests, lopping and pollarding of trees, and whole tree cutting for fuel

[1] We use the spelling Siwalik as per Kumar et al. 2020. Other authors have used Shivalik or Sivalik.

Fig. 3.3 Changing slope, aspect and geology can influence forest type and allow for a patchwork of species and stands. Partitioning of *Cupressus torulosa* (dark green, right side), *Cedrus deodara* (bright green), *Quercus leucotrichophora* (light green, right side), and *Q. floribunda* (brownish, foreground). (*Pic*: *Vishal singh*)

and timber, transformed Himalayan forests long before commercial forestry. As human populations have grown, the degradation, fragmentation and reduction in area of forests has occurred (Singh and Singh 1992; Reddy et al. 2018). In many areas it is difficult to ascertain what the original vegetation would have been. Champion and Seth (1968) also recognised the presence of degradation stages in their classification. Moreover, the scale of deforestation varied. While much of it was local, in many cases, there were regional impacts. For example, the spread of railways in the nineteenth century led to large-scale cutting of sal in India (Tucker 1983). In Nepal, agriculture expanded widely in *Schima-Castanopsis* forests (Stainton 1972). Furthermore, human-induced forest fragmentation disrupted ancient migration routes of flora and fauna, the impacts of which on forest growth are not well understood.

Thus, in the Himalayas, forest vegetation follows multiple and complex gradients (see Table 3.2), which favour some species and disadvantage others. This varied topography leads to a patchwork of altitudinal and moisture combinations, which makes characterising regional patterns more difficult.

Influence of Biogeography

The Himalayas are at the confluence of the Palearctic and Indo-Malayan biogeographic realms (Udvardy, 1975) which exert considerable influence on the vegetation of Himalayan forests. The Palearctic realm brings species familiar to Europe and North Africa that have evolved with stronger seasonality common in more temperate

Table 3.2 Influence of biogeographic variables on Himalayan forests

Variable	Impact
Altitude	Temp drops 0.50–0.67 °C/100 m. forests change from tropical to alpine within 100 km, moving north across the range
Precipitation (amount) *Across the Himalayan arc*	Rainfall typically decreases from east to west as the impact of the monsoonal system decreases
Divide of outer and inner ranges' precipitation (amount) *Rain shadow effect*	Outer ranges get the full force of monsoons and are wet, while the inner ranges are dry being shielded from monsoons by the high Himalayas. North of the Greater Himalayas are rain-shadow areas
Precipitation (seasonality) *Monsoonal* vs. *westerlies impact*	NW Himalayas get significant winter/spring rain due to westerlies (more evenly distributed precipitation); Central and Eastern Himalayas and outer ranges largely monsoonal (strongly seasonal precipitation)
Latitudinal impact	NW Himalayas have latitudes 10° further north compared to the eastern regions, leading to increased seasonality, day-length variation and summer-winter variation
Slope and aspect	North slopes and hill bases are moister with higher diversity and biomass, and different species as compared to south slopes and ridge tops (see Chap. 11)
Rainfall efficiency	Transeau ratio varies. Evapotranspiration is lower at higher elevations, so the same amount of rainfall produces moister vegetation at higher elevations
Biogeographic province	Different biogeographic provinces—Indo-Malayan (southern and eastern) and Palearctic (northern and western) meet in the Himalayas, influencing vegetation in different parts
Human impact	Humans cut down, fragment and degrade forests; they favour certain species intentionally (silviculture) or unintentionally (disturbance—grazing, fire), thereby impacting forests

climes. The influence of the Palearctic is strongest in the northern and western parts of the Himalayas and is characterised by species such as *Cedrus deodara*, *Pinus wallichiana*, *Betula utilis* and *Rhododendron campanulatum*. The Indo-Malayan, which covers much of India, is characterised by a tropical and subtropical climate, and a high level of biodiversity with higher endemism. *Shorea*, which dominates across the lower foothills, is representative of this realm. At higher elevations, species such as *Bauhinia purpurea, Lithocarpus elegans*, *Schima wallichii* and *Duabanga sonneratioides*, are common examples. The forests in the south and east of the Himalayas are characterised by their Indo-Malayan affinities.

Classification Using Biogeography and Rainfall Patterns

The two precipitation regimes prevalent in the Himalayas have been discussed. The southwestern summer monsoon dominates across much of the Himalayas and leads to heavy precipitation from May–June through September. The outer ranges receive the bulk of the rainfall, which progressively recedes moving west and north. The extra-tropical upper Westerlies supply smaller but significant moisture inputs, especially in the winter, in the northwestern Himalayas and the inner ranges. While monsoonal forests are wetter, the seasonality of rainfall is also important in determining forest type.

Based on these affinities and moving from southeast to northwest across the Himalayas, the vegetation can be divided into the following broad divisions.

Type I

Submontane and lower montane zone of broadleaved evergreen tropical and subtropical forests, largely Eastern Himalayan as well as the southern flank of the Himalayas, consisting of wet/moist forests that are strongly monsoonal, with high tree diversity and strong Indo-Malayan affinities.

They are best found in the Northeast Himalayan regions of India, Bhutan and eastern Nepal. Dipterocarps such as *Shorea* characterise the lower reaches, and as altitudes increase, Ericaceae (e.g. *Rhododendron*), Theaceae (e.g. *Schima*) and Fagaceae (e.g. *Castanopsis*, *Quercus*) become more common. Forests are typically diverse, and no single species dominates. For example, in the Buk oak (*Quercus lamellosa*) forests, named for a single species (Type IV; Group 11B, Ci-1b as per Champion and Seth 1968), the Buk oak only constitutes 12.5% of the forest trees. *Castanopsis* and *Schima* are common genera, and various Rhododendron species often find great abundance at higher elevations. Several subtropical and lower temperate species, including Bauhinia's and species such as *Duabanga grandiflora, Lithocarpus elegans, Engelhardtia spicata, Toona ciliata, Magnolia hodgsonii, Acrocarpus fraxinifolius, Michelia cathcartii, Bischofia javanica* and *Careya arborea, are commonly found in these forests. Such forests are predominantly found in the Eastern Himalayas up to the temperate zone. While sal-dominated forests extend into the Western Himalayas, these western sal lack the high diversity that normally typifies Type I forests.*

Type II

Montane forests of the Central and Western Himalayas, and higher elevation forests across the Himalayan arc: These are forests with stronger Palearctic and Tibetan affinities, which grow in drier regions than Type I forests, but can have equally high productivity and growth rates. They are simpler in structure, with lower tree diversity, often forming monodominant stands. Consequently, these are the forests that have been most intensively managed for silviculture.

These Type II forests best represent the Himalayan forests. Not only is their geographical extent greater, but they also more strongly represent species that have evolved and adapted to the Himalayas' montane conditions. While sensitive to moisture, they are adapted to varying edaphic factors, and these trees can often dominate across a greater variation of soil types and parent materials. Soil depth and soil moisture are typically more important than the nature of the parent rock, and many of these forests grow on soils that are immature and lacking a well-developed profile.

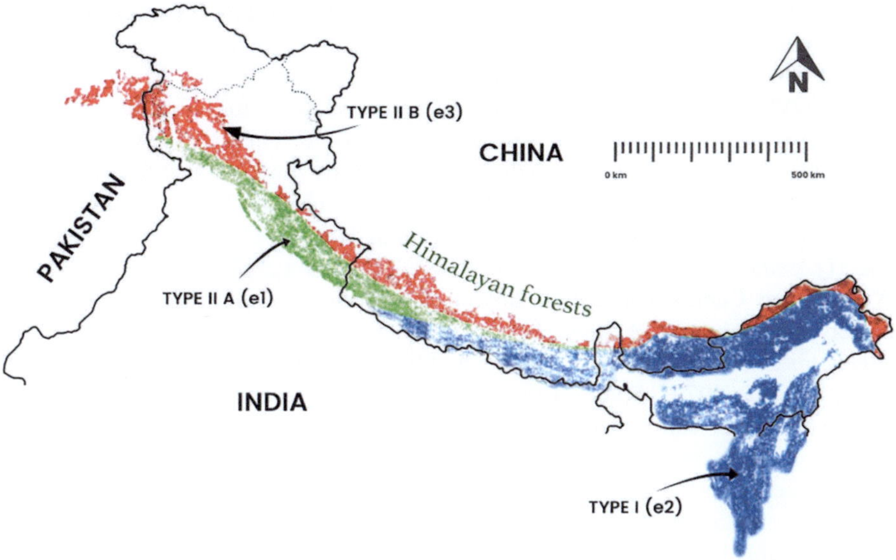

Fig. 3.4 Different forest and leaf types across the Himalayas (boundaries of forest type are representative and indicative)

They can be further subdivided into the following.

II-A: *Forests of the Outer Himalayan ranges,* which are exposed to the direct monsoon thrust. These include broadleaved oaks, alder, laurels, rhododendrons, some winter deciduous forests, as well as some conifers: a Diploxylon pine (*Pinus roxburghii*) and a fir (*Abies spectabilis*). Evergreen species which shed leaves annually and simultaneously leaf out, dominate this eco-division.

II-B: *Forests of the Inner Himalayan ranges*. These are the drier ranges, often in the rain-shadow areas shielded from the direct monsoon thrust. The impact of the westerlies is higher on this vegetation, which is rich in conifers with multiyear leaf lifespans, such as deodar (*Cedrus deodara*), spruce (*Picea*), Haploxylon pines (such as *Pinus gerardiana*, *P. wallichiana*), larch and juniper forests, and some firs (*Abies pindrow*). Also included are winter deciduous forests, both broadleaved (birch, particularly *Betula utilis*, and willows) and a conifer (larch). Some of the most attractive forest types, consisting of tall (up to 60 m) stately trees of deodar, spruce and firs, occur in this division (Champion and Seth 1968). Figure 3.4 broadly illustrates the contours of these Type I and II forests and overlays these with the leaf based classification described below.

Abundance of Monodominant Forests

When it comes to classifying vegetation, we heavily depend on identifying dominant species, which becomes difficult when species richness is high, as is the case in the more mesic Eastern Himalayas (termed above as Type I Himalayan forests).

In the western and central parts of the Himalayas (Type II forests), monodominant stands abound. Interestingly, every elevational zone in this part of the Himalayas has species which can form large, contiguous monodominant stands. In the foothills (up to 1000 m), it is Sal (*Shorea robusta*), which is known for its extreme gregariousness, typically allowing in few other species (Stainton 1972). Chir pine (*P. roxburghii*) can form pure stands in the frequently burned forest between 1000 and 1800 m. At warm temperate altitudes (1500–2100 m), banj oak (*Q. leucotrichophora*) dominates in the Central Himalayas, above which kharsu oak (*Q. semecarpifolia*) is highly successful from about 2500–3200 m. In the Western and Central Himalayas, *Betula utilis* dominates the treeline, while firs, namely *Abies pindrow*, *A. spectabilis* and *A. densa,* can form monodominant stands at/near treelines in the Western, Central and Eastern Himalayas, respectively (Troup 1921; Stainton 1972). Some of these forest types, such as *Pinus roxburghii* and *Cedrus deodara*, can attain valley-level dominance over hundreds of kilometres regardless of local variabilities. In other cases, particular edaphic factors or disturbances cause a single-species stand. *Cupressus torulosa* (Cypress) often owes its monodominance to a steep south-facing slope with limestone rocks, prone to gravitational fall. Such a habitat is inhospitable to most other species. Alder (*Alnus nepalensis*) colonises landslides and forms monospecific stands in areas varying in annual precipitation from 1000 to 5000 mm, both in the eastern and western regions of the Himalayas. *Dalbergia sissoo* and *Acacia catechu* colonise freshly formed substratum along rivers (Champion and Seth 1968).

Thus, dominant tree species that belong to very different families and may be either angiosperms or gymnosperms are a very visible and important component across Himalayan forests (especially Type II forests) across a range of altitudinal and moisture regimes.

This differentiation of forests is also reflected to an extent in the canopy heights of forests across the Himalayas (Fig. 3.5).

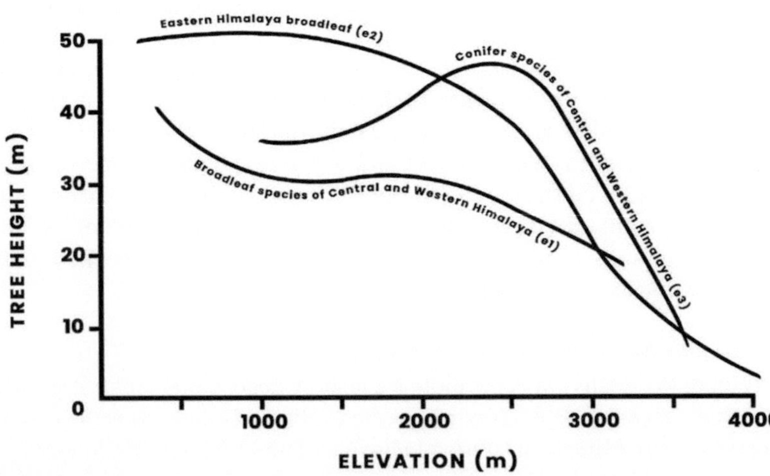

Fig. 3.5 Tree height for different groups of trees at varying altitudes. (Redrawn from Singh and Singh 1987)

Using a Leaf-Based Classification

Leaf characters, such as leaf life span, size, extent of sclerophylly, periodicity of leafing and leaf drop, have been used to classify vegetation (Raunkiaer 1934; Webb 1959; Wu 1980, 1987; Medina et al. 1990; Ashton 2014; Ashton and Zhu 2020). Division of forests into evergreen and deciduous types, or broadleaved and coniferous, has been practiced since the beginning of vegetation classification. Leaf size is known to influence several aspects of tree function, such as rate of transpiration and photosynthesis (Malhado et al. 2009) and is thought to increase along the gradient of increasing precipitation (Givnish 1987). Leaf size and wood density often show an inverse correlation: trees with lower wood density have larger leaves (Wright et al. 2007). Smaller leaves may suffer less herbivory during leaf expansion (Moles and Westoby 2000). Leaf characters directly and indirectly influence community composition and adaptive features and are predominantly associated with the physiognomy of vegetation. Leaf traits are known to determine plant and stand characters (Reich et al. 1992). Leaf life span influences photosynthesis and defensive traits (Matsuki and Koike 2006). Leaf traits vary in response to variation in resource availability (Zhang et al. 2020), and leaf life span and photosynthesis parameters are linked (Vasfilov 2016). Leaf structural traits greatly impact physiological functions (Trueba et al. 2022). There is evidence to indicate that competition between conifers and angiosperm trees is affected by leaf characters (Brodribb et al. 2012), and the adaptive significance of evergreen vs deciduous leaves (Givnish 2002) is related to several other plant traits and habitat conditions. Leaf traits are associated with several ecosystem characteristics (Luo et al. 2005). As an example, needle leaves are known to retain more water, are less damaged by snow, and are more energy-saving than broadleaved (Norris 2018). Conifer leaves also have a higher carbon concentration than leaves of broadleaved species (Ma et al. 2018). Plant phenology is strongly interlinked with ecosystem processes and biodiversity (Gray and Ewers 2021).

In view of this widespread relevance of leaf characters, leaf traits of canopy species can be used to classify Himalayan forests (Singh and Thadani 2024). More importantly, the focus on leaf characters enables us to characterise Himalayan forests at a regional scale and distinguish them from the forests of other regions. We have considered leaf traits such as leaf life span, phenology, leaf-form (e.g. broadleaf and needle-like leaf) and extent of sclerophylly of dominant tree species. Instead of treating conifer leaves as one category, we have divided them keeping in view the adaptive significance. The forest groups, recognised based on leaves, are given in Fig. 3.6 and described below.

Singh and Thadani (2024) divide Himalayan forests into five main types with regard to leaf types, viz. (i) evergreen with about 1 year leaf life span and simultaneous leafing and leaf drop during pre-monsoon (e1); (ii) tropical evergreen with mixed leaf life span (e2); (iii) coniferous type, with leaves of greater than 1-year longevity (e3, generally of 3–6 years); and finally deciduous (d)—forests where the dominant trees lose their leaves in response to unfavourable conditions for part of the

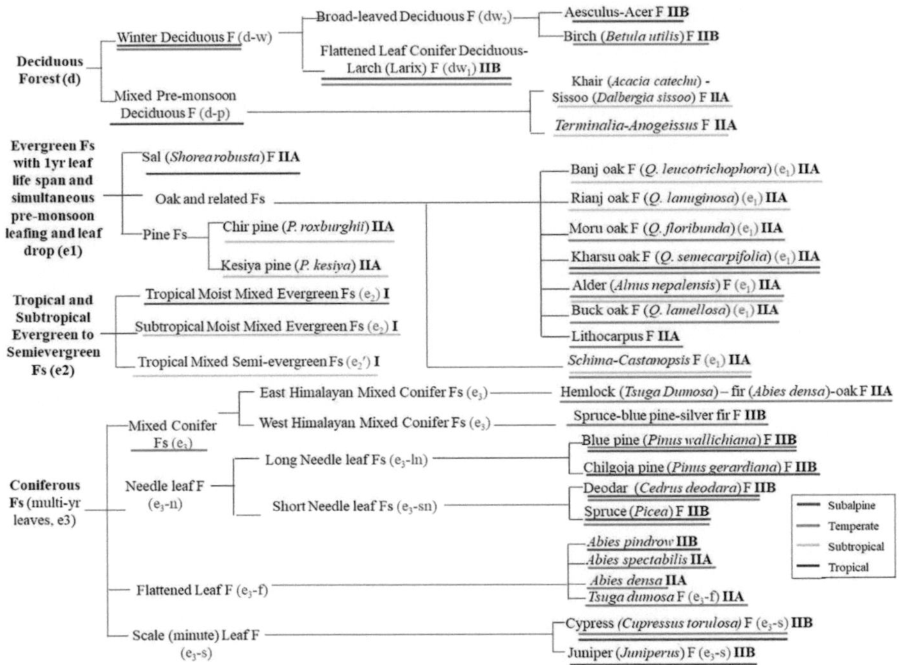

Fig. 3.6 Leaf-based classification of Himalayan Forests—a schematic representation
The Himalayan forests are divided into five main types with regard to tree leaves, viz. deciduous (d) represented by (i) winter deciduous and (ii) drought (pre-monsoon) deciduous; (iii) evergreen with about 1 year leaf life span and simultaneous leafing, and leaf drop during pre-monsoon (e_1); (iv) tropical evergreen with mixed leaf life span, size and other leaf characters (e2); and (v) coniferous type, generally with leaves of several years longevity (3–6 years) (e3). Each of these large groups can be further divided. Winter deciduous is divided into broadleaf deciduous and flattened leaf conifer deciduous, which includes larch. The e_1 type is a heterogeneous group, consisting of mesophyllous sal, sclerophyllous oaks and long-needle-leaved pines. The classification emphasises that conifers also vary considerably in leaf form, and species with different leaf forms vary in adaptation to the environment and photosynthetic rate. Coniferous forests (e3) type can be divided into conifers with needle-like leaves, flattened leaves, scale-like leaves and so on. The colour of underlines indicates altitudinal belts corresponding to tropical, subtropical, temperate and subalpine areas. When a forest occurs in more than one belt, it is underlined with more than one colour. F is for forest (After Singh and Thadani 2024).

year which are further divided into (iv) winter deciduous (dw) and (iv) pre-monsoon deciduous (dp).

This classification stresses the significance of pre-monsoon drought (March-May) when many leafless deciduous species produce new leaves, and many evergreen species replace their old leaves with new leaves. The degree of pre-monsoon drought varies considerably from one year to another, depending on the frequency of rainstorms. Climate change-induced warming may affect the phenology of these species by intensifying pre-monsoon drought.

e1 Species are Central to the Himalayan Forests

The Himalayan forests are best exemplified by evergreen species with concentrated leaf drop and simultaneous leafing during the pre-monsoon (March–May, dry and warm season). These dominate in the Western and Central Himalayas, but are commonly found in the Eastern Himalayas as well. Referred to as e1 type, these include varied growth forms that range from sclerophyllous oaks (e.g. *Quercus leucotrichophora*, *Q. floribunda*), mesophyllous dipterocarps (*Shorea robusta*), an evergreen maple (*Acer oblongum*) and even subtropical needle-leaved pines (*Pinus roxburghii, Pinus kesiya*). While globally pines are associated with leaf (needle) life spans of 2–40 years, subtropical Himalayan chir pine (*P. roxburghii*) has a needle retention of 1 year and comes closer than any other pine to being deciduous (Richardson and Rundel 1998; Zobel and Singh 1997). During March–April, chir pine looks more brown than green, as new needles are still short and emerging. Specific leaf mass varies widely for e1 species from 112 g m^{-2} for *Shorea robusta*, 137–200 g m^{-2} for the oaks and over 300 g m^{-2} for *Pinus roxburghii* (Singh et al. 1994).

While leaf flush primarily occurs in the pre-monsoon, some e1 species have secondary leaf production during or after monsoon to take advantage of favourable growth periods with fresh leaves. However, unlike evergreen tropical rainforests, which have leaf fall throughout the year (Reich 1988), leaf fall is episodic in e1 forests and mostly during the pre-monsoon dry period. In many species from diverse genera, including *Pinus roxburghii* and *Shorea robusta,* which are among the most common Himalayan species (FSI 2021), there is a period when the tree has a sparse leaf cover during the pre-monsoon, and in this respect, they resemble pre-monsoon deciduous trees. The shedding of old leaves and thinning of the canopy during pre-monsoon drought allows these trees to replenish tree water status enough to initiate the growth of young leaves (Zobel et al. 2001). The e1 species occur over a wide range of elevation from foothills (*Shorea robusta*) to treelines (*Quercus semecarpifolia*) but are much more pronounced in the outer Himalayan ranges directly exposed to the monsoon thrust (i.e. in Type II A forests).

This trait, viz. evergreen with a leaf life span of about 1 year, and simultaneous leaf drop and leafing during pre-monsoon, contributes to separating Himalayan forests from other regions of the world. The reasons why such species prevail and often dominate in the Himalayas have been discussed by Singh et al. (1994) and Zobel and Singh (1997). The monsoon pattern of rainfall, characterised by 3–4 months of concentrated rains, followed by long rainless spells during winters and early summer, is a major climatic factor that greatly influences the selection of plant functional types and species. The autumn (post-monsoon season) is highly favourable for plant growth with sunny and long days (>11 hours of daylength through October in Nainital in the Central Himalayas), high relative humidity and soil moisture and night-time temperatures that remain above freezing up to a considerably high elevation. This enables the e1 species to attain peak photosynthetic rate during the fall (post-monsoon) season (Thadani 1999; Tyagi et al. 2023). In contrast, in most temperate regions across the World, the fall/autumn season is when leaves of broad-leaved trees start senescing and decreasing their physiological activities because of cold temperatures and short day lengths (Luo et al. 2017).

In conditions characterised by mild winters with relatively long day length and high relative humidity, species that produce new leaves just before the favourable season for photosynthesis and retain them around the year (e1-leaf type) have a competitive advantage over other species. These evergreen species (e1) dominate over deciduous species as they can assimilate carbon throughout the year, including during the mild winters. By overhauling the entire foliage mass in the pre-monsoon, they have fresh leaves to maximise the photosynthesis by the time rains arrive. Multi-year evergreen species, in contrast, have many old leaves with reduced photosynthetic capacities. It must be noted that tree species with other strategies also co-exist in each forest type and are adapted to different niches. As an example, *Rhododendron arboreum* with a leaf life span that considerably exceeds 1 year is commonly found as an undercanopy species in e1-dominated forests of *Quercus leucotrichophora*.

e2 Dominates in the East (Evergreen to Semi-Evergreen and Tropical to Subtropical)

In lower elevation forests, especially of the Eastern Himalayas, evergreen trees with multiple leaf flushes often dominate. Classified as e2 species, their leaf life span is usually less than 1 year (Athokpam and Garkoti 2013; Athokpam et al. 2014), though some e2 species have longer leaf life spans, often up to 20 months (Boojh and Ramakrishnan 1981). Peak leaf fall usually occurs soon after winter (January–February) (Kikim and Yadava 2001). Among leaf types, this category is the least investigated and less understood. Common e2 trees include *Schima wallichii, Duabanga sonneratioides, Dillenia pentagyna*, *Artocarpus chaplasa, Tetrameles nudiflora* and several species of *Lithocarpus* and *Castanopsis.* At higher elevations, Rhododendrons dominate.

Typically, e2 tree species are not sclerophyllous, have leaves of mesophyll size and occasionally megaphyll. At tropical latitudes, these may be associated with a significant number of deciduous species (Kikim and Yadava 2001), though the general character of the forest remains evergreen.

e3: Evergreen Conifers with Multiple Year Leaf Life Spans

While some pines, notably chir, can have a one-year leaf life span with simultaneous leaf production and leaf fall (e1 type), the majority of conifers follow a distinct pattern with leaves that persist for multiple years (generally 2–6 years). These are categorised as e3 type and include the fir, spruce, cedar, cypress and junipers. While they are found across the Himalayas, the e3 conifers dominate in the inner (northern) ranges at altitudes above 2000 m. They are more prevalent in the Western Himalayas.

Fig. 3.7 Leaf types in e3 conifers. (**a**) Needle type leaf: *Cedrus deodara*. (**b**) Flattened type leaf: *Abies pindrow*. (**c**) Scale type leaf: *Cupressus torulosa*. (Photo credits: Cedrus & Cupressus (R Thadani); Abies (Soham Kacker))

On the basis of leaf morphological characters, e3 type conifers are divisible into three main types: (i) needle-like leaves, such as in *Cedrus* and *Picea*, (ii) flattened leaves (dorsiventrally asymmetric) such as in *Abies, Tsuga, Larix* and *Taxus* and (iii) scale-like leaves as in *Cupressus* and *Juniperus* (family Cupressaceae) with appressed imbricate scale leaves (Fig. 3.7). The mean leaf width to thickness ratio (WTR) of the cross section is less than 2 in needle-like leaves and more than 2 in flattened type (Du et al. 2020). In the pine family, needle-like leaves show a higher photosynthetic rate than flattened leaves (Brodribb et al. 2007; Brodribb and Feild 2008). Needles are suited to sunny and dry environments, whereas flattened leaves are better adapted to shady and humid conditions. As an example, *Pinus krempfii*, which is found in the tropical rainforests of South Vietnam, has flat leaves adapted to the deep shade of rainforests (Brodribb and Feild 2008). In general, needle-like conifer leaves have higher carbon assimilation rates. Studies of leaf anatomy and the three-dimensional organisation of tissues are important to understand plant water use and carbon uptake (Trueba et al. 2022).

Members of the *Cupressaceae,* with scale-like leaves, are adapted through a combination of morphological and physiological means to some of the driest areas. Their appressed small and imbricate leaves, accompanied by reduced xylem-specific conductivity (Brodribb et al. 2007; Pittermann et al. 2012), reduce transpiration losses. Physiologically, the Cupressaceae are most efficient among conifers to be able to survive among the lowest water potentials without cavitations and embolisms that damage leaves of other conifers (Brodribb et al. 2014). Consistent with these attributes, several *Juniperus* species and *Cupressus torulosa* occupy some of the driest sites in the Himalayas. Junipers are the only evergreen species that occur in the cold deserts of Ladakh, with about 100 mm annual precipitation.

Conifers with flattened leaves can occur in a range of moisture regimes. Mention may be made of the two fir species, *A. pindrow* largely associated with dry inner ranges and *A. spectabilis,* a fir that can form extensive stands in the moist outer ranges along much of the Himalayan arc. *A. pindrow* is the main fir of Kashmir valley, a region shielded from the monsoon by Pir Panjal ranges though it can also be found in more monsoonal climes. It occupies lower elevations than *A. spectabilis*, and generally these do not form mixed stands. *Abies spectabilis* (silver fir), along with other conifers with flattened leaves, such as *Tsuga dumosa* (hemlock), generally occur in the outer (southern) ranges and can even occur among the most mesic forests, such as the cloud forest zone of the Himalayas (Miehe et al. 2015).

Conifers with short needles, such as *Picea* and *Cedrus,* largely occur in dry areas, particularly in dry inner valleys. In moist outer ranges, they are unable to compete with conifers like silver fir with flattened leaves, and with broadleaved oaks (Champion and Seth 1968).

The conifers do, however, form mixed stands with broadleaved trees. In the Eastern Himalayas, mixed coniferous forest consists of *Tsuga dumosa* and *Abies densa* in the canopy, and *Lithocarpus pachyphyllus, Quercus lineata, Acer campbelli, Magnolia campbelli, Taxus baccata* and rhododendron in the undercanopy are common. In the west, mixed coniferous forests consist of spruce, silver fir, blue pine and deodar, along with varying mixtures of oaks and deciduous broadleaved species.

Deciduous (d) Type

Deciduous forests are divided into two distinct groups: (1) winter deciduous trees, where leaves are shed in the late fall because of colder temperatures, and (2) pre-monsoon deciduous (March-May), where leaf loss is a response to droughty conditions.

In the Indian Himalayas, there are 727 evergreen tree species and 574 deciduous species (Bhatt et al. 2020). The Western Himalayan forests are richer in deciduous species than evergreen, and even in the eastern forests, Shukla and Ramakrishnan (1982) report 54 deciduous species out of 122 tree species in a subtropical humid seasonal forest. However, deciduous species form far fewer forest types and occupy much smaller forest areas than do the evergreen species. Above 2000 m elevation, Miehe et al. (2015) recognised only three deciduous (winter deciduous) forest types, namely birch (*Betula utilis*) forest, *Aesculus-Acer* forest and larch, a conifer deciduous forest. Champion and Seth (1968) recognised only two viz. birch and larch forests. With the exception of upper elevation birch forests, a characteristic feature of the Himalayas is the absence of large areas under deciduous species in spite of there being a broad floristic base of deciduous trees.

Winter Deciduous Forests

Winter deciduous forests are further divided into broadleaf deciduous species, such as *Betula utilis* (*Himalayan birch*), and into flattened leaf deciduous conifers, which include larch (*Larix griffithii*) (Pinaceae family). These shed their leaves in response to the changing photoperiod and decreasing temperature as winter approaches, which prompts hormonal changes that lead to leaf fall. These trees typically enter a dormant state during winter, and physiological activity is very low. These forests are best seen at higher elevations. *Betula utilis,* for example, dominates treeline areas of inner valleys in the Western and Central Himalayas. At temperate altitudes, several deciduous species can form smaller patches. Common among these are *Aesculus indica* (Himalayan horse chestnut), *Acer* spp. (e.g. *A. caesium* and *A. pictum*—Himalayan maples), *Carpinus viminea* (Hornbeam) and *Fraxinus micrantha.*

Drought Deciduous Forests

Pre-monsoon or drought deciduous trees, on the other hand, shed their leaves in response to prolonged dry periods and drought conditions. Water scarcity triggers the abscission process to help the trees survive dry periods. The lack of soil moisture is usually the trigger to initiate abscission, and hence, many of these trees, if grown in mesic environments, will not completely lose their leaves during the dry period.

Pre-monsoon deciduous forests are most common in the foothills. Examples include *Dalbergia sissoo*, *Acacia catechu* and *Bauhinia variegata*, as well as species of *Terminalia* and *Anogeissus*. The pre-monsoon deciduous and the e1-type evergreen species show a continuum, with several e1-type species (including major species such as *Shorea robusta* and even *Pinus roxburghii*) having a short phase with a distinctly reduced canopy and an almost leafless condition in some sites. For example, in drier areas, *Schima wallichii* has been reported to be deciduous (Boojh and Ramakrishnan 1981), whereas across much of its range, this tree is treated as evergreen (Kikim and Yadava 2001). The same is true for Sal (*Shorea robusta*). In parts of the bhabhar or dry south-facing slopes, sal may be for some time deciduous, while retaining a more evergreen nature in moister areas in the terai or on cooler slopes. While sal has been described as a deciduous species by some authors, in our opinion, it can be considered evergreen, as sal trees typically do not become leafless and remain physiologically active during the period of leaf thinning. Sal has a greater degree of canopy thinning than other e1 species. However, its canopy thinning is on a continuum with chir pine and most oaks, which are regarded as evergreen by all.

Thus, winter and pre-monsoon deciduous differ fundamentally. Winter deciduous (dw) species show a period of physiological dormancy, which is generally not seen in the drought deciduous (dp). The drought deciduous trees are also more responsive to the limiting variable (moisture) for leaf shedding and, as such, are on a continuum with the e1-type evergreen (see Fig. 3.8). The specific leaf area of pre-monsoon deciduous species is lower (9.46 m^2/kg) than that of winter deciduous species (15.2 m^2/kg) (Zobel and Singh 1997).

Occurrence of Evergreen and Deciduous Forests

At mid-elevations, the canopy is usually evergreen. At least in the Western Himalayas, deciduous species are often confined to small pockets along water courses, and in hollows and depressions. Deciduous species form occasional stands amidst evergreen forests by colonising landslide deposits and eroded sites (Suri 1933). The winter deciduous forest, *Aesculus-Acer,* frequently occurs as patches

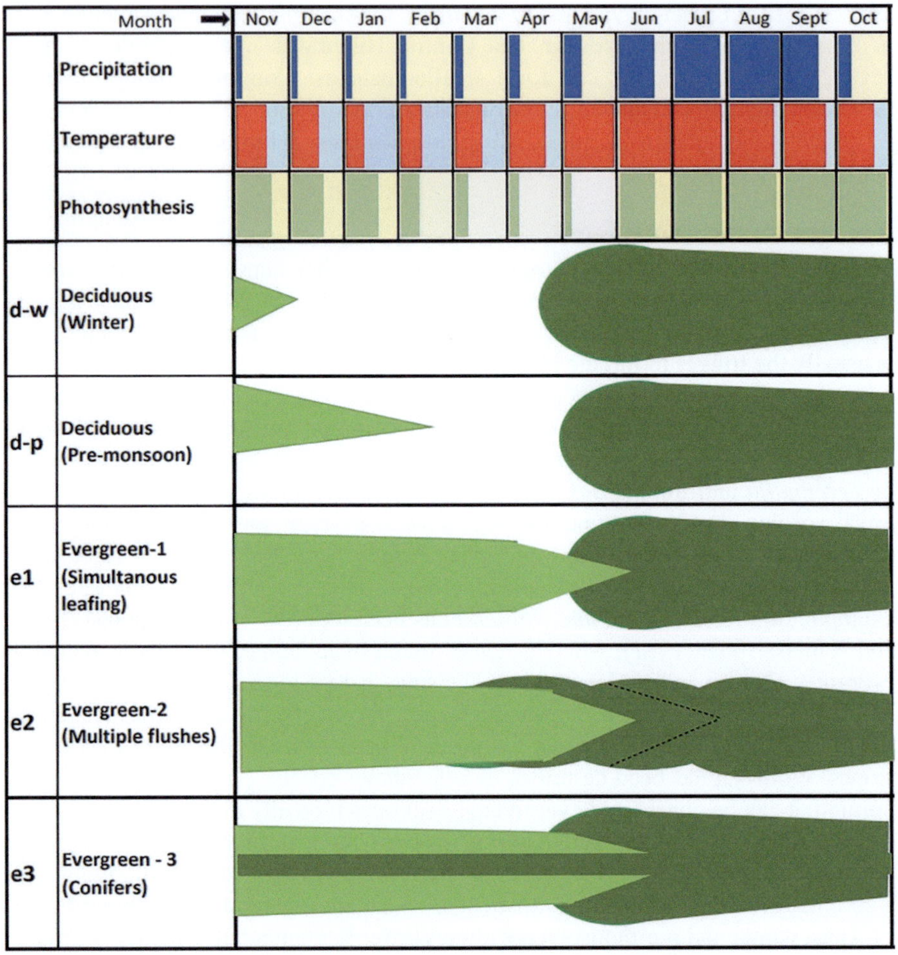

Fig. 3.8 Leaf drop and new flush among broadleaved species: a continuum seen from d-p, e1 and e2 species
Figure showing changing precipitation, temperature and max photosynthesis patterns in a monsoonal forest. The time of leaf drop and leaf flush is shown for species with different strategies. (After Singh and Thadani 2024).

along the water courses on shady slopes, often with boulder fields in the western part (69°30′–82°30′ E), generally beyond the full thrust of monsoon.

In some areas, especially in the dry inner valleys, evergreen and deciduous species occur in mixed patches. Common deciduous species in these forests include *Aesculus indica, Acer caesium* and *A. cappadocicum, Ulmus wallichiana, Cornus macrophylla, Populus ciliata, Carpinus viminea*, *Fraxinus micrantha* and *Sapium insigne* in the Western Himalayas. We refer to these as the Western Himalayan mixed deciduous forest. At higher elevations, birch may occur mixed with oaks such as *Q. floribunda* and *Q. semecarpifolia*. In the Eastern Himalayas, deciduous species include *Magnolia campbellii, M. globosa, Osmanthus suavis, Corylus ferox, Acer caudatum, A. hookeri, A. pectinatum, A. sikkimense, Prunus rufa* and *Sorbus thomsonii* (Stainton 1972). Distributional patterns of these species in Eastern Himalayan mixed winter deciduous forest and their contributions to forest communities need to be better investigated, as data are limited.

Limitations of a Deciduous Strategy in the Himalayas

The deciduous habit has been referred to as an acquisitive leaf strategy, and the evergreen habit as a conservationist strategy (Donovan et al. 2011; Bai et al. 2015). Where resources are not strongly limited, deciduous species have a competitive advantage over evergreen species as they are able to acquire higher carbon amounts at a lower leaf dry mass cost (Reich et al. 1992; van Ommen Kloeke et al. 2012). In the Himalayas, high rates of soil erosion in the monsoon-drenched outer ranges may be limiting in nutrients. While photosynthetic capacity per unit dry weight of leaf is typically lower in evergreen species than in deciduous (Reich et al. 1991; Wright et al. 2005), the advantage which is associated with the longer leaf life span of evergreen species is their higher nutrient retention potential, which gives them a competitive edge in infertile habitats (Pornon et al. 2011).

In the moist outer ranges of the Himalayas, deciduous species are outcompeted by evergreen species with e1 leaf habit, which, much like deciduous species, produce most of their new leaves annually just before the most favourable periods for photosynthesis (moist and warm days). As winters are relatively warm and sunny, these e1 species are also able to fix carbon throughout the winter, which compensates for the higher cost of the evergreen, often sclerophyllous, leaves. Thus, the e1 species, by combining traits of deciduous and evergreen habits, are able to dominate.

In the dry and cold inner ranges, conifers and broadleaf species (such as many *Rhododendrons*) with multi-year leaves outcompete deciduous species from infertile habitats. The inherently resource-conservation traits of these longer-lived leaves help them dominate in resource-poor habitats. Even at the limits of tree growth, at the treeline, species such as *Abies spectabilis* and various rhododendrons dominate, especially when moisture is adequate. Only at the extreme limits of cold and dry climates near the timberline do birch (*Betula utilis*) and some species of *Sorbus* develop communities over large stretches. This is especially in the Western Himalayas, where the climate is more continental, winter temperatures are lower, and the impact of the monsoons is less (Bobrowski et al. 2017). These more inhospitable conditions (the treeline is 500 m lower in the western compared to Eastern Himalayas, and winter temperatures colder due to higher latitude and higher temperature lapse rates (see Chap. 2)) and the presence of higher winter snowfall due to the westerlies (which leads to good pre-monsoon soil moisture) allows birch to outcompete the evergreen species (Schickhoff et al. 2015; Bobrowski et al. 2017).

While the Eastern Himalayas, up to mid elevations, is dominated by evergreens which flush multiple times (e2 types of forests), the outer, moister ranges of the

Central and Western Himalayas are associated with evergreens with a 1-year leaf lifespan (e1). The inner ranges seem to be associated with multiple-year leaves of conifers (e3). Among the important broadleaved species, only a few deciduous species such as *Betula utilis, Aesculus indica* and some species of *Acer* occur in inner ranges (dw). The pre-monsoon drought deciduous (dp) trees form their greatest extent in the drier zones of the Himalayan foothills.

While predicting the impact of climate change on Himalayan forests, this split in forest type into moist outer ranges (where e1-type dominates) and dry inner ranges (where e3-type dominates) has never been analysed. The two regions may respond differently to climate change, as the relative impacts of monsoon and westerlies vary.

Table 3.3 An overview of some of the common Himalayan species, indicating their geographical range in the Himalayas and also providing their altitudinal range and classification based on leaves.

Advantages of a Leaf-Based Classification System

In the Himalayas, the steeply undulating topography leads to dramatic changes in soil depth, solar radiation and even rainfall over short distances of only a few kilometres. The heterogeneous terrain can lead to rapid variations in species mix, making classification of forests unnecessarily complex and of little functional use. The inhospitable terrain also makes ground truthing to correctly identify forest type difficult, and there is increasing dependence on remote sensing methods.

A classification system based on leaf type and phenology meets the suggestions of Running et al. (1995) that classification systems be (a) based on simple, observable, unambiguous characteristics of vegetation structure that are important to ecosystem biogeochemistry and could be measured in the field for validation, (b) remotely sensible so that repeatable reclassifications of existing vegetation will be possible and (c) directly translates into the biophysical parameters of interest to climate and biogeochemical models.

Important to this logic is the explicit separation of climate from the vegetation classification, to allow the classification to be based purely on observable remotely sensed vegetation properties (Singh and Thadani 2024). Leaf longevity, sometimes simply termed evergreen versus deciduous canopy, is an extremely critical variable in carbon cycle dynamics and is important for seasonal albedo and energy transfer characteristics of the land surface. Thus, leaf longevity class defines whether a plant must completely regrow its canopy each year, or merely a portion of it, with direct consequences to ecosystem carbon partitioning, leaf litterfall dynamics, and soil carbon pools (Running et al. 1995).

Using and building on this proposed classification system, it should be possible through simple seasonal measurements (and largely cloud-free seasons are chosen for measurement) using remote sensing methods to differentiate between these various Himalayan forest types (Table 3.4).

Table 3.3 Common tree species of the Himalaya

Species	Family	<-- Western Himalaya Eastern Himalaya -->											Zone	Altitudinal range (m)	Leaf based classification
		NW	Kashmir	Himachal	UK	W. Nepal	E. Nepal	Sikkim	Bhutan	Arunachal	Meghalaya	NE states			
Juniperus squamata														3500-5000	e3-s
Betula utilis													Alpine and Sub-alpine	3000-4300	dw
Rhododendron setosum														3000-5000	e2
Rhododendron campanulatum														2700-4500	e2
Juniperus recurva														2900-4100	e3-s
Juniperus indica (syn J. wallichiana)														3000-4000	e3-s
Abies spectabilis	P													3000-4000	e3-f
Abies densa	P													2900-4000	e3-f
Picea spinulosa	P													2800-4000	e3-n
Quercus semecarpifolia	F													2500-3800	e1
Tsuga dumosa	P													2500-3500	e3-f
Abies pindrow	P													2500-3300	e3
Picea smithiana	P													2400-3400	e3
Pinus wallichiana (syn P. excelsa)	P													2000-3500	e3-n
Pinus gerardiana	P													2000-3000	e3-n
Acer campbellii														1800-3000	dw
Acer oblongum														1800-3000	e1
Cedrus deodara	P												Temperate	1600-3000	e3-n
Quercus floribunda	F													2000-2600	e1
Rhododendron arboreum														1600-3000	e2
Cupressus torulosa														1600-2800	e3
Quercus baloot (syn Q. ilex)	F													1800-2600	e1
Quercus lamellosa	F													1300-2700	e1
Aesculus indica														1400-2600	dw
Lithocarpus pachyphyllus (syn Quercus pachyphylla)	F													1200-2800	e2
Lithocarpus elegans	F													1200-2600	e2
Lyonia ovalifolia														1500-2300	dw
Quercus lanata	F													1400-2400	e1
Quercus leucotrichophora	F													1500-2200	e1
Quercus griffithii	F													1200-2400	e1
Castanopsis indica	F													900-2700	e2
Castanopsis purpurella (syn C hystrix)									Abundant					1000-2200	
Castanopsis tribuloides	F													800-2200	e2
Alnus nepalensis														600-2600	dw
Pinus roxburghii	P													1000-2100	e1
Betula alnoides													Sub-tropical	700-2100	dw
Olea europaea subsp. cuspidata														700-2100	e2
Quercus serrata	F													600-2000	dw
Schima wallichi														500-2000	e2
Pinus kesiya	P													800-1600	e1

(continued)

Table 3.3 (continued)

Species	Family		Zone	Range	Type
Engelhardia spicata			Tropical	500-1800	d
Phoebe hainesiana			Tropical	600-1400	e2
Litsea monopetala			Tropical	200-1200	e2
Terminalia elliptica (syn T. tomentosa)			Tropical	400-1300	dp
Dipterocarpus retusus (syn D. macrocarpus)	D		Tropical	400-1200	e2
Mesua ferrea			Tropical	400-1200	e2
Quercus semiserrata	F		Tropical	300-1300	dw
Senegalia modesta (syn Acacia modesta)	L		Tropical	400-1200	dw
Terminalia arjuna			Tropical	200-1400	dp
Terminalia myriocarpa			Tropical	200-1300	e1
Anogeissus latifolia			Tropical	400-1000	dp
Dalbergia sissoo	L		Tropical	200-1200	dp
Mallotus philippensis			Tropical	200-1200	e1
Duabanga grandifolia (syn D. sonneratioides)			Tropical	100-1200	dp
Ficus spp			Tropical	0-1200	e2
Kayea assamica			Tropical	200-1000	e2
Shorea robusta	D		Tropical	200-1000	e1
Dipterocarpus turbinatus	D		Tropical	200-800	e2
Shorea assamica	D		Tropical	200-700	e2
Senegalia catechu (syn Acacia catechu)	L		Tropical	0-1000	dw
Bauhinia vahlii	L		Tropical	200-700	dw

Family Key		
	P	Pinaceae
	F	Fagaceae
	D	Dipterocarpaceae
	L	Fabaceae (Legumes)

Table 3.4 Differentiating between forest types based on leaf cover in different seasons

Forest type	Nomenclature	Winter (December)	Pre-monsoon (March)	Post-monsoon (October)
Evergreen—Tropical and subtropical	e2	++	++	++
Evergreen—1 year leaf life span, simultaneous leafing and leaf drop	e1	++	+	++
Deciduous (Winter)	dw	–	+	++
Deciduous (Pre-monsoon)	dp	+	–	++
Coniferous (Multi-year life span)	e3	++	++	++

++ full leaf cover; + sparse leaves; – leafless

The months suggested are relatively cloudless months when clear satellite images can typically be obtained. Variations in leaf drop and leaf flush lead to varying canopy cover, which can be used to distinguish these forest types.

A Continuum Based on Moisture

In Himalayan forests up to temperate altitudes, the lack of moisture during the dry pre-monsoon is often limiting to growth. For these forests, the pre-monsoon dry period is phenologically the most important phase (Shukla and Ramakrishnan 1982; Singh and Singh 1987; Kikim and Yadava 2001). Many trees of these forests drop their leaves between November and April. This is not a response to low temperature, but rather low moisture availability. This is a period when rainfall is sparse and trees conserve moisture through low photosynthetic rates and leaf senescence.

However, this is also a common period for leaf renewal to take advantage of the abundant moisture that the summer monsoon rains bring. Photosynthetic rates appear to be high through the monsoon, and often peak in the post-monsoon season. The most common and seemingly efficient response is exemplified by the e1 species, in which nutrient retranslation occurs during the pre-monsoon, and leaf fall and new flush occur almost simultaneously. During this period, photosynthesis rates are low, and moisture loss is also low, so as to conserve water, which is better available for the new flush (Zobel et al. 2001). The timing of the new flush allows the trees to make the best use of the favourable conditions that occur in the monsoon.

The d-p group of trees, which are deciduous in response to the pre-monsoon drought, usually drop their leaves earlier, between October and January. These are often found in areas of higher moisture stress (e.g. at lower altitudes where heat is greater). Leaves are lost 1–3 months before new leaf flush so as to tide over this period of highest water stress.

In mesic environments, such as in the Eastern Himalayas (the e2 forests), the maximum leaf fall is between January and March (Shukla and Ramakrishnan 1982; Kikim and Yadava 2001), the period of the pre-monsoon drought. Leaf flush occurs at multiple times of the year, reflecting the more spread-out periods of favourable photosynthetic gain. These e2 trees also often produce a leaf flush in the pre-monsoon, though additional flushes may occur to take advantage of the favourable post-monsoon conditions (Ralhan and Singh 1987; Kikim and Yadava 2001). Figure 3.8 shows this continuum between these strategies, viz. dp, e1 and e2.

This Himalayan strategy contrasts strongly with trees from most other temperate and alpine zones, where photoperiod or changing day length trigger leaf abscission. While mechanisms of leaf fall are not well studied in the Himalayas, for most common Himalayan species, changes in soil moisture appear to have a role in leaf abscission.

References

Apollo M (2017) The population of Himalayan regions–by the numbers: past, present and future. Contemp Stud Environ Tour:145–160

Ashton PS (2014) On the forests of tropical Asia, lest the memory fade. Harvard University

Ashton PS, Zhu H (2020) The tropical-subtropical evergreen forest transition in East Asia: an exploration. Plant Divers 42:255–280

Athokpam FD, Garkoti SC (2013) Variation in evergreen and deciduous species leaf phenology in Assam, India. Trees 27:985–997

Athokpam FD, Garkoti SC, Borah N (2014) Periodicity of leaf growth and leaf dry mass changes in the evergreen and deciduous species of Southern Assam, India. Ecol Res 29:153–165

Bai K, He C, Wan X, Jiang D (2015) Leaf economics of evergreen and deciduous tree species along an elevational gradient in a subtropical mountain. AoB Plants 7

Bhatt D, Sekar KC, Kumar K (2020) Tree diversity of Indian Himalayan region. GB Pant National Institute of Himalayan Environment, Almora, Uttarakhand. ISBN: 978-93-5408-274-0

Bobrowski M, Gerlitz L, Schickhoff U (2017) Modelling the potential distribution of Betula utilis in the Himalaya. Glob Ecol Conserv 11:69–83. https://doi.org/10.1016/j.gecco.2017.04.003

Boojh R, Ramakrishnan PS (1981) Temperature responses to seed germination in two closely related tree species of Schima, Reinw. Curr Sci 50(9):416–418

Bookhagen B, Burbank DW (2010) Toward a complete Himalayan hydrological budget: spatio-temporal distribution of snowmelt and rainfall and their impact on river discharge. J Geophy Res 115:1–25

Brodribb TJ, Feild TS (2008) Evolutionary significance of a flat-leaved Pinus in Vietnamese rainforest. New Phytol 178:201–209

Brodribb TJ, Feild TS, Jordan GJ (2007) Leaf maximum photosynthetic rate and venation are linked by hydraulics. Plant Physiol 144:1890–1898

Brodribb TJ, Pittermann J, Coomes DA (2012) Elegance versus speed: examining the competition between conifer and angiosperm trees. Int J Plant Sci 173(6):673–694

Brodribb TJ, McAdam SA, Jordan GJ, Martins SC (2014) Conifer species adapt to low-rainfall climates by following one of two divergent pathways. Proc Natl Acad Sci 111(40): 14489–14493

Champion HG (1936) A preliminary survey of forest types of India and Burma. Indian Forestry Record (NS) Silviculture 1(1)

Champion HG, Seth SK (1968) A revised survey of forest types of India. Government of India Press, New Delhi, p 404

Donovan LA, Maherali H, Caruso CM, Huber H, de Kroon H (2011) The evolution of the worldwide leaf economics spectrum. Trends Ecol Evol 26(2):88–95

Du H, Ran JH, Feng YY, Wang XQ (2020) The flattened and needlelike leaves of the pine family (Pinaceae) share a conserved genetic network for adaxial-abaxial polarity but have diverged for photosynthetic adaptation. BMC Evol Biol 20(1):1–12. https://doi.org/10.1186/s12862-020-01694-5

Dudgeon W, Kenoyer LA (1925) The ecology of Tehri-Garhwal. A contribution to the ecology of western Himalayas. J Indian Bot Soc 4:233–285

FSI (2021) Indian state of forest report. Forest Survey of India, Dehradun

Givnish TJ (1987) Comparative studies of leaf form: assessing the relative roles of selective pressures and phylogenetic constraints. New Phytol 106:131–160

Givnish TJ (2002) Adaptive significance of evergreen vs. deciduous leaves: solving the triple paradox. Silva Fenn 36:703–743

Gray RE, Ewers RM (2021) Monitoring forest phenology in a changing world. Forests 12(3):297

ICFRE (2013) Forest types of India: revisited. Indian Council of Forestry Research and Education, Dehradun

Joshi R, Tamang ND (2021) Temperature lapse rate in the Himalaya. In: Singh SP, Singh RD, Gumber S (eds) Interpreting Mountain treelines in a changing world. Chapter 7. Central Himalayan Environment Association and International Centre for Integrated Mountain Development. https://doi.org/10.53055/ICIMOD.787

Kattel DB, Yao T, Yang K, Tian L, Yang G, Joswiak D (2013) Temperature lapse rate in complex mountain terrain on the southern slope of the central Himalayas. Theor Appl Climatol 113:671–682

Kenoyer LA (1921) Forest formations and successions in the Sat Tal Valley, Kumaon, Himalayas. J Indian Bot Soc 2:236–258

Kikim A, Yadava PS (2001) Phenology of tree species in subtropical forests of Manipur in north eastern India. Trop Ecol 42(2):269–276

Kumar A, Patil M, Singh AN (2020) Shivalik, Siwalik, Shiwalik or Sivalik: which one is an appropriate term for the foothills of Himalayas? J Sci Res 64(1)

Luo T, Luo J, Pan Y (2005) Leaf traits and associated ecosystem characteristics across subtropical and timberline forests in the Gongga Mountains, Eastern Tibetan Plateau. Oecologia 142:261–273

Luo Z, Guan H, Zhang X et al (2017) Photosynthetic capacity of senescent leaves for a subtropical broadleaf deciduous tree species Liquidambar formosana Hance. Sci Rep 7:6323. https://doi.org/10.1038/s41598-017-06629-7

Ma S, He F, Tian D, Zou D, Yan Z, Yang Y et al (2018) Variations and determinants of carbon content in plants: a global synthesis. Biogeosciences 15:693–702. https://doi.org/10.5194/bg-15-693-2018

Malhado ACM, Whittaker RJ, Malhi Y, Ladle RJ, Ter Steege H, Butt N et al (2009) Spatial distribution and functional significance of leaf lamina shape in Amazonian forest trees. Biogeosciences 6(8):1577–1590

Matsuki S, Koike T (2006) Comparison of leaf life span, photosynthesis and defensive traits across seven species of deciduous broad-leaf tree seedlings. Ann Bot 97(5):813–817

Medina E, Garcia V, Cuevas E (1990) Sclerophylly and oligotrophic environments: relationships between leaf structure, mineral nutrient content, and drought resistance in tropical rain forests of the upper Rio Negro region. Biotropica 22(1):51–64. https://doi.org/10.2307/2388719

Miehe G (2015) Landscapes of Nepal. In: Miehe G, Pendry CA, Chaudhary RP (eds) Nepal: an introduction to the natural history, ecology and human environment of the Himalayas. Royal Botanic Garden, Edinburgh, Edinburgh

Miehe G, Miehe S, Bohner J, Baumler et al (2015) Vegetation ecology. In: Miehe, G.; Pendry, C.A. and Chaudhary, R.P. (Editors). Nepal: an introduction to the natural history, ecology and human environment of the Himalayas. Edinburgh: Royal Botanic Garden, Edinburgh

Moles AT, Westoby M (2000) Do small leaves expand faster than large leaves, and do shorter expansion times reduce herbivore damage? Oikos 90(3):517–524

Norris L (2018) The functions of needle leaves. Home guides, garden, soil care. SFGATE. Newsletter. 06 December, 2018. https://homeguides.sfgate.com/functions-needle-leaves-82695.html

Ohsawa M (1987) Life zone ecology of the Bhutan Himalaya I. Chiba University, Chiba

Pittermann J, Stuart SA, Dawson TE, Moreau A (2012) Cenozoic climate change shaped the evolutionary ecophysiology of the Cupressaceae conifers. Proc Natl Acad Sci USA 109: 9647–9652

Pornon A, Marty C, Winterton P, Lamaze T (2011) The intriguing paradox of leaf lifespan responses to nitrogen availability. Funct Ecol 25(4):796–801

Ralhan PK, Singh SP (1987) Dynamics of nutrients and leaf mass in central Himalayan forest trees and shrubs. Ecology 68(6):1974–1983

Rana SK, Price TD, Qian H (2019) Plant species richness across the Himalaya driven by evolutionary history and current climate. Ecosphere 10(11):e02945

Raunkiaer C (1934) The life forms of plants and statistical plant geography; being the collected papers of C. Raunkiaer. The life forms of plants and statistical plant geography; being the collected papers of C. Raunkiaer

Reddy CS, Saranya KRL, Pasha SV, Satish KV, Jha CS, Diwakar PG et al (2018) Assessment and monitoring of deforestation and forest fragmentation in South Asia since the 1930s. Glob Planet Chang 161:132–148

Reich PB (1988) Variation among plant species in leaf turnover rates and associated traits: implications for growth at all life stages. In: Lambers H, Poorter H, Van Vauren MMI (eds) Inherent variations in plant growth: physiological mechanisms and ecological consequences. Buckhuys, Leiden, pp 467–487

Reich PB, Uhl C, Walters MB, Ellsworth DS (1991) Leaf lifespan as a determinant of leaf structure and function among 23 Amazonian tree species. Oecologia 86:16–24

Reich PB, Walters MB, Ellsworth DS (1992) Leaf life-span in relation to leaf, plant, and stand characteristics among diverse ecosystems. Ecol Monogr 62(3):365–392

Richardson DM, Rundel PW (1998) Ecology and biogeography of Pinus: an introduction. In: Richardson DM (ed) Ecology and biogeography of Pinus. Cambridge University Press

Running SW, Loveland TR, Pierce LL, Nemani RR, Hunt ER Jr (1995) A remote sensing based vegetation classification logic for global land cover analysis. Remote Sens Environ 51(1):39–48

Sakai A, Malla SB (1981) Winter hardiness of tree species at high altitudes in the East Himalaya. Nepal Ecology 62(5):1288–1298

Schickhoff U, Bobrowski M, Böhner J, Bürzle B, Chaudhary RP, Gerlitz L, Heyken H et al (2015) Do Himalayan treelines respond to recent climate change? An evaluation of sensitivity indicators. Earth Syst Dynam 6(1):245–265

Schmidt-Vogt D, Miehe G (2015) Land Use. In: Miehe G, Pendry CA, Chaudhary RP (eds) Nepal: an introduction to the natural history, ecology and human environment of the Himalayas. Royal Botanic Garden, Edinburgh, Edinburgh

Schweinfurth U (1957) Die horizontale und vertikale Verbreitung der Vegetation im Himalaya. Bonner Geogr Abh 20:1–375. (The Horizontal and Vertical distribution of vegetation in the Himalaya, Bonner Geographical abstracts)

Schweinfurth U (1984) The Himalaya: complexity of a mountain system manifested by its vegetation. Mt Res Dev:339–344

Shukla RP, Ramakrishnan PS (1982) Phenology of trees in a sub-tropical humid forest in North-Eastern India. Vegetatio 49(2):103–109

Singh JS, Singh SP (1987) Forest vegetation of the Himalaya. Bot Rev 53:80–192

Singh JS, Singh SP (1992) Forests of the Himalayas: structure, function and impact of man. Gyanodaya Prakashan, Nainital, 294 pages

Singh SP, Thadani R (2024) A leaf based classification for Himalayan forests. Int J Ecol Environ Sci 50(1):11–31

Singh SP, Adhikari BS, Zobel DB (1994) Biomass, productivity, leaf longevity, and forest structure in the central Himalaya. Ecol Monogr 64(4):401–421

Singh SP, Singh RD, Gumber S (2021) Interpreting mountain treelines in a changing world. Central Himalayan Environment Association and International Centre for Integrated Mountain Development

Stainton JDA (1972) Forests of Nepal. John Murray, London

Suri PN (1933) A study in the ecology and silviculture of the Himalayan Spruce and silver fir with special reference to works in progress in Kulu. In Proceedings Punjab forest conference

Thadani R (1999) Disturbance, microclimate and competitive dynamics of tree seedlings in Banj Oak (*Quercus leucotrichophora*) in Forest of Central Himalayas India. A dissertation presented to faculty of graduate school of Yale University in candidacy for the Degree of Philosophy

Troup RS (1921) The silviculture of Indian trees. Clarendon Press

Trueba S, Théroux-Rancourt G, Earles JM, Buckley TN, Love D, Johnson DM, Brodersen C (2022) The three-dimensional construction of leaves is coordinated with water use efficiency in conifers. New Phytol 233(2):851–861

Tucker RP (1983) The British colonial system and the forests of the Western Himalayas. In: Tucker RP, Richards JF (eds) Global deforestation and the nineteenth century world economy. Duke University Press, Durham

Tyagi V, Singh SP, Singh RD, Gumber S (2023) Chir pine and banj oak responses to pre-monsoon drought across slope aspects and positions in Central Himalaya. Environ Monit Assess 195(2): 258

Udvardy MDF (1975) A classification of the biogeographical provinces of the world. IUCN Occasional Paper no. 18. IUCN, Morges van Ommen Kloeke AEE, Douma JC, Ordonez JC, Reich PB, Van Bodegom PM (2012) Global quantification of contrasting leaf life span strategies for deciduous and evergreen species in response to environmental conditions. Glob Ecol Biogeogr 21(2):224–235

Vasfilov SP (2016) The effect of photosynthesis parameters on leaf lifespan. Biol Bull Rev 6:96–112

Webb LJ (1959) A physiognomic classification of Australian rain forest. J Ecol 47:551e570

Wright IJ, Reich PB, Cornelissen JH, Falster DS, Groom PK, Hikosaka K et al (2005) Modulation of leaf economic traits and trait relationships by climate. Glob Ecol Biogeogr 14(5):411–421

Wright IJ, Ackerly DD, Bongers FJJM, Harms KE, Ibarra-Manriquez G, Martinez-Ramos M, Mazer SJ, Muller-Landau HC, Paz H, Pitman NCA, Poorter L, Silman MR, Vriesendorp CF, Webb CO, Westoby M, Wright SJ (2007) Relationships among ecologically important dimensions of plant trait variation in seven neotropical forests. Ann Bot 99(5):1003–1015. https://doi.org/10.1093/aob/mcl066

Wu ZY (1980) The vegetation of China. Science Press, Beijing. (in Chinese, with maps)

Wu ZY (1987) The vegetation of Yunnan. Science Press, Beijing. (in Chinese, with maps)

Zepner L, Karrasch P, Wiemann F, Bernard L (2020) ClimateCharts.Net—an interactive climate analysis web platform. Int J Digit Earth. https://doi.org/10.1080/17538947.2020.1829112

Zhang XL, Ha BB, Wang SJ, Chen ZJ, Ge JY, Long H, He W, Da W, Nian XM, Yi MJ, Zhou XY (2018) The earliest human occupation of the high-altitude Tibetan Plateau 40 thousand to 30 thousand years ago. Science 362(6418):1049–1051

Zhang YL, Moser B, Li MH, Wohlgemuth T, Lei JP, Bachofen C (2020) Contrasting leaf trait responses of conifer and broadleaved seedlings to altered resource availability are linked to resource strategies. Plants 9(5):621. https://doi.org/10.3390/plants9050621

Zobel DB, Singh SP (1997) Himalayan forests and ecological generalizations. Bioscience 47(11): 735–745. https://doi.org/10.2307/1313096

Zobel DB, Garkoti SC, Singh SP, Tewari A, Negi CMS (2001) Patterns of water potential among forest types of the central Himalaya. Curr Sci 80:774–779

Major Forest Types and Associations in the Himalayas

4

Rajesh Thadani and S. P. Singh

Abstract

This chapter provides an analysis of major forest types and associations across the Himalayan arc, mapping their distribution patterns, dominant species, and ecological characteristics. While drawing from Champion and Seth's (1968) forest classification system as its organisational framework, the chapter extends beyond India to encompass forest formations throughout the entire Himalayan range, including Nepal, Bhutan, and Pakistan. Rather than strictly adhering to the established classification, the authors emphasise important species and genera that define Himalayan forests, providing detailed accounts of their ecological relationships and geographic distributions. The Himalayas host an extraordinary range of forest ecosystems, from tropical evergreen forests in the eastern foothills to high-altitude birch forests at the treeline. While eastern Himalayan forests exhibit greater species richness and structural complexity, in the western Himalayas, forests are characterised by monodominant stands of species like Sal (*Shorea robusta*), chir pine (*Pinus roxburghii*), and various oaks (*Quercus* spp.). Conifers, particularly from the Pinaceae family, dominate higher elevations and inner valleys, with firs, spruce, and deodar being ecologically and economically significant. The chapter highlights how anthropogenic factors have shaped forest composition, with some species expanding due to disturbance while others

R. Thadani (✉)
Centre for Ecology Development and Research (CEDAR), New Delhi, India

Department of Botany, Jai Hind College, Churchgate, Mumbai, India
e-mail: thadani_rajesh@hotmail.com

S. P. Singh
INSA Honorary Scientist, Dehradun, Uttarakhand, India

Centre for Himalayan Studies, Graphic Era (deemed to be) University, Dehradun, Uttarakhand, India

S. P. Singh, R. Thadani (eds.), *Himalayan Forests in a Changing World*,
https://doi.org/10.1007/978-981-95-0978-2_4

face depletion. By examining both forest types and their constituent species, the chapter provides essential context for understanding Himalayan forest dynamics and their sensitivity to environmental change.

Keywords

Himalayan forests · Champion and Seth classification · monodominant stands · biogeographic gradients · Sal · Chir pine · oaks · conifers · cloud forests · treeline ecology · anthropogenic impacts

Introduction

The Himalayas harbour a diverse range of forest formations along a 2400 km arc that epitomise the region's ecological diversity and complexity. In the previous chapter, we made generalisations for this complex and dynamic landscape, so as to classify these forests into broad categories. We described the major biogeographic variables, such as altitude, precipitation (amount and seasonality), position on slope, aspect, latitudinal factors and human impacts that lead to variations in forest types (Table 3.2). We also presented a leaf phenology-based classification system to divide the Himalayan forests into different categories.

As altitude increases, vegetation transitions from tropical, through sub-tropical, then montane temperate broadleaf and coniferous forests, and finally to sub-alpine and alpine scrub and meadows. Precipitation patterns influence these transitions, with moister regions having more dense and stratified canopies and a higher diversity of trees and epiphytes than dryer regions. The impact of the monsoon rains is particularly high in the windward slopes of the outer ranges and diminishes in the inner hills where rain-shadow areas host drier, sparse vegetation.

At lower elevations, below 1000 m, tropical forests characterised by a rich diversity of broadleaf species is common. The eastern part of the arc has rainforest like conditions at these elevations with a luxuriance of evergreen broadleaved species and high diversity. Dipterocarps abound in this belt, best known of which is sal (*Shorea robusta*) which dominates the foothills across much of the central and eastern Himalayas. In the extreme west, where precipitation is very low, it is replaced by a scrub forest of *Olea-Acacia*.

Above these are sub-tropical forests, often composed of chir pine (*Pinus roxburghii*) in the dryer central and western region, and *Schima-Castanopsis* in Nepal. As we ascend into the montane temperate zone, a variety of oaks assert their dominance, in particular banj (*Quercus leucotrichophora*) in the monsoon drenched outer ranges in the central Himalayas. While oaks remain important in the east, oaks such as *Q. lanata* and *Q. lamellosa* do not form pure stands but occur as components of mixed forests. Evergreen species with about 1-year leaf life span occur prominently in the western and central part of the monsoonal ranges and include broadleaved oaks as well as needle-leaved pines (See Chap. 3). North of the outer ranges lie the inner valleys with monsoon shadow forests, which have a domination of tall needle-leaved conifers, which have multi-year leaf life spans.

The monsoon-exposed outer ranges have transeau ratios nearly twice that of the inner monsoon-shadow ranges and species richness is similarly higher in the monsoonal forests compared to the inner ranges. Oaks and rhododendrons become sparse in the inner monsoon shadow valleys, possibly because of their dependence on high monsoon precipitation. In the temperate zones above 2000 m, conifers such as deodar (*Cedrus deodara*), blue pine (*Pinus wallichiana*), and fir (*Abies pindrow*) dominate.

The sub-alpine zones, above 3000 m see an increasing coniferous component especially in the west, and firs dominate along with spruce (*Picea smithiana*). Hardwoods remain relevant and Kharsu oak (*Q. semecarpifolia*), some maples (*Acer* spp) and birch (*Betula utilis*) often dominate at the treeline around 4000 m in the west and central regions, while a variety of rhododendrons occupy the upper reaches particularly around the eastern treelines. *Rhododendrons such as R. campanulatum,* a krummholz species and at lower altitudes *R. arboreum* are quite widespread in much of western and central Himalayas. *Stunted* vegetation, including Junipers also make their presence felt at these altitudes.

The interplay of altitude (which influences temperatures) and precipitation profoundly impacts the distribution and structure of forests. Higher altitudes typically receive less precipitation but are characterised by snow rather than rain, which influences the soil moisture regime and nutrient cycling. Also, lower evaporation at colder, higher altitudes means drought is less severe. Tree species and their associations are finely tuned to these gradients, leading to distinct forest formations.

This chapter delves into characteristic forest formations of the Himalayas, providing brief accounts of their distribution and characteristic species. There is a dichotomy between common forest types and common forest tree species. While the eastern Himalayas have forest associations which cover large geographical areas, the most common Himalayan trees are found in the western regions, often forming contiguous monodominant stands. For example, in the banj oak (*Quercus leucotrichophora*) forests of Kumaon, banj can constitute >70% of the canopy (Singh and Singh 1992). The large tracts of conifer forests in the western Himalayas show similar levels of single species dominance. Such a gregarious nature is lacking in eastern Himalayan associations, and the region is characterised by more diverse multi-species forests. The most common tree species of the Himalayas are thus usually found in the central or western parts and include (in approximate order of importance) *Shorea robusta* (in the foothills), followed by conifers: *Pinus roxburghii, Pinus wallichiana, Abies pindrow, Abies densa, Picea smithiana,* and *Cedrus deodara. Quercus leucotrichophora, Schima wallichii,* and then again, a conifer *Tsuga Dumosa* make up the remainder of the top ten Himalayan trees (see Table 4.1 below for the most common Himalayan trees).

A list of common Himalayan tree families and representative species is given in Table 4.2.

Table 4.1 Dominant trees (by wood volume) of the Himalayas (India, Pakistan, Nepal, Bhutan)

Genus/species	Volume in the Himalayas	Range and common species
Pinus	~700 m cu m[a]	Pines are the most common Himalayan tree and *P. roxburghii* and *P. wallichiana* are the dominant species, each ~ 300 m cu m volume. *P. roxburghii* (chir) dominates in Uttarakhand and western Nepal, while *P. wallichiana* is found across the Himalayan arc. *P. kesiya* is common in pockets of the NE and *P. gerardiana* in parts of the west.
Quercus	~600 m cu m	Oaks best typify Himalayan forests and are a common component across the E-W arc. *Q. leucotrichophora* dominates the central Himalayas and is the most common oak (>100 m cu m). *Q. semecarpifolia* is widespread close to the timberline. Eastern Himalayas have a diverse group of oaks. *Q. griffithii* is among the most common trees of Bhutan
Abies	~400 m cu m	Firs dominate higher altitudes. *A. pindrow* in the western Himalayas (typically below 3300m) and *A. densa* in the east (in particular Bhutan) are the main species. *A. spectabilis* occurs at higher elevations (>3000 m) especially in the central parts of the Himalayas
Shorea	~400 m cu m	In the Siwaliks and foothills of India and Nepal, *Shorea robusta* (sal) is the most common tree species of the Himalayas
Cedrus deodara	>150 m cu m	Deodar is commonly found in the western Himalayas in India and Pakistan
Picea smithiana	>150 m cu m	Spruce is common in the western Himalayas in India, Pakistan, and Nepal
Rhododendrons	~150 m cu m	*R. arboreum* is best known and found across the east-west arc, but this genus is at its most diverse and dominant at high elevations of the eastern Himalaya
Castanopsis	>100 m cu m	This genus is common in the eastern Himalayan forests of India and Bhutan and extends into eastern Nepal.
Schima wallichii	~100 m cu m	Eastern India, Nepal, and Bhutan
Tsuga Dumosa	~85 m cu m	Common in Nepal and Bhutan in particular

Sources: India (FSI 2021); Nepal (DFRS 2015), Bhutan (FMID 2023; Ohsawa 1987), Pakistan (FAO 2005)

[a]Estimated volumes across India, Nepal, Bhutan, and Pakistan in million cubic metres. Species listed when a single species dominates in the Himalayas

Himalayan Forest Types and Associations

We describe below, some of the dominant forest types and associations of the Himalayas. We use the Forest-type (group and sub-group) numbers as given by Champion and Seth (1968) to link our forest types with Champion and Seth's (C&S). For current extent of coverage, data from FSI (2021) is used for Indian

Table 4.2 Common families and representative tree species in the Himalayas

Family	Representative species
Pinaceae	*Abies pindrow* (Fir); *Cedrus deodara* (Deodar); *Picea smithiana* (Spruce); *Pinus roxburghii* (Chir); *Pinus wallichiana* (kail); *Tsuga dumosa* (Hemlock)
Fagaceae	*Castanopsis tribuloides* (Chestnut); *Quercus lamellosa*; *Quercus leucotrichophora* (Banj); *Quercus semecarpifolia* (Kharsu)
Ericaceae	*Lyonia ovalifolia*; *Rhododendron arboreum* (burans); *Rhododendron campanulatum*; *Vaccinium nummularia*
Betulaceae	*Alnus nepalensis* (Himalayan Alder, Utis); *Betula alnoides*; *Betula utilis* (Himalayan Birch); *Carpinus viminea* (Himalayan Hornbeam)
Dipterocarpaceae	*Shorea robusta* (Sal); *Dipterocarpus turbinatus* (Gurjan)
Fabaceae	*Albizia lebbeck* (Siris); *Bauhinia variegata (*kachnar*)*; *Dalbergia sissoo (*shisham*)*; *Senegalia modesta*
Lauraceae	*Litsea monopetala*; *Machilus odoratissimus* (syn *Persea odoratissima*); *Neolitsea pallens*; *Phoebe lanceolata*

forests. For forests of Nepal, the excellent descriptions of Miehe, Miehe et al. (2015) and Stainton (1972), for Bhutan, Ohsawa (1987) and for Pakistan Rasheed et al. (2017) among others form the basis of our descriptions.

We have chosen important Himalayan tree species, genera, or forest types so as to provide as holistic an overview of Himalayan forests as we possibly can in the span of a chapter. Forest types that cover substantial areas or demonstrate prominence in the Himalayas are the focus. We have (Table 4.3) also tried to link these forest associations with Champion and Seth's (1968) classification system (Figs. 4.1, 4.2, 4.3, and 4.4).

Himalayan Rainforests: Moist Tropical Forests of the Eastern Himalaya

(Northern Tropical Wet Evergreen forests, C&S 1B)

Brief Description

Moist tropical forests inhabit some of the wettest areas of the foothills of the Eastern Himalayas. These forests have multilayered canopies, with occasional emergents—often Dipterocarps—projecting above the main canopy. These are tall forests, often over 50 m high with luxuriant growth. The middle storey is dominated by evergreens. Bamboos are common along with canes and epiphytes. The dense canopy leads to poorly developed ground vegetation and grasses are largely absent. Epiphytes and climbers are common. Cauliflory, i.e. fruiting from the main stems, is also common. The more moist of these forests resemble true rainforests.

Climate

These are forests of heavy monsoonal rainfall. Annual precipitation typically exceeds 2400 mm and may exceed 5000 mm in some places. Rainfall in the winters (December—February) is low, often under 20 mm a month, but humidity levels

Table 4.3 Forest types as per Champion and Seth (1968) and Himalayan forest associations described in this chapter

Group	Climate type group	Type group name (as per Champion and Seth)	Himalayan sub-group	Himalayan forest association described in this chapter
I. Moist tropical forest	1	*Tropical wet evergreen forests*	1B	*Himalayan rainforests: Moist tropical forests of the eastern Himalaya*
	2	Tropical semi-evergreen forests	2B	East Himalayan mixed tropical semi-evergreen forests
	3	Tropical moist deciduous forests	3C	Himalayan Sal forests
	4	Littoral and swamp forests	–	*(uncommon in Himalayas)*
II. Dry tropical forest	5	Dry tropical forests	5B	Himalayan Sal forests
	6	Tropical thorn forests	–	*(uncommon in Himalayas)*
	7	Tropical dry evergreen forests	–	*(uncommon in Himalayas)*
III. Montane subtropical forests	8	*Sub-tropical broadleaved hill forests*	8B/C1	*Schima-Castanopsis*: Eastern Himalayan moist sub-tropical forests
	9	Sub-tropical pine forests	9C1	Chir pine: Sub-tropical pines
	10	Sub-tropical dry evergreen forests	10 C1	Olea and Senegalia
IV. Montane temperate forests	11	Montane wet temperate forests	11B/C1	*Quercus lamellosa*: Cloud forests of the Eastern Himalayas
	12	Himalayan moist temperate forests	12C1a	Banj oak: Moist lower temperate forests
			12C2a	High altitude oaks: *Quercus semecarpifolia* forests
			12/C1c	Deodar (*Cedrus deodara*)
			(12C1f)	Pinus wallichiana
			(12C12b)	Fir forests
	13	Himalayan dry temperate forests	13/C2	Deodar (*Cedrus deodara*)
			13/C4	Pinus wallichiana
V. Sub-alpine forests	14	Sub-alpine forests	14/C2/2S1	Pinus wallichiana
			14C1	Fir forests
			14C2	Birch forests
VI Alpine scrub	15	Moist alpine scrub	15C1	Birch/Rhododendrons
			15E1	Rhododendron scrub
	16	Dry alpine scrub		*(Largely scrub, dwarf shrubs and not really forests)*

Source: Champion and Seth (1968)

Fig. 4.1 Foothill forests (**a**) Tropical semi-evergreen (Assam); (**b**) *Shorea robusta*, Surkhet (600 m), Nepal. (Photo: (**a**) Rajesh Thadani; (**b**) Vishal Singh)

remain high. Cooler temperatures in these months reduce evapotranspiration. The mean annual temperature in these forests lies between 22–26 °C, though minimal temperatures can fall to 5–10 °C. Maximum temperatures rarely exceed 37 °C. These forests grow on soils that retain moisture quite well.

Occurrence

These are found in the plains of Assam and on low hills and gentle slopes up to around 1000–1200 m. The lower slopes of Arunachal Pradesh facing the

Fig. 4.2 Lower temperate forests around 1800 m. Chir pine (**a**) and Banj oak (**b**) dominate in the central Himalayas. Chir is better known as a sub-tropical species, but can form monodominant stands well into the lower temperate zone. (Photo credit: Rajesh Thadani)

Brahmaputra valley (such as the districts of East and West Siang and Lohit), southern edges of Sikkim and Bhutan, Meghalaya and small fragments in Nagaland and hills of other Northeastern States, such as in the Patkai hills. While not truly Himalayan, they cover about 4000 km^2 in the Himalayan foothills.

Dominant Trees and Associates

These are forests of high diversity. The emergent canopy is dominated by large dipterocarps such as *Dipterocarpus retusus* (syn *D. macrocarpus*) and *Shorea assamica*. *Mesua ferrea* (Calophyllaceae) and *Castanopsis indica* (Fagaceae) are often important trees. The second storey, also 30 m high includes *Prasoxylon excelsum* (syn. *Dysoxylum procerum*) (Meliaceae); *Terminalia myriocarpa* (Combretaceae); *Elaeocarpus sphaericus*[1] (Elaeocarpaceae), *Baccaurea sapida* (syn. *B. ramiflora;* or Burmese grape); *Vatica lanceifolia* (Dipterocarpaceae) (locally abundant in upper Brahmaputra valley); *Terminalia bellirica*[2] (Bahera; Combretaceae) *Garcinia, Myristica*, and *Eugenia.*

[1] The fruit of *Elaeocarpus* yields Rudraksh beads used in rosaries and necklaces.

[2] *Terminalia bellirica* is a common deciduous tree found across the Indian subcontinent and is a one-third component of the common ayurvedic remedy triphala.

Fig. 4.3 Upper temperate forests >2500 m. (**a**) Mixed stand of Kharsu oak (*Quercus semecarpifolia*) and Maple (*Acer caesium*). (**b**) Deodar (*Cedrus deodara*) forest in Himachal Pradesh. (Photo: (**a**) Soham Kacker and (**b**) Rajesh Thadani)

Fig. 4.4 High altitude forests above 3000 m. (**a**) Blue pine in Niti Valley (Uttarakhand). (**b**) *Betula utilis* in Kashmir forming the treeline on a north-facing slope, with dense Rhododendron undergrowth, firs in the background. (Photo credit: (**a**) Gopal S Rawat and (**b**) Zafar Reshi)

Bamboos of various genera including *Dendrocalamus hamiltonii* and *Bambusa pallida* are common and can grow 15–18 m tall. Shrubs include *Clerodendrum* and *Ixora*. Many of these species such as *Mesua ferrea*, *Prasoxylon excelsum*, *Kayea assamica* (Assam ironwood) represent the Indo-Malayan biogeographic zone.

In the Upper Assam Valley (C&S 1B/C2), evergreen forests are dominated by *Mesua* and *Kayea*. In southern Assam, on the lower slopes of the Cachar hills (C&S 1B/C3) and Khasi—Jaintia hills of Meghalaya are the Cacher Tropical Evergreen forests rich in *Palaquium* (*Palaquium polyanthum*) (family Sapotaceae), *Diospyros toposia* (Family Ebenaceae), and *Dipterocarpus turbinatus* among others.

In areas of moist evergreen and semi-evergreen forests, heavy cutting or other degradation can lead to short lived euphorbiaceous trees being established with species such as *Macaranga denticulata*, *Mallotus philippensis,* and *Trema orientale* dominate in the early stages. Canebrakes (C&S 1B/E1) may be found in areas.

Human Impacts

Much of the original growth of massive Dipterocarps has long since been destroyed for timber. Much of the Khasi and Jaintia hills were home to jhum cultivators and the original forests have given way to secondary forests and these jhum impacted areas have a greater component of *Schima wallichii*, *Artocarpus heterophyllus* (Jackfruit, family Moraceae) (ICFRE 2013).

East Himalayan Mixed Tropical Semi-Evergreen Forests (C&S 2B)

Brief Description

This is an intermediate group of forests that lie between "Tropical Wet Evergreen" and "Moist Deciduous forests" with traits that are transitional between the two. While the forests have a closed canopy, with a height of 25–35 m, the canopy is less dense than in the evergreen forests and undergrowth is denser. The barks of trees in these forests tends to be thicker and rougher than trees of wet-evergreen forests. These forests are found in poorer and lower rainfall sites than the Tropical wet evergreen forests, and can be found in patches in Assam and the Northeast.

Climate

Rainfall ranges from 1500 to 3000 mm and the dry season is short. These forests are found in the plains and lower foothills below 1000 m.

Occurrence

While mostly found around Assam, along the southern banks of the Brahmaputra river and in Cachar, these extend into the foothills of the northeastern states, and the southern boundaries of Sikkim. This is an important forest type in terms of extent and covers over 35,000 km^2 of land area (FSI 2021).

Dominant Trees and Associates

These are rich and diverse forests. In general, Dipterocarps while present, are less common than in the moist tropical forests. Among the evergreens, *Syzygium*,

Cinnamomum, *Artocarpus,* and members of the Magnoliaceae are common. Among deciduous tree species, *Terminalia*, *Tetrameles* and *Stereospermum* are common representatives. In areas where fires are common, *Shorea robusta* may occur. An understorey rich in Lauraceae is typical of these forests with species such as *Phoebe* and *Machilus*.

In North Bengal and adjacent parts of Assam and Bhutan, Assam Valley Semi-evergreen forests (C&S 2B/C1) are found. Trees such as *Phoebe hainesiana*[3] (Lauraceae) and *Syzygium formosum* (syn *Eugenia formosa*) (Myrtaceae) and *Beilschmiedia* species are common in Northern Bengal (Kalimpong). In other parts of Kalimpong, *Mesua ferrea* can occasionally form over 50% of the canopy. In the Aka hills in Arunachal (near the Bhutan/ Assam borders) and northern Assam, in addition to *Phoebe* and *Beilschmiedia spp, Cinnamomum* and *Machilus* are abundant. The second storey has some *Quercus*, *Litsea,* and *Artocarpus*. In other forests of this semi-evergreen group, more wet evergreen components such as *Mesua ferrea* are found, while in the lower hills, *Quercus lamellosa* along with *Castanopsis* may extend downwards into this forest type. In these sub-Himalayan tracts, *Schima wallichii* and *Bauhinia purpurea* are found on lower slopes. *Terminalia myriocarpa*, *Syzygium*, *Tetrameles,* and *Phoebe* spp. are also common in this forest type.

An important species of these lowland forests is *Tetrameles nudiflora*, which dominates in many parts of Assam and the Northeast. This is a fast-growing deciduous tree, the only species of the genus (monotypic genus). The family Tetramelaceae is confined to the Indo-Malayan region with only two genera, *Tetrameles* and *Octomeles*, both of which are monotypic. Tetrameles is often found in tracts that have been jhumed for long periods.

The Cachar forests, near Silchar in Assam were known for their diversity and large area they covered, though tea and rice cultivations have since depleted these forests. Here Dipterocarps such as *Dipterocarpus turbinatus* and *D. tuberculatus* are common along with *Duabanga grandiflora* and *Artocarpus chama* (Syn *A. chaplasha*). A mix of similar species, viz. *Duabanga—Terminalia—Pterospermum* can also be seen as a seral stage in sub-Himalayan tracts of Arunachal and in the Naga and Patkai hills and is classified under moist deciduous forests by Champion and Seth (1968) (3/1S2).

As evident from the above descriptions, these forests are very diverse and plant associations vary greatly. Collectively the semi-evergreen forests of the northeast encompass most of the species found in Wet evergreen to Moist deciduous forests.

Human Impacts

As in the Wet evergreen forests, cane and bamboo brakes can dominate these areas for varying periods post forest clearing and human disturbance. Large swathes of these forests have also been depleted and cleared for tea cultivation. The tea gardens

[3] As per POWO *Phoebe hainesiana* is a syn for *Phoebe bootanica*. However, *P. hainesiana* continues to be used in current literature so we are using this name.

of Cachar are famous for their strong, full-bodied black teas. Wet rice cultivation has also expanded along the Brahmaputra valley at the cost of native forests.

Sal Forests of the Himalayas
(C&S 3C and 5B)

Brief Description

Sal (*Shorea robusta*) is a major forest type of the Himalayas which dominates the lowland forests almost across the entire Himalayan arc from 76 to 93°E. While a tree of the monsoonal foothills and Siwalik's (up to 1200 m), it forms a band so extensive from east to west that it is the most common tree species of the Himalayas. Sal trees are tall and handsome, attaining heights of 35 m or more in favourable conditions.

Sal has been classified by many authors as a deciduous species as it sheds its leaves during the dry season, typically from February–May, and new leaves flush between April and early July. In areas where the climate is humid, such as parts of the eastern Himalayas and where soils are more moist such as the Terai, Sal may exhibit a semi-evergreen nature. We have (Chap. 3) included sal as an e1 species (evergreen with 1-year leaf life span) given its nature of flushing in the pre-monsoon so as to benefit from the moist season ahead.

Climate

Sal forests are the dominant vegetation in the North Indian Tropical Moist Deciduous forests (C&S 3C). They can also be found commonly in the Northern Tropical dry deciduous forests (C&S 5B). Monsoon rainfall is the main form of precipitation in the Sal zone, and it can vary widely from 800 mm in the west to over 2500 mm in the east. Typically, sal forests dominate in areas where annual rainfall is between 1300 and 2000 mm much of it in the 4–5 monsoon months. The dry season can vary from almost 8 months at the western edge of its range to under 5 months in the east (Miehe et al. 2015).

Mean temperatures vary between 21 and 27 °C though minimum and maximum can range from freezing (0 °C) to over 40 °C. Sal is a tropical dipterocarp sensitive to frost, and hence usually does not grow much above 1000 m elevation.

Occurrence

Sal occurs along the Himalayan foothills and Siwaliks from Himachal Pradesh in the Western Himalayas to Meghalaya and the Brahmaputra valley in the East. Pockets of Sal forests can be seen beyond, especially in the West—for example in Jammu and Mirpur in Jammu and Kashmir. Along the Himalayan belt, sal is best seen in the central Himalayas in the Uttarakhand foothills and in terai regions of Nepal.

Sal can form monodominant stands across varied formations, from the marshy alluvial soils of the Terai to the dry and coarse pebble and gravel deposits of the Bhabhar, and the shallow unconsolidated sedimentary rocks of the Siwaliks. Along river valleys, sal forests can penetrate deep into the lesser Himalayas. These forests are notable in not just the range of substrate types they occupy, but also a remarkably uniform composition across this broad belt despite significantly varying precipitation regimes.

Sal is not found on recently formed alluvium or on waterlogged soil, and along riversides it is often replaced by tropical deciduous riverain forest. As per Stainton (1972), if "in the spring one flies over these forests, it is easy to see the dominance of the Sal, for at that time of year, the Sal is in flower and its sprays of white flowers and fresh green leaves are clearly visible from the air. Its dominance is broken only along the streams and rivers where the leafless trees of the deciduous riverain forest can be seen." Close to the river's edge and on newly-formed gravels, there are often strips of forest composed of *Acacia catechu* and *Dalbergia sissoo*. In time though the soil on these riverside terraces may mature sufficiently for Sal to succeed and replace the deciduous species.

In Champion and Seth's (1968) description of "North Indian Tropical Moist deciduous forests" (Sub group 3C), so important is the Sal forest that these authors divide this broad group "into two categories, Sal forests, and the rest." As a timber species, Sal is the most important and abundant tree both in India (FSI 2021) and in Nepal (DFRS 2015). In Nepal, Sal alone constitutes close to 20% of the stem volume and total forest biomass (DFRS 2015) with an area that extends over about 11,500 km^2 (TISC 2010). In the Himalayan foothills in India, Sal occurs across 4800 km^2 in the Western Himalayas and an additional 1450 km^2 in the Eastern Himalayas (FSI 2021), for a total of almost 18,000 km^2 across the Himalayas. In addition, Sal also occurs as a component of mixed forests, especially in the eastern Himalaya.

Sal is not exclusively a Himalayan tree and considerable additional areas of Sal are seen across the Indian subcontinent in areas of moderate rainfall. Two distinct contiguous forest patches are found across India—The north Indian Sal extends into the plains of Uttar Pradesh in addition to the Himalayan belt, while a large belt of Sal also can be found in east and central India stretching from Bangladesh to Madhya Pradesh and covering an estimated over 100,000 km^2 (Gautam and Devoe 2006). Overall, Sal is easily the most common tree of India constituting 10.9% of the timber volume (FSI 2021).

Sal is an aggressive, extremely gregarious species, and rarely found as a component of any other type of forest. This great timber species has high coppicing power, resistance to burning, and is adaptable to a variety of soil and site conditions. Deeply weathered sandy or silty colluvial soils, often formed through mass wasting processes, are common. In deep, rich soils, Sal forests can reach close to 50 m tall, though 25–35 m tall stands are more common. In the Siwalik hills, Sal trees are often under 15 m as soils are shallow.

Dominant Trees and Associates

The dominance of Sal and uniformity of these forests leads to generally poor floral diversity in Sal forests (Stainton 1972). Associates include various species of *Terminalia, Anogeissus latifolia, Adina cordifolia, Lagerstroemia parviflora,* and others. In the western alluvial, Sal forests *Mallotus philippensis* is a common associate. *Bauhinia vahlii* (Fabaceae) is a common climber. In the very moist eastern forests *Schima wallichii* is a common associate. A lower canopy with *Mallotus philippensis*, *Litsea monopetala*, *Phoebe,* and some oaks may be found. Bamboos such as *Dendrocalamus hamiltonii* are found in the understorey along with shrubs such as *Desmodium*, *Indigofera*, and *Leea*.

In the Siwaliks, species such as *Anogeissus latifolia* (Combretaceae) and *Terminalia elliptica (syn T. tomentosa)* are common associates while in the Bhabhar *Terminalia elliptica* may associate with *Lagerstroemia parviflora* (Lythraceae), and *Mallotus philippensis* (Euphorbiaceae) forms the second storey. In the Terai, *Syzygium cumini* (Myrtaceae, Jamun) and *Mallotus nudiflorus* (syn *Trewia nudiflora*) are common associates.

The Sal of the Bhabhar, while not as luxuriant, is certainly more abundant. As the Bhabhar is waterless for many months owing to its unique geology (see Chap. 1: Introduction), habitations and forest removal for farming is less common than in the Terai, and many Sal forests have survived. A narrow band of Sal can be found stretching across the Bhabhar, its dominance broken only along streams and rivers by deciduous riverain forest, and in dips and moist gulleys by tropical evergreen forest. Epiphytes and climbers are scarce. In some places there is a dense green mass of shrubs and seedlings beneath the trees, but in others, the forest is carpeted only with dry brown leaves. Sal forests are commonly burned in the spring, and this hinders the growth of a shrub understorey. In the far eastern foothill forests of Assam and Arunchal, where rainfall is high, *Sal* gives way to other Dipterocarps, such as *Dipterocarpus turbinatus* and *D. tuberculatus*.

Overall, while Sal is not a tree of the mountains, it is nonetheless the most common species of the Himalayan zone due to its tenacity and ability to dominate along the foothills and adjacent plains across much of the Himalayan arc.

Human Impacts

Sal forests are managed for timber and are also a source of a variety of non-timber forest products (NTFPs) valued by local communities. This tree is used for fuel-wood, fodder, litter for compost fertiliser, gums and resins, and many other products.

In the Terai, the region where it grows to its most massive proportions, many Sal forests were cleared for agriculture, especially post 1947 in India and a little later in Nepal. This has led to considerable depletion and fragmentation. The best examples of Terai Sal can be found in national parks such as Corbett (Uttarakhand, India), Dudhwa (Uttar Pradesh, India), and Chitwan National Park (Nepal).

The value of *Shorea robusta* as one of the main timber trees of India led to it being favoured silviculturally, and during the era of colonial forestry, it was managed to expand its distribution and dominance over competing species (Champion and Seth 1968). Controlled fire and moderate grazing have been used as silvicultural tools to encourage Sal regeneration, but when uncontrolled they can damage and deplete these forests. Fire was once considered the only weapon available to foresters for controlling weeds (Champion and Osmaston 1962). Controlled burning was also prescribed to reduce the load of dead leaves which hinder regeneration (Troup 1921) and to remove some competing species. Fire was used to maintain Sal forests as the climax type in wet Sal forest regions (Jacob 1941) without which these get converted into mixed broadleaved forests (Gautam and Devoe 2006). However, fire also leads to an increase in the number of herbs and shrubs, especially palatable plants, which leads to higher grazing.

Grazing was considered effective in checking the growth of palatable weeds such as *Imperata cylindrica* which otherwise inhibit the establishment of Sal seedlings (Rowntree 1940; Sarkar 1941). However, uncontrolled grazing reduces the litter content of soil, leads to a dramatic fall in sapling density, and depletes soils of nutrients leading to failure of Sal regeneration.

***Schima-Castanopsis*: Eastern Himalayan Moist Sub-tropical Forests** (e2 Type)(C&S 8B/C1)

Brief Description

In the eastern Himalayas, east of central Nepal and extending to Arunachal and beyond, *Schima wallichii* (Family Theaceae) and various species of *Castanopsis* (Family Fagaceae) commonly associate and form a forest type that may dominate the sub-tropical zone.

Climate

The Schima-Castanopsis forests are found in areas of high monsoonal rainfall, from 1500 to 5000 mm (Miehe et al. 2015). The dry season is short, typically no more than 3 months, and it occurs in the winters (usually Nov–Feb). Champion and Seth (1968) included these forests in "Northern Subtropical broadleaved Hill forests" (Sub-group 8B).

The mean annual temperature ranges from 15 to 20 °C, and temperatures are relatively stable from April to October. Temperatures rarely ever drop below freezing or rise above 30 °C in the middle of this zone (1100–1500 m).

Occurrence

Schima-Castanopsis forests are found from 800 to 2000 m and the dominance of this association is greatest in Eastern Nepal and Sikkim, though *Schima-Castanopsis* forests can be found from central Nepal to Arunachal. This is a zone of heavy cultivation and often these forests have been reduced to coppiced stumps or a few scattered trees. At one time though these forests would have dominated across east central Nepal around the Arun and Tamur valleys and the lower altitudes south of Annapurna (Stainton 1972).

These forests may abut with the Sal zone at their lower margins, and *Schima wallichii* extends lower into the tropical zone. On dryer sites, and moving into western Nepal, the *Pinus roxburghii* forests outcompete this association.

Dominant Trees and Associates

The *Schima-Castanopsis* forest is an evergreen multi-storeyed forest with a closed canopy, dense undergrowth, tall, fleshy-leaved forbs and ferns (Miehe et al. 2015). Canopy height can reach 50 m under favourable conditions. These forests are best

Table 4.4 The dominant species of *Schima-Castanopsis* forests

Species	Altitudinal range (metres)	Latitudinal range	Notes
Schima wallichii	250–1800	83–100°E	Abundant and among the ten most common Himalayan trees
Schima khasiana	900–1800	East of 88° E	Very restricted distribution, e.g. Shillong plateau
Castanopsis indica	600–1800	83–100°E	Can be found in wet tropical forests
Castanopsis hystrix	1500–2450	83–100°E	The largest of the Castanopsis, and the only species common in temperate altitudes
Castanopsis tribuloides	600–2200	78–100°E	Found on dryer sites and west into the Kumaon Himalaya

seen in eastern Nepal and Sikkim where the two genera may constitute >50–60% of the canopy. In Sikkim, *Schima wallichii* is the most abundant forest tree followed by Castanopsis species.

A brief description of the main species that define this associations is given in Table 4.4.

A luxuriant growth of epiphytes and lianas are found in this forest type. *Engelhardia* may associate in the canopy while various Magnolias and some deciduous trees may be found scattered. Lauraceous trees and shrubs prevail in the understoreys with *Ilex* spp., *Camellia kissi*, *Maesa chisia*, *Elaeagnus umbellata (*syn *E. Parvifolia)*, *Viburnum coriaceum*, *Eurya* spp., and *Symplocos* spp.

Moving further east, Fagaceae, viz. *Quercus*, *Lithocarpus,* and *Castanopsis* are often present and especially abundant at the higher elevations in Arunachal Pradesh and the north-eastern states where they intermix with other temperate genera such as *Alnus*, *Prunus* and *Betula*. In the Shillong plateau, *Schima khasiana* may be locally abundant. These forests are generally evergreen, though some deciduous species may occur and occasionally, such as in the case of *Betula*, form pure associations. In the Khasi hills, oaks such as *Quercus griffithii* and *Q. semiserrata* are common, and *Pinus kesiya* has expanded its range in disturbed areas.

Schima wallichii can descend to tropical altitudes where rainfall averages over 3000 mm and is a component of the Assam Valley Tropical wet evergreen forests (C&S, 1B/C1). However, it typically occurs as a part of sub-tropical forests. It may be found in pure pockets growing gregariously in Assam, and is the most common tree in the outer hills of western Bhutan up to 1300 m, but less common in Eastern Bhutan (Troup 1921).

To their west, as the climate gets drier in central Nepal, these forests give way to the western Himalayan oaks, viz. *Quercus lanata*, *Q. leucotrichophora* and also chir pine—*Pinus roxburghii*. The eastern Himalayan moist sub-tropical forests see a close mixing of Palearctic and Indo-Malayan vegetation.

Human Impacts

In the Indian eastern Himalayas, jhum cultivation has largely replaced the original flora. In the Khasi hills for example, this forest type was replaced by pine over a half century ago (Champion and Seth 1968). Khasya Pine (*Pinus kesiya*) was originally an edaphic climax restricted to well drained and dryer sites but has since expanded to replace the lower Schima-Castanopsis forests due to excessive anthropogenic disturbance. In Nepal, these forests are often reduced to coppiced stumps or a few scattered trees (Miehe et al. 2015).

Chir Pine Forests (Sub-tropical Pines)

(C&S 9 C1)

Brief Description

Chir pine (*Pinus roxburghii*) is the most abundant Himalayan tree. This three-needled pine (three needles per dwarf or spur shoot) is an aggressive early successional, capable of forming pure stands across large tracts, especially in the central Himalayas. Across India it is exceeded in wood volume only by sal and teak (FSI 2021) both of which are largely found in the plains. It is also among the most common trees of Nepal. Chir stands are adapted to frequent low intensity ground fires, and the periodicity and intensity of fires considerably influences these forests. Chir forms the canopy, and often no undercanopy or shrub layer will be present in forests subject to frequent fires. Chir is among the only pine species with a leaf life span of 1 year, and it appears partially deciduous during March–April when brown old leaves prevail over new green leaves. Tree height is greatly influenced by site quality, but it can grow to 40 m on favourable sites though heights of 20–30 m are common.

Climate

Mean annual temperatures in the chir zone are typically between 15 and 20 °C. Maximum summer temperatures can exceed 40 °C at lower altitudes, while minimums of −5 °C and lower occur in the upper reaches. Rainfall is largely from the Southwest monsoon (June-Sept). The range of rainfall is considerable and can vary from 900 mm in rain shadow regions to over 2500 mm in the monsoon drenched outer hills.

Occurrence

P. roxburghii forests are found across the Himalayan arc from Western Pakistan (72° E) through Bhutan (91°E), and cover almost 14,000 km^2 in the Indian Himalayas alone. They are among the most extensive forests dominated by a single species in the Himalayas. Chir forests descend to below 600 m on northern ridges where soils are thin and can be found as high as 2300 m on southern slopes. Heavy snow can damage this tree and may limit upward altitudinal extent (Thadani 2023). Chir is absent from the valleys of Kashmir where monsoons are weak. The centre of its distribution is Uttarakhand, though Chir forests cover large areas in Himachal and the dry upper valleys of Western Nepal, (Stainton 1972).

Dominant Trees and Associates

Known as long-needled pine, Chir has 25–30 cm long needles. An older synonym, *Pinus longifolia* ("long-leaves") well reflected this trait. Chir is a gregarious species that dominates its environment. Oaks are the most common associates, especially *Quercus leucotrichophora* in the central Himalayas. Yet, this oak and pine often do not truly intermix. The pine occupies ridges, south slopes, or sites with shallow soils, while the oaks dominate on northern aspects, mid and lower slopes, and sites with deeper soils. An understorey of oak under a pine canopy may occur in protected sites where oak is reclaiming space it originally occupied.

In the higher altitude oak zone (C&S 9/C1b), an ericaceous second-storey of *Lyonia ovalifolia* (*angyar*) and *Rhododendron arboreum* is often found along with *Myrica esculenta* (*kaphal*) and *Pyrus pashia,* though these species are often remnants of an older oak canopy removed as a result of human disturbance. At lower altitudes, as in the Siwalik (C&S 9/C1a), Chir associates with *Terminalia, Mallotus, Syzygium cumini,* and may be contiguous with tropical *Shorea robusta*. An understorey of shrubs such as *Carissa spinarum (syn C. opaca)* (*jangli karaunda*), *Dodonaea viscosa*, *Berberis aristata,* and *Rubus ellipticus* are common if fires are not too frequent.

Ground fires, often intentional, are common in the spring or early summer (pre-monsoon), and lead to a rich growth of grass in the monsoon suitable for grazing. In the winter, this grass dies out and the soil appears bare until fallen needles cover the surface in the early spring, often to be burnt again and the cycle repeats annually in many Chir forests, and occasionally more frequently. If fires are excluded, a shrub layer develops and burnt coppices of oak sprout and grow at a rapid rate. Chir seedlings also develop a thick corky bark within a few years, which protects them from at least cool ground fires. The oaks in contrast need longer protection, usually well over a decade to be able to escape even from a cool ground fire.

Chir has often been thought of as an invasive species of disturbed environments, and this is true in many areas where Chir replaces oak forests due to excessive human disturbance (Thadani 1999). Dudgeon and Kenoyer (1925) referred to Chir as a xerophytic edaphic climax in a broadleaved sclerophyllous formation. Champion and Seth (1968) however refer to Chir as a genuine climatic climax in many sites including quartzite formations, scarp slopes, and schists between 1200 and 1800 m.

Human Impacts

Human disturbance benefits this pine. Biotic influence has extended the range of Chir to many areas where broad-leafed sclerophyllous trees, such as the oaks, would normally occur. The Chir-banj association is among the best described and studied in the Himalayas, and as chronic disturbance and forest degradation has increased (See Chap. 9), Chir has encroached onto many oak sites (Singh and Singh 1992; Thadani 1999). Chir is also an efficient coloniser when agricultural fields are abandoned, and especially in dryer regions, abandonment of agricultural terraces often leads to even aged Chir stands. Protection from fire, lopping, and litter removal will lead to the reestablishment of oaks, which can outcompete Chir under natural conditions in better quality sites.

Similar Forest Types

East of Bhutan, and especially in the Khasi, Naga, and Manipur hills, Chir pine is replaced by *Pinus kesiya,* another three-needled long leaved pine. Interestingly much of eastern Nepal is devoid of any pine, as *P. roxburghii* diminishes rapidly and *P. kesiya* is absent. This Khasi pine forms structurally similar forests (Assam sub-tropical pine forests; C&S 9/C2) between 800 and 1600 m in the Khasi and Naga hills and at sites where rainfall typically exceeds 1800 mm. Associates of this pine include *Quercus griffithii*, *Q. serrata*, *Castanopsis spp*, *Acer,* and others.

Subtropical Dry Evergreen Forests of *Olea* and *Senegalia* (C&S 10C1)

Brief Description

These are often thorny, low scrub-like forests of small-leaved evergreen trees and shrubs. While one of the 16 forest types of Champion and Seth (1968), they are relatively unimportant in the Himalayas and make up <1% of forests.

Climate

These are monsoonal forests, but rainfall is low, typically 500–1000 mm. A long hot and very dry season in the spring and summer is followed by some monsoon rains. Some frost may occur in the winter.

Occurrence

Found up to 1000 m in the Bhabhar and Siwalik hills as well as the foothills of the western Himalayas, especially in Himachal and Kashmir.

Dominant Trees and Associates

Olea europaea subsp. *cuspidata* (syn *Olea cuspidata*), and *Senegalia modesta* (syn. Acacia modesta) are common. Chir pine may be found in the upper reaches. A scrub of *Dodonaea viscosa* occurs on more degraded sites and hot sites.

Quercus lamellosa and Cloud Forests of the Eastern Himalaya (Montane Wet Temperate forests (C&S, sub-type 11B/C1b))

Brief Description

These are species rich, closed wet evergreen forests of medium height, but with trees of large girth. The large branching crowns of trees are festooned with epiphytes.

Frequent mist and cloud contribute significant occult moisture, which helps maintain a lush and diverse vegetation, with not only numerous tree species but also a rich diversity of epiphytic mosses, ferns, fruticose lichens, and orchids. Due to their ability to draw moisture from damp winds and clouds, these forests are of considerable hydrological significance and their presence can enhance streamflow and help maintain high baseflows in downstream rivers (see Chap. 12, Forest hydrology).

Buk oak, a common occurrent of the eastern cloud forests, is the tallest oaks of the Himalayas and can reach 40 m. This broad crowned tree grows best on shady humid slopes. Its massive acorns, up to 7.5 cm in diameter (Troup 1921), are highly valued by wildlife as a food.

Climate

These are Monsoonal forests, with rainfall usually above 2000 mm. Dark monsoon mists characterise these forests that have been described as cloud forests by Miehe et al. (2015). Winters are dry, with December–January being the driest months. There is good rain much of the rest of the year. The cooler winter temperatures minimise evapotranspiration and allow for a damp, epiphyte rich environment to prosper round the year. The mean minimum temperature is around freezing, and summer temperatures rarely exceed 30 °C. However, there is considerable variation with altitude.

Occurrence

These are Eastern Himalayan forests commonly found from eastern Nepal through Sikkim and north Bengal, Bhutan, and Arunachal. They occur at elevations of about 1750–3000 m.

The *Eastern Himalayan wet temperate forests (11B/C1)* are among the most extensive forest type in the Indian Himalayas (FSI 2021). Including the Buk oak and Lauraceous forest sub-types, it covers well over 18,000 km^2 in the Indian eastern Himalayas, much of it in Arunachal Pradesh. In addition, these forests cover considerable areas in eastern Nepal and Bhutan.

Dominant Trees and Associates

These are evergreen forests with high diversity. No single tree dominates unlike in the Western Himalayas. As per Champion and Seth (1968) (Type 11B/C1b), *Quercus lamellosa* only constitutes 12.5% of the eponymously named buk oak forests in Kalimpong, West Bengal.

Fagaceae dominate the canopy, though Lauraceae are abundant. Common fagaceous dominants include *Quercus lamellosa* (Buk oak) and at higher altitudes, *Lithocarpus pachyphyllus* (syn *Quercus pachyphylla*), which typify these forests. Other common oaks include *Q. glauca*, *Q. acutissima*, and the Indian chestnuts *Castanopsis tribuloides* and *Castanopsis hystrix*. Lauraceous species such as *Machilus edulis*, *Cinnamomum bejolghota* (syn *C. obtusifolium*), *Litsea* and *Beilschmiedia* are abundant, but often form part of the second storey.

A diverse assortment of deciduous trees occurs but they typically form an insignificant part of the canopy. A few canopy trees such as *Acer campbellii* are locally abundant in Bhutan and Sikkim. The layers below the canopy are evergreen. Dwarf bamboos (particularly *Arundinaria*) commonly occur in the understorey. At higher altitudes, rhododendrons become important. At lower altitudes *Schima* and *Engelhardtia* are important. Magnolias are common through this forest type, and *Magnolia doltsopa* (syn *Michelia doltsopa*) and *Magnolia campbellii* are canopy trees which produce conspicuous large showy flowers in the early spring. Despite their prominence when flowering, Magnolias are usually not a significant part of the canopy, though they can be common in the Naga Hills (11B/C2). In several areas, a dense under canopy of rhododendrons and laurels occurs, rendering the canopy sparse.

The forest composition changes in drier areas. For example, in the driest part of its range *Quercus lamellosa* will occur sandwiched between *Quercus lanata* (syn Q. *lanuginosa*) and *Q. semecarpifolia* (Stainton 1972), but these species will disappear when rainfall is heavy, and in its purest form *Quercus lamellosa* forests are dark and gloomy with a rich diversity of epiphytes hanging and clinging to its branches.

The ecological equivalent of these cloud forests in the Western Himalayas are the *Quercus floribunda* forests of Uttarakhand and western Nepal, and high elevation *Q. semecarpifolia* forests. While low temperatures and high rainfall give these forests a cloud forest like character, the longer dry seasons these forests experience would make us hesitate in labelling these as cloud forests.

Human Impacts

The forests have been extensively disturbed by shifting cultivation (Jhum) in the northeastern Himalayas. Human induced fires are common due to jhum. At higher altitudes, considerable grazing pressure is evident as well as potato cultivation.

Banj Oak: Moist Lower Temperate Forests
(C&S 12C1a)

Brief Description

Quercus leucotrichophora (syn *Q. incana*) forms among the best-known forest types in the central Himalayas (C&S12/C1a). Locally known as Ban/ Banj oak, its specific epithet alludes to the white (leuco) hair (tricho) on the leaf underside. Banj well represents the western Himalayan oaks (See Boxes 4.1 and 4.2) and is among the ten most common trees in India (FSI 2021), second only to Chir among Himalayan trees.

Banj is an e1 type tree and has leaves with a generally 1-year lifespan, which are shed in the spring, around the same time as new leaves flush. Re-translocation of nutrients from the senescing leaves is low and the nutrient rich leaf litter rapidly decompose in the warm moist conditions of the monsoon, which allows for the building of a good soil, rich in humus in banj forests.

Box. 4.1: Two Centres of Western Himalayan Oaks in Uttarakhand

Uttarakhand appears to be centre of the western Himalayan oaks: *Q. leucotrichophora, Q. floribunda,* and *Q. semecarpifolia* all occur most commonly and show their greatest size in this state. At Cheena-Kilbury (2200–2600 m) in Nainital, four oaks (*Q. lanata* being the fourth), can be seen growing within a 3 km transact, unusual oak diversity for the Western Himalayas. Old forest stands of three of these species can be found in the Chopta transect (2200–3000 m) in Garhwal viz., *Q. leucotrichophora, Q. floribunda,* and *Q. semecarpifolia*. The three oaks along with two firs, birch and rhododendrons make Chopta-Tungnath a unique transect in Himalayas.

Box 4.2: Confusion at a Himalayan Scale: Naming the Oaks

The oaks are among the best known and most valued genera of the Northern Hemisphere, but their nomenclature has been fraught with issues and confusion abounds. There have been some eight major classification attempts of the genus *Quercus* over the past two centuries, most based entirely on morphological characters. A recent classification (Denk et al. 2017) uses molecular phylogenetic techniques.

Some older workers such as DeCandolle (1863) and Bentham and Hooker (1880) regarded *Quercus* and *Lithocarpus* to be a single genus (as cited in Soepadmo 1968) though others (Orsted 1871; Camus 1936–1954) treated the two as distinct genera. Soepadmo (1968) accepts this differentiation and expounded on the many differences between the two genera. Today *Quercus* and *Lithocarpus* are well accepted as distinct genera (POWO 2024). Thus, the Himalayan tree commonly referred to as *Quercus pachyphylla* in literature (Champion and Seth 1968), is now accepted as *Lithocarpus pachyphyllus* (POWO 2024).

The taxonomic status of *Cyclobalanopsis* was even more intensely debated. This group is found in subtropical regions of southeast and east Asia, and in the eastern Himalayas, and consists of evergreen species found in mesic habitats (Ohsawa 1987). In older classifications, this subgenus *Cyclobalanopsis* was given prominence and raised to a genus level by some authors such as Orsted (Denk et al. 2017). While other authors disagreed with this, it was only the modern molecular phylogenetic techniques which helped settle this debate. Denk et al. (2017) demote *Cyclobalanopsis* to a section level along with section *Cerris* and *Ilex*—all part of Subgenus *Cerris*. This includes several of the east Himalayan oaks which are now considered part of the genus *Quercus* (See Fig. 4.5). Oaks of section cyclobalanopsis are found on moist north slopes across the eastern Himalayas and include:

(continued)

Box 4.2 (continued)

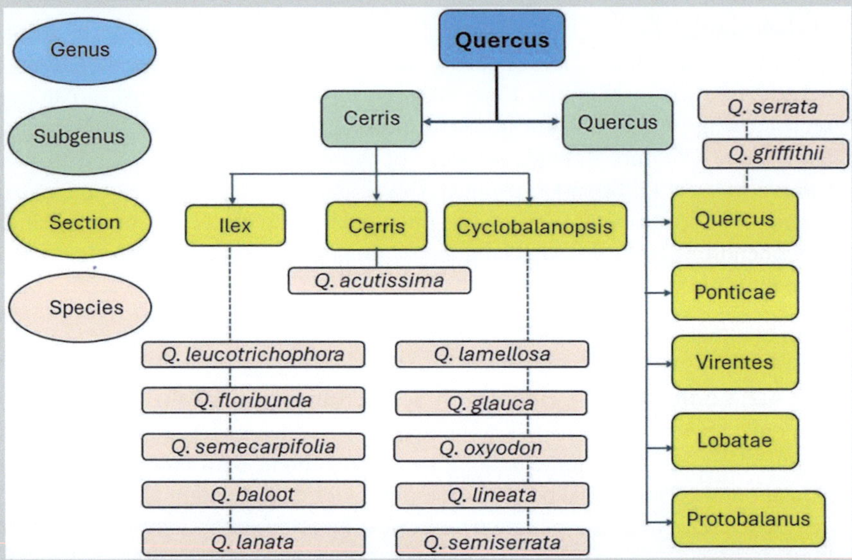

Fig. 4.5 Classification of Genus Quercus and common Himalayan species. (Redrawn using Denk et al. 2017 and Yang et al. 2021)

Quercus oxyodon (Syn *Q. lineata var. oxyodon*; syn: *Cyclobalanopsis oxyodon*).
Quercus semiserrata (Syn *Cyclobalanopsis semiserrata*).
Quercus lamellosa (Syn *Cyclobalanopsis lamellosa*).

These oaks, along with the closely allied genera *Lithocarpus* form a complex with nomenclature depending on the authority being given primacy. We follow the modern taxonomic treatments (Denk et al. 2017; POWO 2024) wherein *Lithocarpus* is a distinct genus and *Cyclobalanopsis* is an older name of species now included in genus *Quercus*. The majority of Himalayan oaks belong to subgenus Cerris, though some fit into subgenus Quercus (Fig. 4.5).

Oaks of the Western Himalayas are of the Ilex section while those of the Eastern Himalayas are predominantly of section Cyclobalanopsis as per Denk (2017). Most Himalayan oaks are evergreen, with diffuse porous wood, though a few species such as *Q. serrata*, *Q. griffithii,* and *Q. acutissima* are deciduous with ring porous or semi-ring porous wood anatomy and distinct growth rings (Gupta and Gupta 2020).

(continued)

Box 4.2 (continued)

While nomenclatural issues based on taxonomic treatments are understandable and being resolved using modern molecular techniques, in the Himalayas many of the common species are referred to by multiple Latin names leading to confusion and bad science. While some of this confusion originates from legitimate name changes required as a result of the use of specific epithets that were already in use, much of this is a result of sloppy researchers. For example, Thadani (2023) details the confusion surrounding the nomenclature of Banj oak (*Quercus leucotrichophora* A. Camus)—the most common and well researched Himalayan oak. Early researchers knew this species as *Quercus incana* Roxb. While Camus proposed *Q. leucotrichophora* in 1935, it was only after Bahadur (1975) gave clear justification and a need for the name change that *Q. leucotrichophora* became widely used. In recent years, *Q. oblongata* D.Don came into use likely due to an error in outdated taxonomic databases. In parts of the Himalayas, the old name *Q. incana* is still used despite this being a completely different species—the American bluejack oak.

Multiple Latin names are used for a diverse array of Himalayan oaks. Table 4.5 lists some of the common Himalayan oaks, many of which have older or incorrect synonyms. A few additional species are reported from the Himalayas, but are not listed as they are very uncommon—such as *Q. rex*; *Q. gomeziana*, *Q. mespifolia*, *Q. blakei* (syn *Q. patkoiensis*), *Q. thomsoniana*, *Q. helferiana* (for e.g. see Lahiri et al. 2017).

Table 4.5 Nomenclature of some common Himalayan Oaks

	Correct Latin name	Incorrect names	Notes on incorrect names	Common name
Himalayan oaks of subgenus Cerris, section ilex				
1	*Quercus leucotrichophora* A. Camus	*Q. incana* Roxb.	Used before 1980. Still often used	Banj or ban; Grey oak
		Q. oblongata D.Don	Incorrect name listed in some older databases.	
2	*Quercus floribunda* Lindl. ex A. Camus	*Q. dilatata* Lindl. ex A.DC	Older name, used until the 1940's; still used in Pakistan	Tilonj; Green oak
		Q. himalayana Bahadur	Published by Bahadur (1971). Not accepted	
3	*Quercus semecarpifolia* Sm	*Q. obtusifolia* D.Don	Incorrect name	Kharsu; Brown oak
4	*Quercus lanata* Sm	*Q. lanuginosa* D.Don	Commonly used old name	Rianj

(continued)

(continued)

Box 4.2 (continued)

Table 4.5 (continued)

	Correct Latin name	Incorrect names	Notes on incorrect names	Common name
5	*Quercus baloot* Griff.	*Q. Ilex* Linn.	*Q. Ilex* was used by Troup (1921) and the usage continues to date	Balut
Himalayan oaks of subgenus Cerris, section Cyclobalanopsis				
6	*Quercus glauca* Thunb.	Correct name generally used	Wide distribution, common along water courses	Phaliat
7	*Quercus lamellosa* Sm.	Correct name generally used	The most important species of eastern Nepal and Darjeeling hills	Buk oak
8	*Quercus semiserrata* Roxb.	*Cyclobalanopsis semiserrata* (Roxb.) Oerst.	Cyclobalanopsis has been discarded now	Khasi oak
9	*Quercus oxyodon* Miq.	*Cyclobalanopsis oxyodon*/Q. Lineata	See above and in text	–
10	*Quercus lineata* Blume	Correct name generally used	Eastern Himalayan species. Common companion of Buk oak	–
Himalayan oaks of subgenus Cerris, section Cerris				
11	*Quercus acutissima* Carruth	Correct name generally used	Reported in the extreme eastern Himalaya	
Himalayan oaks of subgenus Quercus, section Quercus				
12	*Quercus griffithii* Hook.f. & Thomson ex Miq.	Correct name generally used	Very common in Bhutan and Arunachal	Paisang
13	*Quercus serrata* Murray	Correct name generally used but sometimes confused with *Q. serrata* Roxb (which is now known as *Castanopsis indica*)	An eastern Himalayan deciduous oak, planted in the West Himalayas in the 1990's	Konara; Manipuri banj
Nomenclature changes at genus level				
14	*Lithocarpus elegans* (Blume) Hatus. ex Soepadmo	*Quercus spicata Bonpl.*	Earlier regarded as Quercus	

(continued)

(continued)

Box 4.2 (continued)

Table 4.5 (continued)

	Correct Latin name	Incorrect names	Notes on incorrect names	Common name
15	*Lithocarpus pachyphyllus (Kurz) Rehder*	*Quercus pachyphylla*	Troup (1921) refers to this as one of the most common Eastern Himalayan "oaks"	
16	*Castanopsis acuminatissima (Blume) A.DC.*	*Quercus acuminatissima (Blume) A.DC/ Quercus lineata*	Reported but not a Himalayan species, found in Myanmar	

Compiled from: POWO (https://powo.science.kew.org/); WFO (http://www.worldfloraonline.org/); Troup (1921); Singh and Singh (1992); Lahiri et al. (2017); Champion and Seth (1968); Thadani (1999); Flora of Pakistan (le gacy.tropicos.org)

Banj is a strong coppicer, and regeneration and small trees with a diameter of up to 20–30 cm coppice well if cut, and the resulting regrowth can be fairly rapid. This makes banj somewhat resilient to fires and occasional human disturbance.

Climate

Banj is a tree of the lower montane temperate zone. It dominates in the outer Himalayas where the impacts of monsoon are strong. Maximum temperatures rarely exceed 35 °C, and winter minima are usually above -5 °C. While snow is not uncommon in the banj zone, it is usually light and does not stay for long.

Occurrence

Quercus leucotrichophora forms a closed canopy, 20–25 m tall on favourable sites, at altitudes between 1700 and 2200 m. In moist situations, it may be found considerably lower. For example, in Shahpur range near Palampur (Himachal Pradesh), a dense thicket of bank oak was seen by this author (Thadani) at below 800 m altitude. Troup (1921) reports natural banj in Mothronwala, (Dehradun) at 600 m elevation.

While it is best exhibited in the central Himalayas (particularly in Uttarakhand), it extends from the monsoonal regions of Pakistan (Iqbal et al. 2019), through western Nepal (up to ~ of 80°E) though small patches are reported in the far eastern Himalayas. Banj forests in India and Pakistan cover 6000 km^2. Like other oaks, it is absent from the Kashmir valley, which is shielded from the monsoon thrust by the high Pir Panjal ranges.

While found in west Nepal, in central Nepal, it is replaced by the ecologically and morphologically similar *Q. lanata*. Banj is not reported from Sikkim or Bhutan (Stainton 1972) though it has been reported in the Assam region and Burma. Possibly though this is a case of misidentification. This oak can grow on a wide variety of geological formations including shale, gneiss, mica-schist, quartzite, and limestone.

Dominant Trees and Associates

Banj oak forms monodominant stands that can constitute 70% or more of a stand (Singh and Singh 1992). In favourable sites, it can completely colonise and dominate abandoned fields forming over 90% of the canopy (Thadani 1999). Above 2000 m and in moist areas, it mixes with *Quercus floribunda* (tilonj oak). Deodar and blue pine (*Pinus excelsa*) are common associates at higher elevations. *Quercus glauca* is abundant along shady ravines and water courses while in the east it mixes with *Q. lanata*.

Rhododendron arboreum, *Lyonia ovalifolia (syn Pieris ovalifolia), Pyrus Pashia,* and *Myrica esculenta* are common associates growing below the oak canopy. As these make poor fodder and fuel, they are often unharmed when human disturbance is high. Various Lauraceae such as *Machilus* spp. and *Litsea* are commonly found in Banj forests. A shrubby layer with *Berberis*, *Indigofera*, *Viburnum,* and *Desmodium* is common.

In moist hollows, such as in damp ravines and along streams, a mix of deciduous species may be found and trees are clothed in mosses, and during the monsoon with epiphytic ferns and even orchids. Deciduous associates include *Carpinus viminea*, *Aesculus indica*, *Prunus cerasoides*, *Juglans regia, Acer caesium,* and *Pyrus pashia*.

Its most common associate is however Chir pine (*Pinus roxburghii*). The banj-Chir association is perhaps the best studied in the Himalayas (Singh and Singh 1992) with Chir favouring dryer ridges and steep southern slopes with thin soils, and banj occupying the more favourable sites with deeper soils.

Human Impacts

The banj zone is favourable for human settlement and cultivation. Banj yields a variety of useful biomass products from an excellent fuelwood (with high calorific value), high quality leaf litter (used to make compost fertiliser), and an abundantly available leaf cattle fodder. Consequently, the tree is extensively lopped and the forest floor swept clean of leaves, which inhibits acorn germination. Repeated lopping leads to the tree losing its vitality and becoming prone to parasitic mistletoe and rot, and over a period of time entire hillsides have become denuded of banj oak due to overuse by humans. Such chronic disturbance allows for encroachment by Chir pine (see Chap. 9 in Volume 2 of this book).

While heavy disturbance is unfavourable, moderate disturbance can favour banj. In moist areas, abandonment of agricultural fields when oak rootstock is present can give rise to almost pure stands of banj oak. Occasional ground fires may benefit oak regeneration. While the aboveground part of young oaks is killed, the rootstock is resistant to fire and will sprout back with great vigour and outcompete most competition.

Similar Forest Types

Quercus lanata (syn *Q. lanuginosa*) forms localised patches in the eastern parts of the banj range starting from Kumaon. It becomes dominant in central Nepal, initially increasingly in moister aspects of *Q. leucotrichophora* forests, and east of 83°E it is found on drier southern aspects while moister aspects are occupied by oaks such as *Q. lamellosa*.

In moister habits and between a narrow band of about 1900–2400 m *Q. floribunda* co-occurs with banj and can form extensive monodominant stands in the central Himalaya.

Quercus Semecarpifolia: The High-Altitude Oak
(C&S 12C2a)

Brief Description

Commonly known as Kharsu or brown oak, this is a large, spreading, evergreen, light-loving oak which can grow to 35 m on favourable sites, though is usually under 20 m tall.

Climate

Rainfall ranges 800–1600 mm in the Kharsu zone, with significant winter snowfall. This is a cold temperature oak and mean maximum temperatures are below 22 °C in the warmest month, while in the coldest month at higher altitudes, mean minimum temperatures are about −5 °C. Absolute minimums may drop below −10 °C but the ground does not remain frozen.

Occurrence

This is an oak of the upper temperate zone with a broad geographical distribution. *Quercus semecarpifolia* can be found from the moist monsoonal slopes of the SE Hindu Kush in Afghanistan (72°E) to the rain shadow regions of Bhutan (90°E) and China. As with other oaks, Kharsu is absent in Kashmir (See Boxes 4.3: The Missing oaks of Kashmir and 4.4), and as per Stainton (1972) also absent from Sikkim. It is typically found at altitudes of 2400–3600 m. This oak is common on southern exposures, on steep rocky slopes.

> **Box 4.3: The Missing Oaks of Kashmir**
> Oaks are among the most common and diverse genera of Himalayan trees and a component of most Himalayan forests. In the western Himalayas *Quercus leucotrichophora*, *Q. floribunda*, and *Q. semecarpifolia* are important canopy

(continued)

Box 4.3 (continued)

components often forming monodominant stands from about 1500 m to over 3000 m. In Nepal *Q. lanata* gains importance as *Q. leucotrichophora* declines, though *Q. semecarpifolia* remains important at the treeline. In eastern Nepal, *Q. lamellosa* gains significance and further east in Bhutan, *Q. griffithii*. In the eastern Himalayas, the more diverse flora includes other Fagaceae including *Castanopsis* and *Lithocarpus*, which together with *Quercus* are an important component of the upper sub-tropical and temperate forests.

Kashmir is unusual in lacking oaks. Oaks are absent from the forests of the Kashmir valley which lies to the north of the Pir Panjal range, despite the apparent suitability of climatic conditions for its growth. Vishnu-Mittre (1963) stated "*Among several anomalies in the flora of the Kashmir valley, the oaks present the most vexing problem.*" Numerous botanists dating back to Royle (1839) have reported the absence of oaks from Kashmir valley. However, this was not always the case.

Puri (1945) found abundant fossil leaves of oak and reported the Fagaceae to be the most important fossil family in the Lower Karewa Series of Kashmir. Karewas are lacustrine deposits, seen as flat-topped mounds found in the valley of Kashmir and formed from the filling of ancient lakes. They are a rich source of fossils dating back to the Pleistocene. Oak leaves constituted nearly half of the entire fossil plant material of the Karewas. A great abundance of five common Himalayan oaks were found in five localities. Interestingly while *Quercus leucotrichophora* (syn *Q. incana*) and *Q. glauca* were found at two higher altitudes (~3000 m), *Quercus semecarpifolia*, *Q. floribunda* (syn *Q. dilatata*), and *Q. baloot* (syn *Q. Ilex* L) were found at lower altitudes of 2000–2700 m. This is anomalous as *Q. semecarpifolia* is normally found at significantly higher elevations than *Q. leucotrichophora* and could possibly indicate misidentification at the species levels.

The presence of oaks in the Kashmir valley during the Pleistocene and their absence today has been explained by the uplift of the Pir Panjal which acted as an effective barrier and prevented the monsoon from entering the valley, resulting in low precipitation. However, evidence of pollen grains of oaks in the Post glacial period (~ 8000 years BP) discredits this hypothesis simply because the uplift of the Pir Panjal occurred millions of years earlier. Also, sufficient low altitude passes lie in the Pir Panjal to allow for the movement of oaks during postglacial warming periods.

Further, given the wide range of requirements of the various oak species, it is unlikely that any drying or cooling event would be detrimental for all the oaks. *Q. baloot* grows in dry almost arid climes, while *Q. floribunda* is at its best in moist and humid climes. Climate stations across Kashmir valley indicate sufficient precipitation (from 500 mm in Pulwama to 2000 mm in Sonmarg) for oaks to thrive.

(continued)

Box 4.3 (continued)

One plausible explanation is that the last ice age led to the local extinction of oaks and this genus has largely not yet succeeded in recolonising the lost ground (Stainton 1972; Singh and Singh 1992). Also, evergreen oaks are likely less competitive at more northern latitudes as winter day length is very short.

Vishnu-Mittre (1963) reported that in addition to the exotic *Q. robur*, local oaks such as *Q. semecarpifolia* and *Q. floribunda* were found in small patches in the Kashmir valley and were regenerating naturally. It is unclear if these patches were planted or were natural relict populations, though it would appear that the populations are recent plantations. However, in the 60 years since, there seems to have been no expansion in oak territories in Kashmir. Iqbal et al. (2017) for example report the presence of only two species of oak in the Kashmir valley. More common is the English oak (*Q. robur*), which was undoubtedly introduced during the colonial period and is not native to the region. It is believed that the Maharaja's of Kashmir procured *Quercus robur* planting material from England in the mid-nineteenth century. It is found most commonly around Srinagar, in parks, along roadsides and appears to have adapted well to local conditions. These authors also report the occasional presence of holm oak (*Quercus ilex*). It is likely this is mis-identified and is actually *Quercus baloot* which is native to the western Himalayas. It appears likely that these too have been planted, though the possibility of small natural populations of baloot oak cannot be precluded. In either case, these are uncommon and it would appear that natural populations of oak in Kashmir are absent.

Today, while the Kashmir valley lacks oaks, in the Rajouri district of Jammu and Kashmir, the monsoonal climate leads to flora very similar to Uttarakhand with *Q. leucotrichophora* dominating and an abundance of *Q. floribunda*, *Q. semecarpifolia,* and *Q. glauca* (Jazib 2023). Further north in disputed Kashmir (regions referred to as AJK or POK by Pakistan and India, respectively), these same oaks are common along with *Q. baloot* (Gillani et al. 2023) which is also found westwards into Afghanistan.

The lack of oaks remains an unresolved puzzle of the Kashmir Himalayas. The higher latitudes which make evergreens less competitive, non-monsoonal climate, and reluctance of this genus to disperse far, may be reasons that help explain this absence, but as yet there is no definitive proof to convincingly resolve this conundrum of the missing oaks.

Box 4.4: Winter Hardiness of Himalayan Trees

The treeline is a zone of inhospitable terrain and temperatures (See Chap. 1 of Volume 2 on Treelines). The Himalayan region has the highest and most diverse treeline in the world (Shi and Wu 2013) and trees can be found at altitudes well over 4500 m (Miehe et al. 2007). Most Himalayan species of high altitudes however are not capable of resisting temperatures below −30 °C which is significantly warmer than the cold thresholds of many temperate trees. This is possibly due to the higher humidity and more moderate cold temperatures of Himalayan winters (Sakai and Malla 1981). The winter hardiness of some high elevation Himalayan species is given below using data from eastern Nepal (Table 4.6).

Table 4.6 Xylem freezing resistance of various Himalayan trees

	Freezing resistance (°C)	Sampling altitude (in m)
Deciduous		
Betula utilis	−40 to −50	3300–3800
Acer campbellii	−30	3300
Salix denticulata	−70	3300
Salix daltoniana	−70	3300
Evergreen broad leaved trees		
Quercus semecarpifolia	−15	3000
Quercus lanata	−8	2250
Pieris formosa	−15	3300–3450
Rhododendron campanulatum	−25 to −35	3800
Rhododendron anthopogon	−30 to −50	3500–3800
Conifers		
Abies spectabilis	−22 to −30	3300–3880
Larix potanini	−22 to −25	3300
Tsuga dumosa	−20	3300
Pinus wallichiana	−20	3450
Juniperus recurva	−30	3300–3880

Data from Sakai and Malla (1981)

Dominant Trees and Associates

At the upper end of its range silver fir (*Abies pindrow*) is the most common associate and mixed forests of fir and kharsu oak are common, giving way to birch and fir near the treeline. *Pinus wallichiana* and *Picea smithiana* are also common associates. While kharsu forms the treeline in some areas, especially areas with monsoonal climate, more commonly at the upper end of its range it transitions into sub-alpine forests of *Abies*, *Betula,* and *Rhododendron*. At the lower end of its limit *Q. floribunda* is common. On humid sites, it often transitions to *Tsuga* forests while at the dryer end of its range *Pinus wallichiana* becomes abundant. The understorey is rich in lauraceous and sclerophyllous trees and shrubs.

Human Impacts

Kharsu is lopped extensively for its excellent fuelwood and to make charcoal. The leaves are an abundantly available fodder during the dry season (Rawat et al. 2022). While permanent habitations are rare at elevations where Kharsu grows, the alpine meadows that pastoralists frequent are close to the kharsu zone leading to its considerable degradation.

Like many oaks, kharsu regeneration benefits from moderate disturbance. In highly disturbed areas, heavy lopping lowers flowering and acorn production (Shrestha 2003) and grazing causes considerable seedling mortality and depletes regeneration (Rawat et al. 2022). However, in undisturbed areas an excessively thick litter layer and high herbaceous cover is also deleterious to regeneration (Shrestha 2003).

Deodar (*Cedrus deodara*) Forest
(Moist Temperate Forests C&S 12/C1c & Dry Temperate Forests 13/C2b)

Brief Description

Deodar, also known as the Himalayan Cedar (*Cedrus deodara*) is among the most magnificent of the Himalayan trees. Tucker (1982) refers to deodar as the "great tree of the mountains", whose dense and fine-grained wood was ideal for construction, as it is durable and resistant to pests such as termites. The tree has a straight bole and grows tall to 30–40 m, but large specimens can be over 50 m in height. Deodar needles have a multi-year lifespan (3–6 years) and are 2.5–5 cm long. The vast forests of deodar around Shimla were once considered inexhaustible and used extensively for building colonial infrastructure and railway lines starting in the 1860's, and deodar forests were among the first to be cut on a large scale by colonial foresters.

Climate

Deodar occurs across a range of rainfall and elevations. In the moist temperate zones at elevations of 1700–2500 m, rainfall may exceed 1800 mm. In the inner rain shadow valleys, precipitation is low—typically under 1000 mm and as low as 450 mm. Here deodar occurs between 2100 m and 3250 m. Deodar is found in a range of soils and geological formations but avoids poorly drained soils. It is susceptible to fire, particularly when young.

Occurrence

Deodar forests cover over 5500 km^2 and are a component both of moist and dry temperate forests (group 12 and 13 of Champion and Seth). These can be found as far west as Afghanistan and are an important forest type in Pakistan, Kashmir, Himachal, and Uttarakhand (especially Garhwal).

The most majestic and largest trees are found in moist temperate forests, especially in Himachal. However, deodar is geographically most widespread in the dry temperate forests of inner valleys and rain-shadow areas. This species does best in areas where there is significant winter precipitation and long-lasting snow cover. Champion and Seth (1968) speculate that the lack of winter snowfall is a significant factor in the absence of deodar eastwards. *Cedrus deodara* is sparsely found in the outer ranges where monsoon rainfall and humidity are high as it outcompeted by oaks. Here it will dominate only on the steep, dry, south-facing scree slopes and limestone cliffs with strong up-valley winds where it can outcompete monsoon forest species (Miehe et al. 2015). Deodar finds its greatest dominance in areas where summer drought occurs. It is a light demanding conifer, tolerant of desiccating conditions.

Being the most valuable timber tree of Himalayan forests, it has been planted extensively since British times and large trees are seen growing well outside its normal range.

Dominant Trees and Associates

Deodar is capable of forming large tracts of almost pure forests. Blue pine (*Pinus wallichiana*), spruce (*Picea*), and fir (*Abies pindrow*) are common associates at upper elevations, while in the more humid end of its range evergreen oaks (*Q. floribunda* and *Q. semecarpifolia*) as well as *Rhododendron arboreum* can be found under the deodar canopy. Among shrubs, *Viburnum*, *Indigofera*, *Rosa,* and *Daphne* are commonly found. At its western periphery in Pakistan, *Quercus baloot* and *Olea* spp. may be found in the undercanopy.

Human Impacts

Most of the massive forest cuttings from 1860 to 1900 were in the deodar forests. This led to a dramatic reduction in the extent of deodar, and most of the accessible and good quality ancient trees were logged. Silviculturists gradually learned that the deodar is not only slow growing but difficult to propagate (Tucker 1982) and many old deodar forest lands were planted over by other species.

Today, this species is revered and protected. Planting of deodar for aesthetic reasons, around temples, and also for its timber values has led to humans increasing its spread and distribution in the past half century. This tree is rarely logged or disturbed by local communities. Occasional illegal logging however continues across its range.

Blue Pine (*Pinus wallichiana*) Forest: A Strong Coloniser
(C&S 12C1f; 13C4;14C2/2S1)

Brief Description

Pinus wallichiana or blue pine (syn *P. excelsa*), locally known as Kail, is a five-needled pine that can grow up to 50 m in height. It forms even-aged, monospecific, single storeyed stands across a range of altitudes and across the Himalayan arc. Blue pine is a light-demanding pine which, like Chir, thrives on disturbance and is often considered invasive. As per Champion and Seth (1968) "*Pinus wallichiana* is the universal coloniser of vacant sites." In moist temperate forests, where it is commonly found, it occurs where oaks and other conifers previously dominated.

Together with *Alnus nepalensis*, blue pine is among the fastest-growing forest trees of the region (tree ring widths of up to 1 cm/year in Bhutan at 2400 m). This is a haploxylon or soft pine. Haploxylon pines have softer, less resinous wood, a single vascular bundle per needle and are better adapted to colder and high-altitude conditions. In contrast, pines such as chir and *P. kesiya* are diploxylon or hard pines characterised by thick resinous wood, two vascular bundles per needle and a greater tolerance to fire and drought (Singh et al. 2023).

Climate

Blue pine occurs across a broad range of precipitation regimes from 300 to 2600 mm and is a major species of monsoon-shadow areas. In the northwest Himalayas, much of the precipitation may occur as winter rain, while in the inner valleys of Bhutan in the east, monsoons dominate. Blue pine forests in the Hindu Kush and NW Himalayas are covered with snow for several months, but east of 83°E, snow is not a significant ecological factor (Miehe et al. 2015).

Occurrence

Blue pine is found across much of the Himalayan arc, from Afghanistan (70°E) to Bhutan (92°E). However, it changes from southern monsoon-exposed slopes in the western Himalayas to the rain shadow regions of the Inner Valleys east of 83°E (Miehe et al. 2015). It is poorly represented in the central Himalayas, especially Kumaon, where it transitions from outer, monsoonal slopes, to the dry inner Himalayas of the eastern part of the arc.

In the Pir Panjal range, it descends as low as 1600 m. It has an upper distribution limit of around 3600 m in Central Nepal as well as in the arid zones of Ladakh. The highest altitude recorded for this species is 4400 m in Pinus-Betula woodlands of Dolpa. In sub-alpine areas also, it is found in disturbed zones where earthslides or occasional fires disturb existing vegetation. In undisturbed forests, it is gradually replaced by fir and other climax species.

In the Indian Himalayas, *Pinus wallichiana* occupies over 4500 km^2 and ranks among the top ten trees in India in terms of wood volume (FSI 2021) and is among the only species to straddle three different sub-types of Champion and Seth (Himalayan moist temperate (12C1), dry temperate (13C4), and sub-alpine forests (14/2S1). It is the most common pine both in Bhutan and Pakistan as well an important species in Nepal, making it among the most well distributed and dominant of Himalayan species (see Table 4.1).

Dominant Trees and Associates

In Dolpa, it grows together with *Abies pindrow* and *Picea smithiana* at lower altitudes and *A. spectabilis* and *Betula utilis* at higher altitudes. In Ladakh, at altitudes of 3000–3600 m it associates with *Pinus gerardiana*, *Abies spectabilis,* and some *Betula utilis* as well as *Rhododendron campanulatum* occasionally scattered in the lower storey.

Natural stands are widely spaced and have a thick litter layer of needles with open to sparse shrub and herbaceous layers, mostly fire-tolerant species (such as *Pteridium aquilinum*) or thorny and aromatic weedy species (*Xanthoxylum*, *Rubus*, *Mahonia*, *Berberis*) or Ericaceous shrubs. Ground fires are often used as a tool to enhance grazing potential, in which case tropical grasses prevail and the thick needle layer is absent. However, while it can survive fires when it is older, this tree does not have the fire tolerance of Chir pine.

Human Impacts

In areas where extensive cutting or human disturbance has occurred, almost pure stands of *Pinus wallichiana* (Low level blue pine forests—C&S 12/C1f) occur, often replacing native oaks. These blue pine forests tend to colonise all vacant and disturbed sites due to their prolific ability to seed. They occur on sites where oaks or mixed conifers would normally form climax forests and are often an earlier seral community. Pure stands of *P. wallichiana* may originate on abandoned terraces in dryer sites (C&S 12/2S1) or where shifting cultivation has stopped.

As the timber is considered fair, blue pine forests in North Pakistan in particular have been unsustainably felled in recent years and the species has diminished in these regions.

Fir Forests of the Himalayas
(C&S 12C2b; 14C1)

Brief Description

Firs (*Abies* spp) are an important component of Himalayan forests in the upper temperate and sub-alpine zones (above 2400 m) (see Box 4.5: Conifers of the Himalayas). Three main species are found of which *Abies pindrow* is the most common (FSI 2021), though *A. densa* and *A. spectabilis* also have significant ranges. These firs can be difficult to differentiate, and considerable confusion exists at the species level.

Box 4.5: The Conifers: The Dominant Trees of the Himalayas

The conifers are an ancient group of cone-bearing seed plants that arose more than 300 million years ago. Along with cycads, Ginkgo, and gnetophytes these collectively form the Gymnosperms.

Conifers are typically single-stemmed, evergreen trees with separate male and female reproductive structures. Leaves are simple, and most are needle- or scale-like (Neale and Wheeler 2019) and persist for many years. A few genera (such as Larix) shed their leaves annually.

(continued)

Box 4.5 (continued)

Conifers tend to occur on soils and climates suboptimal for plant growth. Almost all conifers have elaborate mycorrhizal symbionts (Farjon 2018) and these associations with fungi help conifers persist in adverse conditions by increasing the surface area of their roots and greatly increase water and nutrient uptake.

Conifers dominate in Boreal climatic zones and cover vast tracts of land in North America, Eurasia, and mountain ranges such as the Himalayas. They are less conspicuous at lower latitudes and in the southern hemisphere though diversity is highest in these areas. The island of New Caledonia near Australia has 43 conifers, while all of Europe has only 41 species of conifers (Farjon 2018). Unlike the northern conifers, the tropical and southern conifers do not tend to form forests or monodominant stands; they typically occur scattered among other trees in diverse ecosystems.

Today there are six (Neale and Wheeler 2019) or seven (Farjon 2018) extant conifer families, which together constitute only about 615 species. This is not a large number, and the total species count of all conifer families barely exceeds that of the single angiosperm genus viz. the oaks! The Genus *Quercus* alone is estimated to have over 500 species (Nixon 2006; Denk et al. 2017). Despite their dominance in the northern hemisphere, the conifers are thought of as a group in decline (Neale and Wheeler 2019) as they have gradually ceded ground to angiosperms. Most existing conifer families arose during the Jurassic (Farjon 2018) during which time (~200–145 million years BP) there may have been as many as 20 conifer families and up to 20,000 species, most of which are now extinct (Farjon 2008).

In India, conifers are rarely found outside the Himalayas. As per Farjon (2018), while India has a rich fossil record of Gondwanan conifers and other gymnosperms, these are from the Cretaceous or even earlier, and most have gone extinct. Almost all conifers found in India today are confined to the Himalayas and came from the north—i.e. they are of Laurasian origin.

Among conifers today, family *Pinaceae* is the most species-rich (with approx. 230 species in 11 genera worldwide) and easily the most important economically. This is the most important tree family of the Himalayas and half the most common species of Himalayan trees belong to Family Pinaceae including the pines (*Pinus roxburghii*, *P. wallichiana*), firs (*Abies pindrow*, *A. densa*, *A. spectabilis*), deodar (*Cedrus deodara*), and spruce (*Picea smithiana*). Other common Himalayan pinaceae include *Tsuga Dumosa* (hemlock) and *Larix griffithii* (Himalayan larch).

The conifer family *Cupressaceae* is also common in the Himalayas and includes *Juniperus indica*, *J. communis,* and *Cupressus torulosa* (surai), while *Taxus wallichiana* is a well-known example of family Taxaceae. Conifers are especially abundant in the Western and Central Himalayas where they are the tallest trees and can exceed 50 m in height.

Climate

Firs avoid the driest zones. *A. pindrow* grows in areas above 1000 mm annual rainfall, while the other species need higher precipitation. *A. densa* is usually found where precipitation >2000 mm. The low evapotranspiration rates in the fir zone lead to higher Transeau ratio's especially at the upper part of the range even when total precipitation is low. Temperatures are cold across much of the fir zone and minimum temperatures may fall below −20 °C, and night temperatures can remain well below freezing for several months of the year. Fires are rare in fir forests, but can be quite destructive if they occur.

Occurrence

These firs do not mix much, but due to the similarity of their morphology there are often problems in identification. *Abies pindrow* is the most common fir and the tallest. It can exceed 50 m in favourable sites and is found as far west as Afghanistan, with significant distribution in Pakistan, Kashmir, and Himachal. In Pakistan and Kashmir, it is likely the only fir (see Fig. 4.6). It can be found from 2000 m up to the treeline at 3500 m in the upper Indus catchment in Kashmir and Pakistan. It is less common in Uttarakhand and tapers off in Western Nepal.

Abies spectabilis is found at higher altitudes from 3000 to 4200 m from Himachal through Nepal. It is uncommon west of Himachal Pradesh and may be largely absent in Kashmir and Pakistan. While its altitudinal range overlaps with *A. pindrow*, the two rarely mix. When they co-occur, as they do in Uttarakhand and western Nepal, *A. spectabilis* is found in a zone above *A. pindrow*. This range of this species ends in Sikkim and it is not found further east.

Fig. 4.6 Confusion over fir species nomenclature: A picture (left) taken by HG Champion of sub-alpine fir and birch forests in Sonmarg (Kashmir) in c. 1964 identified the fir as *A. spectabilis*. The picture on the right taken over 50 years later shows the same region, but fir is identified as *A. pindrow*. The snow cover seems largely unchanged in this site influenced by the westerties. Tree cover is also largely unchanged (Photo courtesy: Zafar Reshi)

Table 4.7 Common firs of the Himalayas

S. no.	Species	Occurrence	Latitude and Altitude	Distinguishing features
1	*Abies pindrow*	The most common tree of temperate altitudes in the western Himalayas. It is most common in monsoon-shadow regions.	Western Himalayas from 70° to 82°E. Possibly the only fir of Pakistan and Kashmir, it also dominates in Himachal. Usually 2000–3000 m.	No prominent stomatal bands underneath the needles which are longer; twigs denser (see Chap. 5 on forest phenology)
2	*Abies spectabilis* (syn *A. webbiana)*	It occurs at elevations above *A. pindrow* and the two are not known to mix.	Across Himalayan arc but common from Himachal through Nepal. Usually found at 3000–4200 m.	Needles are flat, with two white stomatal bands underneath
3	*Abies densa*	The most common tree of Bhutan and among the most common Himalayan trees.	Eastern Himalayas, east from Sikkim ~88°E. Dominant in Bhutan. From 2800–4300 m	Needles are slightly shorter and thicker compared to *A. spectabilis*; the two are often confused. Prominent white stomatal bands on the underside
4	*Abies delavayi*	Scarce across much of the Himalaya	Occurs east from Arunachal and into China, northern Myanmar and Vietnam from 3000 to 4200 m	Needles are shorter and more densely packed compared to other Himalayan firs, but this too has prominent white stomatal bands

Abies densa (Bhutan fir) is the common fir of Sikkim, Bhutan, and the eastern Himalayas. This is easily the most common tree species of Bhutan (Ohsawa 1987; FMID 2023) and commonly found above 3000 m across the Northeast. It is unclear if *A. densa* and *A. spectabilis* intermix, and being very similar it is often difficult to distinguish them. While *A. spectabilis* is found West of approx. 87–88°E longitude, *A. densa* is typically found East of this longitude. Table 4.7 provides some details of these common firs.

Dominant Trees and Associates

The firs most often exist as a significant component of various forest associations. The Western Mixed Coniferous forest (12/C1d) is among the most common forest types of the temperate zone of the Western Himalayas and extends over some 7500 km^2. These are among the most attractive forests of the Himalayas with *Abies pindrow*, *Pinus wallichiana* and Spruce (*Picea smithiana*) and deodar with canopy heights that can reach over 50 m. Various oaks (esp. *Quercus floribunda*),

Maples, and Birch can form the second storey. Dwarf bamboos (*Arundinaria falcata*, *Thamnocalamus spathiflorus*) are often found in the understorey. In Uttarakhand in particular, the Himalayan Upper oak-fir forests (12/C2b) between 2600 and 3400 m have a canopy typically of fir and spruce (*Abies pindrow* and *Picea smithiana*) which overtops oaks (*Quercus semecarpifolia* and sometimes *Q. floribunda*). At higher elevations birch's (*Betula alnoides* and *B. utilis*) can be an important component with *Abies pindrow* forming the canopy and oak or birch forming a second storey.

In the eastern Himalayas the upper temperate zone (2400–3000 m) may have *Abies densa* associated with *Tsuga Dumosa* and a second storey of *Quercus pachyphylla* and other oaks, maples, and magnolias. In parts of Arunachal, *Abies delavayi* is found, but this fir is more common in China and south east Asian mountains east of the Himalayas, from 2700 to 3400 m.

At higher reaches, in the sub-alpine zone and near the tree-line, forests are fairly similar across the Himalayan arc with fir, birch and rhododendrons being the three main genera. In the Northwest pure crops of fir may be found—with *Abies pindrow* reaching the tree line in Kashmir and Pakistan, and *Abies spectabilis* across much of the central Himalayan zone. Birch (*Betula utilis*) trees grow in-between and *Rhododendron* can form a dense undergrowth and completely dominate in protected hollows. *Quercus semecarpifolia* is occasionally found mixed in these forests. In the Eastern Himalayas, the component of rhododendron increases. While in the west, *R. campanulatum* is the most common rhododendron at upper altitudes, but in the east this genus is diverse and dominates the treeline. Sub-alpine forests of the eastern Himalayas also have significant components of fir and birch, viz. *Abies densa* and *Betula utilis* (C&S 14/C2) along with some juniper (*Juniperus indica*, syn *J. wallichiana*).

Human Impacts

Fire, while occasional, can have devastating impacts on fir forests which are not adapted to fires. In many regions, felling of fir has caused considerable localised depletion such as in Pakistan for timber, and in Nepal for roofing shingles. Browsing by ruminants kept by pastoralists can severely impact regeneration. At high elevations, once these forests are cleared, they are difficult to re-establish due to unimpeded snow avalanches. Also, young fir seedlings show best survival and growth when under a shaded canopy.

Thus, while human disturbance is low at the altitudes where firs grow, this genus is quite susceptible to disturbance.

The Himalayan Rhododendrons
(C&S 15C1, Component of 12C1)

Brief Description

Two Ericaceous members, viz. *Rhododendron* and *Lyonia* are common Himalayan genera. Both find prominence in a wide range of oak forests. *Rhododendron* is among the most widespread Himalayan genera, comparable in distribution to the oaks and the pines and a genus known for its showy and prominent flowers. Yet other than the Alpine scrub forest types, it barely finds mention in Champion and Seth's (1968) forest types. Early botanists however gave great importance to this genus. Wallich, the Father of Nepalese Botany, stated that the genus Rhododendron "*claims the highest ranks among the plants*" of Nepal (Hooker, 1849).

A large number of Rhododendrons are found across the Himalayas as a component of numerous forest types. They are found from the upper sub-tropical to the alpine elevations and across the Himalayan arc. However, the majority of Rhododendrons are not trees. Nonetheless given their importance, we describe this genus and a few of its species.

Rhododendron arboreum (Burans)

This is a small to medium sized gnarled tree, 8–15 m tall, and with a DBH that can reach 1.5 m. It has showy red flowers which are produced in great abundance in the early spring. The bright red on a background of dark green leaves makes the burans in bloom a very attractive sight. The tree was described to western botanists as early as 1796 by Captain Hardwicke in the Siwaliks (Hooker 1849) and intrigued western botanists as unlike shrubby rhododendrons they had encountered, this species took a tree form. Burans is often found as an understorey tree in oak forests. As oak forests are prone to heavy disturbance (lopping, cutting, litter removal), often the oak canopy gets degraded and removed, leaving behind a forest rich in rhododendron. Oaks do not regenerate under its deep shade, but at higher elevations *Abies* can survive low light and eventually overtop and shade out this species. The ground is covered in a layer of slowly decomposing leaves.

Climate

Annual precipitation ranges from 850 to 2000 mm and this species occurs from the sub-tropics to the sub-alpine. The temperatures can range from a maximum of over 36 °C in the lower part of its range, to −20 °C in the upper part of its range.

Occurrence

Rhododendron arboreum is usually found from 1200 to 3400 m across the Himalayan arc from Swat in Pakistan in the West, to Northern Myanmar in the East (latitudinal range 72–98°E). While most common around Uttarakhand, where it is the third most common tree species (FSI 2021), it is abundant in the Northeast. Interestingly *Rhododendron arboreum* is the State Tree of Uttarakhand in the central Himalayas and the State Flower of Nagaland in the Eastern Himalayas.

The Flora of Bhutan (Grierson and Long 1991) recognises four intraspecific taxa that show different altitudinal ranges (*R. arboreum* ssp. *arboreum*, 1200–3000 m; *R. arboreum ssp. cinnamomeum* 2800–3400 m; R. *arboreum v. roseum* 1400–4000 m; *R. arboreum ssp. delavayi* 1700–3400 m). However, the last is now recognised as its own distinct species, *Rhododendron delavayi* (Franch.) (POWO 2024). The extend of the distribution of this species can be gauged by another subspecies, *Rhododendron arboreum subsp. Zeylanicum* which is found as far south as the hills of Sri Lanka.

Dominant Trees and Associates

Rhododendron arboreum is a common understorey tree of oak forests. However, it is frequently found with conifers such as chir pine, particularly where disturbance has removed the oaks. At higher elevations, it is found under fir.

Human Impacts

This species is not grazed by cattle (Hooker 1849) and it regenerates well after fires. While occasionally lopped for fuelwood, it is typically left unmolested and hence may increase its dominance in forests that are heavily disturbed.

High Altitude Rhododendrons and Those of the Eastern Himalaya

The spectacular variety of rhododendrons in the Eastern Himalayas piqued the interest of Sir Joseph Dalton Hooker, the famous taxonomist and biogeographer (and author of "The Flora of British India"). His friendship with Charles Darwin and his curiosity about plant distribution led him to a deeper examination, which led to the book "*Rhododendrons of the Sikkim-Himalaya*", considered a seminal contribution to botanical literature.

In the almost perennially wet slopes from E Nepal (88°E) to upper Myanmar (98° E), at 3000–4000 m (lower at very humid sites), *Rhododendron hodgsonii* forests are found at altitudes similar to *Abies densa,* but in more humid seasonally waterlogged conditions. The extraordinarily large leaves (up to 30 cm long and 10 cm wide) at such a high-altitude are a consequence of the mesic conditions. In these forests, *Abies* and *Betula* are absent due to the high humidity and presence of epiphytic bryophytes which form large cushions up to a metre in diameter. Common associates of *R. hodgsonii* include *Rhododendron falconeri* as well as *R. grande* (in eastern Nepal) and *R. kesangiae* (in Bhutan) along with cold-deciduous *Sorbus* and *Acer* spp.

At even higher altitudes, from about 3800–4500 m, *Rhododendron* thickets are found, often on shaded North exposures. Precipitation here can be quite high and can reach 4000 mm on some sites. While rhododendron diversity is high, typically only one-two species are found together and could include *R. aeruginosum*, *R. wightii,* and *R. bhutanense* in Bhutan; *R. wallichii*, *R. fulgens,* and *R. campylocarpum* in E Nepal. At the highest altitudes, above 4500 m, dwarf thickets often under 50 cm high are found, which may be of *R. anthopogon*, *R. nivale,* and *R. setosum*.

At higher elevations, across much of the Himalayan arc *Rhododendron campanulatum* forms extensive thickets in sub-alpine forests above 3000 m. This can be a shrub or small tree, up to 5 m high. Found as far west as Pakistan, it is more common east of Himachal and all the way to Sikkim. From Himachal to Western Nepal, *R. campanulatum* forms monospecific thickets ("Krummholz") which decrease in height with altitude, from 3 to 4 m at lower elevations to dwarf shrubs a metre high at their upper distribution limit. Multi-stemmed *Sorbus microphylla* sometimes overtops these thickets. The attractive flowers of this small tree commonly known as the bell rhododendron, range from white to light pink, and are the state flower of Himachal Pradesh.

Birch and the Himalayan Timberline
(C&S 14C1, C2; 15C1)

(A) *Betula utilis*
Brief Description

Betula utilis, the Himalayan silver birch is a deciduous species of high altitudes. Locally called *Bhojpatra*, its whitish bark was used to make a paper for ancient writing. It has bright green papery leaves, which turn golden-yellow in the autumn before they fall in October. Typically, 8–12 m tall, with a diameter rarely exceeding 0.5 m and an open canopy, it can grow to 20 m in favourable environments (Ghimre et al. 2020). In the geologic past, during the Pliocene, this birch may have occurred at altitudes of ~1800 m (Vishnu-Mittre 1963), but today it is usually found over 3000 m.

Climate

Bhojpatra grows in areas of annual precipitation from 450 to 1700 mm, but this is a tree best seen in inner valleys and non-monsoonal belts. Around Safed Koh (70°E) it thrives in areas of heavy snowfall during winter and spring (450 mm), with low summer rainfall (70 mm) (strongly non-monsoonal). Similarly, in upper Hunza (Ladakh) forests receive 200–500 mm snowfall and 80 mm of monsoon rain. However, it also grows in the Central Himalayas in monsoonal zones. *Betula* forests of S Tibet also receive only summer rainfall.

This is a tree of high altitudes, and maximum temperatures rarely exceed 18 °C in the summer, and sub-zero temperatures with heavy snow is common for many months.

Occurrence

Betula utilis is common at higher elevations in the inner Himalayas and is distributed along the entire Himalayan arc from the Karakoram and Afghanistan to Burma. It also extends into China and western Tibet and can be found from about 70–95°E. The species is most dominant in a belt above the fir forests, at elevations above 3000 m to the timberline at ~4300 m. This birch is tolerant of long-lasting snow cover and snow deformation.

Dryer conditions and less competition in the Western Himalayas allow this tree to form extensive and gregarious stands (Singh and Singh 1987). In the Nanda Devi biosphere reserve in Uttarakhand for example it is the dominant tree at high altitudes with an IVI of 217–267 and a basal cover of 22–29 m^2/ha (Maletha et al. 2020) forming almost pure stands. The harsh, more continental climes in the West/North-western Himalayas with warmer summers and more severe winters, and a longer duration of snow cover, favours *B. utilis* (Bobrowski et al. 2017). This dominance of birch in the western Himalayas is also related to higher soil moisture at the beginning of the growing season due to snowmelt, and soil moisture during the pre-monsoon period (March–May) (Bobrowski et al. 2017).

In the east Himalayas, the dominance of this species weakens considerably. The lower amplitude of seasonal temperature variations towards the more oceanic

Eastern Himalayas leads to the birch becoming less competitive, and it is displaced by coniferous species and rhododendrons as the principal treeline species.

While birch is more competitive in the western Himalayas, the birch treeline is located at lower elevations (3600–3800 m) compared to the eastern Himalayas where treeline birch can occur as high as 4500 m. This is related to the much colder winter temperatures of the west—the mean temperatures of the coldest month at 3900 m is −13 °C in the Karakorum compared to −5 °C in southeast Tibet and Yunnan (Bobrowski et al. 2017)—as well as the shorter growing season, and longer duration of snow cover in the Western Himalayas.

Birch forests thrive on humid, shady slopes with deeply weathered soils, and are uncommon on south-facing slopes, in particular in the more continental Western Himalayas, where they tend to dominate on moist, north facing slopes. In the eastern Himalayas, birch is more sparsely distributed but may be found in dryer environments presumably because of the lack of competition with thicket forming plants such as rhododendrons (Bobrowski et al. 2017). In timberline sites in Sikkim, Singh et al. (2021) report the absence of *B. utilis* from upper altitudes and a dominance of rhododendrons and silver fir (*Abies densa*).

Dominant Trees and Associates

Betula utilis along with firs dominates the sub-alpine forests throughout the Himalayas. In Western Nepal, from 3000–3500 m, *B. utilis* is often associated with firs (*Abies pindrow*), spruce (*Picea smithiana*), *and Pinus wallichiana*, but from 3600 to 4200 m it forms a belt above the firs and is gregarious (Miehe et al. 2015). In many areas an undergrowth of *Rhododendron campanulatum* is common.

In the Western Himalayas, *B. utilis* is an integral part of the sub-alpine birch-fir forests (14/C1b) where it associates with *Abies pindrow, Rhododendron campanulatum* and *Quercus semecarpifolia*. These forests are best seen near the tree line in Kashmir and Himachal. In the central Himalayan region A. *spectabilis, Pinus wallichiana* and *Picea smithiana* are associated, and in patches Kharsu oak.

In the Eastern Himalayas (14/C2), *B. utilis* associates with *Abies densa*, *Juniperus wallichiana,* and various rhododendrons. The importance of rhododendrons increases in the east.

Human Impacts

The specific epithet, *utilis*, refers to the many uses of the different parts of the tree. The whitish bark, which is smooth and shining with horizontal lenticels, has numerous thin papery layers that exfoliate in broad horizontal rolls and was used in ancient times for writing scriptures and texts. In parts of Kashmir, it was the main source of writing paper. As per legend, the bark was used as clothing by attendants of Lord Shiva. This bark also has important medicinal properties, and biological actives from this species have anti-inflammatory, anti-bacterial properties and are effective against Cancer and HIV (Wani and Pant 2021). It is widely used in the Ayurveda and Unani systems of medicine. In Ladakh and Kashmir, it is used to treat anaemia and blood diseases (Ghimire et al. 2020).

The leaves are lopped as cattle fodder. The wood makes a good firewood and is used for building in the inner Himalayas. Overextraction for fuelwood and fodder and the removal of bark threatens this tree. Being a slow growing species, the species is considered vulnerable.

(B) Other Species of Family Betulaceae

The Family Betulaceae has other significant species across the Himalayas. Among birches, *Betula alnoides,* is found at lower elevations east from Himachal to Myanmar at altitudes of 1500–3000 m. This birch is found sporadically mixed with oaks, as well as conifers, and is gregarious when conditions favour its natural regeneration—on landslips and sites where soils have been newly exposed. Seedlings are often found on road cuttings though the regeneration cannot withstand shade.

More common and important is *Alnus nepalensis*, a common native species widely used in forestry and agro-forestry, especially in the eastern Himalayas. Similar in longitudinal range to *B. alnoides*, this alder occurs from 1000 to 2500 m. It is suited to exposed sites and regenerates well in freshly exposed, landslide affected, degraded, and highly disturbed sites and is often planted to check landslips and reduce erosion (Sharma et al. 1998). It grows best along streams, where it forms fast growing monospecific cohorts with a closed canopy up to 30 m tall (Miehe et al. 2015). *Alnus* is the most prominent actinorhizal genus in the Himalayas, i.e. it can form a symbiotic relationship with the nitrogen fixing actinomycete genus Frankia. This ability to fix environmental nitrogen and increase soil fertility has helped this tree become important for agroforestry systems and land restoration. In the north east, alder is often cultivated as a shade tree in cardamom plantations. Conditions created by Jhum cultivation (slash and burn) favour this species. Given its ability to fix nitrogen, and fast turnover of leaves, encouraging *Alnus nepalensis* in jhum fields can help shorten the fallow period of a jhum cycle in addition to providing harvestable poles.

While *A. nepalensis* is commonly planted in the western Himalayas, native to the region is *Alnus nitida* which is found occupying a similar niche from West Nepal through Pakistan at altitudes of 1000–2700 m.

Conclusions

In the preceding sections, we have introduced the reader to some of the important forest associations and dominant tree species from across the Himalayan arc. By describing these major forest types and their constituent species (Fig. 4.7), we aim to provide a foundation for understanding Himalayan forests and their distribution patterns. This framework serves as an essential context for subsequent chapters in exploring the dynamics and challenges facing Himalayan forest ecosystems.

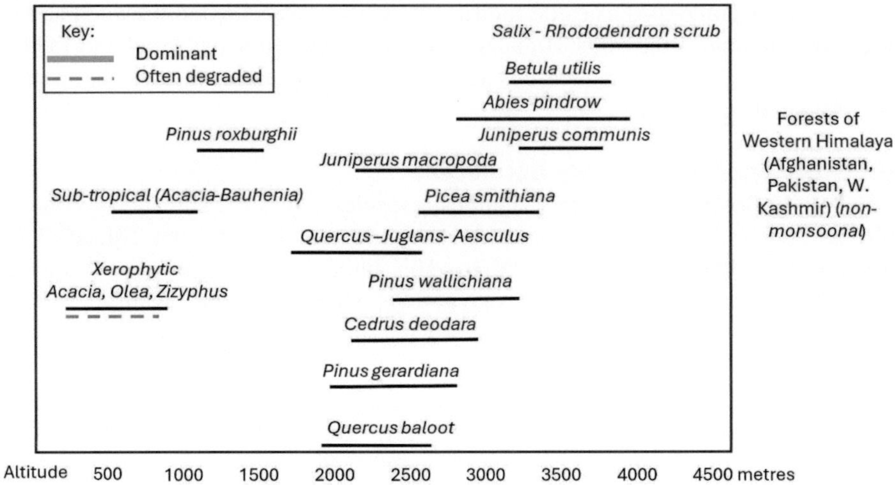

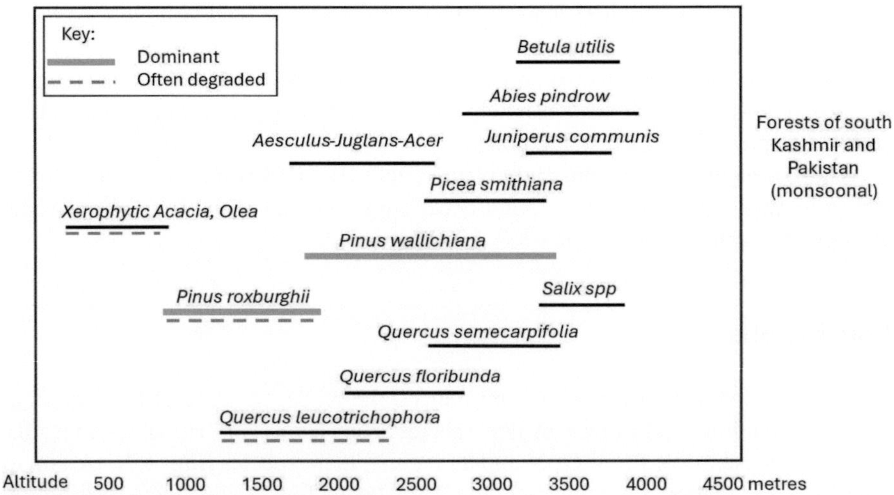

Fig. 4.7 Tree distribution diagrams

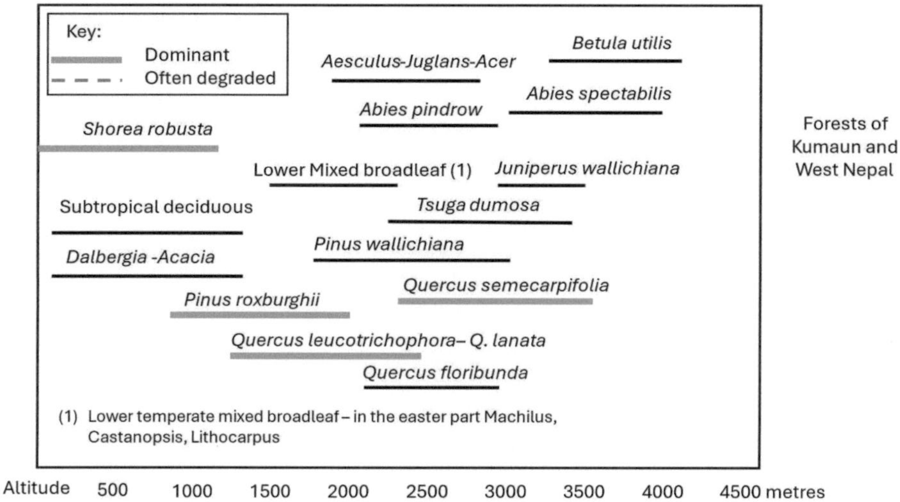

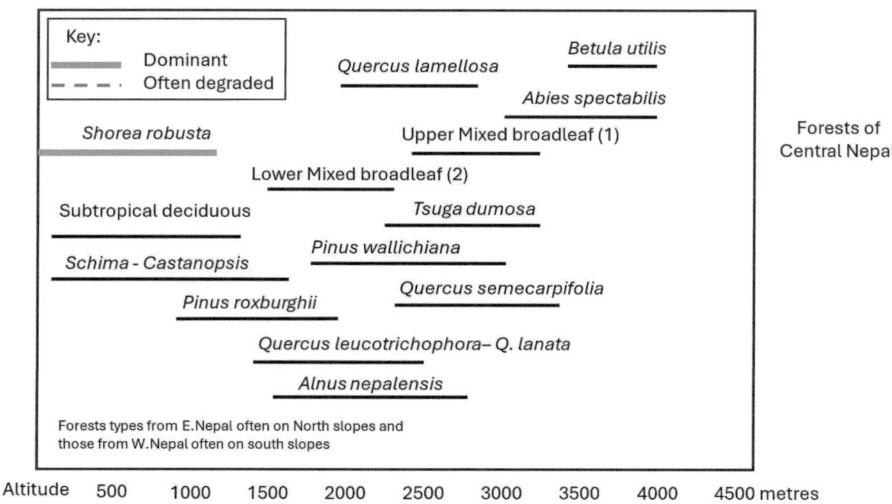

Fig. 4.7 (continued)

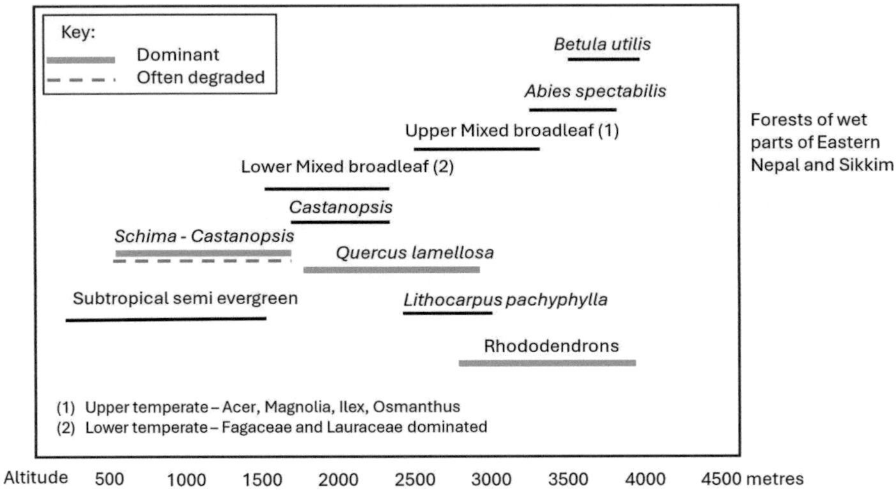

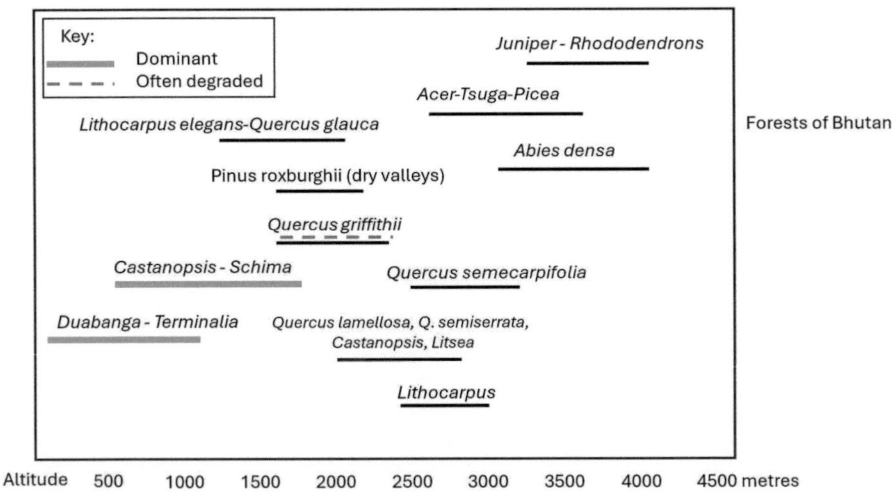

Fig. 4.7 (continued)

Commentary: Comparative Ecology of Two Dominant Tree Families of Tropical and Temperate Asia

Mark S. Ashton
The Forest School at The Yale School of Environment, Yale University, New Haven, CT, USA

The Fagaceae and Dipterocarpaceae are good examples of convergent evolution. They comprise unrelated tree taxa that dominate forests in very different wet-to seasonally dry climates—temperate and tropical—but have several ecological similarities. The best example in the Fagaceae are the oak-dominated forests of Laurasia (North America, Europe, and Asia); while the best example in the Dipterocarpaceae of Gondwanan origin, is the genus *Shorea* of SE Asia. The two families meet at the Himalaya, between temperate and tropical, and between Laurasia and Gondwana.

Both families have speciose tree genera, especially *Quercus* (oaks) in Fagaceae and *Shorea* (meranti, lauan) in Dipterocarpaceae. These genera comprise the main canopy structure and composition of their respective forest types. Some of their ecological similarities include:

1. Supra-annual masting events to satiate seed predators.
2. Poor seed dispersal.
3. Recalcitrance, whereby seeds germinate immediately or after a brief period of stratification (winter).
4. Seed that germinates post mast year in abundance in the forest understorey and survive for a period under low light conditions as advance reproduction.
5. Reliance on a large canopy level disturbance for regeneration release to become saplings as a new stand of trees.
6. Both families, especially the genera *Quercus* and *Shorea*, exhibit speciation across topographic, elevation, and disturbance gradients that facilitate formation of complex associations (see Fig. 4.8).
7. Differentiation among species within a genus occurs with slight differences in the physiology, morphology, and anatomy that they have in common. For example, both families have simple alternate arranged leaves. Leaves of species within a genus are similar but differ in their size, thickness, and arrangement making the species with smaller, thicker leaves, and smaller stomata more water-use efficient than conspecifics that have larger dimensions.
8. It is not a surprise therefore that species separate by site. As an example, the water-use efficient species is restricted to grow on the drier soils, while the more water-wasteful species grows faster and is more competitive on the wetter soils.
9. Lastly, these families are ectomycorrhizal, with a dominance on old weathered, nutrient poor soils.

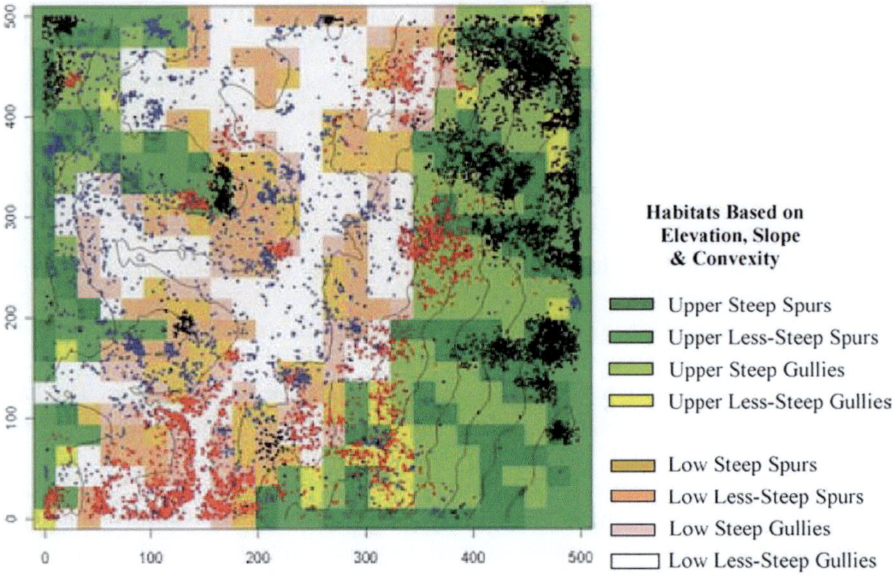

Fig. 4.8 A stem map of three tree species in a spatially explicit 25 Ha plot in mixed dipterocarp forest of southwestern Sri Lanka. Distribution patterns of (c) *Shorea worthingtonii* (black) found predominantly on upper steep spurs, *Shorea trapezifolia* (blue) found predominantly on the low-elevation less-steep spurs, and *Shorea megistophylla* (red) found predominantly on the low-elevation less-steep gullies (from Gunatilleke et al. 2006). If a spatially explicit stem map of equal area was available for the Himalayas, the distribution patterns for oak would look similar between ridges and gullies.

Fagaceae and Dipterocarpaceae in the Himalayas

The Temperate Forests of the Middle Hills

There are over 500 species of oak in the world. The epicentre for oak species diversification is North America with over 160 species. However, there are more than 35 species of oaks in the greater south Asian realm, of which about 14 are in the Himalayan region. Forests in the western Himalaya can become monodominant to oak. Eastern Himalayan forests are more diverse in tree species, and oak as a genus does not usually dominate but represents a part of the mixed broadleaf forests. On the other hand, *Castanopsis* is a tree genus in the Fagaceae, whose epicentre of diversification, comprising nearly 140 species, is in South and Southeast Asia. About 11 species are found in the eastern Himalaya.

Studies elsewhere have demonstrated that forests with many species of oaks have diversified through niche partitioning—with species having differing degrees of shade tolerance and therefore successional status; and occupying different sites across the forest topography in relation to hydrology and soil fertility (Cavender-

Table 4.8 Differences in leaf morphology, anatomy, and photosynthesis between two Himalayan oaks growing in high light and moist soils conditions (Thadani et al. 2009)

Trait	*Q. leucotrichophora*	*Q. floribunda*	Significance
1. Leaf area (cm^2)	30 ± 1.9	11 ±0.5	$p < .0001^{***}$
2. Leaf dry weight (g)	0.42 ± 0.03	0.16 ± 0.01	$p < .0001^{***}$
3. Leaf thickness (μm)	173 ± 8	229 ± 7	$p = .0008^{**}$
4. Leaf mass/area (mg/cm^2)	13.8 ± 0.2	14.6 ± 0.4	$p = .0773$
5. Palisade thickness (μm)	68 ± 5	95 ± 3	$p = .0016^{*}$
6. Spongy thickness (μm)	71 ± 5	104 ± 5	$p = .0018^{*}$
7. Cuticle thickness (μm)	3.3 ± 0.3	2.5 ± 0.1	$p = .0859$
8. Size of palisade cell ($μm^2$)	420 ± 34	488 ± 34	$p = .1942$
9 Palisade cell length/breadth	4.66 ± 0.18	4.70 ± 0.21	$p = .8815$
10. Stomatal length (μm)	19.5 ± 0.1	17.9 ± 0.2	$p < .0001^{***}$
11. Stomatal density (stoma/mm^2)	439 ± 15	661 ± 27	$p < .0001^{***}$
12. Photosynthesis (μmol/m^2/s)	8.0 ± 0.3	12.2 ± 0.4	$p < .0001^{***}$
13. Conductance (mmol/m^2/s)	0.14 ± 0.01	0.23 ± 0.01	$p < .0001^{***}$
14. Leaf toughness (g/mm^2)	144 ± 7	156 ± 8	$p = .2563$
15. Chlorophyll (SPAD units)	43.7 ± 1.5	42.9 ± 0.8	$p = .7062$

Fishers PLSD, $^{*}p < .01$, $^{**}p < .001$, and $^{***}p < .0001$

Bares 2019). While some studies have been done in the Himalayas, more are needed to investigate oak species differentiation in shade tolerance and topographic/elevation affinity. For instance, leaf morphology and physiology studies show *Q. floribunda* to demonstrate greater levels of plasticity and to be more opportunistic of light environments but less drought tolerant as compared to *Q. leucotrichophora* (Table 4.8). These along with *Q. semecarpifolia* differentiate out broadly across the middle hills of the western Himalaya in relation to elevation from 1500 m to 3300 m. *Q. leucotrichophora* dominates lower elevations, *Q. floribunda*'s range is in the middle and *Q. semecarpifolia* can be found at the highest elevations. What exact mechanisms and adaptations promote their separation need further study. In addition, little to no parallel work has been done for *Castanopsis* or *Lithocarpus*, underscoring a dire need for further research on this genus.

The Tropical Lowlands and Hill Forests

There are over 196 species of *Shorea* with the epicentre of species differentiation being across the equatorial tropical wet seasonal climates of Borneo and peninsular Malaysia. In India Dipterocarpaceae are not nearly as diverse but are well represented. There are 7 species (four genera) recognised for north-east India; 11 species (three genera) in the Andaman Islands; and 13 species (four genera) in the Western Ghats (Ashton 1982). There are about three species of *Shorea (Anthoshorea)* of which two are now rare and found in the wetter northeastern regions on the border with Myanmar. Similarly, there are about 11 species of *Dipterocarpus* in India, of which 5 can be found in the foothills of the Himalayan northeast and/or the Andamans (Ashton 2014). The most widespread is *Shorea robusta* (sal) which dominates in the Siwaliks. The tree is a shade-intolerant generalist, that can be its own forest type, forming monodominant stands on very seasonal, weathered, nutrient poor soils where fire is prevalent. An enormous amount of research has focused on Sal silviculture, primarily because it is one of the most important economic trees across India. But little research has focused on the ecology of the remaining *Shorea* (*Anthoshorea*) species (*S. assamica*, *S. roxbughii*) in the wetter remnants of India's northeast forests (Nagaland, Arunachal Pradesh) where the genus has potentially differentiated across succession and topography. Likewise, no studies have investigated the potential habitat differentiation of *Dipterocarpus* species from the northeast. But studies elsewhere in South Asia (e.g. Sri Lanka) demonstrate that like *Quercus*, *Shorea* and *Dipterocarpus* are genera having species that differ in their topographic, elevational, and successional affinities.

References

Ashton PS (1982) Dipterocarpaceae. Flora Malesiana 9:237–552

Ashton PS (2014) On the forests of tropical Asia: lest the memory fade. University of Chicago Press

Bahadur KN (1975) A name change for *Quercus incana* Roxb is inevitable. Indian Forester 101(1): 99–102

Bobrowski M, Gerlitz L, Schickhoff U (2017) Modelling the potential distribution of Betula utilis in the Himalaya. Global Ecol Conserv 11:69–83. https://doi.org/10.1016/j.gecco.2017.04.003

Camus A (1936–1954) Les Chênes. Monographie du genre Quercus. Tome I, II, II. (Volumes I–III). Texte. Paul Lechevalier, Paris

Cavender-Bares J (2019) Diversification, adaptation, and community assembly of the American oaks (*Quercus*), a model clade for integrating ecology and evolution. New Phytol 221:669–692

Champion H, Osmaston FC (1962) E. P. Stebbing's the forests of India, vol IV. Oxford University Press, London

Champion HG, Seth SK (1968) A revised survey of Forest types of India. Government of India Press, New Delhi, p 404

Denk T, Grimm GW, Manos PS, Deng M, Hipp AL (2017) An updated infrageneric classification of the oaks: review of previous taxonomic schemes and synthesis of evolutionary patterns. In: Oaks physiological ecology. Exploring the functional diversity of genus Quercus L, pp 13–38

DFRS (2015) State of Nepal's Forests. Forest Resource Assessment (FRA) Nepal, Department of Forest Research and Survey (DFRS). Kathmandu, Nepal

Dudgeon W, Kenoyer LA (1925) The ecology of Tehri-Garhwal. A contribution to the ecology of western Himalayas. J Indian Bot Soc 4:233–285

FAO (2005) Global Forest Resources Assessment, Country Report; Pakistan

Farjon A (2008) A natural history of conifers. Timber Press, Portland

Farjon A (2018) The Kew review: conifers of the world. Kew Bull 73:1–16

FMID (2023) National forest Inventory Volume I: State of Forest Report, Bhutan

FSI (2021) Indian state of Forest report. Forest Survey of India, Dehradun

Gautam KH, Devoe NN (2006) Ecological and anthropogenic niches of Sal (Shorea robusta Gaertn. f.) forest and prospects for multiple-product forest management–a review. Forestry 79(1):81–101

Ghimire N, Kunwar RM, Hussain W, Abbasi MA, Bussmann RW, Paniagua-Zambrana NY (2020) Betula utilis D. Don Betulaceae. In: Ethnobotany of the Himalayas. Springer International Publishing, Cham, pp 1–11

Gillani NA, Khan MQ, Hussain K, Sabir S (2023) Agro-morphological diversity of evergreen Oaks (Quercus) in Western Himalayas, Azad Kashmir, Pakistan. Pure Appl Biol 13(2):156–174

Grierson AJC and DG Long (1991) Flora of Bhutan, Vol 2(1) Royal Botanic Garden, Edinburgh https://archive.org/details/dli.pahar.3642/mode/2up. Accessed July 5, 2024

Gunatilleke CVS, Gunatilleke IAUN, Esufali S, Harms KE, Ashton PMS, Burslem DF, Ashton PS (2006) Species–habitat associations in a Sri Lankan dipterocarp forest. J Trop Ecol 22(4): 371–384

Gupta P, Gupta S (2020) Wood anatomy of Indian oaks, with reference to systematic, ecological and evolutionary perspectives. Nord J Bot 38(4)

Hooker JD (1849) The rhododendrons of Sikkim-Himalaya, 2nd edn. Reeve, Benham and Reeve, London

ICFRE (2013) Forest Types of India Revisited. Indian Council for Forestry Research and Education, Dehradun. India, pp 484

Iqbal J, Dutt V, Ahmad H, Bhat G, Ganie M, Wagay S et al (2017) Status, distribution and utilization of oak (Quercus spp.) in western Himalayan (Kashmir valley). Indian Forester 143(1):1–5

Iqbal M, Bilal SA, Iqbal A, Almzeb M, Saeed M, Mehreen R, Bibi S, Naz F, Ali M, Muntaha S (2019) Distribution of Quercus leucotrichophora at different location and elevation in Jandool valley, northern Pakistan. Acta Ecol Sin 39(6):438–442

Jacob MC (1941) Sal regeneration de novo: a rejoinder. Indian Forester 67:337–346

Jazib MJ (2023) Community composition, population structure and regeneration potential of tree species in oak-dominated mixed forests of Rajouri district in Jammu and Kashmir, India. Sustainab Agric Food Environ Res 11:1–13

Lahiri S, Das M, Mukherjee A (2017) A contribution to the study of Quercus L. in the Himalayas. Indian J Scient Res 7(2):15–20

Maletha A, Maikhuri RK, Bargali SS (2020) Criteria and indicator for assessing threat on Himalayan birch (B. Utilis) at timberline ecotone of Nanda Devi biosphere reserve: a world heritage site. Environmental and Sustainability Indicators, Western Himalaya. https://doi.org/10.1016/j.indic.2020.100086

Miehe G, Miehe S, Vogel J, La D (2007) Highest treeline in the northern hemisphere found in southern Tibet. Mt Res Dev 27(2):169–173

Miehe G, Miehe S, Bohner J, Baumler et al (2015) Vegetation ecology. In: Miehe G, Pendry CA, Chaudhary RP (eds) Nepal: an introduction to the natural history, ecology and human environment of the Himalayas. Royal Botanic Garden, Edinburgh

Neale DB, Wheeler NC (2019) The conifers. In: The conifers: genomes, variation and evolution. Springer, pp 1–21. https://doi.org/10.1007/978-3-319-46807-5

Nixon KC (2006) Global and neotropical distribution and diversity of oak (genus Quercus) and oak forests. In: Ecology and conservation of neotropical montane oak forests. Springer, Berlin, Heidelberg, pp 3–13

Ohsawa M (1987) Life zone ecology of the Bhutan Himalaya I. Chiba University, Chiba (Japan)

Ørsted AS (1871) Bidrag til Kundskab om Egefamilien. Kongl Danske Vidensk Selsk Biol Skr 5 naturvidensk math Afd 6:331–538 (As referenced in Denk et al. 2017)

POWO (2024) Plants of the World Online. Facilitated by the Royal Botanic Gardens, Kew. Published on the Internet; http://www.plantsoftheworldonline.org/ Retrieved 02 June 2024

Puri GS (1945) The genus Quercus in the Karewa deposits of Kashmir, with remarks on the oak forests of the Kashmir Valley during the Pleistocene. Proc Indian Acad Sci Sect B 22:232–256

Rasheed F, Yaqoob S, Ahmad AH (2017) Forest types of Pakistan. In: Text book of applied forestry. University of Agriculture Faisalabad, pp 17–28

Rawat B, Rawat JM, Purohit S, Singh G, Sharma PK, Chandra A et al (2022) A comprehensive review of Quercus semecarpifolia Sm.: an ecologically and commercially important Himalayan tree. Front Ecol Evol 10. https://doi.org/10.3389/fevo.2022.961345

Rowntree JB (1940) Grazing versus burning as an aid to Sal regeneration. Indian For 66:645–653

Royle, JF (1839) Illustrations of the Botany and other branches of natural history of the Himalayan Mountain and of the Flora of Cashmere. WH Allen and Co, London. https://www.google.co.in/

books/edition/Illustrations_of_the_Botany_and_Other_Br/DQHQjbwsvpMC?hl=en&gbpv=1&pg=PP11&printsec=frontcover. Accessed Jan 2025

Sakai A, Malla SB (1981) Winter hardiness of tree species at high altitudes in the east Himalaya, Nepal. Ecology 62(5):1288–1298

Sarkar MM (1941) A short note on the Sal reproduction in the upper Bhabar tract, at Raimatong, Buxa division, Bengal. Indian For. 67:23–29

Sharma E, Sharma R, Pradhan M (1998) Ecology of Himalayan alder (Alnus nepalensis D. Don). Proc Indian Natl Sci Acad Part B 64:59–78

Shi P, Wu N (2013) The timberline ecotone in the Himalayan region: an ecological review. In: High-altitude rangelands and their interfaces in the Hindu Kush Himalayas, vol 108

Shrestha BB (2003) Quercus semecarpifolia Sm. in the Himalayan region: ecology, exploitation and threats. Himal J Sci 1(2):126–128

Singh JS, Singh SP (1987) Forest vegetation of the Himalaya. Bot Rev 53:80–1902

Singh JS, Singh SP (1992) Forests of Himalaya. Gyanodaya Prakashan, Nainital, p 294

Singh SP, Singh RD, Gumber S (2021) Interpreting mountain treelines in a changing world. Central Himalayan Environment Association and International Centre for Integrated Mountain Development. https://doi.org/10.53055/ICIMOD.787

Singh SP, Gumber S, Singh RD, Pandey R (2023) Differentiation of diploxylon and haploxylon pines in spatial distribution, and adaptational traits. Acta Ecol Sin 43(1):1–10. https://doi.org/10.1016/j.chnaes.2021.07.007

Soepadmo E (1968) A revision of the genus Quercus L. Subgen. Cyclobalanopsis (Oersted) Schneider in Malesia. Gardens' Bulletin, Singapore, pp 355–427

Stainton JDA (1972) Forests of Nepal. John Murray, London

Thadani R (1999) Disturbance, microclimate and the competitive dynamics of tree seedlings in banj oak (*Quercus leucotrichophora*) forests of the central Himalaya, India. Doctoral dissertation, Yale University

Thadani R (2023) Snow damage in a Himalayan forest maintains the dominance of evergreen oaks. Biotropica 55:737–741. https://doi.org/10.1111/btp.13236

Thadani R, Berlyn GP, Ashton MS (2009) A comparison of leaf physiology and anatomy of two Himalayan oaks in response to different light environments. J Sustain For 28(1-2):74–91

TISC (2010) Forest and Vegetation types of Nepal. Govt. of Nepal. Ministry of Forests and Soil Conservation. Tree Improvement and Silviculture Component. Kathmandu, Nepal

Troup RS (1921) Silviculture of Indian trees. Clarendon Press

Tucker RP (1982) The forests of the Western Himalayas: the legacy of British colonial administration. Forest & conservation history. J For Hist 26(3):112–123. https://doi.org/10.2307/4004579

Vishnu-Mittre (1963) Oaks in the Kashmir Valley with remarks on their history. Grana 4(2): 306–312. https://doi.org/10.1080/00173136309436753

Wani ZA, Pant S (2021) Betula utilis D. Don: an ecologically and economically important timberline species of Himalayan region in jeopardy. Bot Rev:1–15

Yang Y, Zhou T, Qian Z, Zhao G (2021) Phylogenetic relationships in Chinese oaks (Fagaceae, Quercus): evidence from plastid genome using low-coverage whole genome sequencing. Genomics 113(3):1438–1447

Forest Phenology

5

S. P. Singh

Abstract

This chapter is largely based on leafing, leaf drop, flowering, and fruiting phenophases both at the species and community level. Of the 54 Central Himalayan tree species, 15 were deciduous (d) and 39 were evergreen (e^1) with about one-year life span, with simultaneous pre-monsoon (March–May) leafing and leaf shedding (e^1 type). The e1 type was prominent also among shrubs (23 out of 39 species). Among the seasons, pre-monsoon was phenologically most active, with sum of phenological score 104, the others scoring <30. Multiple leafing is more common in the humid Eastern Himalayas (82% in Meghalaya than Central Himalayas, 40%). In general, much of leaf expansion and leaf maturation occur before the arrival of monsoon, which enables species to take full advantage of favorable growth conditions. The average leaf size of all species of the Central Himalayas, 69.5 cm^2/leaf, is 1.6 times larger than the global average, 42.5 cm^2/leaf, based on 1389 species. Across the 30 study species, bud survival was 42–45% but it varied widely and is still poorly known. The duration of stable leaf population on twigs on average was 2 months for deciduous species and 4 months for evergreen. The average leaf mass loss that accompanied leaf senescence was 26.2% for evergreen species and 31.6% for deciduous species. Leaf mass per unit area was higher in treeline species than in those of lower areas (e.g., 128 gm^{-2} in the tree line deciduous species, *Betula utilis,* compared to 70–111 gm^{-2} in lower deciduous species). Treeline species showed a higher nutrient resorption percentage than lower forest species. Interestingly, *Pinus*

S. P. Singh (✉)
INSA Honorary Scientist, Dehradun, Uttarakhand, India

Centre for Himalayan Studies, Graphic Era (deemed to be) University, Dehradun, Uttarakhand, India
e-mail: spsecology@gmail.com

S. P. Singh, R. Thadani (eds.), *Himalayan Forests in a Changing World*,
https://doi.org/10.1007/978-981-95-0978-2_5

wallichiana which occurs both in outer and inner ranges resembles *P. roxburghii* in leaf phenology, showing deciduous tendency during the pre-monsoon season. Generally, autumn and early winter are the period of seed fall in Inner Himalayan conifers, and fallen seeds remain on ground for varying periods during winters and pre-monsoon months, before germinating during May to September.

Keywords

Central Himalayas · Deciduous · Evergreen · Phenophases · Treeline · Leaf senescence

Introduction

In forests, the phenophases which are generally studied are bud burst, leaf flushing, leaf senescence and leaf drop, flowering, fruiting, and seed dispersal (Davi et al. 2011). In several European countries, recording of phenological events dates back thousands of years. Charles Moren, a Belgian botanist, was the first to use the term phenology in the eighteenth century to describe the second flowering event that occurred as a result of unusual warming. Leaf emergence (flush) phenology has been given particular importance, as it has a direct controlling impact on major processes such as primary productivity, carbon storage, nutrient cycling, and water balance (Gray and Ewers 2021).

Visual observation is a common method to study plant phenology, however, in recent years, with technological development several new ways of studying phenology are being used. They range from the use of remote sensing techniques, phenocams, to drones and collection traps. Accelerations and dendrometers are being used to detect changes in aboveground tree mass due to leaf emergence and leaf drop (Gougherty et al. 2018). Micro-coring is being used to collect samples to measure xylem development and correlated leaf flush events (Michelot et al. 2012; Puchalka et al. 2017). Eddy covariance flux towers have been installed at many places to seasonally measure CO_2 fluxes and estimate changes in gross primary productivity and ecosystem respiration. The use of fixed camera and eddy covariance on towers and the application of drones to take field observations have increased in recent years.

Monitoring forest phenology in a changing world is drawing widespread interest in ecology (Gray and Ewers 2021). Examples of phenological changes include more flowering and fruiting along forest edges than inside the forests (D'Eça Neves and Morellato 2012), and advanced flowering due to global warming (Herrerías-Diego et al. 2006), resulting in an increased asynchrony between pollinators' visits and flowering (Visser and Holleman 2001). New molecular tools have been developed to improve the understanding of phenology (Gray and Ewers 2021).

The global warming seems to have already raised the upper elevational limit of the tree presence. For example, recently *Quercus pannosa* (an evergreen oak) was found up to 4270 m in SE Himalayas (Yang et al. 2020), which could be the highest elevation of occurrence for any oak species in the world. Flowering and fruiting events have been found to occur 15–20 days earlier in a disturbed and fragmented forest in comparison to an undisturbed and continuous forest.

A citizen science approach could play a significant role in studying phenology, particularly in regions like the Himalayas, where trained human resource is limited, terrain is difficult, and often the sites where plants occur are varied and far-flung.

The Nature of Phenological Studies in the Himalayas

Phenological studies in the Himalayas are based on a few forest types and isolated individual efforts. Troup (1921) gave considerable importance to phenology in his famous book on Silviculture of Indian trees (3 volumes). Research papers on forest phenology in the Himalayas started appearing only in the early 1980s. The first published research papers were on a subtropical humid forest of the Indian Eastern Himalayas, IEH (Boojh and Ramakrishnan 1981; Shukla and Ramakrishnan 1982). They emphasized that many trees flowered when they were leafless, and the majority of species fruiting during monsoon had fleshy fruits. Subsequently, similar observations were made on the community-level phenology of subtropical broadleaved forests of Manipur (Kikim and Yadav 2001). Recently, leaf phenology of evergreen and deciduous species in wet tropical forests of another region (Silchar) of IEH was analyzed (Athokpam et al. 2014). Some of the major early studies on community phenology across a series of principal forest types along a wide elevational gradient (foothill to over 2200 m) are due to the efforts of the Kumaun University ecology group (Ralhan et al. 1985a, b; Negi and Singh 1992; Negi 2006). Their publications gave importance to leaf development, starting from bud break to leaf expansion, leaf mass accumulation and leaf senescence, and leaf life span.

In recent years, the flowering of *Rhododendron* species and the impact of climate change have drawn the attention of researchers (Khanduri et al. 2008; Gaira et al. 2014; Ranjitkar et al. 2014; Singh 2014a, b; Negi et al. 2021). Poudyal et al. (2004) and Tewari et al. (2016) analyzed relationship between phenology and tree water status. Negi et al. (2022) reported the effect of warmer years as earlier bud break and leafing in the semi-deciduous and evergreen forests of this region. Recently, a study was conducted on treeline species phenology of the Indian Central Himalayas, ICH (Singh et al. 2018).

For this chapter, we largely depend on the studies carried out (i) in the Kumaun region (a part of Indian Central Himalayas, ICH) across the major forest types from foot hills to about 2200 m elevation, and (ii) subtropical broadleaf forests and tropical moist forests from the Indian Eastern Himalayan (IEH) of India. These studies shed light on community-level phenology, comparison between evergreen and deciduous trees, and seasonal patterns of leaf expansion, leaf mass changes, and senescence. In community-level phenology, the focus is on the seasonal changes in the number of species in leafing, leaf drop, flowering, and fruiting. As for dry inner ranges, there is a dearth of recent data, so we used the classical work of Troup (1921) to summarize the phenological behavior of some conifers which occur widely in inner ranges.

The chapter also briefly dwells upon treeline phenology, based on the Tungnath site of ICH. Thus, in ICH, we have phenological studies for the entire elevational range from foothills to treelines. Research on the effect of climate change on plant phenology is rather scanty and is limited to a few species.

A Regional Scenario Based on a Sequence of Forests: Central Himalayas

To find out the seasonal pattern of phenological activities across the forest communities of the Central Himalayan region, we aggregated the data given in Ralhan et al. (1985a, b) on six forest types along the elevational transect, 350–2150 m in Kumaun Himalayas. In each, forest trees, shrubs, and herbs were considered for assessing the importance of season or month for phenological activities at the regional scale. Among the various forests studied, sal (*Shorea robusta*) and oak (*Quercus* spp.) forests had crown density >80%, pine forests were mixed oak forests, and pure pine forests were open type with crown density between 40% and 60%. The pine (*P. roxburghii*) forest encounters frequent human-ignited fire, while burning is moderate in sal forest and pine-mixed oak forest and negligible in oak-dominated forests. Shrubs were suppressed in frequently burned pine forest stands, and they were abundant in oak-dominated forests and intermediate in the remaining forest types. Of the 54 tree species encountered across the six forest types of the elevational transect (studied by Ralhan et al. 1985a, b), 39 were evergreen and 15 deciduous. In oak-dominated forests occurring in the higher parts of the elevation gradient, 86–100% tree species were evergreen, compared to 40–45% species in lower elevation forests. There were 39 shrub species across the forests, of which 13 were deciduous, 23 evergreen with concentrated leaf drop (e^1), and 3 evergreen with continual leaf drop (e^2). In the small group of evergreen shrubs with continual leaf drop, leafing and leaf drop are staggered over a long period of the year.

While trees and shrubs were investigated for leaf drop, leaf flushing, flowering, and fruiting, herbs were studied for seasonality in net primary production. In each forest, the number of species in a given phenophase was counted to identify the seasonal peak for that phenophase. Across the six forest communities and nine phenophases (four phenophases each for trees, shrubs and one for herbs), there were 54 primary species peaks (the main large peak) (Fig. 5.1). In certain phenophases, secondary peaks also occurred because some species showed multiple leafing and leaf drop (occasionally also multiple flowering).

To get an idea of the scale of monthly phenological activities, scores were given to the months/ seasons when phenological peaks occurred (when the highest number of species were displaying that phenophase). The score was 3 to each primary peak, 1 to each secondary peak, and none to others. The scores were summed up for months/ seasons to express their overall phenological significance at the regional

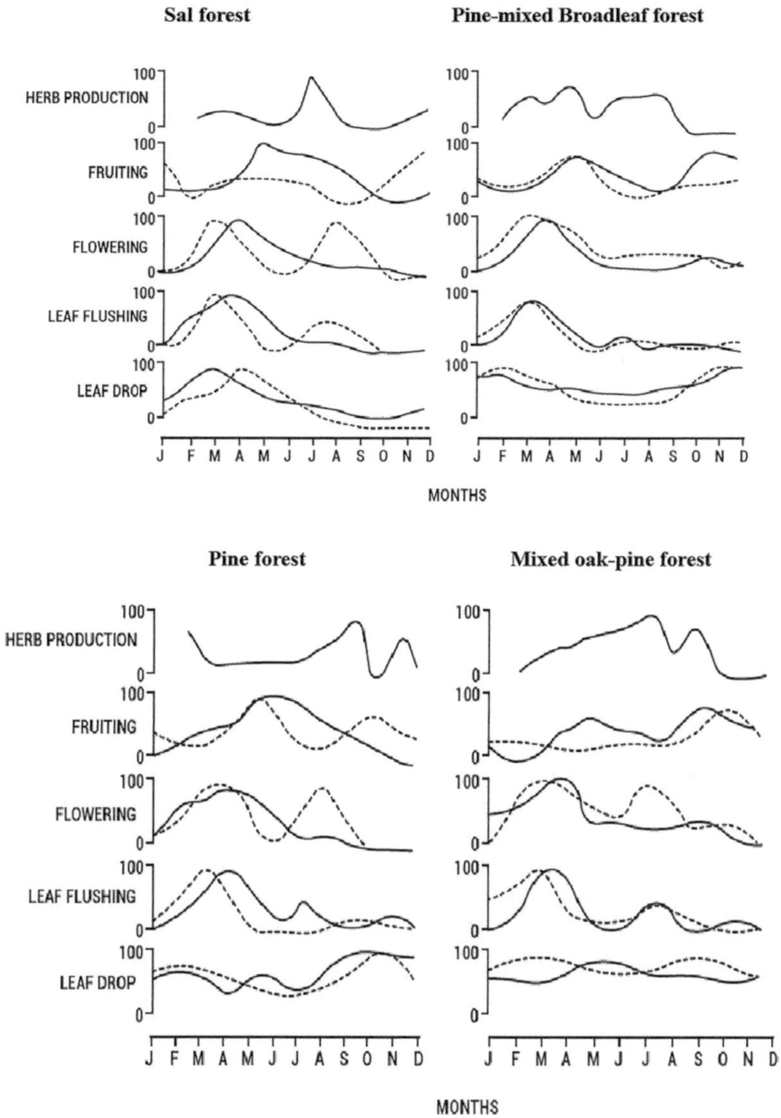

Fig. 5.1 Representation of periodicity of major phenological activities in six major forest types of Kumaun, Uttarakhand Himalayas (between 28°43′–31°27′ N and 77°34′–81°02′ E, a part of Central Himalaya). The peak value (the highest number of species in the given phenophase, such as leaf fall) in each case is taken as 100, and all other values are computed with respect to that. Broken curves are for shrubs and solid curves for trees. Herb production is in terms of g m^{-2} $month^{-1}$. The average altitudes of the study forests were as follows: sal (*S. robusta*) 350 m, pine (*P. roxburghii*)-mixed broadleaved forest 1350 m, chir pine forest 1750 m, mixed banj oak (*Q. leucotrichophora*) 1850 m, mixed oak-rianj (*Q. lanata*) dominated forest 2150 m, and mixed oak-tilonj (*Q. floribunda*) dominated forest 2150 m. (Developed from Singh and Singh 1992)

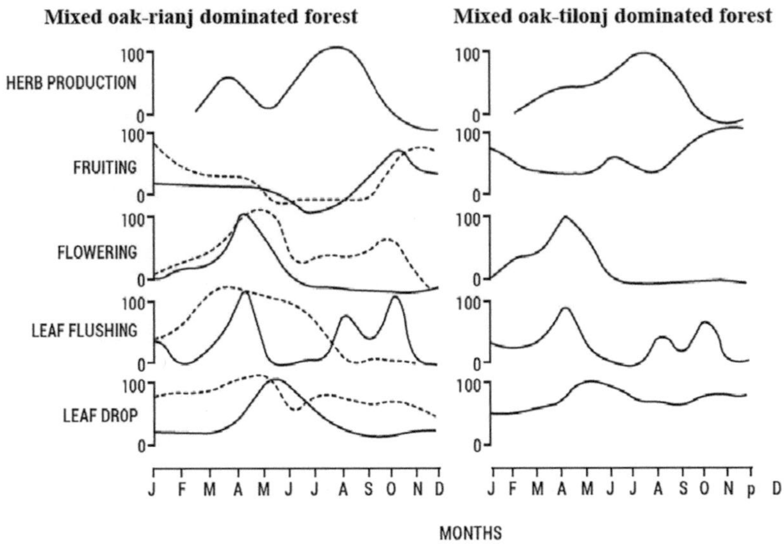

Fig. 5.1 (continued)

level. The phenological score at the regional scale (six forest types) was 104 for March alone, and 140 for the pre-monsoon (March–May), and 26 each for monsoon (mainly July–August) and autumn (October–November) (Fig. 5.2). Thus, for the basic phenophases, pre-monsoon the driest period is the "busiest" time, when many species of the six forests leaf out, shed leaves and flower. During this period, the canopy is set for renewal, and the forest floor is under a thick carpet of fresh leaf litter. Thus, during pre-monsoon forests have a spectrum of green canopy and brown ground floor, supporting myriad life forms of herbivores, detritivores, and pollinators. Presumably, trees take the risk of producing leaves during the driest part of the year in many parts of the world to avoid herbivores (Viera et al. 2016). The peak time when trees contain mature fruits and seeds varies more across the months, while leafing and flowering durations overlap at the regional level. Herbaceous biomass peaks at the end of monsoon, just before autumn.

Community-Level Phenological Patterns by Forest Types

Phenological studies by individual forests of the central Himalayas indicate that the evergreen woody species showing a concentrated pre-monsoon (March–May) leaf drop and simultaneous leafing (e^1), dominate ecologically in the region. Such species can also be referred to as "leaf exchanging" type of species (Longman and Jenik 1974). By the time, the monsoon arrives, these forests with fully developed foliage are ready to take full advantage of the favorable conditions of the warm and wet period. Leafing during the warm and dry period is prevalent also in the eastern

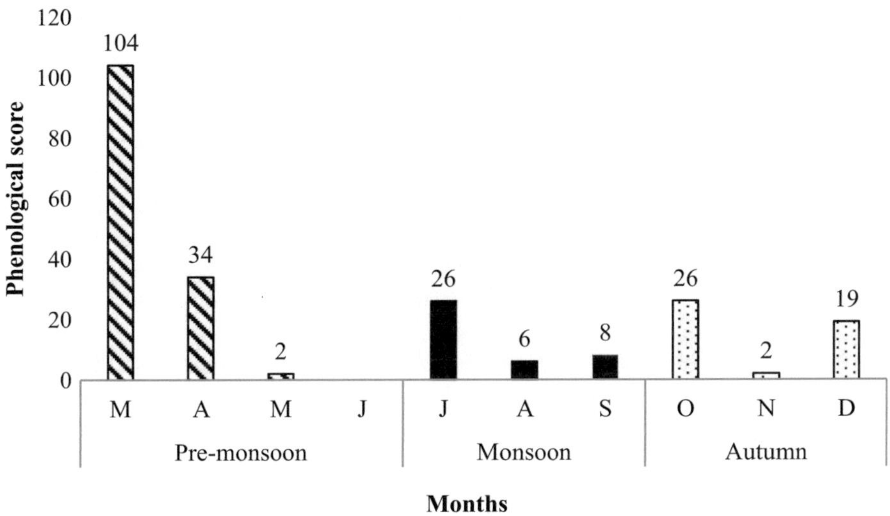

Fig. 5.2 Phenological scores based on four common phenophases, namely leafing, leaf drop, flowering, and fruiting, and six Central Himalayan forests between 300 and 2200 m. The score considers all four phenophases, both for trees and shrubs, across the six forests. In addition to these, biomass changes in the herb layer were considered. Both primary (score of 3 for each phenophase) and secondary peaks (1 score for each phenophase) were considered. The secondary peaks occur because some species produce leaves or flowers more than once, or a group of species leaf out at a different time than when the majority of species produce leaves. The first month is March (M) and the last December (D)

Himalayan trees, but they drop their leaves earlier and over a longer period from December to March. The trees might also flush up well before the monsoon rain and suffer from drought to avoid insect herbivory (Puchałka et al. 2017).

In these evergreen forests (e^1), smaller secondary peaks for leafing during the warm-wet season (July–August) and autumn (October) also occur because of the multiple leafing in some species, which are more common in the eastern forests (82% species) than in the Central Himalayan forests (40% species). The percentage of species with multiple leafing is significantly greater in the overstory than in the understory. It appears that the lower availability of light limits the multiple leafing in the undercanopy environment. Several advantages are attributable to multiple leafing: (i) it enables the species to utilize efficiently the favorable spells of potentially unfavorable seasons, such as winter, (ii) it ensures the presence of some leaves in case the leaves of the primary flush are damaged, (iii) it distributes the demand for nutrients in time, and (iv) it promotes staggered leaf drop and staggered nutrient release, thus minimizing the leakage of nutrients.

The leaf longevity for broadleaf species, on average, is slightly longer for the eastern Himalayan trees (average for 12 dominant species = 20 months) than for the Central Himalayan trees (average of six dominant species = 13 months). Phenological observations on individual leaves in Kumaun Himalayan forests indicate that the leaf expansion phase is longer for overstory species (average 6 weeks) than for understory species (average, 4.2 weeks), and for shrubs it is shortest (average, 2.5 weeks) among the three growth forms (Singh and Singh 1992). Climate warming accompanied by moisture is expected to lengthen leaf life span.

The flowering peak also occurs during the pre-monsoon, however, another equally large flowering peak is formed by a group of shrubs, either during the rainy season or in autumn. The overlap between leafing and flowering durations at the community level in all forests indicates that trees in the region have enough carbon reserves to sustain both the vegetative and reproductive activities simultaneously. However, in the year of seed masting of oaks the carbon demand of seed crops stops or reduces leaf flushing (Negi et al. 1996).

Among the trees of the Central Himalayan forests, the three forests of relatively lower elevations show the peak for the standing crop of mature fruits during the warm and dry period (May) (Ralhan et al. 1985a). In these forests, seed germination is synchronized with the commencement of the rainy season. On the other hand, in the forests of higher elevations the peak for the standing crop of mature fruits is formed during autumn. In several of these species, seeds have partial dormancy and staggered seed germination. The shrubs of a great majority of Central Himalayan forests show the peak standing crop of mature fruits in autumn (Singh and Singh 1992).

Notwithstanding the differences in the temporal pattern of peaks, the majority of species of these forests show lengthy fruit retention, and fruits of one or the other species are available to wildlife during much of the year. Food-rich seeds of oaks, such as *Q. leucotrichophora* and *Q. lanata*, and *Aesculus indica* are available on trees or ground throughout the winter and spring (Ralhan et al. 1985a), while the other oaks, *Q. floribunda* and *Q. semecarpifolia,* are laden down with acorns during monsoon months.

Among the herbs, three major categories with regard to growth time were discernible (Singh and Singh 1992): (i) spring perennials common across all forests growing rapidly during spring; (ii) rainy season herbs consisting of a mixture of annuals and perennials, showing peak growth after that of the woody species; and (iii) autumn or winter season perennial herbs growing actively when woody species enter the dormant phase. Such a time-separation in the growth behavior of herbs has several implications in regard to the functioning of the ecosystem. The spring perennials, which are able to maintain rapid initial growth at the expense of food stored in the perennating organs, by restricting their growth at the start of the frost-free period, are able to escape the competitive effects of woody species. These herbs release nutrients as their shoots decompose rapidly during summer, which become available to other herbs. The niche of rainy season herbs is characterized by greater adaptability to deep shade, which is associated with complete development of the canopy, and the cloudiness that prevails from the end of June to the end of September.

The monsoon- and autumn-flowering shrubs play a key role in supporting pollinators as flowers are scarce during that time of the year. For example, the honey bee *Apis cerana* heavily depends on the shrubs such as *Berberis asiatica* and *Pyracantha crenulata* for pollen and nectar during autumn and early winter. Compared to trees, shrubs generally start their photosynthetic activity earlier. Thus, by shifting their growth activity to a relatively less favorable period (low temperatures and solar radiation), they minimize the competitive interaction with trees. Notably, a certain quantity of green tissue is always present in forest ecosystems.

In all forests, the peak microarthropod population follows the litterfall and moisture, that is, in the beginning of the rainy season (Singh and Singh 1992). This is the period when the peak rate of weight loss in decomposing litter occurs. The microbial population, on the other hand, peaks by the end of monsoon, i.e., in September/October, when the ground is still quite moist and days are sunny. Microarthropods which generally occupy the interface between soil and litter decompose the litter physico-chemically and biologically.

Despite differences in the elevation and physiognomy, the Central Himalayan forests occurring between 300 and 2200 m elevations show a marked phenological similarity to one another. Presumably, pre-monsoon (March–May) temperatures at lower elevations are not high enough to curtail bud burst in woody species. At higher elevations, on the other hand, the temperatures are not low enough to hasten dormancy during winters. At these latitudes, day lengths are quite long even during winters.

To sum up, these ecosystems seem to be geared to maximize the advantages of the warm and moist monsoon period (June–September): (1) The foliage of woody species is renewed yearly before the monsoon sets in. Since monsoon commences earlier in the eastern Himalayas, the species of the eastern Himalayan forests renew canopy earlier than in the western Himalayan forests (Boojh and Ramakrishnan 1981; Shukla and Ramakrishnan 1982); (2) Possibly, the maturation of foliage prior to rainy season minimizes the losses due to insects, which too are abundant during the warm and wet period. In other words, the trees flush up during pre-monsoon drought not only to maximize the advantages monsoon rain but also to avoid insect herbivory; (3) The arrival of monsoon following the concentrated leaf drop ensures a rapid decomposition of litter as the wet and warm monsoon season promotes the growth of soil microbes and other biota; (4) The advantage of warm and moist condition is so much that the seeds of annual herbs and many woody species germinate during the monsoon months, despite the disadvantage of the low light intensity at forest floor and highly competitive environment for the growth of herbs and seedlings of woody species; and (5) The rapid buildup of biomass in the forest during monsoon months ensures immobilization of nutrients released through rapid decomposition, thus protecting nutrients against the losses due to run off and soil erosion. Because of heavy rainfall, the majority of nutrients would otherwise be leached out or lost in surface runoff.

A study based on major Central Himalayan forests between 300 and 3300 m indicates that of the 29 tree species, 12 are mesophyllous (>40 cm^2; 40% of total), 11 notophyllous (20–20 cm^2; 40%), and 6 microphyllous (<20 cm^2). Mesophyllous species occur mostly in lower elevations (e.g., *S. robusta, Bauhinia* spp., *Sapium insigne, Mallotus philippensis*), and the leaves of two smaller categories occur in higher elevations. Leaf area generally decreases with decreasing temperature.

Leaf size is also affected aspect. For example, on the central Himalayas transect, the leaf area on average is greater on the warmer south (S) slope aspect (50.5 ± 18.8 cm^2) than on the cooler north (N) aspect (47.8 ± 18.4 cm^2, Singh and Negi 2018a), indicating that temperature, rather than moisture is the main constraint to leaf growth that begins soon after winters when temperature are still low. Notably, the leaf size was larger on the S aspect also for *Q. floribunda* and *Machilus duthiei* which generally occur on the N aspect. Of the four Himalayan oaks (evergreen), *Q. floribunda* is microphyllous, *Q. leucotrichophora* and *Q. semecarpifolia* notophyllous, and *Q. lanuginosa* mesophyllous.

Furthermore, the average leaf area is greater for canopy species (73.7 cm^2/leaf) than undercanopy species (40.6 cm^2/leaf) (Singh et al. 1994; Singh and Negi 2018b). In view of shade adaptation, the undercanopy species are expected to have larger leaves. It seems that shade in combination with shortages of nutrients and water promotes the selection of species with smaller size in the forest undercanopy. Between evergreen and deciduous, the average leaf size in the latter is significantly larger: 41.4 ± 14.8 and 107.9 ± 51.7 cm^2/leaf, respectively (Negi and Singh 1992). The average leaf area of all the 26 species (69.5 cm^2/leaf) of the transect is 1.63 times larger than the global average leaf size, 42.5 cm^2/leaf (based on 1389 tree species, Maynard et al. 2022). The average leaf area of the western Himalayan oaks is 23.1 cm^2, which is similar to the average for evergreen oaks of the world (19.2 cm^2 of 45 species, Sancho-Knapik et al. 2021).

Among the 11 shrub species studied in the Central Himalayan elevation transect, 9 were microphyllous, 1 was notophyllous, and 1 mesophyllous. Why leaves in forest shrubs are smaller than of the trees? Since expansion period in small leaves would be short, they are less likely to be affected by herbivory, which is likely to be more common near the ground.

Bud Recruitment, Mortality, and Shoot Growth

Whatever little is known about the phenology of buds for the Himalayan trees is due to a study of Negi (2006) on 30 tree species of Central Himalayan forests. They generally start recruiting vegetative buds about 1 month after the shoot emergence, and continue to recruit buds for months: 2–3 months in 17 species and 4–6 months in the rest of 13 species. However, in *Sapium insigne, R. arboreum,* and *P. roxburghii*, buds begin to be recruited during the monsoon, several months after the shoot formation.

The bud recruitment mostly progresses from the shoot base to the apex, except for a few basal leaves where buds are not formed. The number of buds per shoot ranges from 1.9 in *P. roxburghii* to 14.5 in *Salix alba*. Across the 30 species, the numbers of bud recruitment and leaf are positively correlated ($r = 6.975$, $p < 0.01$). Bud mortality mostly occurs during autumn and winter and is completed well before leaf fall during pre-monsoon months.

Species vary in the length of period when their bud populations are relatively stable, from 13 (*Populus ciliata*) to 30 weeks (*Prunus cerasoides* and *Engelhardtia colebrookiana*), and both extreme values are of the deciduous species. The average values of deciduous and evergreen species were similar for the duration of stable bud population (21.4 weeks for evergreens and 21.7 weeks for deciduous) and bud survival populations (42.2% and 43%, respectively, for evergreen and deciduous, Negi 2006). Across the species, bud survival ranged from 23% in *Myrica esculenta* (an undercanopy tree species) to 84% in *P. roxburghii* (Fig. 5.3). So the pine employs a conservative strategy—a small bud population accompanied by a high survival rate.

In deciduous species, the bud mortality peaks in and around March, when they produce new leaves. It seems that a sizeable proportion of buds do not break and die out. In comparison, in evergreens bud mortality occurs over a long period and generally before the production of new leaves. For example, it occurs from November to May in *Q. lanata*, August to January in *Q. floribunda*, October/November to February in *Q. leucotrichophora* and in *P. roxburghii*, and January to May in *R. arboreum* (Fig. 5.3). In other words, the deciduous species retain the bud stock intact until the time of bud break which triggers mass bud mortality. In contrast, evergreen species gradually reduce the stock of buds during winters and are left with a limited stock of buds to produce new leaves. In this respect, *S. robusta* largely follows the pattern of e[1] species.

Average leaf area per shoot in deciduous species is (587.8 ± 174.4 cm^2), more than twice as much as in evergreen species (246.7 ± 35.7 cm^2). A similar difference was found in average shoot length between deciduous (19.3 cm) and evergreen species (11.2 cm). However, the average time period of shoot elongation in the two groups was similar (18 weeks for evergreen and 20 weeks for deciduous species).

Comparison between chir pine and banj oak, the major forest species of central Himalayas shows that they have the same period of shoot elongation, March to October (but most growth occurring in the first 2 months) and similar shoot length (12–15 cm), despite differing considerably in shoot thickness, the shoot diameter being the 18.8 mm in chir pine and 3.2 mm in banj oak (in *Q. floribunda* it is 2.8 mm, the lowest among all species). Across all the broadleaved species, shoot diameter ranges between 3 and 7 mm, with an average of 5.6 mm for deciduous and 4.6 mm for evergreen species.

After the monsoon, as soil moisture depletes, shoots experience shrinkage, ranging from 7% in *Q. floribunda* to 28% in *S. insigne*. While oaks have very dense wood (~900 kg m^3), *S. insigne* is known for its light wood (~438 kg m^3). The average shrinkage of shoots was significantly more in deciduous species (18.9%) than in evergreen species shoots (14.2%), presumably because of the difference in wood density.

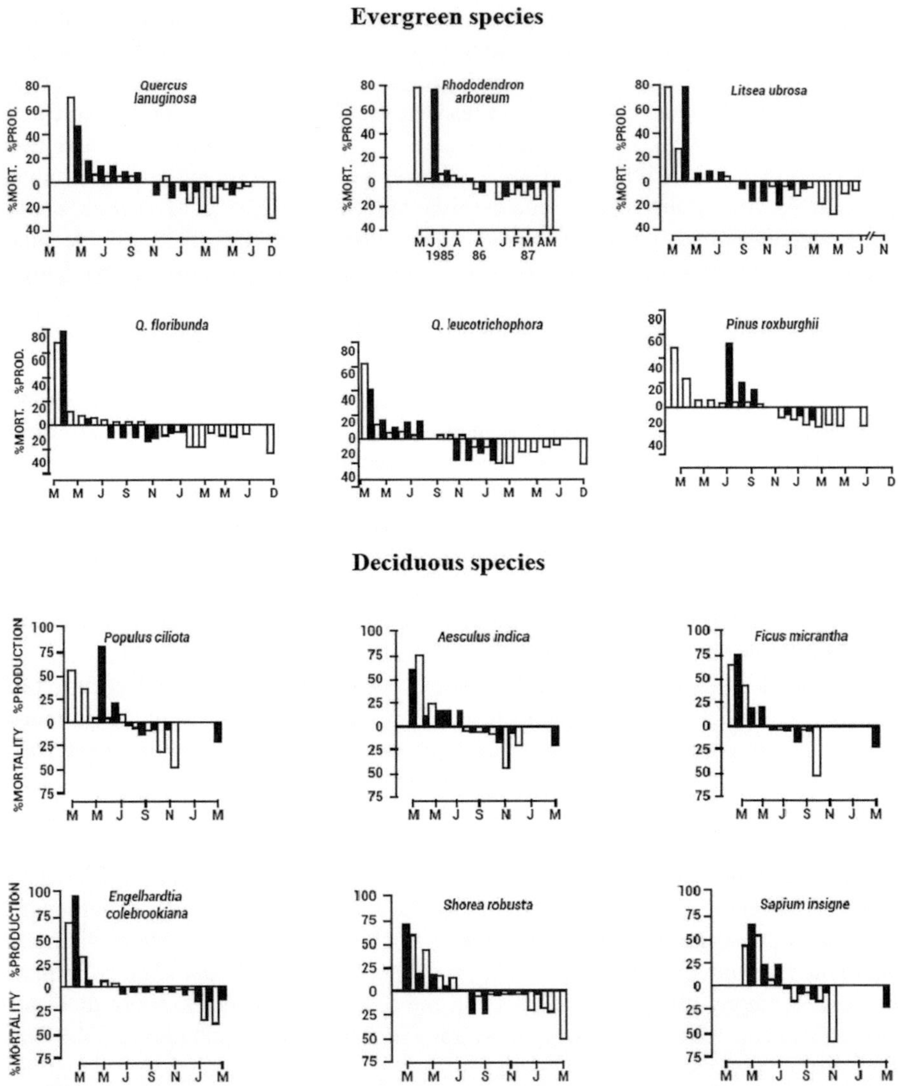

Fig. 5.3 Periodicity of the recruitment and mortality of leaves (open bars) and vegetative buds (solid bars) in the shoots of forest tree species of the Indian Central Himalayan (ICH) region (from Negi 2006). Only some representatives are included. Values are average/percentages of the total number of buds/leaves produced and dried out (mortality) in the period indicated. Months are M: March; M: May, J: July, S: September, N: November, J: January, M: March

Environment Cues for Shoot Elongation

What kind of environmental cues do the Himalayan trees require to induce shoot growth in a year? In a dry tropical climate, water stress may suppress shoot growth till the onset of rainy seasons (Frankie et al. 1974; Longman and Jenik 1974). The Himalayan species including those of the tropical belt like *S. robusta* leaf out well before (2–3 months before) the onset of monsoon. Their growth initiation is synchronized with the rising temperature in March–April, despite water stress (Zobel et al. 2001). In higher elevations, around 3000 m the temperature requirement for shoot growth may be met in the second half of pre-monsoon, May and June.

In all the dominant forest species and several others, leaf initiation triggers shedding of old leaves, thus resulting in a marked thinning of the canopy and reduction in water demand. However, in lower elevation forests, trees may begin to shed leaves long before leafing. A study in Nepal on five tree species showed that wood moisture in all five species decreased as they put forth new leaves during pre-monsoon, indicating withdrawal of stem water (Poudyal et al. 2012). In a year of severe drought (2000) when buds of *Q. lanata* were seen dying, the wood moisture was reduced to 58.6% and hydraulic conductivity to 4.01×10^{-6} kg s^{-1} MPa m^{-2} compared to 24.2×10^{-6} kg s^{-1} MPa m^{-2} in normal years (Poudyal et al. 2004). Wood moisture also serves some species to produce additional leafing. For example, the multiple leafing species, *Castanopsis indica* in Nepal, is known to put forth leaves until November (Poudyal et al. 2012).

A Comparison of Leaf Size, Leaf Longevity and Growth Form

A summary of averages of evergreen (11 species) and deciduous tree species (15 species) based on typical Central Himalayan forests is given in Fig. 5.4. Except for *P. roxburghii,* all the evergreen species are broadleaved, largely with one-year leaf life span. They include three oaks (*Q. leucotrichophora, Q. floribunda,* and *Q. lanata*), sal (*Shorea robusta*), and several undercanopy species, such as *Machilus duthiei, Ilex dipyrena,* and *Litsea umbrosa*. The deciduous category includes both winter deciduous (leafless during winters) such as *Aesculus indica, Sapium insigne, Populus ciliata,* and *Pyrus pashia,* and pre-monsoon deciduous (drought deciduous) such as *Bauhinia variegata* and *Engelhardtia colebrookiana.* Of the 26 tree species, the leaves of 14 species have entire margin (e.g., *R. arboreum, Bauhinia variegata, S. robusta*) and 12 have leaves with serrated (e.g., *Q. leucotrichophora* and *Q. lanata*) or spiny margin (e.g., *Q. floribunda* and *Ilex dipyrena*). Broadleaved species with serrated margins largely occur in cooler forests of temperate regions (in tropical rainforests they are almost absent). Evidences are there that early season photosynthesis when temperatures are low in temperate regions photosynthesis is concentrated on serrated margins (Baker-Brosh and Peet 1997; Royer and Wilf 2006). However, rhododendrons which prevail in the cold environment of treelines have leaves with entire margins. Rhododendrons have curvature diversions for

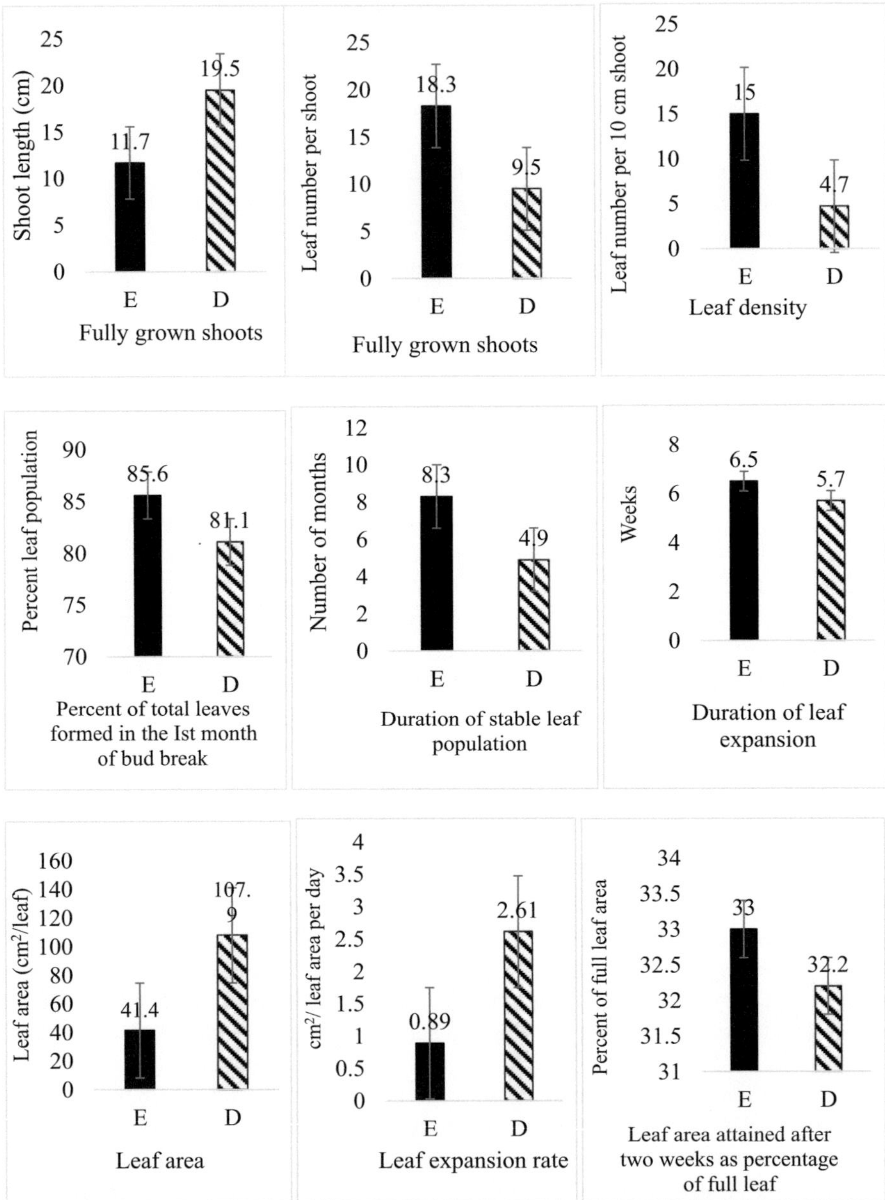

Fig. 5.4 Comparison of evergreen (e[1] species: filled histogram, also indicated as E) and deciduous species (15 species, hatched histogram also indicated as D) of Central Himalayan forests for various leaf-related characters. (Based on Negi and Singh 1992. The values are averages of each group)

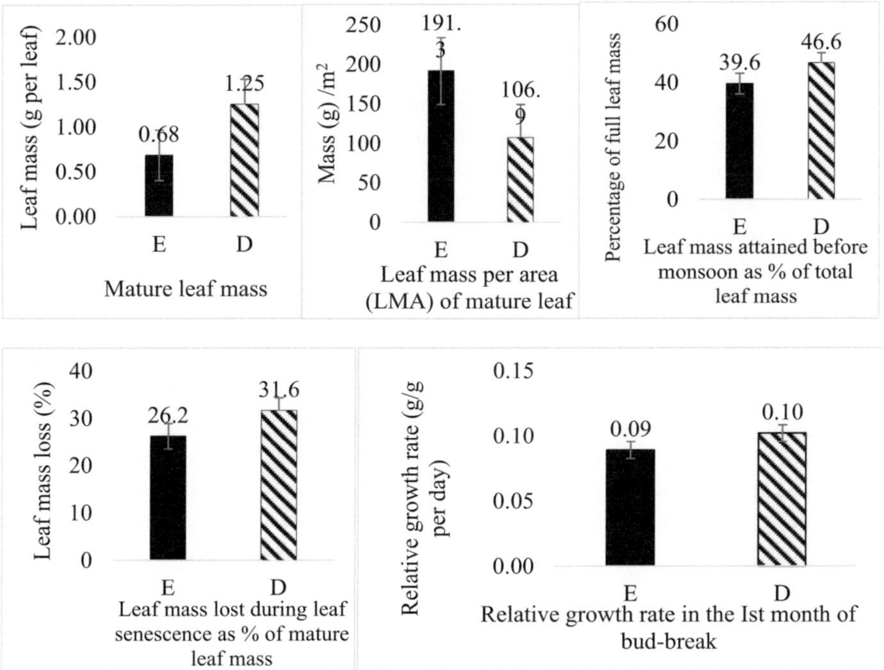

Fig. 5.4 (continued)

efficient nastic and tropic leaf movements that increase cold-hardiness and drought resistance (Wang et al. 2020).

On average, the deciduous species have longer shoots with sparser leaves than evergreen species (Negi and Singh 1992). The leafy shoots of evergreen species have a relatively longer period of stable leaf population (on average, 4 months, compared to about 2 months for deciduous species). On average, the deciduous species have larger leaves with lower specific leaf mass than evergreen species. So, construction cost per unit area is expected to be greater in evergreen species than deciduous species of the region (Givnish 1987).

At an individual leaf level following generalizations can be made (Negi and Singh 1992). The average leaf area of deciduous species (107.9 cm^2 $leaf^{-1}$) is more than twice that of evergreen species (41.4 cm^2 $leaf^{-1}$). The average of all tree species is 79.2 cm^2 $leaf^{-1}$, which is about twice that of the average tree leaf area (42.5 cm^2 $leaf^{-1}$) globally (Maynard et al. 2022). In both groups, leaves are fully expanded before the arrival of the monsoon. The average percentage of total leaf mass that the species makes before the arrival of monsoon is 39.6% for evergreen species and 46.6% for deciduous species. The period of increasing leaf mass culminates before the end of monsoon in deciduous species, while in evergreen species it continues late into the monsoon.

The average period of steady state when leaf mass is stable (the period when leaf mass is over 90% of the peak mass) in evergreen (4 months) species is roughly twice as long as for deciduous species (2 months). After the phase of a steady state mass, the leaves gradually lose mass as they senesce. This cycle of seasonal changes leading to leaf drop within a year occurs for both deciduous and e1 evergreen species of the Central Himalayas. This pattern is different from that of the evergreen species of temperate regions, which keep on accumulating mass in the first growing season (Mooney and Rundel 1979).

However, the decrease in leaf mass from a senescing leaf does not mark the terminal drop in its physiological activities. Species with senescing leaves can retain a considerable fraction of the photosynthetic capacity of mature leaves. In *Liquidambar formosana,* senescing leaves are reported to retain 42% of the net photosynthesis rate of mature leaves (Luo et al. 2017). Considerably high winter carbon assimilation (5.4–8.8 μ mol CO_2 m^{-2} s^{-1}) has been reported for subtropical evergreen broadleaved species, presumably because of climate warming (Sun et al. 2010). An understanding of the photosynthetic capacity of senescing leaves is important to evaluate the competitive success between evergreen and deciduous species in the Himalayas, particularly in view of the prolonged growing season under the influence of climate change. Banj oak leaves have been found to be active photosynthetically even a few hours before they are shed (Singh S.P. personal observation). It seems that these evergreen broadleaved species utilize the entire year to fix carbon, despite shedding leaves yearly.

The average leaf mass loss before leaf fall was significantly ($p < 0.05$) less in evergreen species (26.2%) than in deciduous species (31.6%). Of the 15 deciduous species, 9 shed leaves and remain leafless during winters (e.g., *Aesculus indica* and *Fraxinus micrantha*). The remaining 6 become leafless later on, largely in response to water stress (e.g., *Bauhinia variegata* and *Engelhardtia colebrookiana*). The drought deciduous species seem to occupy an intermediate position between winter deciduous and evergreens with about a one-year leaf life span (e^1). Interestingly, the drought deciduous species (e.g., *Bauhinia variegata, Grewia optiva,* and *Engelhardtia colebrookiana*) shed leaves and become leafless for a short period, and then leaf out during the drought itself. The leafless period seems to enable them to revive water potential enough to leaf out again while water stress is still high (Poudyal et al. 2004). The average leaf life span of drought deciduous species is about 324 days, compared to 224 days for winter deciduous species (Table 5.1).

Since leaf mass buildup in deciduous species is more synchronized with pre-monsoon, they are likely to be more adversely affected by pre-monsoon drought which could intensify due to the warming of the global climate. So, deciduous species are not only likely to be deprived of winter-time carbon assimilation, but also seem to be more exposed to the vagaries of pre-monsoon climatic conditions than evergreen species.

The sparsely distributed large leaves should enable deciduous species to project a greater leaf area outside the canopy shade. This character is particularly associated with species of open early successional environments where the canopy is multilayered (Wheat 1980; Boojh and Ramakrishna 1982). Deciduous species such as

Table 5.1 Indian Central Himalayan (ICH) tree species (600–2200 m, Kumaun Himalaya): average leaf life span (LLS) by leaf categories derived from Negi (2006) (range in days is given in parenthesis)

Leaf category	Leaf life span (LLS in days)
Evergreens of about one-year leaf longevity, with pre-monsoon leaf drop and simultaneous leafing- e^1	370.1 days ($n = 11$; 364–423)
Pre-monsoon deciduous (Dp)	324.2 days ($n = 6$; 316–351)
Winter deciduous (Dw)	224 days ($n = 7$; 179–295)

Here, e^1 also includes *S. robusta* and *Alnus nepalensis* which are treated as deciduous by some. *R. arboreum* is evergreen broadleaved with a clearly more than 1-year leaf life span (LLS), 637 days, so is not included in e^1

Populus ciliata and *Sapium insigne* are early seral species (Champion and Seth 1968), and some of them are restricted in distribution to moist sites along water courses, where bare sites are frequently formed due to erosion and landslides. Such species seem to be outcompeted by evergreen species from most of the other areas. Because the difference between favorable and non-favorable periods (winters) is not large with respect to temperature and day length in this region of intermediate latitudes, evergreen species (e^1 type) which produce new leaves yearly as well as use the entire year for carbon assimilation are favored (Table 5.1).

Treeline Phenology: How It Differs from That of Forests?

Phenological growing period is of critical importance with regard to treeline species (Körner et al. 2023). For this section, we heavily depend on a five-year study of five species of the Tungnath site of ICH (Singh and Negi 2023). Located at 3334 m elevation, the Tungnath site is a wet site (>4000 mm). The growing season average temperature (2016–2020) at Tungnath treeline was 9.6 °C (Joshi et al. 2023), which is distinctly warmer at present than 6.7 ± 0.8 °C growing season mean temperature with which treelines are associated worldwide (Körner and Paulsen 2004).

All major tree growth forms that occur in the Himalayas are found in the Tungnath treeline: broadleaved deciduous (e.g., *B. utilis*), broadleaved evergreen (e.g., *Q. semecarpifolia*), and conifers (e.g., *Abies spectabilis*) with multi-year leaves. *Q. semecarpifolia* (kharsu oak) which is e^1 type, is unable to retain the character of concentrated leaf drop and simultaneous leafing. In treeline, its leafing is spread over 3 months from May to July, and leaf drop over the period of 4 months from December to early March. In other treeline species of ICH also, leafing and leaf drop are separated in time (Fig. 5.5). In *A. spectabilis* and *R. arboreum*, leaf drop is an event of the autumn to winter period, while in the birch (*B. utilis*) leaf fall starts a bit earlier and renders the trees leafless in October. Interestingly, in

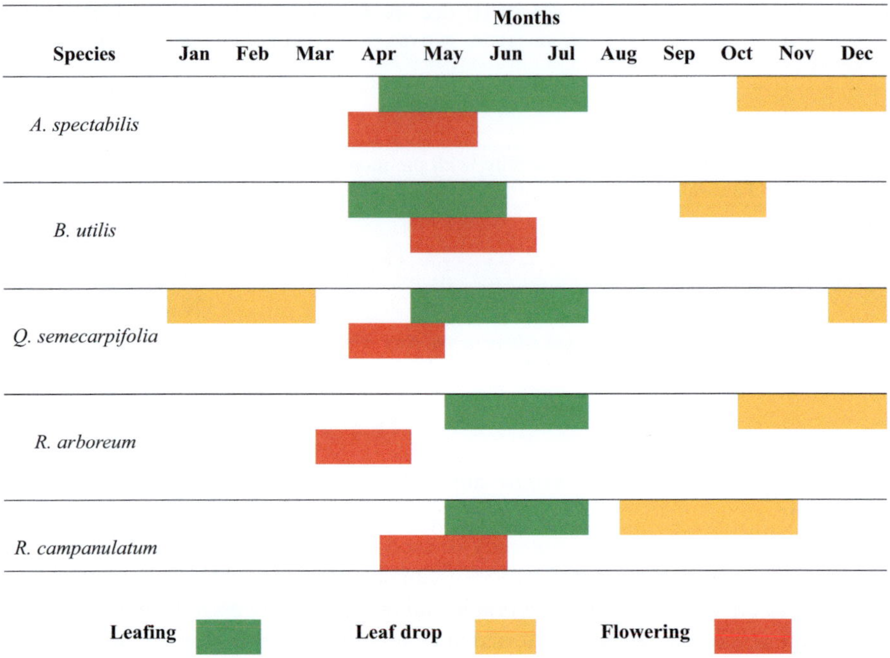

Fig. 5.5 Phenological calendar of various phenophases for the five treeline species (mean across 2016–2017, 2017–2018, 2018–2019, 2019–2020, and 2020–2021 phenological cycles). (Derived from Singh and Negi 2023)

R. campanulatum which has leaf life span of about 2 years leaf drop begins in the middle of monsoon rain (in August).

How long leaf populations is stable in the treelines? In fir (*A. spectabilis*), the population of densely packed narrow leaves (148.3 leaves per first-year shoot of 10.2 cm lengths) remains relatively stable throughout the first year. In contrast, leaf population is stable for only two monsoon months (July–August) in the deciduous birch, and, for 4 months (August–November) in evergreen (e^1) kharsu oak (*Q. semecarpifolia*). In the two rhododendrons, leaf populations are stable for 8–10 months.

Leaf size varies widely from 0.33 cm^2 per leaf (microphyllous) in the fir to 54.9 cm^2 in *R. campanulatum* (mesophyllous). The remaining three species are notophyllous. Leaf size is much larger (macrophyllous) in another tree rhododendron, *R. hodgsonii* (e.g., in Sikkim 314 cm^2 per leaf; Basnett and Devy 2021) occurring in Arunachal Pradesh, Sikkim, Bhutan, and Nepal.

Because of the year-to-year variation in temperature, the growing period of a species differs considerably. During the 5 years of study, Singh and Negi (2023) found that in the warmer year, 2017 the mean early growing period temperature was 8.33 °C, while in 2020, a cold year, it was lower 6.59 °C. Because of this, the growth advanced by 2–3 weeks in the warm year and delayed by a similar margin in the cold year (Singh and Negi 2023).

The month in which treeline species attained peak leaf mass was August in *Q. semecarpifolia*, October in *R. campanulatum,* and September in the other three species.

In the Tungnath treeline, the leaves of fir were microphyllous (0.33 cm^2 per leaf), those of kharsu oak (26.9 cm^2) and birch (24.2 cm^2) notophyllous, and the other two rhododendrons mesophyllous (41.2–54.8 cm^2). Interestingly, the leaf size of deciduous species, birch, was hardly different from that of evergreen oaks. To survive in a treeline environment, leaves need structural enforcement, and in such an environment, persistence, rather than an increase in net carbon assimilation, is important. In ICH treelines, leaf mass per area (LMA) ranges from 128 (*B. utilis*) to 251.5 g m^{-2} (*A. spectabilis*). An LMA of 128 g m^{-2} for a deciduous leaf is much higher than those of tropical wet forests of Silchar (mostly <70 g m^{-2}). The krummholz rhododendron, *R. campanulatum* with a leaf life span of 2 years has tough textured leaves, with LMA 213.5 g m^{-2}, which is far higher than those of any other broadleaf forest species of the region, and of *R. campanulatum* individuals occurring below treelines.

Interestingly, all evergreen species including multi-year evergreens *A. spectabilis* commonly lose some mass during winters, when conditions are stressful. Generally, tree leaf size decreases with elevation. However, as referred to earlier, species with macrophylls are found in all elevational bands in the Himalayas. For example, species with large leaves occur in tropical wet forest of Silchar (*Dipterocarpus turbinatus* 313 cm^2 $leaf^{-1}$), in mid-elevations *Quercus lamellosa* (16–45 × 6–15 cm) around 2000 m in eastern Himalayas (Sikkim, Bhutan, and eastern Nepal) and *R. hodgsonii* (17–24 × 6.5–10 cm, 314 cm^2 $leaf^{-1}$) (3500–3745 m). All of them, however, occur in very moist sites. *R. hodgsonii*, in fact, occurs almost in waterlogged conditions (Miehe et al. 2015). It seems that high moisture overrides the limiting effect of low temperature on leaf size in treelines. The leaves of *R. hodgsonii* are much thicker (0.74 mm) than those of the other 9 rhododendrons (0.24–0.60 mm) that were studied in Sikkim by Basnett and Devy (2021), but less dense, and of intermediate LMA (155.6 g m^{-2}, compared to 122.1–213.8 g m^{-2} in other 9 species). Large and thick leaves are expected to have a thick boundary layer and reduced exchanges with the atmosphere, which would keep leaves warm by insulating from the atmosphere characterized by low temperature, strong radiation, and strong wind (Kudo 1996).

Overlap Between Leafing and Flowering

There was a broad overlap in the duration of leafing and flowering in the birch and silver fir, while they were separate in *R. arboreum,* and the overlap between them was limited in *R. campanulatum* and the oak (*Q. semecarpifolia*). In the rhododendrons and oak, flowering occurred earlier at relatively lower temperatures, and leafing time coincided with warmer temperatures. The non-overlap between flowering and leafing may enable species to optimally allocate resources to reproduction and vegetative growth (Körner 2012). The separation between the two phenophases (leafing and flowering) allowed rhododendrons and the oak to rapidly produce leaves as temperatures were higher during June–July when they produced leaves.

Eastern Himalayan Forests

In subtropical broadleaved forests of Meghalaya (IEH), tree species initiate leafing relatively early in December and continue to produce leaves (up to March) (Boojh and Ramakrishnan 1981; Shukla and Ramakrishnan 1982). Deciduous species (18 species) and evergreen species (44 species) together shed leaves from winter (December to February) to early pre-monsoon months (March–April). The main features of their phenology are as follows: (i) Leafing and leaf fall periods are affected by the prevalence of multiple leafing species, which are 82% of the total tree species in the Meghalaya forests, compared to 40% in ICH forests (Kumaun). In these subtropical broadleaved forests of Meghalaya, average leaf longevity is 20 months (average 12 species), compared to 13 months for ICH forests (Singh and Singh 1992). (ii) Flowering in leafless trees (majority in March) was quite common in the subtropical forests of the eastern Himalayas (Shukla and Ramakrishnan 1982). (iii) A majority of species that bear fruits during monsoon had fleshy fruits.

In another study on the tree phenology of subtropical mixed forest conducted in Manipur (24°45′–24°51′ N latitude and 93°48′–93°55′ E longitude, between 865 and 1785 m, annual precipitation 1390 mm), Kikim and Yadava (2001) adopted a community-level phenology approach. Here, the rainfall is monsoonal, about 82% of the annual rainfall of 1511 mm occurring from April to October, and winter is the driest time in the year. In this study, leaf fall, leafing, flowering, and fruiting of 32 important tree species were analyzed (19 canopy species (Fig. 5.6) and 13 undercanopy species), of which important species were *Lithocarpus fenestrata, L. dealbata, Schima wallichii, Q. serrata, Castanopsis tribuloides, Engelhardtia spicata,* and *Albizzia chinensis*. Of the 32 species, 20 were evergreen and 12 deciduous. While evergreen species were equally divided between canopy and undercanopy layers, 9 out of 12 deciduous species belonged to the canopy layer. The forest was diverse, with the IVI of each species remaining below 75. At the forest level, leaf drop starts from November and continues during much of the subsequent dry period, with a peak in February (in terms of number of species) (Fig. 5.6). Leafing follows it, and peaks in April (28 species in leaf flushing). A secondary peak of leafing occurs in August–September (Fig. 5.6), as multiple leafing is quite common in canopy species (e.g., *B. alnoides, Castanopsis tribuloides, Lithocarpus* spp., *Q. serrata*). Leaf flushing before the monsoon to maximize the advantage of warm and moist conditions seems to be a common feature in much of the Himalayas under the impact of monsoon. A prominent secondary peak of leafing

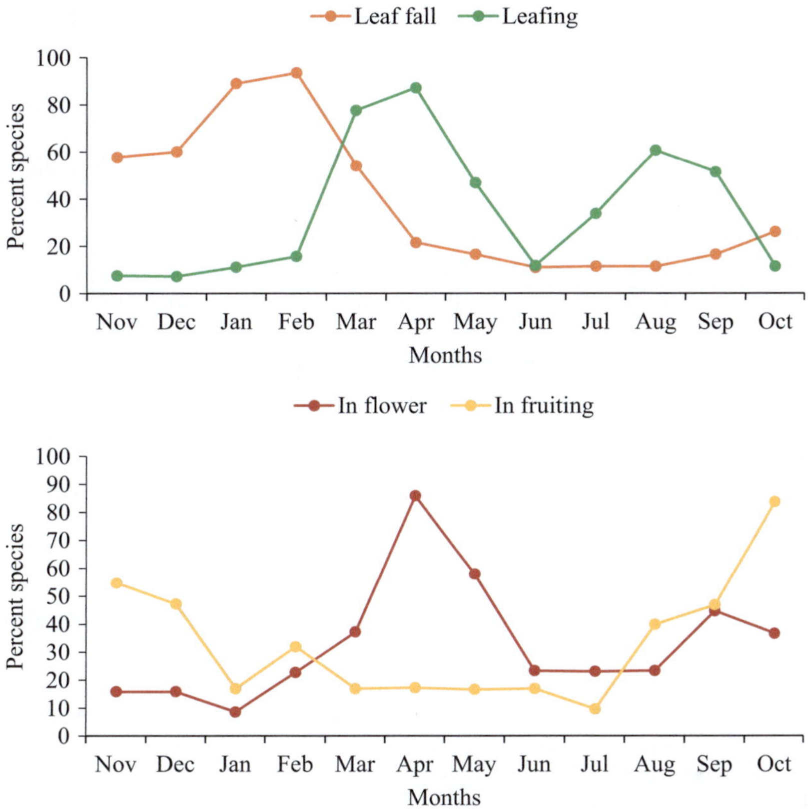

Fig. 5.6 Percent canopy species (19 canopy species) in different phenophases in subtropical mixed forests of Manipur (24°45′–24°51′ N latitude and 93°48′–93°55′ E longitude, between 865 and 1785 m, annual precipitation 1390 mm). Secondary peak is largely due to multiple activities (Derived from Kikim and Yadava 2001)

during monsoon (Kikim and Yadava 2001) is associated with subtropical eastern Himalayan forests.

In subtropical mixed forests of the eastern Himalayas, species vary considerably in leaf shedding time, and hence, the process of leaf exchange at the community level is not apparent like ICH forests, where dominant species are characterized by concentrated leaf drop and simultaneous leafing. In these subtropical forests, flowering also shows two species peaks, one in April and the other, smaller one in September. It appears that several species produce flowers while leafing out.

The seasonal pattern of flowering is affected by the presence of species with extended flowering activity. Among the canopy species, *Alnus nepalensis, Castanopsis tribuloides,* and *Schima wallichii* flower for 10 weeks, while undercanopy species *Eurya japonica, Saurauia punduana* bear flowers for 28 and 38 weeks, respectively. Kikim and Yadava (2001) observed that while evergreen species mostly flower after completing leaf production, the deciduous species vary considerably; (i) they may flower before leafing (e.g., *Bauhinia purpurea* and *Engelhardtia spicata*); (ii) flower simultaneously with leafing (e.g., *Emblica officinalis*); (iii) flower soon after leafing (included many species, e.g., *Albizzia chinensis, Celtis australis, Quercus serrata*); and (iv) flower quite some time after leafing (e.g., *B. alnoides, Q. griffithii*).

Fruiting shows one peak in October, soon after the end of the wet period. The species takes a distinctly shorter time in fruit maturation (1–5 months) than oak forests of Central Himalayas (up to 22 months). Generally, seeds lie on the ground during much of the dry period of winter and pre-monsoon. Seeds of many species have winter dormancy, broken by the arrival of the warm and humid monsoon season.

In brief, in comparison to the community-level phenology of typical ICH forests, the subtropical broadleaved forests of the Indian Eastern Himalayan states have a greater tendency for having secondary leafing and flowering peaks during the wet part of the year. Furthermore, leaf shedding occurs over a long period, starting from dry winter which could be partly related to higher species richness of both deciduous and evergreen species in subtropical mixed forests.

Tropical Wet Broadleaf Forest, Silchar, Indian Eastern Himalayas (IEH)

In the tropical wet forest of Silchar, Assam, the average leaf size based on 19 important species (13 evergreen and 6 deciduous species) is 93.6 cm^2, which is significantly larger than the average leaf size of ICH forest trees (65 cm^2 per leaf; mean of 30 important species). In this tropical wet forest, leaf size ranges from 13 cm^2 in *Syzygium syzygioides* to 317 cm^2 in *Dipterocarpus turbinatus*. The same in Central Himalayan forests is 0.29 cm^2 in *Abies spectabilis* to 181.3 cm^2 in *S. robusta* (Singh et al. 1994). With regard to Raunkiaer's classification (1934) of leaf size, in tropical wet forest, 1 species is macrophyllous (*Dipterocarpus turbinatus*), 14 mesophyllous (45–182.25 cm^2), 2 notophyllous (20.25–45 cm^2), and 2 microphyllous (2.25–20.25 cm^2). In the central Himalayas, the Raunkiaer's leaf size distribution is as follows: 40% mesophyllous, 40% notophyllous, and 20% microphyllous.

In 8 of the 14 evergreen species of the tropical wet forest, leaf life span (LLS) is less than 300 days (205–296 days). Despite the short leaf life span of individual leaves, these evergreen species never become leafless as leaves are produced several times a year. The evergreen species with clearly more LLS than 1 year are *Dipterocarpus turbinatus* (509 days), *Palaquium polyanthum* (504 days), and *Podocarpus neriifolius* (489 days). The average leaf mass per unit area (LMA) is 77.4 g m^{-2} for evergreen species and 48.4 g $m^{-2}$for deciduous species (Fig. 5.7). These values are much lower than those of ICH forest species (191.3 g m^{-2} and 106.9 g m^{-2}, respectively, for evergreen and deciduous species) (Table 5.2). The average LMA of tropical wet forest of Silchar (67.7 g m^{-2}) is in fact, lower than that of tropical rain forest (73 g m^{-2}) and graminaceous species (72 g m^{-2}) at global level (Table 5.2).

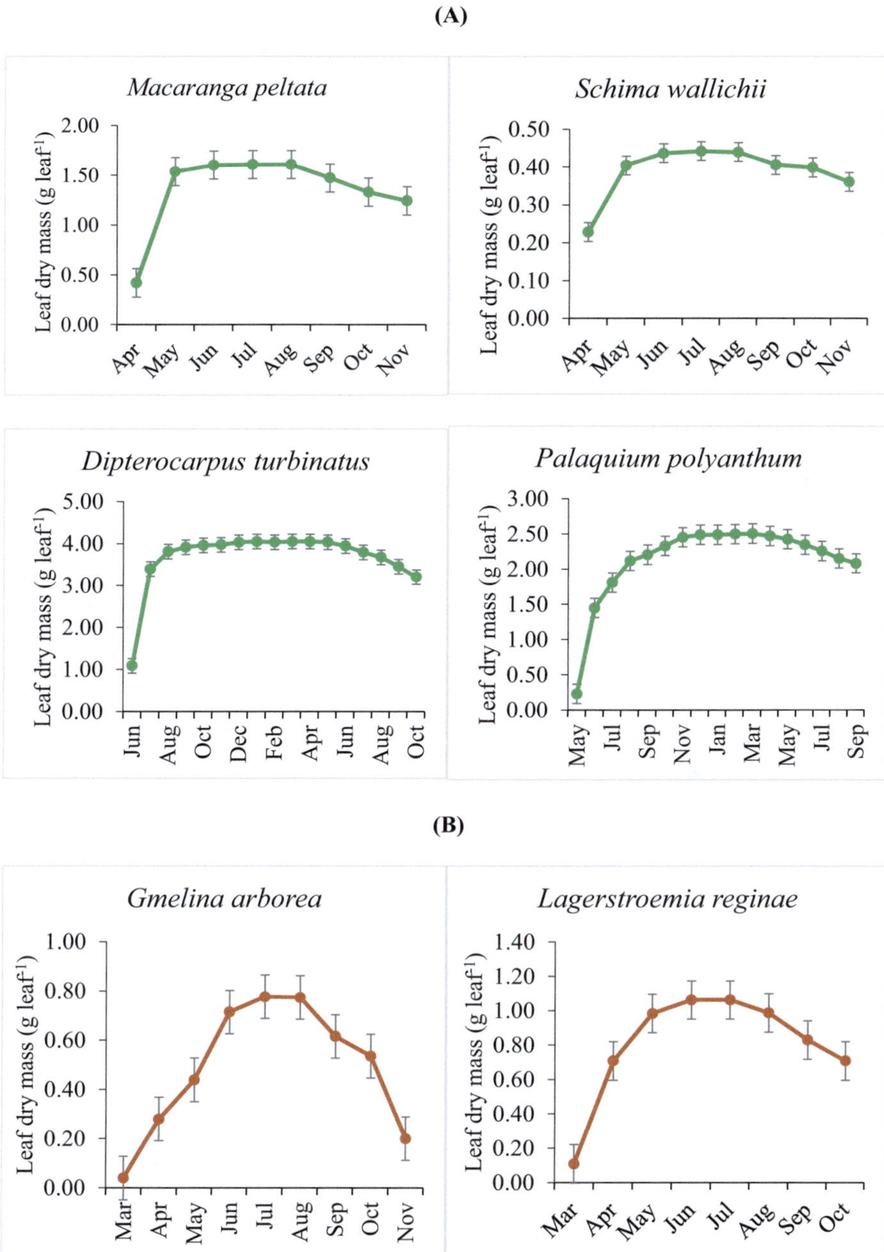

Fig. 5.7 Monthly changes in leaf dry mass in a year of important representative species of the tropical wet forest of Silchar, Assam. (**a**) Evergreen species, and (**b**) deciduous species. (Adapted from Athokpam et al. 2014)

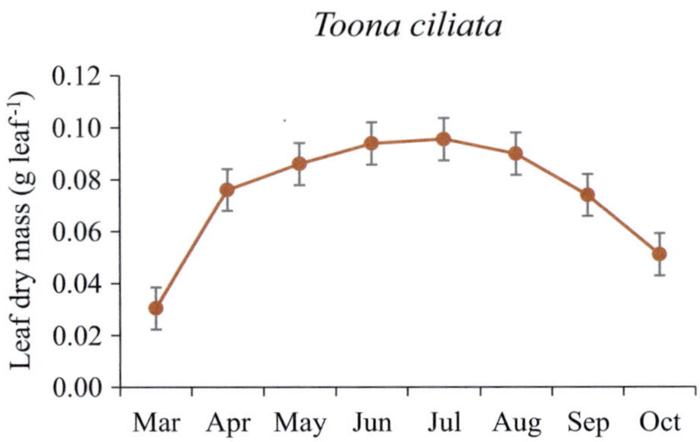

Fig. 5.7 (continued)

Table 5.2 Comparison of leaf mass per unit area (LMA; g m^{-2}) of the Himalayan regions (Singh et al. 1994; Negi 2006; Athokpam and Garkoti 2013) with various global groups

Group	Leaf mass area (LMA) (g m^{-2})
Himalayan	
Broadleaved Evergreen	
Evergreen Central Himalayan forest	191.3
Deciduous Central Himalayan forest	106.9
Evergreen conifer forests	
Evergreen tropical wet forests, Silchar, Assam	77.4
Deciduous tropical wet forests, Silchar, Assam	48.4
Global	
Evergreen gymnosperms	227
Evergreen shrubs	161
Evergreen broadleaved trees	106
Deciduous trees	75
Deciduous shrubs	73
Herbs	60
Gramineae	72
Tropical rainforests	73
Tropical deciduous forests	83
Temperate forests	81
Woodlands	129
Shrublands	166
Deserts	186

Source: Poorter et al. (2009)

The LMA values of Central Himalayan trees are more than those of temperate trees (81 g m^{-2}) and deciduous trees (75 g m^{-2}, global average) of the world. In general, a high leaf density and high leaf thickness contribute to higher LMA values.

Growth, light, temperature, submergence, nutrient, and water availability affect LMA variously (Box 5.1).

Box 5.1: Factor Affecting Leaf Mass per Area (LMA)

Since LMA is a key driver of leaf and canopy reflectance, we need to understand the factors which influence variation in LMA. It depends on two leaf components, leaf density (LD) and leaf thickness or leaf volume-to-area ratio (LVA). For both evergreen and deciduous species, variations in LVA are positively dependent on mesophyll volume per unit area (VA or thickness). However, with regard to evergreen leaves positive relationship of LVA with VA of epidermis, vascular and sclerenchyma tissues, and air spaces is also found to affect LMA (de la Riva et al. 2016). The higher LMA in evergreens is largely due to a greater leaf thickness, caused by a greater volume of mesophyll and air spaces. However, within each group (deciduous and evergreen), the higher LMA is due to higher leaf density, which is caused by higher volumetric fractions of mesophyll and lower fractions of air species and epidermis (Villar et al. 2013). The leaf carbon concentration is positively related to mesophyll VA in deciduous species, and with VA of vascular and sclerenchyma tissues in evergreens. LMA and LD increased with increasing water stress.

In Silchar forests, evergreen (96.2 cm^2) and deciduous species (91.5 cm^2) have similar mean leaf sizes, largely because some of the evergreen species like *Dipterocarpus turbinatus* (317 cm^2, a macrophyll), *Palaquium polyanthum* (169 cm^2), and *Gmelina arboreum* (177 cm^2) have quite large leaves.

Even some deciduous species in the central Himalayas retain leaves during much of the winter and have a leaf life span of around 300 days. Because of these, the LMA values for deciduous species in the Central Himalayas are higher than the global average (Poorter et al. 2009). Several species with high LMA are known to have relatively large structural investment, which reduces herbivory and physical and environmental stresses (Givnish 1988). Such leaves can last longer. Deciduous species seem to invest more in symplastic components, which allows them to resorb a larger fraction of mass.

However, in some other phenological details the Silchar forest species resemble the species of the Central Himalayas. For example, in a majority of species leaf expansion was completed well before the onset of monsoon, i.e., in June or earlier in both regions.

Evergreen species in tropical wet forests of Silchar lose mass and senesce during the winter of the year in which they are initiated. However, their leaf mass loss is less (18.3% of mature leaf mass) than in deciduous species (30.3% of mature leaf mass). In comparison, evergreen species of temperate forests continue to accumulate mass throughout the first growing period (Mooney and Rundel 1979). Average rate of leaf expansion decreased with elevation, the value being much higher (2.65 cm^2/day) in Silchar than in ICH forests (0.91 cm^2/day) and in the timberlines (0.73 cm^2/day).

In brief, relatively low LMA regardless of growth form and rapid leaf expansion clearly separate tropical wet forest species of IEH from those of the ICH.

At full leaf expansion, species generally attain 60% of the total leaf dry mass. However, they also vary considerably, the percentage being close to 98% for *Macaranga peltata* and 47.9% of the gymnosperm, *Podocarpus neriifolius*.

In tropical wet forests of Silchar, some of the deciduous species attain full leaf mass by the beginning of early part of monsoon, and begin to lose mass (senescence) before the end of monsoon rains (Fig. 5.7). This should be seen in view of the fact that in Silchar rains begin in March, several months earlier than in the areas of Central Himalayan forests (Athokpam and Garkoti 2013).

In Silchar, deciduous species become leafless soon after autumn, while some of the Central Himalayan deciduous species retain leaves during most of the winter season.

Inner Dry Valley Forests

In the dry inner Himalayan ranges, conifers (e.g., *C. deodara, P. wallichiana,* and *Picea smithiana*) form extensive forests. They, however, also occur in outer ranges, with the full thrust of monsoon rain. Little is known about what extent their phenologies are influenced by the lack of the full thrust of the monsoon. As described in Troup (1921), these conifers leaf out during pre-monsoon (March–June), and shed leaves by the end of June, except *C. deodara* which sheds its needles in autumn after a considerable gap.

In *P. wallichiana* (blue pine), new needles appear during March–April (Fig. 5.8) and attain full length by August–September. Leafing seems to induce a rapid leaf shedding, and trees give a somewhat bare appearance during May–June (Troup 1921) as do *P. roxburghii* trees during March–April. The bare appearance (Troup 1921) during the pre-monsoon occurs because old leaves are shed, and new leaves are only partially developed. Similar to *P. roxburghii,* the leaf life span (LLS) in *P. wallichiana* is about 12–15 months. Despite having the contrasting associations with monsoon (*P. roxburghii* is distributed only in monsoon-dominated areas while *P. wallichiana* largely occurs in areas away from monsoon thrust), these two pines show a similar "deciduous tendency." *P. wallichiana* also has several early successional traits: an early age of (10–12 years of age) of seed production, high frequency of seed production (5 in 13 years), and ability to colonize bare sites. *P. roxburghii* is also an effective colonizer. There is a need to investigate how blue pine responds phenologically to differences in the conditions of inner and outer ranges.

P. gerardiana has similar phenology; except that leafing and leaf shedding are delayed by a month or so compared to *P. wallichiana* possibly because it occupies higher elevations. The timing of leafing and leaf drop in *Picea smithiana* is similar to that of *P. wallichiana* (Fig. 5.8). *A. pindrow* too has similar leaf phenology except that the duration of leafing and leaf fall is longer. *C. deodara* puts forth new leaves

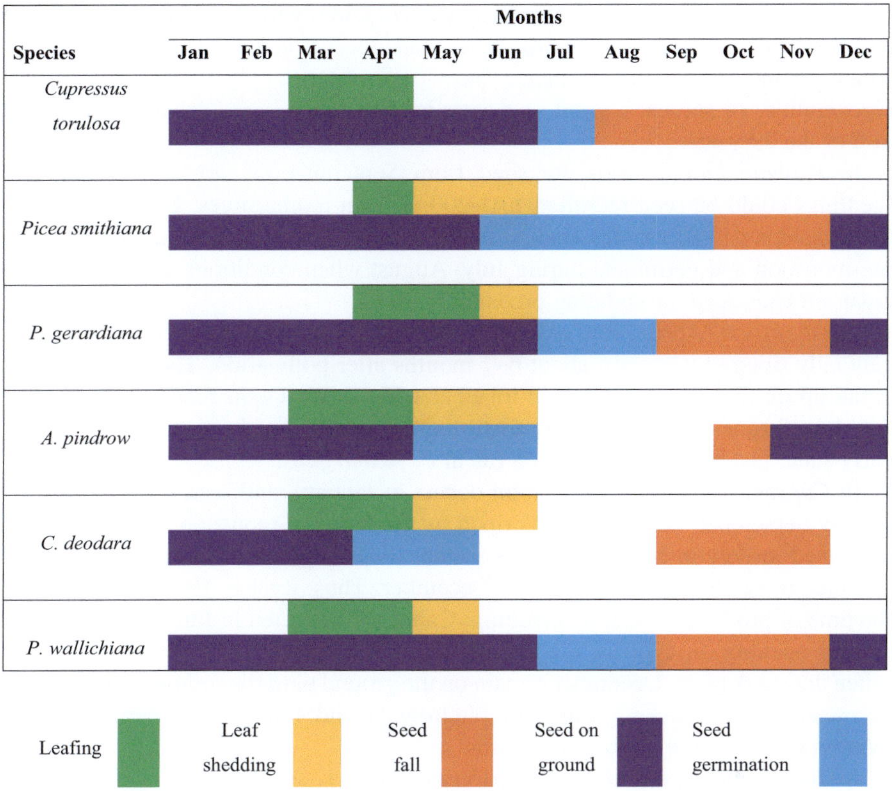

Fig. 5.8 Pattern of leafing, leaf shedding, seed fall, seed on ground, and germination in some evergreen conifer species which occur extensively in dry inner valleys. (Data developed from information given in Troup 1921)

during March–April, but sheds leaves a few months later during autumn when its seed cones are mature (Fig. 5.8).

In *P. gerardiana,* pollination takes place by July when female cones are full-sized, and seeds ripen in September–October of the year it is pollinated.

In deodar (*C. deodara*) cones ripen from September to November, about 12.5–13.5 months after the first appearance. Seeds remain on the ground for about 6 months (Fig. 5.8) and germinate in the following spring, March–April, earlier than other conifers, and well before the monsoon. Snow-melt water is important for seedling establishment, particularly where the monsoon is weak. Though seed fertility is high (72–98%), the oily seeds of deodar lose vitality soon, so cones are collected immediately before they open. Seeds are winged and disperse effectively. Good seed years in deodar are infrequent.

In spruce, pollination occurs from mid-April to mid-May, and by July, seed cones attain full size, while still green. The seed cones ripen in October–November, 6–7 months after pollination, and the seeds begin to germinate from the end of May to

September. Thus, seed germination in spruce coincides with the monsoon as does in the case of *Q. floribunda*. However, spruce also occurs abundantly in inner valleys when the monsoon is weak, hence snow-melt water supply is important for its regeneration. In spruce, a good seed year is uncommon, good seeding being only in 6 of the 29 years.

In *P. wallichiana,* seeds are shed from September to early November, and seedlings could be seen recruited 170–200 m from mother trees. Seeds remain on the ground for 8–10 months (longer than for oaks) throughout winter and ensuing pre-monsoon and germinate during July–August when conditions are moist. Snow cover protects seeds from being predated by birds.

Pre-monsoon (April–May) is the time of pollination in pindrow fir, and its cones generally ripen by October, about 6–7 months after pollination. The ripened cones break up on the tree and then fall together with scales. Good seed years are more frequent (7 times in 20 years) in this fir than in spruce. Seeds germinate in May to early June, just before the monsoon (as in *P. roxburghii*).

In *Cupressus torulosa.* Pollination occurs in January, and seed cones are of full size by September–October. Seeds ripen by the next pre-monsoon in April and are contained loosely in cones. As cones soften, seeds begin to be shed in August and continue to be shed until November–December. Thus, it takes 19–22 months from the time of production to seed shedding. Cones are collected in June and kept in the sun for opening. In cypress, the first seeding takes place at 15 years of age. Seeds fallen during August–December remain on the ground until the following July, when germination commences. Thus, like pindrow fir and spruce, seed germination in cypress is linked with monsoon months.

In *Juniperus macropoda* too, seed germination is synchronized with monsoon months. Seeds are shed after the monsoon, from the end of September to November. Subsequently, seeds remain on the ground from the end of winter to summer.

The conifers associated with inner ranges are divisible into two groups, one in which seed germination is synchronized with typical monsoon months, July–September (mostly July–August), and the other in which seeds germinate before the monsoon. While cypress, juniper, pines, and spruce belong to the first category, deodar and pindrow fir represent the second category. How these meet their water requirement for seed germination when they occur in areas shielded by high ranges for monsoon is not well documented.

All the conifers shed seeds during autumn, except cypress in which seed fall extends from August to December. The fallen seeds remain on the ground during winter and pre-monsoon, varying in period from three in deodar to 7 months in cypress.

To conclude, there is a need to investigate phenology of these conifers, which are largely associated with dry inner Himalayan ranges, where monsoon is weak. Except for pines, they have multi-year leaves, and able to develop some of the most handsome forests in Himalayan despite the dry environment. They have orthodox seeds, which remain lying on ground for several months, and tend to wait for monsoon or melt water to germinate.

Periodicity of Primary Production

Data collected from the carbon tower indicated that September/October was the time of lowest atmospheric carbon concentration. Monsoon season is the time when most of the biomass accumulation occurs. For example, most of the species achieve much of the leaf expansion before the monsoon commences, but leaves gain in dry matter largely during the monsoon.

A study based on the monthly harvest of seedlings of *Q. floribunda* indicates that of the annual net production, 84% occurs from July to October (Singh and Singh 1992). About 75% of above aboveground increment and 80% of the foliage increment occur from July to September. The period from July to October also accounts for more than half (57%) of annual root production, with a sizeable root production (24%) occurring from November to December. Though pre-monsoon is the time of leaf and shoot production, monsoon is the time when most of mass accumulation including of leaves occur.

Timing of Atmospheric CO_2 Concentration Over Canopy of Foothill Forests

Based on two eddy covariance (EC) flux towers, Watham et al. (2020) provide variations in atmospheric CO_2 concentration over the canopies of two deciduous forests of dry and moist sites located in the Western Himalayan foothills. The major species of moist forests were *S. robusta, Terminalia tomentosa,* and *Syzygium cumini,* and those of dry forests were *Dalbergia sissoo, Acacia catechu,* and *Holoptelea integrifolia.* At these sites, the values of mean annual precipitation (1178–1106 mm) were similar. In both CO_2 concentrations were similar and peaked in July (398.4–415.6 ppm). However, the lowest CO_2 concentration occurred in May (363.8 ppm) in the moist forest and in September in the dry forest (359.7 ppm).

Diurnal changes show that at both sites the maximum CO_2 concentration (445.3 ppm) was recorded at 06.00 am in the morning and the minimum (434.3 ppm) at 3.30 pm. In a dry deciduous forest, NDVI peaked (MODIS product) in September, which is the time when CO_2 concentrations were the lowest. In the dry forest, the NDVI was the lowest in March when the canopy of many tree species was the thinnest.

Climate Change Impact

Information on climate change impact on forest phenology is limited to a few studies, pertaining to changes in the flowering of rhododendrons and differences in leafing and leaf shedding, and growth period of major forest species.

Using the remote sensing technique, Khanduri et al. (2008) showed that in temperate latitudes of the Himalayas growth period of forest species (based on

650 species) has increased. The average decadal advancement in the growth period was estimated at 1.9 days, and the average delay of autumn events at 1.4 days.

A study on the effect of the recent decades of climate warming on growing season length in Himalayan evergreen species with about one-year leaf life span (leaf-exchanging type) showed that while leafing advanced in them, it did not result in an increased growing season length (Negi et al. 2021). The study included seven major evergreen tree species occurring between 300 and 2200 m, namely *S. robusta, P. roxburghii, Myrica esculenta, Q. leucotrichophora, Q. floribunda, R. arboreum,* and *Machilus duthiei*. The phenological measurements were made two times at an interval of about three decades (1985–1987 and 2014–2016) during which the mean atmospheric temperature had increased by 1.10 °C and annual precipitation had decreased by 432 mm. The average rate of advancement in the bud break time of the seven species was 12 days, i.e., 0.2 day/year ($p < 0.01$). Since bud break induced leaf drop, leaf fall was also advanced (at the rate of ~0.4 day/year), and the decrease in average leaf life span (from 410 ± 30.74 to 400 ± 29.16) was not significant. On average, the leaf drop during about three decades advanced by 12 days, the advancement being of only 4 days in the tropical sal and of 16–17 days in banj oak and rhododendron, occurring in a warm temperate area. Furthermore, the climate warming did not increase the rate of leaf formation, the average percentage of total leaf population attained in the first month of leafing being 84 ± 3.14 and 87 ± 1.28, respectively, in 1985–87 and 2014–16. In contrast to these leaf-exchanging species, in deciduous tree species leafing not only advanced but leaf life span also increased due to climatic warming.

On the basis of real-time field observations, generalized additive modeling, and herbarium records, Gaira et al. (2014) showed that flowering in *Rhododendron* species advanced because of the increase in maximum winter and post-monsoon and annual mean maximum temperatures. An elaborate study on flowering phenology of *R. arboreum* was carried out at two sites one in Yunnan, China (2150–2850 m) and the other in Nepal (2976–3597 m). In both observations were made in phenological monitoring plots along an elevational transects.The experiment showed that winter temperature was the main determinant of flowering events, both initial flowering and peak flowering. Soil moisture was the next important variable with regard to the initiation of flowering (Ranjitkar et al. 2013). At both sites, trees of lower elevations flowered earlier than those of higher elevations, and flowering period decreased with altitude. In both sites, flowering synchrony among the individual trees was high, which is expected to enhance cross-pollination and genetic diversity. The high genetic diversity, in turn, may enhance the capacity of *R. arboreum* to adapt to stressful conditions of high elevations. Interestingly, unlike many species of cold regions, *R. arboreum* responds to heat throughout the winter, and it has no chilling requirement (Ranjitkar et al. 2013). Soil moisture promotes the initiation of flowering by increasing nutrient uptake (see also Nord and Lynch 2009). Other soil factors, such as litter cover, pH, and organic matter, can cause only minor deviations. In brief, global warming is likely to advance flowering and adaptability of the *R. arboreum*, which in turn may result in its range expansion. Tewari et al. (2016), on the other hand, found that early flowering in *R. arboreum* was triggered by higher tree Ψ (>−0.05 MPa), rather than temperature. According to them, floral bud size increased with a rise in soil Ψ, not by temperature.

Conclusions

In brief, the three principal forest divisions in the Himalayas, namely the moist tropical forests of lower elevations, mid- and high-elevation forests of moist outer ranges (largely studied in Uttarakhand), and those of dry inner valleys, differ in tree forms which dominate them. In all of them, species predominantly produce leaves during the dry period, preceding monsoon rain and or the time of snow-melt. However, leaf drop periodicity varies considerably among the three divisions. In mid-elevations of outer ranges, e^1 species (with about one-year life span) with simultaneous and concentrated leafing and leaf drop during pre-monsoon (March–April) dominate. The pre-monsoon leaf drop is common also in subtropical forests of IEH, but it is relatively earlier, and staggered over a longer period which includes late winter and early pre-monsoon months. In coniferous forests of inner ranges, leaf fall either immediately follows leafing or occurs after a gap of a few months. While winter deciduous forests occur in isolated patches without mixed, evergreen species, drought deciduous species are quite common in lower elevations, and may form extensive forests. The drought deciduous species shed leaves at the end of winter or subsequently during the beginning of the pre-monsoon drought, and then leaf out during the drought itself. In a way, e^1 evergreens represent the stage next to that of drought deciduous species in the continuum from deciduousness to evergreenness. The e^1 species undergo a limited thinning in the canopy without trees becoming leafless (Fig. 5.8). In fact, some of them such as *S. robusta, Alnus nepalensis,* and *Acer oblongum* are referred to both as deciduous and evergreen (we refer to them as evergreen, as in the Himalayas, they rarely become leafless). Thus, in the Himalayas forests are divisible on the basis of the degree of canopy thinning during the pre-monsoon. This applies not only to broadleaved species but also to pines. Only multi-year leaf-bearing conifers and some broadleaved evergreen forests are not affected by these changes in the canopy. The treeline phenology is relatively more heterogeneous. Generally, the species tend to leaf out late, up to July when temperatures are adequately warm and frost-free; however, they vary a lot in leaf shedding phenology. Broadly, autumn and winter together constitute the time of leaf shedding. In low and mid-elevational evergreen forests, multiple leafing is common and increases from west to east.

The time of flowering is generally the same as that of leafing in e^1 type of evergreens. However, there is a strong tendency to produce flowers in the phase of leaflessness in subtropical forests of IEH. Shrubs in e^1 type of forests tend to form another flowering peak during monsoon and autumn. In treeline species, the tendency to minimize overlap between leafing and flowering is apparent, possibly because in the stressful treeline environment, trees may not accumulate enough resources to support both phenophases simultaneously.

A larger leaf size (high percentage of mesophylls) separates the moist lower elevation forests from those of mid and higher elevation forests. Contrary to this, LMA increases from lower to higher elevations. The low LMA in tropical wet forests of Silchar clearly separates it from the forests of the mountainous Himalayas. As for leaf size, moisture can override the effect of elevation, as species with macrophylls in the Himalayas occur in excessively moist forests ranging from lowland tropical wet forests to treeline. In general, for a given growth form LMA tends to be higher in leaves of the Himalayan trees.

Leaf phenology, along with LMA and anatomical characters of leaves, could help considerably in understanding the forest responses to climate change. There is a need to build on this work done earlier on the structural and functional features of forest species (Singh et al. 1994). There is a strong need to study forests of inner ranges, particularly to examine how they have adapted to the lack of monsoon. Leaf phenology investigation in relation to ecosystem functioning can be quite rewarding in understanding ecosystem responses to climate change.

References

Athokpam FD, Garkoti SC (2013) Variation in evergreen and deciduous species leaf phenology in Assam, India. Trees 27:985–997. https://doi.org/10.1007/s00468-013-0850-8

Athokpam FD, Garkoti SC, Borah N (2014) Periodicity of leaf growth and leaf dry mass changes in the evergreen and deciduous species of Southern Assam, India. Ecol Res 29:153–165

Baker-Brosh KF, Peet RK (1997) The ecological significance of lobed and toothed leaves in temperate forest trees. Ecology 78(4):1250–1255

Basnett S, Devy SM (2021) Phenology determines leaf functional traits across *rhododendron* species in the Sikkim Himalaya. Alp Bot 131:63–72. https://doi.org/10.1007/s00035-020-00244-5

Boojh R, Ramakrishnan PS (1981) Phenology of trees in a sub-tropical evergreen montane forest in North-East India. Geo Eco Trop 5(3):189–209

Boojh R, Ramakrishnan PS (1982) Growth strategy of trees related to successional status II. Leaf dynamics. For Ecol Manag 4(4):375–386

Champion HG, Seth SK (1968) A revised survey of the forest types of India. Government of India Publication, Delhi

D'Eça Neves FF, Morellato LPC (2012) Efeitos de bordanafenologi a de árvoresemflorestasemidecídua de altitude na Serra do Japi, SP. In: Vasconcellos Neto J, Polli PR (eds) Novos Olhares da Serra do Japi, 1st ed

Davi H, Gillmann M, Ibanez T, Cailleret M, Bontemps A, Fady B, Lefèvre F (2011) Diversity of leaf unfolding dynamics among tree species: new insights from a study along an altitudinal gradient. Agric For Meteorol 151(12):1504–1513. https://doi.org/10.1016/j.agrformet.2011.06.008

de la Riva EG, Olmo M, Poorter H, Ubera JL, Villar R (2016) Leaf mass per area (LMA) and its relationship with leaf structure and anatomy in 34 Mediterranean woody species along a water availability gradient. PLoS One 11(2):e0148788

Frankie GW, Baker HG, Opler PA (1974) Comparative phenological studies of trees in tropical wet and dry forests in the lowlands of Costa Rica. J Ecol:881–919

Gaira KS, Rawal RS, Rawat B, Bhatt ID (2014) Impact of climate change on the flowering of Rhododendron arboreum in central Himalaya, India. Curr Sci 106(12):1735–1738

Givnish TJ (1987) Comparative studies of leaf form: assessing the relative roles of selective pressures and phylogenetic constraints. New Phytol 106:131–160

Givnish TJ (1988) Adaptation to sun and shade: a whole-plant perspective. Funct Plant Biol 15(2): 63–92

Gougherty AV, Keller SR, Kruger A, Stylinski CD, Elmore AJ, Fitzpatrick MC (2018) Estimating tree phenology from high frequency tree movement data. Agric For Meteorol 263:217–224

Gray REJ, Ewers RM (2021) Monitoring forest phenology in a changing world. Forests 12(3):297. https://doi.org/10.3390/f12030297

Herrerías-Diego Y, Quesada M, Stoner KE, Lobo JA (2006) Effects of forest fragmentation on phenological patterns and reproductive success of the tropical dry forest tree Ceiba aesculifolia. Conserv Biol 20(4):1111–1120

Joshi R, Tamang ND, Sambhav K, Mehra C, Bisht BS, Singh SP (2023) Temperature lapse rate in climatically different Himalayan treeline environments: regional analysis of patterns, seasonality, and variability. In: Singh SP, Reshi ZA, Joshi R (eds) Ecology of Himalayan treeline ecotone. Singapore, Springer Nature Singapore, pp 51–73

Khanduri VP, Sharma CM, Singh SP (2008) The effects of climate change on plant phenology. Environment 28(2):143–147. https://doi.org/10.1007/s10669-007-9153-1

Kikim A, Yadava PS (2001) Phenology of tree species in subtropical forests of Manipur in north eastern India. Trop Ecol 42(2):269–276

Körner C (2012) Alpine treelines: functional ecology of the global high elevation tree limits. Springer Science & Business Media

Körner C, Paulsen J (2004) A world-wide study of high-altitude treeline temperatures. J Biogeogr 31(5):713–732

Körner C, Möhl P, Hiltbrunner E (2023) Four ways to define the growing season. Ecol Lett 26(8): 1277–1292. https://doi.org/10.1111/ele.14260

Kudo G (1996) Intraspecific variation of leaf traits in several deciduous species in relation to length of growing season. Ecoscience 3(4):483–489

Longman KA, Jeník J (1974) Tropical forest and its environment. Longman Group Ltd

Luo Z, Guan H, Zhang X et al (2017) Photosynthetic capacity of senescent leaves for a subtropical broadleaf deciduous tree species *Liquidambar formosana* Hance. Sci Rep 7:6323. https://doi.org/10.1038/s41598-017-06629-7

Maynard DS, Bialic-Murphy L, Zohner CM et al (2022) Global relationships in tree functional traits. Nat Commun 13:3185. https://doi.org/10.1038/s41467-022-30888-2

Michelot A, Simard S, Rathgeber C, Dufrêne E, Damesin C (2012) Comparing the intra-annual wood formation of three European species (*Fagus sylvatica, Quercus petraea* and *Pinus sylvestris*) as related to leaf phenology and non-structural carbohydrate dynamics. Tree Physiol 32(8):1033–1045

Miehe G, Pendry CA, Chaudhary RP (eds) (2015) Nepal: an introduction to the natural history, ecology and human environment of the Himalayas. Royal Botanic Garden, Edinburgh

Mooney HA, Rundel PW (1979) Nutrient relations of the evergreen shrub, *Adenostoma fasciculatum*, in the California chaparral. Bot Gaz 140(1):109–113

Negi GCS (2006) Leaf and bud demography and shoot growth in evergreen and deciduous trees of central Himalaya, India. Trees 20:416–429

Negi GCS, Singh SP (1992) Leaf growth pattern in evergreen and deciduous species of the Central Himalaya, India. Int J Biometeorol 36:233–242

Negi AS, Negi GCS, Singh SP (1996) Establishment and growth of *Quercus floribunda* seedlings after a mast year. J Veg Sci 7(4):559–564

Negi GCS, Singh P, Singh SP (2021) Atmospheric warming-associated phenological earliness does not increase the length of growing season in Himalayan trees. For Sci 67(6):694–700

Negi GCS, Joshi S, Singh P, Joshi R (2022) Phenological response patterns of forest communities to annual weather variability at long-term ecological monitoring sites in Western Himalaya. Trees For People 8:100237. https://doi.org/10.1016/j.tfp.2022.100237

Nord EA, Lynch JP (2009) Plant phenology: a critical controller of soil resource acquisition. J Exp Bot 60(7):1927–1937

Poorter H, Niinemets Ü, Poorter L, Wright IJ, Villar R (2009) Causes and consequences of variation in leaf mass per area (LMA): a meta-analysis. New Phytol 182(3):565–588
Poudyal K, Jha PK, Zobel DB, Thapa CB (2004) Patterns of leaf conductance and water potential of five Himalayan tree species. Tree Physiol 24(6):689–699
Poudyal K, Jha PK, Zobel DB (2012) Role of wood water properties and leaf dynamics in phenology and response to drought in evergreen Himalayan tree species. Ecoprint 19:71–84
Puchałka R, Koprowski M, Gričar J, Przybylak R (2017) Does tree-ring formation follow leaf phenology in Pedunculate oak (Quercus robur L.)? Eur J For Res 136(2):259–268
Ralhan PK, Khanna RK, Singh SP et al (1985a) Phenological characteristics of the tree layer of Kumaun Himalayan forests. Vegetatio 60:91–101. https://doi.org/10.1007/BF00040351
Ralhan PK, Khanna RK, Singh SP et al (1985b) Certain phenological characters of the shrub layer of Kumaun Himalayan forests. Vegetatio 63:113–119. https://doi.org/10.1007/BF00044061
Ranjitkar S, Luedeling E, Shrestha KK et al (2013) Flowering phenology of tree rhododendron along an elevation gradient in two sites in the eastern Himalayas. Int J Biometeorol 57:225–240. https://doi.org/10.1007/s00484-012-0548-4f
Ranjitkar S, Kindt R, Sujakhu NM, Hart R, Guo W, Yang X, Shrestha KK, Xu J, Luedeling E (2014) Separation of the bioclimatic spaces of Himalayan tree rhododendron species predicted by ensemble suitability models. Glob Ecol Conserv 1:2–12
Raunkiaer C (1934) The life forms of plants and statistical plant geography. Clarendon, London
Royer DL, Wilf P (2006) Why do toothed leaves correlate with cold climates? Gas exchange at leaf margins provides new insights into a classic paleotemperature proxy. Int J Plant Sci 167(1): 11–18
Sancho-Knapik D, Escudero A, Mediavilla S, Scoffoni C et al (2021) Deciduous and evergreen oaks show contrasting adaptive responses in leaf mass per area across environments. New Phytol 230(2):521–534. https://doi.org/10.1111/nph.17151
Shukla RP, Ramakrishnan PS (1982) Phenology of trees in a sub-tropical humid forest in North-Eastern India. Vegetatio 49(2):103–109
Singh N (2014a) Flowering phenology of tree *Rhododendron arboreum* along an elevation gradient in different sites of Kumaun Himalayas. Int J Sci Nat 5(3):572–576
Singh SP (2014b) Attributes of Himalayan forest ecosystems: they are not temperate forests. Proc Indian Natl Sci Acad 80(2):221–233
Singh P, Negi GCS (2018a) Seasonal changes in leaf area and leaf mass of dominant trees of Western Himalayan forests. Indian For 144(2):182–190
Singh P, Negi GCS (2018b) Treeline species phenology: shoot growth, leaf characteristics and nutrient dynamics. Trop Ecol 59(2):297–311
Singh P, Negi GCS (2023) Phenological response of treeline ecotone tree species to global warming in Western Himalaya. In: Singh SP, Reshi ZA, Joshi R (eds) Ecology of Himalayan treeline ecotone. Springer, Singapore. https://doi.org/10.1007/978-981-19-4476-5_15
Singh JS, Singh SP (1992) Forests of Himalaya: structure, functioning and impact of man. Gyanodaya Prakashan, Nainital
Singh SP, Adhikari BS, Zobel DB (1994) Biomass, productivity, leaf longevity, and forest structure in the central Himalaya. Ecol Monogr 64(4):401–421
Singh P, Arya V, Negi GCS, Singh SP (2018) Expansion of *Rhododendron campanulatum* krummholz in the treeline ecotone in Tungnath, Garhwal Himalaya. Trop Ecol 59(2):287–295
Sun XL, Xu YF, Ma LY, Zhou H (2010) A review of acclimation of photosynthetic pigment composition in plant leaves to shade environment. Chin J Plant Ecol 34(8):989
Tewari A, Bhatt J, Mittal A (2016) Influence of tree water potential in inducing flowering in *Rhododendron arboreum* in the central Himalayan region. iForest 9:842–846. https://doi.org/10.3832/ifor1525-008
Troup RS (1921) The silviculture of Indian trees, vol I–III. Clarendon Press, Oxford
Viera M, Caldeira MVW, Rovani FFM, Castro KC (2016) Ecological and environmental aspects of nutrient cycling in the Atlantic Forest, Brazil. In: Tropical forests: the challenges of maintaining ecosystem services while managing the landscape, pp 113–117

Villar R, Ruiz-Robleto J, Ubera JL, Poorter H (2013) Exploring variation in leaf mass per area (LMA) from leaf to cell: an anatomical analysis of 26 woody species. Am J Bot 100(10): 1969–1980. https://doi.org/10.3732/ajb.1200562

Visser ME, Holleman LJ (2001) Warmer springs disrupt the synchrony of oak and winter moth phenology. Proc R Soc Lond Ser B Biol Sci 268(1464):289–294

Wang H, Nilsen ET, Upmanyu M (2020) Mechanical basis for thermonastic movements of cold-hardy Rhododendron leaves. J R Soc Interface 17(164):20190751

Watham T, Srinet R, Nandy S, Padalia H, Sinha SK, Patel NR, Chauhan P (2020) Environmental control on carbon exchange of natural and planted forests in Western Himalayan foothills of India. Biogeochemistry 151:291–311

Wheat D (1980) Sylleptic branching in *Myrsine floridana* (Myrsinaceae). Am J Bot 67(4):490–499

Yang Y, Sun H, Körner C (2020) Explaining the exceptional 4270 m high elevation limit of an evergreen oak in the South-Eastern Himalayas. Tree Physiol 40(10):1327–1342

Zobel DB, Garkoti SC, Singh SP, Tewari A, Negi CMS (2001) Patterns of water potential among forest types of the central Himalaya. Curr Sci 80:774–779

Tree Nutrient Concentration and Forest Nutrient Cycling

6

S. P. Singh

Abstract

This chapter on nutrients is based on a series of Indian Central Himalayan (ICH) forests along an elevation gradient from foothills (300 m) to the treeline (>3300 m). In general soil properties changed monotonically (e.g. soil bulk density decreased) with elevation, but for subtropical chir pine forest patterns differ. Generally nutrient concentration in tree components was found in the following order: leaves >twigs >branches/bark>coarse roots >bole (stem) wood. Bark was 1.3 times (Ca) to 1.59 times (P) richer in nutrient concentration than stem wood. Nitrogen concentrations in stem wood of Himalayan forests are generally on higher side of the range given in the literature, making it a large nutrient reservoir. The prevalence of significant positive correlations in nutrient concentrations among the tree components across species suggests a great deal of functional coordination of nutrient allocations across disparate types of tissues. Thus, a species having high Ca concentration in leaf is likely to have high concentration of Ca in stem wood. The average N:P ratio in tree components across the ICH species ranged between 16.7 in coarse roots and 24.9 in twigs. Across the ICH forests total N mass in the ecosystem (vegetation + forest floor litter +30 cm deep soil) varied about two-fold from 5,079 kg ha^{-1} in chir pine to 11,768 kg ha^{-1} in tilonj oak forest, located at 2000 m. The percentage of total nitrogen mass in a forest ecosystem that occurs in vegetation,'decreases rapidly above 2600 m, from about 34.5% in kharsu oak forest to 7.6% in treeline birch forest and 2.71% in krummholz of *R. campanulatum*. Phosphorus (P) mass in

S. P. Singh (✉)
INSA Honorary Scientist, Dehradun, Uttarakhand, India

Centre for Himalayan Studies, Graphic Era (deemed to be) University, Dehradun, Uttarakhand, India
e-mail: spsecology@gmail.com

S. P. Singh, R. Thadani (eds.), *Himalayan Forests in a Changing World*,
https://doi.org/10.1007/978-981-95-0978-2_6

vegetation ranged about 3.5-fold from 13.2 kg ha^{-1} in the treeline forest of *R. campanulatum* to 462 kg ha^{-1} in tilonj oak-dominated forest. Generally, herbs accounted for significantly higher percentage of nutrient uptake than shrubs, largely because of the greater production and nutrient concentration of herbs than shrubs. If we exclude the values of treeline forests and some exceptionally high values of old-growth oak forests, the N return through litterfall ranges between 56.1 and 83.3 kg ha^{-1} yr^{-1}. Canopy interception of precipitation is higher in conifer forests (e.g. 34.1% is *Cedrus deodara* forest) than broad-leaved evergreen forests (e.g. 24.8% in *Q. leucotrichophora* forest. In the ICH forests, the nutrient input through precipitation is in the following order: Ca > K > Mg > N > P. The nutrient turnover time in these forests is short, closer to that of tropical forests than temperate forests: it ranged between 1.22 and 4.35 years across all forests and nutrients. The average resorption of N in the Himalayan species was 50.3%, based on 71 species. To conclude, holding large amount of nutrients the Himalayan forests is of great conservation value.

Keywords

Himalayas · Nutrient concentration · Deciduous · Evergreen · Throughfall

Introduction

Nutrient availability is among the most important factors, which affects biomass production in a forest ecosystem (Viera et al. 2016). In fact, studies on forest nutrient cycling originated from early concerns that forest harvesting was unsustainable in a long term because of the depletion of soil nutrients associated with biomass removal (Johnson and Turner 2019). The connection between nutrient availability and forest productivity is well-established (Switzer and Nelson 1972; Attiwill and Adams 1993). Plant growth and carbon sequestration processes directly and indirectly require at least 16 elements. Of them, N, P, and K are particularly important: N is a vital component of enzymes and chlorophyll (Sardans et al. 2015); P is key component of NADP, ATP, nucleic acids, and phosphates (Deng et al. 2015); K contributes to stress-resistance (Xing et al. 2022). Among others, Ca an essential secondary nutrient is a part of every plant cell, and Mg (another essential secondary nutrient) is required in chlorophyll to capture energy.

Forest nutrient cycling can be divided into three components (Fig. 6.1): geochemical cycling that represents inputs and outputs of nutrients in the ecosystem; biogeochemical cycle that represents nutrients transfer between the plants and soil; and biochemical cycling that is about internal nutrient cycling, such as resorption of nutrients from senescing leaves to perennial organs of trees (Switzer and Nelson 1972). Nutrients come into geochemical cycling through precipitation and dry depositions, biological fixation and rock weathering, and go out in stream water, in a gaseous form, and harvest of biomass.

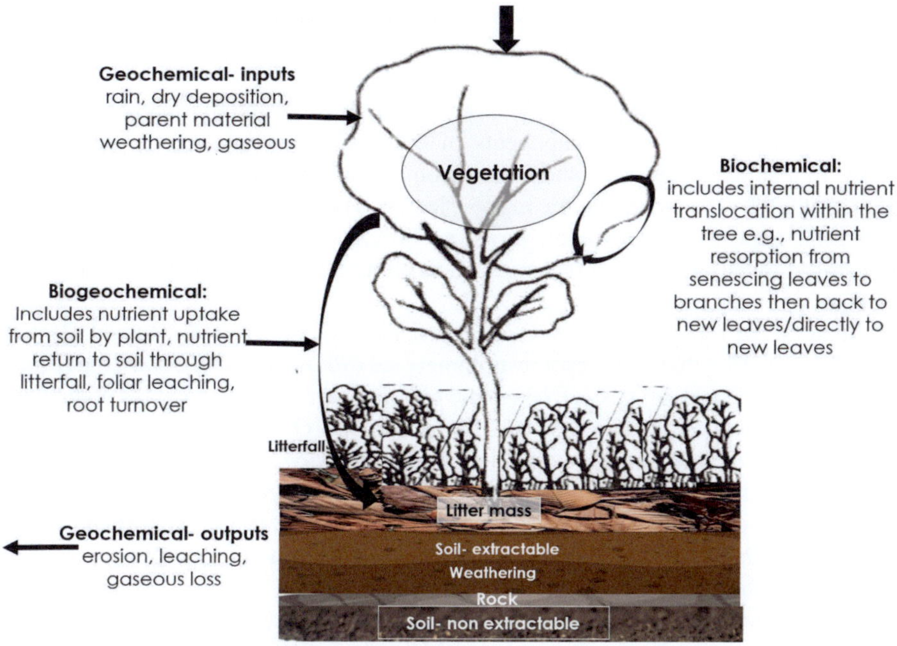

Fig. 6.1 A framework of nutrient cycling in forest ecosystems. (Derived from Johnson and Turner (2019)). The tree is of banj oak (*Quercus leucotrichophora*), a common oak of Indian Central Himalaya

The biogeochemical cycle in a forest involves nutrient transfer between vegetation (e.g. through litterfall) and soil (plants absorb nutrients from soil). In other words, it involves nutrient uptake from soil by plants, and nutrient return to soil in various ways; such as leaching, in leaf litter fall, root mortality, and mortality of individuals. Biochemical cycling represents resorption of nutrients inside the plants, mostly from older to younger tissues (e.g. from senescing leaves to new shoots).

Site level studies on nutrient cycling generally use the average nutrient concentration of different species that occur on a site (Chen et al. 2013); however, within a site nutrient concentration may vary by an order of magnitude between species (Kraft et al. 2008). Furthermore, a large-scale pattern of plant stoichiometry has been largely analyzed around leaves (Han et al. 2011; Tian et al. 2018; Han et al. 2005; and Reich and Oleksyn 2004), through tree leaves and stems have different roles in a forest ecosystem functions (Tang et al. 2018). Leaf is of critical importance for nutrient metabolism, while stem provides long-term storage for nutrients that are of critical importance for internal nutrient-cycling (Heineman et al. 2016; Yan et al. 2016). While leaf nutrient concentration is commonly used to indicate plant nutrient status, nutrient concentration of tree stems (boles) draws little attention, though tree stems contain much larger amount of nutrients than leaves (Yang et al. 2014; Zhao et al. 2016; Fortunel et al. 2012; and Kleyer and Minden 2015). Trees allocate biomass and nutrients between leaves and stems or branches according to environment and growth requirements (Yang et al. 2014; Zhao et al. 2016; Fortunel et al. 2012; and Kleyer and Minden 2015).

Leaf, being the most active component has distinctly higher nutrient concentrations than other components of trees, such as bole (stem) and coarse roots. In a study based on N and P stoichiometry of 153 Chinese terrestrial plants, Han et al. (2005) found geometric mass of leaf, N, P and N:P ratio 18.6, 1.21 mg g^{-1} and 14.4, respectively. However, variations in leaf and stem stoichiometry may be influenced by environmental factors, such as climate and topography (Sardans et al. 2017).

Most of the forests on which this chapter is based are from the Indian Central Himalayan region that includes forests namely sal (*Shorea robusta*), chir pine (*Pinus roxburghii*), mixed banj oak (*Q. leucotrichophora*)-chir pine, mixed oaks dominated by *Q. floribunda*, *Q. lanata*, and *Kharsu* oak forest (*Q. semecarpifolia*), silver fir (*Abies pindrow*), and deciduous species; *Aesculus indica*, maple forest (*Acer* spp.), and birch forest (*Betula utilis*) and a rhododendron (*R. campanulatum* that often goes beyond treelines (Chaturvedi 1983; Khanna 1986; Rawat 1983; Garkoti and Singh 1994; and Adhikari et al. 1995). They together cover common forest types, occurring from the foothills to the upper elevational limit of forests. Sal, oaks, pine, silver fir, birch, and rhododendrons make prominent forest types over a large part of the Himalayas even outside the Indian Central Himalayas (ICH).

The study parameters of nutrient cycling in this chapter include nutrient concentration in plants, nutrient storage in major ecosystem components, nutrient uptake in primary production, nutrient return from vegetation to soil, nutrient resorption (NuR) from senescing leaves, and nutrient turnover on forest floor. For this chapter we made a meta-analysis of coordination among nutrient concentrations in different tree components, such as leaves, twigs, and stem wood, and nutrient ratios to find out generalizations about the Himalayan forests.

The study elevational transects occur in the outer central Himalayan ranges, exposed to full thrust of monsoon, generally with 1500–2500 mm annual precipitation. Conditions tend to become increasingly moist with elevation as evapotranspiration loss decreases with decreasing temperature. That is why forest biomass values around 500 t ha^{-1} are quite common between 2000 and 2800 m in several mature and old-growth oak (e.g. *Q. floribunda* and *Q. semecarpifolia*) and fir (*A. pindrow*) forests. However, these high values are partly because of lower human disturbance than in lower elevations. The forest biomass declines sharply only above oak-silver fir (*A. pindrow*) belt (with biomass of around 500 t ha^{-1}) around 2600–2700 m. It is 308 t in a maple forest (~2800 m), 177 in birch forest (treeline, around 3000 m), and 40.5 t in rhododendron krummholz (>3000 m). The study includes Nainital transect in the lower part (<2200 m) and Pinder Valley transect in the higher part (>2200 m).

Climate change is perceptible during last three decades, with mean air temperature increasing at the decadal rate of 0.38 °C. The mean annual precipitation is indicated to decrease from 2192 mm in 1985–87 to 1760 mm in 2014–16 (Singh and Negi 2023).

Soil Along the Elevation Transect

Several distinct patterns for soil (30 cm depth) properties are evident along the elevation gradient. Mean annual soil temperature declines from 22.4 °C in foothill sal forest to 14.6 °C in tilonj (*Q. floribunda*)-dominated mixed-oak forest at about 2200 m. Between these two end points, the mean annual temperature is 19.0 °C in chir pine forest, and 16.3 °C in banj oak forest. Soil was most moist in tilonj forest (15.5%) and the driest (10.8%) in chir pine forest (values of pre-monsoon, around April–May). Coarse materials (larger than soil particles), such as pebbles and cobbles were not found in the deep alluvial soil of sal forest. Among others the coarse material content was distinctly higher in chir pine forest (32.77 ± 5.16%) than other forests and declined with decreasing relative importance of chir pine in the forest composition (0 to 20.6 ± 0.65%). The soil color ranged from dark brown to yellowish brown, and texturally soil was loamy sand in the sal forest of foothills and sandy loam in the remaining forests occurring in mountains (Table 6.1). Percentage of sand decreased and silt increased significantly with elevation, particularly in 20–30 cm soil layer. However, sand was the dominant (61.33 ± 2.89–82.67 ± 0.67%) component of soil texture in the entire transect. Similar observation has been made for Garhwal forests (Sharma et al. 2010). Soil bulk density in Nainital transect declined significantly ($p < 0.01$) with elevation from 1.14 ± 0.07 in sal seedling-growth forest to 0.73 ± 0.06 g cm^{-3} in mixed-oak tilonj-dominated forest occurring between 2000 and 2200 m (Khanna 1986).

Soil water holding capacity and soil moisture increased with elevation, but for chir pine forest in which they were the lowest. Soil pH was almost the same (around 6.7) in the lower half of the elevation transect, but for chir pine forest in which it was lower (5.67). In higher sites the pH was around 6.

Total organic carbon (C) and total nitrogen (N) concentration increased significantly with elevation ($p < 0.05$). Available P increased with elevation, and at each forest site it decreased rapidly with soil depth. Similar patterns in relation to elevation for N and P have been reported for soil of Garhwal forests (Kumar et al. 2021). Exchangeable K, Mg, and Ca concentrations increased with elevation with a marked dip in chir pine forest soil.

In general, soil properties changed monotonically, but for the mid-elevational chir pine forest, which differed from both lower and higher elevation forests for several parameters. Low pH, high percentage of coarse particles, low soil moisture, low nutrient concentrations, and high C:N ratio separated the chir pine forest from other forests (Chaturvedi 1983; Khanna 1986; and Singh and Singh 1992). The average banj oak forest soil is richer than chir pine forest soil in total N (2.6 vs 2.1 g kg^{-1}), organic carbon (3.2 vs 2.6 g kg^{-1}), available P (1.6 vs 0.08 g kg^{-1}), and total K (1.0 vs 0.6 g kg^{-1}) (Chaturvedi 1983; Rawat 1983). Greater differences across the forests within the elevation range have been reported by Singh and Negi (2023). For example in chir pine forest soil N concentration was about 80% lower than in oak forests (Singh and Negi 2023).

A comparison of two adjacent sal forests, old-growth sal forest and seedling coppice forest shows that the management practices can bring about marked changes

Table 6.1 Physico–chemical properties of soil in different study forests

Parameter		Sal (*S. robusta*) old-growth (350 m)	Sal seedling coppice forest (385 m)	Pine-mixed broadleaf forest (1700 m)	Pine (*P. roxburghii*) forest (1750 m)	Mixed oak (*Q. leucotrichophora*)-pine forest (1850 m)	Mixed oak-rianj oak (*Q. lanata*) dominated forest (2100 m)	Mixed oak-tilonj oak (*Q. floribunda*) dominated forest (2200 m)
Coarse matter (pebbles and cobbles) %	Average ±	–	–	20.60	32.77	11.77	12.17	17.30
	Error	–	–	±0.65	±5.16	±0.41	±3.05	±5.70
Bulk density (g/cm^3)	Average ±	1.02	1.14	0.94	0.78	0.91	0.77	0.73
	Error	±0.03	±0.07	±0.06	±0.07	±0.09	±0.08	±0.06
Sand (%)	Average ±	81.00	78.83	82.67	63.67	61.33	63.33	65.33
	Error	±0.58	±1.88	±0.67	±1.45	±2.89	±0.67	±1.76
Silt (%)	Average ±	11.33	14.50	10.67	24.67	25.67	24.00	24.00
	Error	±1.33	±2.29	±0.67	±0.67	±1.76	±1.15	±1.15
Clay (%)	Average ±	7.67	6.67	7.33	11.67	13.00	12.67	10.67
	Error	±0.88	±0.88	±0.67	±0.88	±0.87	±1.76	±2.40
Moisture (%)	Average ±	13.00	13.67	20.00	12.50	21.50	17.17	25.00
	Error	±0.58	±2.05	±1.80	±1.80	±5.30	±2.03	±3.06
Water holding capacity (%)	Average ±	46.43	53.67	61.30	41.10	55.50	66.17	60.20

	Error	±3.72	±9.18	±5.70	±1.78	±0.95	±7.87	±8.86
pH	Average ±	6.77	6.70	6.60	5.67	6.10	6.03	6.27
	Error	±0.03	±0.06	±0.06	±0.27	±0.10	±0.03	±0.03
Organic carbon (%)	Average ±	0.99	0.51	2.38	2.95	1.90	2.50	2.55
	Error	±0.05	±0.09	±0.59	±0.53	±0.78	±0.86	±0.76
Total organic matter (%)	Average ±	1.70	0.88	4.09	5.04	3.28	4.61	4.40
	Error	±0.08	±0.16	±1.01	±0.90	±1.34	±1.31	±1.31
Total nitrogen (%)	Average ±	0.13	0.07	0.19	0.20	0.20	0.25	0.24
	Error	±0.01	±0.01	±0.06	±0.04	±0.07	±0.11	±0.09
C:N ratio	Average ±	7.79	6.81	13.81	15.24	9.32	9.92	9.88
	Error	±0.54	±0.37	±2.47	±0.69	±0.76	±0.39	±0.47
Available Phosphorus (%)	Average ±	0.005	0.004	0.007	0.007	0.008	0.009	0.008
	Error	±0.001	±0.001	±0.001	±0.001	±0.002	±0.002	±0.002
Exchangeable Potassium (Meq%)	Average ±	0.29	0.26	0.36	0.25	0.47	0.58	0.41
	Error	±0.08	±0.09	±0.11	±0.05	±0.16	±0.20	±0.14

(continued)

Table 6.1 (continued)

Parameter		Sal (*S. robusta*) old-growth (350 m)	Sal seedling coppice forest (385 m)	Pine-mixed broadleaf forest (1700 m)	Pine (*P. roxburghii*) forest (1750 m)	Mixed oak (*Q. leucotrichophora*)-pine forest (1850 m)	Mixed oak-rianj oak (*Q. lanata*) dominated forest (2100 m)	Mixed oak-tilonj oak (*Q. floribunda*) dominated forest (2200 m)
Exchangeable Calcium (Meq %)	Average ±	10.17	14.64	15.52	8.62	12.92	12.94	12.74
	Error	±1.12	±0.57	±0.76	±0.80	±2.39	±1.44	±1.25
Exchangeable Sodium (Meq %)	Average ±	0.25	0.15	0.19	0.26	0.32	0.32	0.18
	Error	±0.09	±0.02	±0.07	±0.05	±0.07	±0.04	±0.05
Exchangeable Magnesium (Meq%)	Average ±	2.98	2.28	2.95	0.95	3.29	5.31	5.16
	Error	±0.66	±0.32	±0.57	±0.19	±0.54	±1.09	±0.51

Largely from Khanna (1986)
The values are averages of three soil depths (i.e. 0–10, 10–20 and 20–30 cm). Altitude is indicated in meters

in soil. Old-growth sal forest had much larger biomass, >700 t ha^{-1} compared to ~400 t ha^{-1} in sal-seedling coppice forest. The latter had young, dense, and even-aged trees, established by managed burning and coppicing. Lower nutrient concentration in sal-seedling coppice forest might be due to recent burning and subsequent soil erosion and higher nutrient uptake by rapidly growing young trees. The two mixed oak forests of higher elevations (2100–2200 m), which had different dominant oak species differed significantly in several soil parameters. The soil of tilonj oak (*Q. floribunda*)-dominated forest had a higher proportion of finer particles like silt and clay than that of the rianj oak (*Q. lanata*)-dominated forest. The tilonj-dominated oak forest had also significantly higher soil moisture and organic carbon, but lower exchangeable K and Mg.

The range of pH along the elevation transect was between 5.4 and 6.8 (20 out of 25 values between 6.0 and 6.8), and it tended to decrease with elevations. While variation in soil properties influences vegetation, difference in soil properties is also caused by species-specific characters, such as nutrient concentration in senesced leaves (Viera and Schumacher 2009; Carnol and Bazgir 2013).

Soil in a Higher (Tungnath) Elevation Transect

The Tungnath elevation transect (2100–3300 m) is largely covered with oaks, and Himalayan silver firs, with sporadic forest patches of broad-leaved deciduous species, such as *Acer* spp., and *Aesculus indica* interspersed among them. The timberline consisted of birch (*B. utilis*), rhododendrons, *A. spectabilis,* and evergreen oak, *Q. semecarpifolia*. Rawal (2022) has distinct changes in soil properties of this transect. Since evapotranspiration decreased with altitude, soil moisture increased significantly with elevation ($p < 0.01$). The water holding capacity (WHC) also increased (from 42.8% to 60.6%) in top layer (0–15 cm) with elevation ($p < 0.05$), as it did in Nainital elevation gradient. Soil of the transect was sandy (67.7%–72.8%) and the sand content increased with elevation ($p < 0.05$). Clay in top soil ranged from 6.22% to 7.67%. The soil pH decreased with elevation, but within a limited range, while soil organic carbon increased with elevation.

Nutrient Concentrations and Correlations Among Tree Components

Mean nutrient concentrations in ICH forest were mostly higher in leaves (that are metabolically more active), compared to other tree components, including that of Ca that is reported to be higher in stem bark than in leaves in some forests, such as pre-montane wet tropical forests of Panama (Jones et al. 2019). Overall, in ICH tree species nutrient concentration generally occurs in the following order: leaves>twigs>branches/bark>coarse root/bole (stem) wood. In ICH trees, the bark nutrient concentrations were distinctly higher than that of stem wood as follows: 1.43 times higher for N, 1.59 times for P, 1.42 times for K, and 1.3 times for both Ca and Na (Table 6.2). A similar pattern was observed in other studies for N,

Table 6.2 Average nutrient concentration (mg g^{-1}) and range across the species in different tree components in the forests of an Indian Central Himalayan elevation transect (300–3000 m)

Nutrient		Nutrient concentration (mg g^{-1})					
		Leaf	Twig	Bark	Branch	Bole	Coarse root
N	Average	**18.52 ± 0.74**	**10.85 ± 0.57**	**6.27 ± 0.55**	**6.65 ± 0.44**	**4.42 ± 0.32**	**4.53 ± 0.20**
	Range	13.4 (*R. arboreum*) - 27.1 (*Acer cappadocicum*)	6.7 (*R. barbatum*) - 17.6 (*Q. semecarpifolia*)	3.7 (*Acer caesium*)– 10.2 (*S. robusta*)	3.4 (*R. barbatum*)- 12.3 (*Q. semecarpifolia*)	2.1 (*R. barbatum*)- 6.5 (*Q. semecarpifolia*)	2.9 (*R. barbatum*)- 6.3 (*Q. semecarpifolia*)
P	Average	**1.00 ± 0.11**	**0.54 ± 0.05**	**0.43 ± 0.07**	**0.43 ± 0.05**	**0.27 ± 0.04**	**0.34 ± 0.03**
	Range	0.47 (*A. indica*) - 2.20 (*Acer cappadocicum*)	0.22 (*Lyonia ovalifolia*) - 0.90 (*Q. lanata*)	0.16 (*Symplocos chinensis*)-1.05 (*P. roxburghii*)	0.14 (*R. arboreum*)- 0.86 (*P. roxburghii*)	0.08 (*Aesculus indica*)- 0.70 (*P. roxburghii*)	0.11 (*S. chinensis*)- 0.60 (*Acer cappadocicum*)
K	Average	**6.72 ± 0.47**	**4.64 ± 0.37**	**2.72 ± 0.34**	**3.07 ± 0.30**	**2.38 ± 0.24**	**2.43 ± 0.19**
	Range	0.60 (*Acer caesium*) -11.40 (*Acer cappadocicum*)	3.00 (*R. campanulatum*) -7.90 (*Q. lanata*)	0.70 (*Acer cappadocicum*)-6.80 (*S. robusta*)	1.50 (*Acer cappadocicum, Symplocos chinensis*)- 6.40 (*P. roxburghii*)	1.20 (*Symplocos chinensis*)– 4.90 (*Q. lanata*)	1.40 (*S. chinensis*)- 4.00 (*P. roxburghii*)
Ca	Average	**12.49 ± 0.94**	**9.87 ± 0.69**	**7.80 ± 1.06**	**7.15 ± 0.68**	**5.99 ± 0.70**	**5.11 ± 0.43**
	Range	5.20 (*R. arboreum*) -20.0 (*Q. semecarpifolia*)	3.60 (*R. arboreum*) – 14.80 (*Ilex dipyrena*)	2.70 (*R. arboreum*)- 13.80 (*Q. floribunda*)	2.60 (*R. arboreum*)- 12.40 (*Q. semecarpifolia*)	1.70 (*R. arboreum*)- 14.10 (*Q. semecarpifolia*)	1.70 (*R. arboreum*)- 8.50 (*Q. leucotrichophora*)
Na	Average	**0.48 ± 0.03**	**0.34 ± 0.03**	**0.28 ± 0.04**	**0.26 ± 0.02**	**0.21 ± 0.02**	**0.20 ± 0.02**
	Range	0.25 (*P. roxburghii*) -0.77 (*Mallotus philippensis*)	0.18 (*P. roxburghii*) – 0.60 (*S. robusta*)	0.13 (*A. caesium*)- 0.64 (*S. robusta*)	0.16 (*A. caesium*)- 0.47 (*S. robusta*)	0.11 (*A. caesium*)- 0.46 (*S. robusta*)	0.09 (*Ilex dipyrena*)- 0.40 (*S. robusta*)

Derived from various sources: Khanna (1986), Singh and Singh (1992), Garkoti and Singh (1994), and Adhikari et al. (1995)

P, and K both in natural forests, and plantations of *B. pubescens, Q. robur, Eucalyptus globulus, E. nitens,* and *Populus* spp. by Rodríguez-Soalleiro et al. (2018).

Investigating variation in stem wood nutrients across 106 tree species, Heineman et al. (2016) found four-fold variation in N, and > 30-fold variation for the other nutrients, which are much more than observed for the ICH tree species (3.9-fold for N, 8.7-fold for P, 4.1-fold for K, 8.3-fold for Ca, and 4.1-fold for Na). It is difficult to find out whether narrower variations for ICH tree species were because of smaller sample size (~20 species for different nutrients), or it is a character of the ICH forests. Apart from the large number of tree species in Panama, they occurred across a wide range of soil fertility, which might have caused marked changes in wood-nutrient concentration, particularly P (Heineman et al. 2016). Values (mg g^{-1}) from some other tropical forests are reported as: N, 0.4–6.2, Ca, 0.6–11.5, P 0.019–0.668, and K 0.16–4.50) (Becker et al. 2012; Martin et al. 2014).

Stem wood (bole) contains much of the nutrient mass of trees, yet controls over stem nutrient concentrations and their functional roles are inadequately known (Heineman et al. 2016). In the ICH tree species mean nitrogen concentrations in stem wood are relatively higher than those reported for tropical forests of Panama; however, the upper side values of the ranges are similar: the values (in mg g^{-1}) being 6.30, 0.70, 5.10, and 10.10, respectively, for N, P, K, and Ca in ICH forests and 5.8, 0.68, 3.12, and 12.6, respectively for N, P, K, and Ca in the trees of Panama forests (Table 6.3). Overall, the nutrient concentrations of the Himalayan trees fall within the range of values reported for some of the similar forests in the literature. Given the significance of stem in forest nutrient cycling, detailed and precise estimations are required on the nutrients of wood and bark at different stages of tree growth across the species of different growth forms and sites of the Himalayas.

A study on natural forests, and forest plantations (five species *B. pubescens, Q. robur,* two species *Eucalyptus, Populus* spp.) of NW Spain indicated following ranges of N concentration in different tree components (mg g^{-1}): leaves 15.7–25.1;

Table 6.3 A comparison of nutrient concentrations in bole (stem wood) between tree species of Indian Central Himalayan forests (distributed along an elevational gradient, 3000–3300 m, compiled from Khanna 1986; Garkoti and Singh 1994; and Adhikari et al. 1995) and the tropical forests of Panama (based on 106 tree species sampled across six lowland and four montane forest plots, Heineman et al. 2016)

	Species mean and range of nutrient concentration (mg g^{-1})			
	Himalayan forests		Panama forests	
Nutrient	Mean	Range	Mean	Range
Nitrogen (N)[a]	4.6	2.1–6.30	2.5	1.3–5.8[a]
Phosphorus (P)[b]	0.30	0.09–0.70	0.11	0.01[a]–0.68
Potassium (K)[c]	2.70	1.3–5.10	0.49	0.061–3.12
Calcium (Ca)[d]	6.30	1.70–10.10	2.08	0.27–12.61
Magnesium (Mg)[e]	–	–	1.62	0.76–5.77

[a]N 0.4–6.8 mg g^{-1} in neotropical forests, [b]P from 0.019–0.67 mg g^{-1}, [c]K 0.6–4.5 mg g^{-1}, [d]Ca 0.06–11.5 mg g^{-1}, and [e]Mg 0.08–1.29 mg g^{-1} (Meerts 2002, 93 species from tropics and temperate region; Martin et al. 2014; Becker et al. 2012)

twigs 4.5–10.9, bark 1.6–7.0, wood 0.9–6.0, and roots 3.8–6.4 (Rodríguez-Soalleiro et al. 2018). These values are generally similar to those of ICH. Xing et al. (2022) showed that in Chinese terrestrial plants average N concentration was 22.15 mg g^{-1} in leaves, 7.21 mg g^{-1} in stem, and 6.8 mg g^{-1} in roots. As for P, the average concentration was 1.22 mg g^{-1} in leaves, 0.86 mg g^{-1} in stem, and 0.27 mg g^{-1} for roots. The mean for K concentration was 25.82 mg g^{-1} for leaves, 7.21 mg g^{-1} for stems and. 3.29 mg g^{-1} for roots.

Coordination and Correlation Among Tree Components

Significant N correlations between nutrient concentrations of tree components (leaf, bole wood, bark, branches, twigs, and coarse roots) across the tree species are given in Fig. 6.2. Across the ICH species N concentrations were significantly correlated

2a. Nitrogen (N)

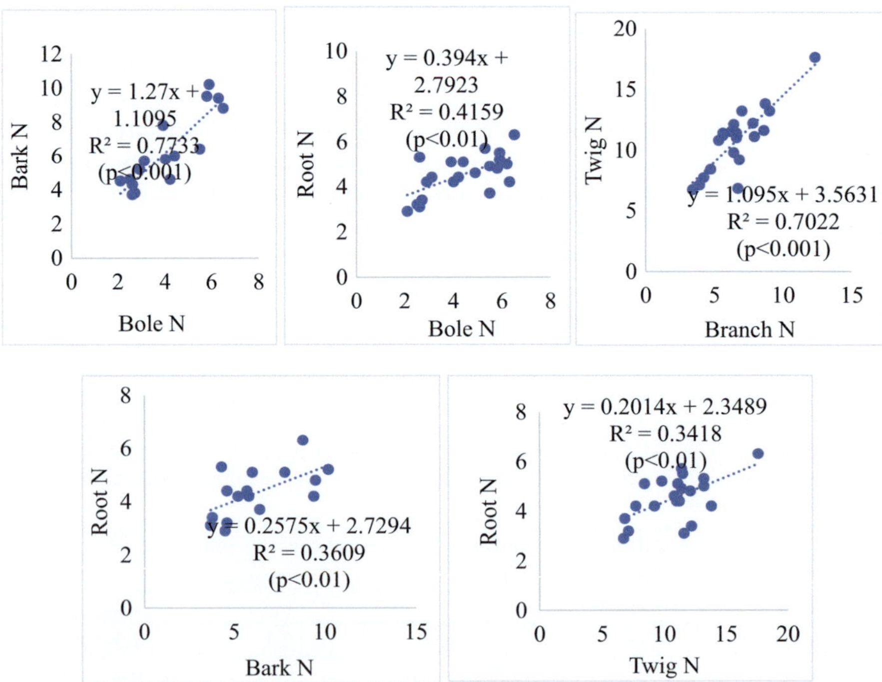

2b. Phosphorus (P)

Fig. 6.2 Correlations in nutrient concentration (mg g^{-1}) between tree components across tree species of Indian Central Himalayas (ICH). Each dot represents a forest species. (**a**) Nitrogen (N); (**b**) Phosphorus (P); (**c**) Potassium (K); (**d**) Calcium (Ca); and (**e**) Sodium (Na). Only significant correlations between tree component nutrient concentrations are shown. (Derived from data given in Khanna (1986), and others shown in earlier section)

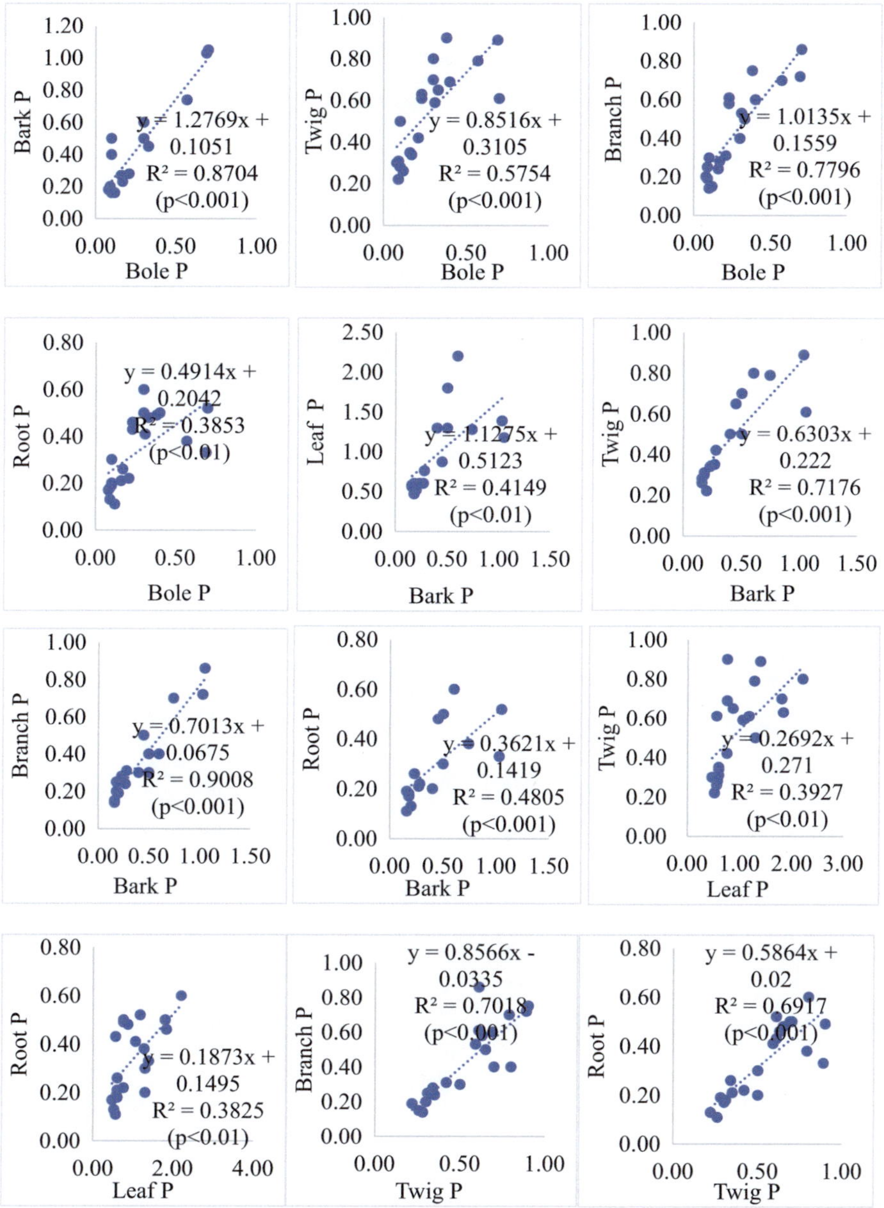

Fig. 6.2 (continued)

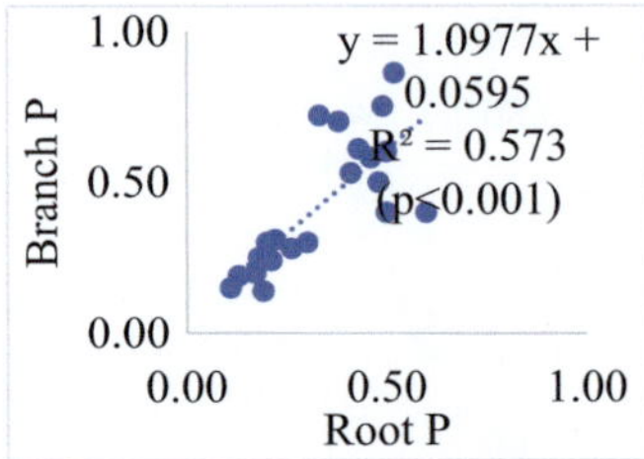

2c. Potassium (K)

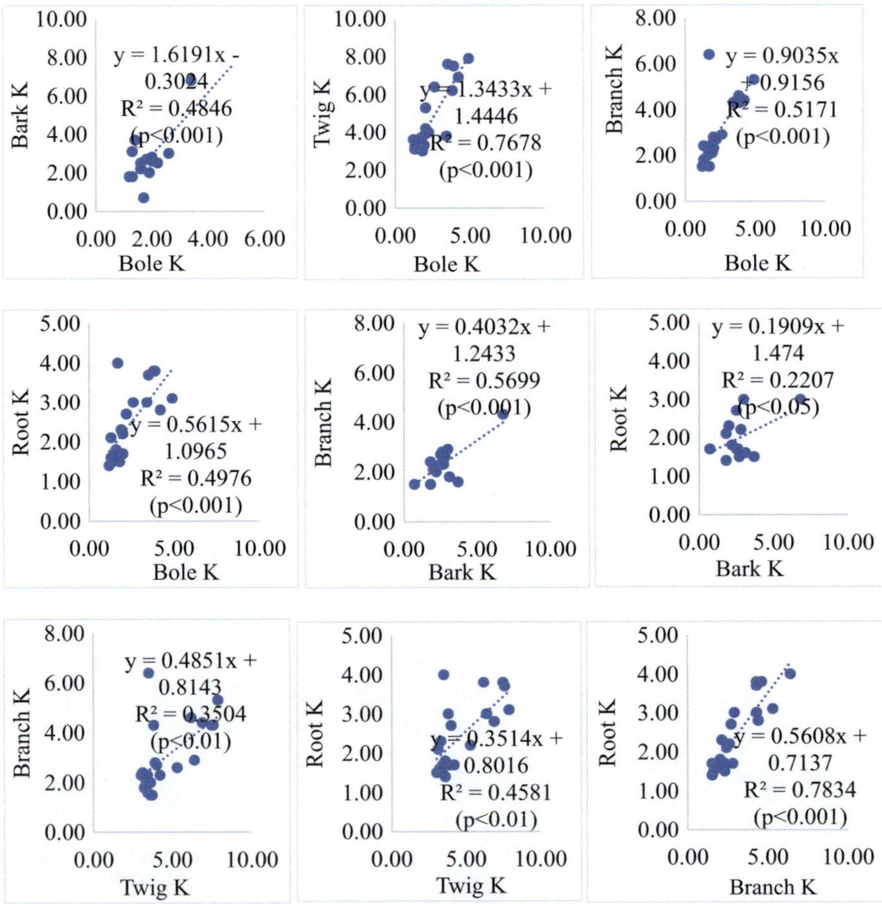

2d. Calcium (Ca)

Fig. 6.2 (continued)

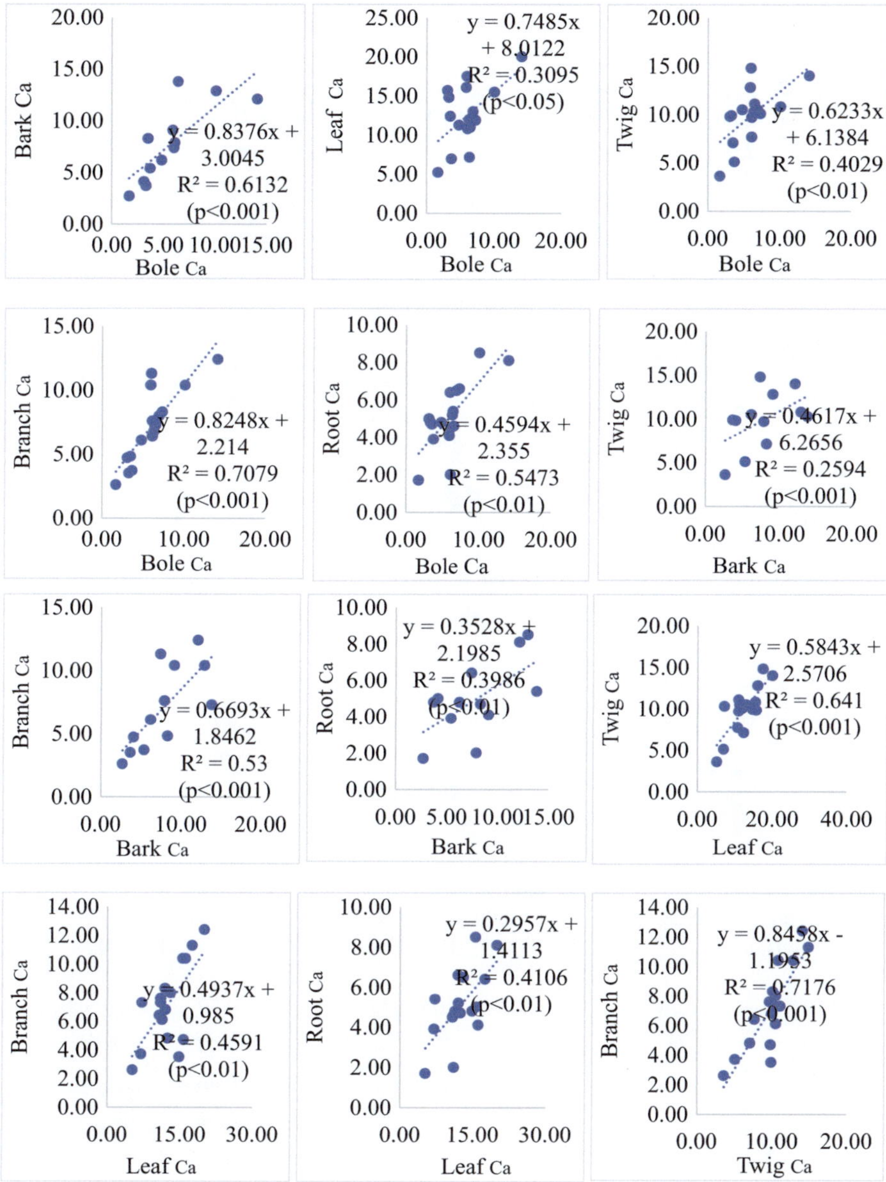

Fig. 6.2 (continued)

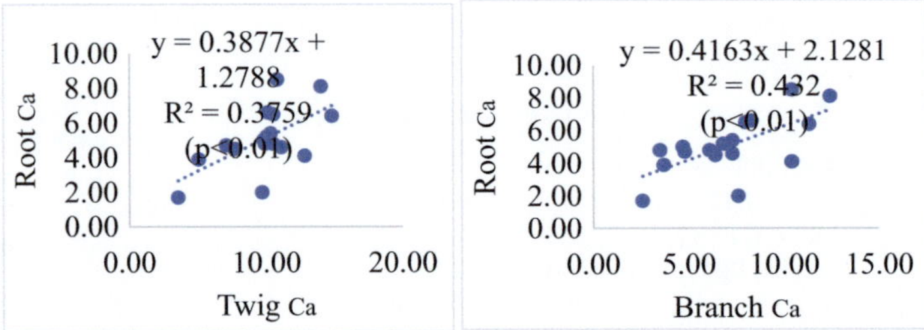

2e. Sodium (Na)

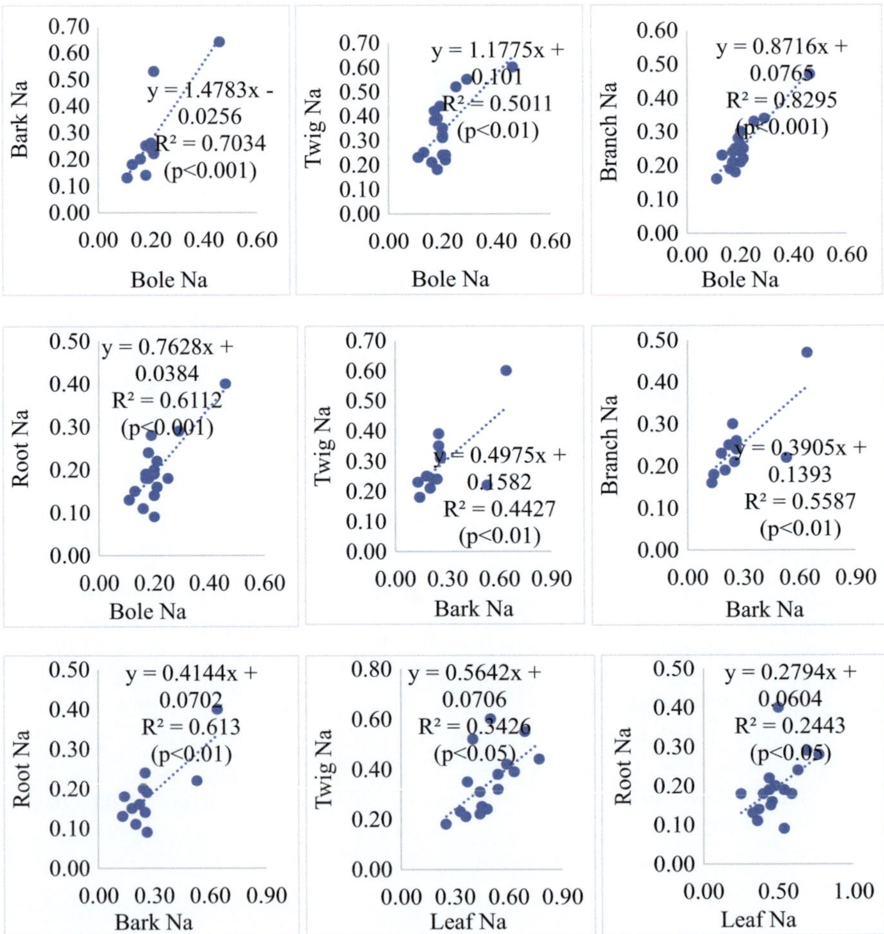

Fig. 6.2 (continued)

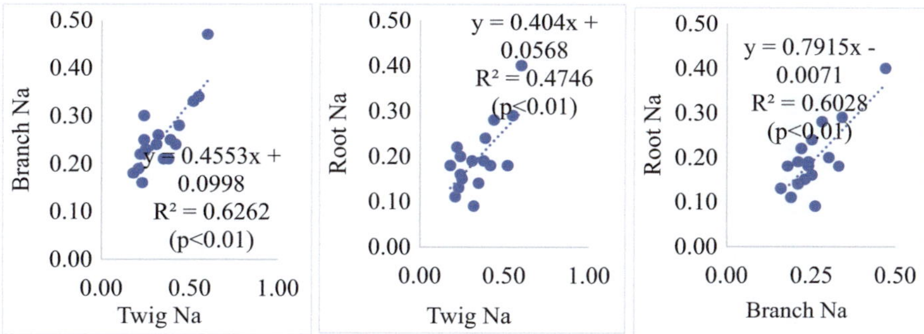

Fig. 6.2 (continued)

between bole wood and bark ($p < 0.001$; $R^2 = 0.773$), bole wood and coarse roots ($p < 0.01$; $R^2 = 0.416$), branch and twigs ($p < 0.001$; $R^2 = 0.702$), bark and coarse roots ($p < 0.01$; $R^2 = 0.360$), and twigs and coarse roots ($p < 0.01$; $R^2 = 0.341$) (Fig. 6.2a). Leaf N concentration was not correlated with that of any other parts. P concentrations among the tree components were relatively more correlated than N concentrations (Fig. 6.2a, b). Bole K concentration across the study species was strongly correlated with that of bark ($p < 0.001$; $R^2 = 0.485$), branch ($p < 0.001$; $R^2 = 0.517$), twig ($p < 0.001$; $R^2 = 0.769$), and coarse root ($p < 0.01$; $R^2 = 0.497$). Among all nutrients, positive inter-tree component correlations for Ca concentration were most common (Fig. 6.2d). The bole and bark Na concentrations were correlated ($p < 0.01$) with those of bark, twig, branch, and roots; and bole, branch, twig and roots respectively (Fig. 6.2e).

The prevalence of significant positive correlations in nutrient concentrations among the tree components across the species suggests a great deal of functional coordination of nutrient allocations across disparate types of tissues (Jones et al. 2019). Thus, a species having high Ca concentration in leaf is likely to have high concentration of Ca in stem wood. However, the extent to which nutrient concentrations in a tree component was correlated with those of the other components varied to an extent. For each nutrient concentration, 15 correlations between tree components were possible, or a total of 75 correlations for all the nutrients, with 15 for each. Of the 75 instances, significant correlations occurred in 53 instances, with 5 for N, 13 for P, 9 for K, 12 for Na, and 14 for Ca. It suggests that Ca concentration in most tree components vary similarly across the species despite them being different in structure, longevity, and other morphological characters. Some tree components resembled more with each in nutrient concentration than other components. The number of significant correlations across the species was higher for bole wood, bole bark, branches, and twigs (14–16) than for leaves and coarse roots (8–9). The link between leaf-nutrient and woody components was discernible. It shows that if Ca concentration is high in leaves of species, it would be high also in woody components, while in case of N, the components vary a great deal. In several deciduous species, as an example, the high leaf N concentration did not ensure high N concentration in bole wood.

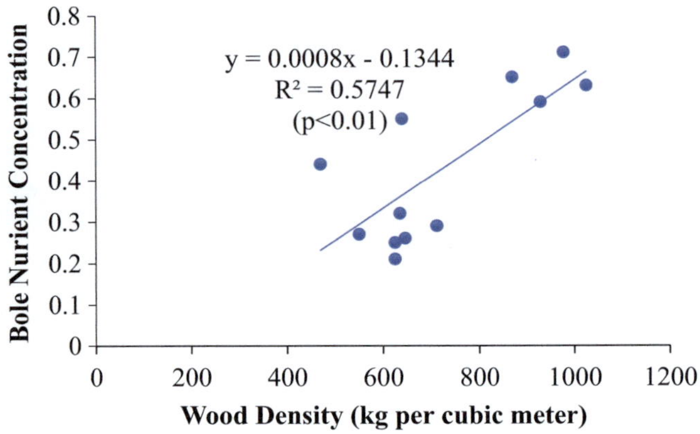

Fig. 6.3 Relationship between bole wood density and nitrogen concentration across the Indian Central Himalayan tree species

Some previous observations are that leaf and stem wood (bole) nutrient concentrations are not closely related across the species of angiosperms (Pietsch et al. 2014; Zanne et al. 2015). However, Heineman et al. (2016) showed that across the Panamian tropical forest species (mostly angiosperms) tree components were significantly positively correlated. Observations on ICH tree species show mixed findings. As for Ca, leaf concentration was positively correlated with all other tree components, but for N leaf concentrations were generally not correlated with those of other tree components. While bole wood and bark N concentrations were correlated, they did not show relationship with leaf N concentration.

The bole bark N concentration values were mostly between those of leaves and boles. Bole wood N concentration in oaks and sal, both known for highly dense wood (>900 kg m^{-3}), was high, 5.2–6.3 mg g^{-1}. In sal of eastern Nepal (SEN) stem wood N concentration (5.7 mg g^{-1}) was similar to that of ICH sal. A positive relationship between wood N and wood density was found in ICH tree species (Fig. 6.3). Martin et al. (2014) reported 14-fold variations in wood N concentration among 59 Panamian hardwood species (0.4–5.9 mg g^{-1}) and the concentration was positively correlated with wood density.

Patterns in Plant Nutrient Concentrations Along an Elevation Gradient

Nitrogen (N) concentration, both in bole (stem wood) ($p < 0.01$) and bark ($p < 0.05$) decreased toward the higher end of elevation gradient (Fig. 6.4a). Contrarily, leaf N concentration increased at the higher end of the elevation gradient, largely because of the presence of some deciduous species: *Acer* spp., *Aesculus indica*, and *B. utilis*. However, the changes along the elevation gradient were

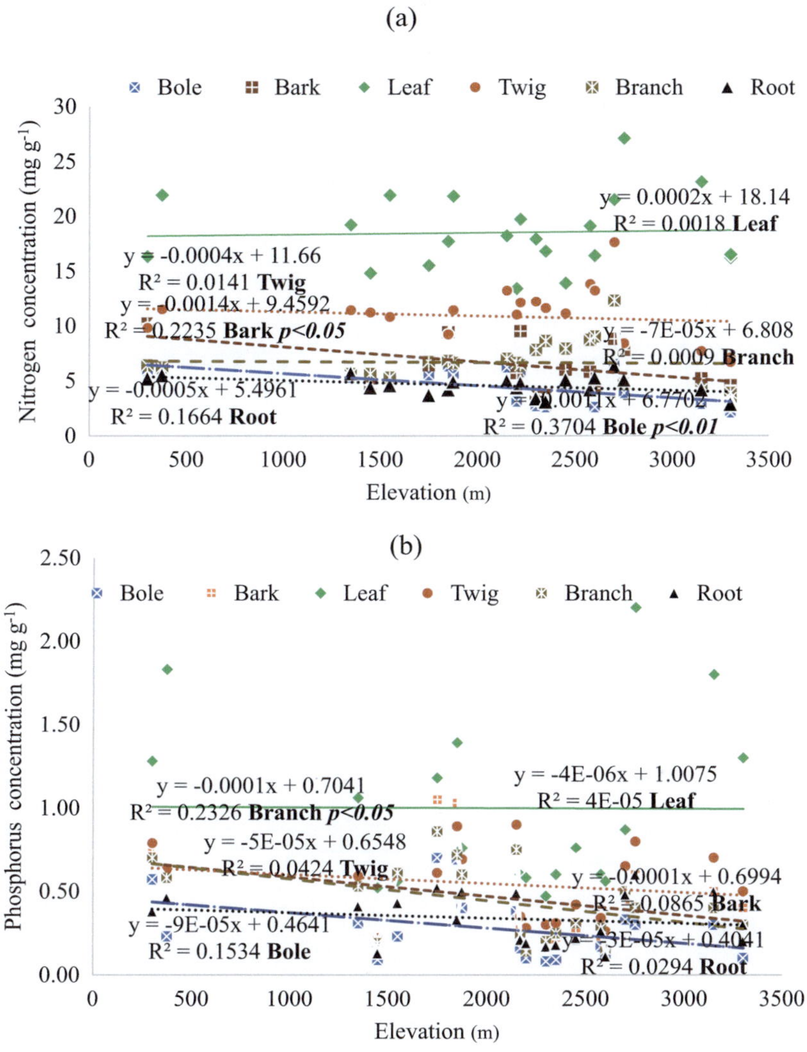

Fig. 6.4 Nutrient concentration in different tree components (viz., bole, bark, leaves, twig, branch, and root) of forests along elevation gradient (300–3300 m) in the Indian Central Himalayas (ICH). (**a**) Nitrogen (N); (**b**) Phosphorus (P); (**c**) Potassium (K); (**d**) Calcium (Ca); and (**e**) Sodium (Na). Forest types are shown in relation to elevation. The forests associated with the study sites were as follows: Sal (*S. robusta*) old-growth forest (350 m), Sal (*S robusta*) new-growth forest (385 m), chir pine (*P. roxburghii*) forest (1700–1750 m), banj oak (*Q. leucotrichophora*) chir pine forest (1850 m), banj oak (*Q. leucotrichophora*) forest (1900 m), rianj oak (*Q. lanata*) forest (2100–2150 m), tilonj oak (*Q. floribunda*) forest (2000–2200 m), horse-chestnut (*Aesculus indica*) forest (2300 m), silver fir (*A. pindrow*) forest (2500 m), kharsu oak (*Q. semecarpifolia*) forest (2650 m), maple (*Acer* spp.) forest (2750 m), birch (*B. utilis*) forest (3150 m), and rhododendron krummholz (*R. campanulatum*) forest (3300 m); elevation (mid-point of the range) indicated at the site. (Developed from data given in Khanna (1986), Singh and Singh (1992), Adhikari et al. (1995), and Garkoti and Singh (1994))

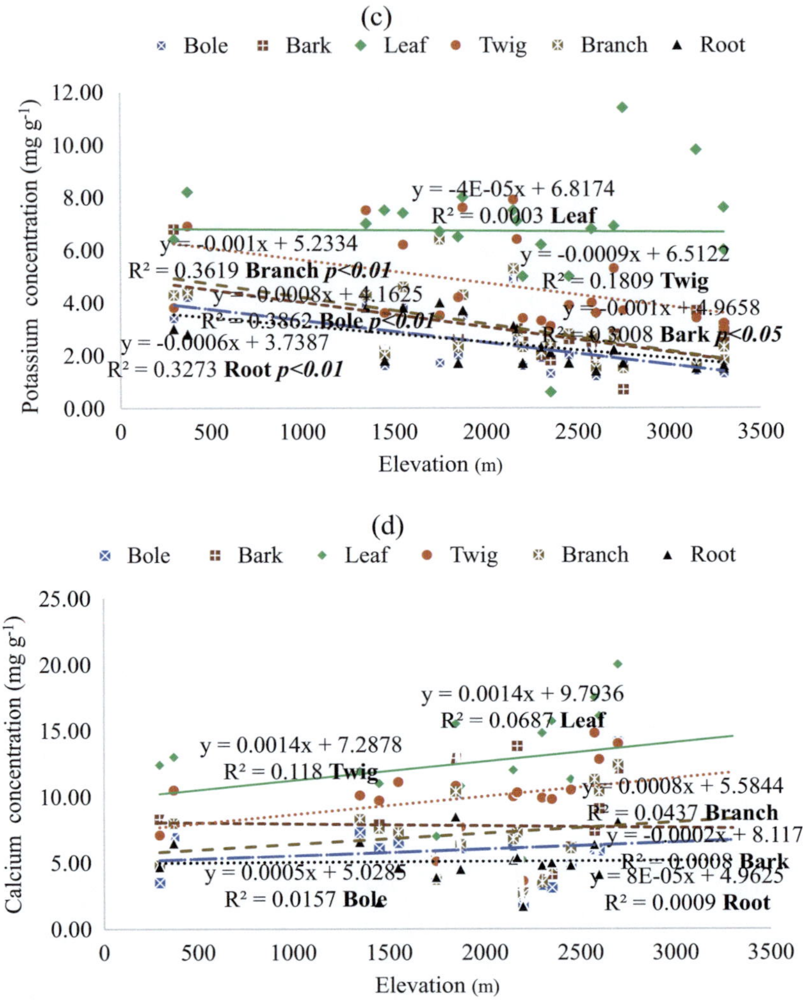

Fig. 6.4 (continued)

insignificant. In other components, N concentration was almost invariant in relation to elevation.

Phosphorus concentration in tree components tended to decrease with increasing elevation, though insignificantly (Fig. 6.4b), except in branch ($p < 0.05$). By and large, N and P concentrations had similar patterns of changes along the elevation gradient. But in leaves, in all tree components, K concentration decreased with elevation (Fig. 6.4c). Generally, Ca concentration increased with elevation (Fig. 6.4d); however, correlation was insignificant. Na concentration decreased with elevation in all the tree components (Fig. 6.4e); however, the decline in Na concentration was significant only in leaf. In brief, N and P showed moderate

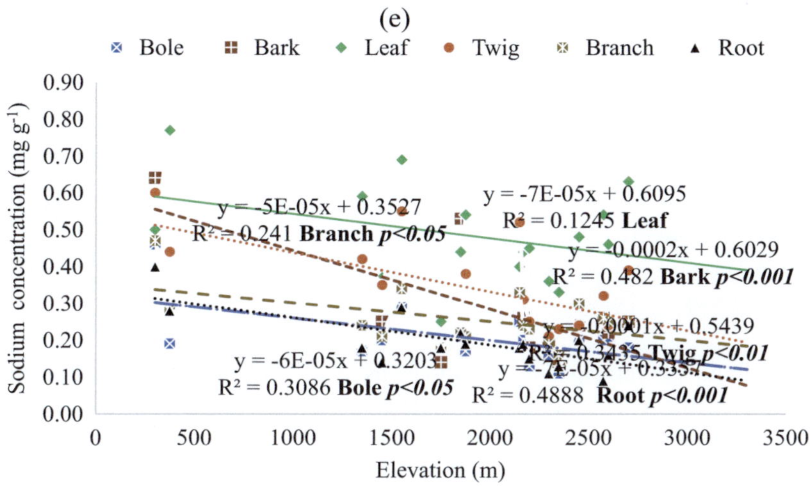

Fig. 6.4 (continued)

changes in relation to elevation. Leaf's nutrient concentration generally increases toward the upper side of elevation gradient largely because of the presence of several deciduous species. Concentrations of K and Na decrease and that of Ca increases from low to high elevation.

Nutrient Ratios in Tree Components

Plant growth is affected by N:P ratio of plant organs. Reich and Oleksyn (2004) showed that N:P ratios in tropical regions are affected by variations in P, while in higher latitudes with temperate climatic conditions, it is N-controlled. Globally, the mean N:P ratio in leaves is 11 (Reich and Oleksyn 2004) and in China's flora 14.4 (Han et al. 2005). The average N:P ratio in tree components across the ICH species ranges between 16.7 in coarse roots and 24.9 in twigs (Table 6.6). The N:P ratios in ICH trees are similar to or higher than those reported for natural forests and plantations of NW Spain (15.5 for leaves and 9.7 in twigs; Rodríguez-Soalleiro et al. 2018). However, in trees of both regions N:P ratios are well within the range suitable for plant growth. It may be pointed out that the high activity level of leaves depends on nutrient uptake by fine roots, the production of which depends on carbohydrates produced by leaves (Chen et al. 2020). Different N:P ratios have been suggested with regard to growth limitation of plants. In a study based on plants of wetlands of Europe N:P ratios less than 14 were considered to be N-limited and N:P ratios higher than 16 to be limited by P (Verhoeven et al. 1996). Some researchers consider that N:P ratios less than 10 are limited by N, and N:P ratios more than 20 are limited by P (Güsewell 2004). However, N:P ratios can vary widely among and within species (Phoenix et al. 2003). Güsewell and Koerselman (2002), as an

Table 6.4 Comparison between evergreen and deciduous woody species for N and P concentrations and mass ratios (Güsewell 2004)

	Nutrient concentration (mg g^{-1})		
	N	P	N:P ratio
Worldwide			
Evergreen woody	13.7	1.02	13.4
Deciduous woody	22.2	1.60	13.9
Mediterranean (Greece)			
Evergreen woody	10.3	0.62	16.6
Deciduous woody	23.0	1.53	15.1

example found that intraspecific variation was more (34%) than the variation among the species occurring at a site (25%). In response to variation in N and P supply N:P ratios in a species may vary 50-fold (Güsewell 2004). Though generally evergreen species have lower N and P concentrations in their tissues, than deciduous species, these plant groups do not differ in N:P ratios (Table 6.4).

Across the ICH species N:P ratios varied widely from 7.9 to 46.7 (5.9-fold) in bole, 6.1 to 35.2 in bark (about six-fold), 11.7 to 40.6 (3.5-fold) in leaves, 10.3 to 55.7 (5.4-fold) in twigs, 7.8 to 60.0 (7.7-fold) in branches, and 7.1 to 48.2 (6.8-fold) in roots (Table 6.5). On an average, deciduous species of ICH had higher N:P ratio (19.2–28.7) than evergreen species (15.2–22.6). Using the criterion of N:P ratio (below 10 limited by N and above 20 limited by P), we can say the growth of pine could be limited by N availability. However, more species seem to be affected by P-limitation. It includes *R. arboreum* at one of the two sites, *Symplocos chinensis, Lyonia ovalifolia* at one of the two sites, *Q. floribunda* at one of the two sites, *Aesculus indica, Acer caesium*, and *Ilex dipyrena*. Phosphorus-limitation is more common among deciduous species. With the exception of *Q. floribunda* at one site, the oaks do not appear to be affected by either N or P availability. In litter both in leaf and woody litter, N:P ratios were lower than in green leaves and bole wood, respectively, indicating that more N was resorbed than P.

The litter:nutrient concentrations (mg g^{-1}) across the six Nainital forests (part of ICH) are reported to range as follows (Mehra et al. 1985): N-leaf litter from 10.7 in chir pine to 16.3 in tilonj oak-dominated forest; and woody litter 1.6 in chir pine to 3.2 in sal- as well as tilonj oak-dominated forests (Table 6.6), P in leaf litter from 0.77 in mixed-oak-chir pine forest to 1.28 in tilonj oak-dominated forest and in wood litter from 0.19 in chir pine forest to 0.33 in tilonj oak-dominated forest; K in leaf 6.0 in sal forest to 7.1 in rianj oak-dominated forest, and in woody litter from 0.8 in chir pine-broad-leaved forest to 1.3 in rianj oak-dominated forest; and Ca in leaf litter from 10.2 in chir pine forest to 14.7 in rianj oak-dominated forest and in woody litter from 2.9 in sal forest to 9.0 in rianj oak-dominated forest (Table 6.6). Both in leaf and wood litter nutrient concentrations were lower than in mature leaves and bolewood. The leaf and wood concentrations across the forests were significantly correlated to for N ($p < 0.05$) and P ($p < 0.01$).

In leaf litter and wood litter of ICH forests the nutrient concentrations were lower than in green leaves and bole wood (Table 6.5), presumably because of nutrient

Table 6.5 N:P ratios in tree components across the species of Indian Central Himalaya

Species	Bole	Bark	Leaf	Twig	Branch	Root
S. robusta	10.35	13.78	12.73	12.41	9.14	13.68
P. roxburghii	7.86	6.10	13.14	11.15	7.79	7.12
Q. leucotrichophora	9.13	9.13	12.73	10.34	9.44	12.73
Q. semecarpifolia	19.70	19.56	24.71	27.08	24.60	13.13
R. arboreum	31.00	35.63	33.62	55.71	62.86	27.89
R. arboreum	18.62		11.75	16.42	11.79	13.33
Symplocos ramosissima	21.30		39.11	17.70	8.69	10.70
Symplocos chinensis	21.67	26.88	29.29	50.77	60.00	48.18
Lyonia ovalifolia	46.67	23.00	28.46	50.91	42.63	36.15
Lyonia ovalifolia	24.00		17.58	20.91	10.77	11.58
A. pindrow	20.95	21.43	18.29	26.43	25.48	23.18
Q. floribunda	12.89		14.80	16.13	10.00	8.89
Q. floribunda	44.38	35.19	32.83	44.86	45.42	32.86
Aesculus indica	33.75	21.11	38.09	40.67	39.00	20.00
Acer caesium	28.89	20.56	28.00	37.42	34.40	17.22
Acer cappadocicum	13.00	13.00	12.32	10.50	11.75	8.50
B. utilis	9.67	10.40	12.83	11.00	10.50	8.40
R. campanulatum	25.00	9.20	12.46	14.20	13.00	10.67
R. barbatum	21.00	11.25	12.69	13.40	11.33	14.50
Myrica esculenta	13.75		28.68	16.52	11.00	9.80
Q. lanata	16.32		23.95	14.67	9.33	10.20
Q. glauca	17.10		18.11	19.32	10.57	13.90
Ilex dipyrena	23.53	25.22	31.83	40.59	31.07	16.15
Mallotus philippensis	25.65		11.97	18.25	10.69	11.96
Average ±SE	**21.51**	**18.84**	**21.67**	**24.89**	**21.72**	**16.70**
	±2.07	**±2.26**	**±1.90**	**±3.02**	**±3.48**	**±2.07**

Derived from Khanna (1986), Singh and Singh (1992) Garkoti and Singh (1994), and Adhikari et al. (1995)

Table 6.6 Nutrient concentration in different litter components (Mehra et al. 1985)

	N (mg/g)		P (mg/g)		K (mg/g)		Ca (mg/g)	
Forest type	Leaf	Wood	Leaf	Wood	Leaf	Wood	Leaf	Wood
Sal forest	15.3	3.2	1.11	0.27	6.00	1.00	11.80	2.90
Pine-mixed Broad-leafed forest	13.3	2.5	0.92	0.23	6.20	0.80	10.90	3.80
Pine forest	10.7	1.6	0.81	0.19	6.10	1.40	10.20	4.80
Mixed oak-pine forest	12.5	2.6	0.77	0.23	6.80	1.20	12.30	4.40
Rianj-dominated mixed oak forest	16.2	2.7	1.02	0.26	7.10	1.30	14.70	9.00
Tilonj-dominated mixed oak forest	16.3	3.2	1.28	0.33	7.00	1.20	14.40	5.80
Average± SE	**14.05 ± 0.92**	**2.63 ± 0.24**	**0.99 ± 0.08**	**0.25 ± 0.02**	**6.53 ± 0.20**	**1.15 ± 0.09**	**12.38 ± 0.75**	**5.12 ± 0.87**

resorption. However, a greater difference in woody litter N and K concentrations may indicate that samples were not comparable adequately.

Singh and Negi (2023) showed that the mean N:P ratio of eight common species of Central Himalaya was 14.7 in mature leaves and 13.9 in senesced leaves, indicating that on an average both N and P were resorbed to a similar extent. The N:P ratio in leaves was the highest at bud break, whereafter it declined and then stabilized for varying periods. It was almost invariant in chir pine needles, generally remaining between 17 and 18 regardless of phenophases. In banj oak it decreased from April to October when the leaves were expanding and gaining mass, whereafter N:P ratios stabilized at around 13. *Q. floribunda* and *Machilus duthei* also followed this pattern. In general, the N:P ratio was lower at senescence stage than at mature stage of leaf, but not significantly.

As for N:K ratios in ICH forests (Table 6.7), the mean across the species for a tree component ranged from 1.92 in bole wood to 3.78 in leaves. An N:K ratio higher

Table 6.7 N:K ratios in tree components across the species of Central Himalaya

Species	Bole	Bark	Leaf	Twig	Branch	Root
S. robusta	1.74	1.50	2.55	2.58	1.49	1.73
P. roxburghii	3.24	–	2.31	1.94	1.05	0.93
Q. leucotrichophora	3.15	3.62	2.72	2.19	2.96	2.47
Q. semecarpifolia	3.25	3.14	3.12	3.32	4.73	2.86
R. arboreum	1.94	2.28	3.90	4.59	3.83	3.12
R. arboreum	1.54	–	2.20	1.86	1.61	2.10
Symplocos ramosissima	1.29	–	2.96	1.74	1.15	1.21
Symplocos chinensis	2.17	2.39	4.56	3.67	6.00	3.79
Lyonia ovalifolia	2.63	2.09	1.95	3.11	4.05	2.61
Lyonia ovalifolia	0.91	–	2.23	1.58	1.02	1.00
A. pindrow	2.20	2.31	2.78	2.85	2.82	3.00
Q. floribunda	1.14	–	2.71	1.59	1.23	1.07
Q. floribunda	2.73	3.17	2.77	2.45	3.76	2.30
Aesculus indica	1.42	1.90	2.89	3.70	3.71	1.48
Acer caesium	2.00	2.06	28.00	3.74	3.58	1.48
Acer cappadocicum	2.29	11.14	2.38	2.27	3.13	3.00
B. utilis	2.07	1.41	2.36	2.26	2.63	2.80
R. campanulatum	1.39	1.70	2.13	2.37	1.70	2.13
R. barbatum	1.62	1.45	2.75	2.09	1.89	1.81
Myrica esculenta	1.57	–	2.73	1.50	1.53	1.32
Q. lanata	1.27	–	2.43	1.67	1.32	1.61
Q. glauca	1.36	–	2.74	1.52	1.30	1.50
Ilexdipyrena	1.82	2.32	2.81	3.45	3.22	1.56
Mallotus philippensis	1.40	–	2.67	1.67	1.41	1.96
Average ±SE	**1.92 ±0.14**	**2.83 ±0.62**	**3.78 ±1.06**	**2.49 ±0.18**	**2.55 ±0.28**	**2.03 ±0.16**

Derived from Khanna (1986), Garkoti and Singh (1994), and Adhikari et al. (1995)

Table 6.8 N:P and N:K ratios in different litter components

	N:P ratio		N:K ratio	
Forest type	Leaf	Wood	Leaf	Wood
Sal forest	13.78	11.85	2.55	3.20
Pine-mixed broadleaf forest	14.46	10.87	2.15	3.13
Pine forest	13.21	8.42	1.75	1.14
Mixed oak-pine forest	16.23	11.30	1.84	2.17
Rianj-dominated mixed oak forest	15.88	10.38	2.28	2.08
Tilonj-dominated mixed oak forest	12.73	9.70	2.33	2.67
Average ± SE	**14.38 ± 0.58**	**10.42 ± 0.50**	**2.15 ± 0.12**	**2.40 ± 0.31**

Derived from Mehra et al. (1985)

than 2 indicates that K is limiting to plant growth (Pan et al. 2020). Among the ICH trees, the average N:K ratio across the species was distinctly higher than 2.0 that was found in leaves and bark (Table 6.7). Stem wood had relatively higher N:K ratio in *P. roxburghii* (3.24), *Q. semecarpifolia* (3.25), and *Q. leucotrichophora* (3.15), while it was markedly lower in several undercanopy species namely, *Lyonia ovalifolia* (0.9), *Symplocos ramosissima* (1.29), *Mallotus philippensis* (1.40), *R. arboreum* (1.54), and *Myrica esculenta* (1.57). Across the tree components of all species, N:K ratios ranged from 0.93 in coarse roots to 4.73 in branches (Table 6.7). In the Spanish forests the average N:K ratio is reported to occur between 1.76 in twigs and 2.99 in leaves. Thus, the tree components of the two regions may suggest that N:K ratios occur within a limited range. The N:Ca ratios of Central Himalayan trees indicate that more Ca concentrations than N occur in bole, bark, branches, and roots, while in leaves the mean N:Ca ratio is higher, 1.44. In twigs, the average N:Ca ratio is close to 1 (1.09). In Spanish forests, N:Ca ratio was between 1.22 in bark and 4.78 in leaves. Among the Himalayan tree species N:Ca ratios in leaves range from 1.14 in *L. ovalifolia* to 2.02 in *Myrica esculenta*. In bole it ranged from 0.73 in *Q. glauca* to 1.69 in *S. robusta*.

The litter N:K ratios on an average increased in wood, from 1.92 to 2.40 (showing more decline of K) (Table 6.8), but decreased in leaf litter (indicating more resorption of N), from 3.78 to 2.15 (Table 6.8).

Standing State of Nutrient Mass in Forests

Nitrogen

Across the ICH forests total N mass in the ecosystem (vegetation + forest floor litter +30 cm deep soil) varied about two-fold from 5079 kg ha^{-1} in chir pine to 11,768 kg ha^{-1} in tilonj oak forest, located at 2000 m (Khanna 1986; Garkoti and Singh 1994; and Adhikari et al. 1995). Large N mass at an ecosystem level was also found in kharsu oak forest (11,048 kg ha^{-1}), and rianj oak forest (10,165 kg ha^{-1}) (Table 6.9), occurring between about 2200 and 2700 m, partly because the forests

Table 6.9 Standing state of nutrients in vegetation, forest floor litter and soil (0–30 cm depth) across the Indian Central Himalayan (ICH) forests: from sal to tilonj oak in Nainital transect sites, and from horse-chestnut forest to rhododendron forest of Pinder valley transect

Forest types with elevation in parentheses	Nutrient	Nutrient (kg ha^{-1}) storage in ecosystems components (% of total nutrient storage in parentheses)			
		Vegetation (trees)	Litter	Soil	Total forest
Sal (*S. robusta*) old-growth (350 m)	N	4455 (52.31)	54 (0.63)	4007 (47.05)	8516
	P	403 (72.22)	3 (0.54)	152 (27.24)	558
	K	2511 (93.17)	15 (0.56)	169 (6.27%)	2695
	Ca	3123 (33.23)	51 (0.54)	6224 (66.23)	9398
Sal new-growth (385 m)	N	2940 (53.90)	57 (1.04)	2458 (45.06)	5455
	P	262 (66.50)	3 (0.76)	129 (32.74)	394
	K	1655 (90.29%)	15 (0.82)	163 (8.89)	1833
	Ca	2155 (17.63)	52 (0.43)	10,017 (81.95)	12,224
Chir pine (*P. roxburghii*)- Site I (1700 m)	N	1161 (19.14)	110 (1.81)	4794 (79.04)	6065
	P	132 (41.12)	11 (3.43)	178 (55.45)	321
	K	628 (70.64)	151 (16.99)	110 (12.37)	889
	Ca	854 (16.49)	131 (2.53)	4195 (80.98)	5180
Chir pine (*P. roxburghii*)—Site II (1750 m)	N	1105 (21.76)	10 (0.20)	3964 (78.05)	5079
	P	199 (47.27)	4 (0.95)	218 (51.78)	421
Banj oak (*Q. leucotrichophora*)-pine (1850 m)	N	2716 (34.36)	99 (1.25)	5090 (64.39)	7905
	P	264 (56.53)	4 (0.86)	199 (42.61)	467
	K	1217 (82.79)	29 (1.97)	224 (15.24)	1470
	Ca	3424 (31.68)	109 (1.01)	7276 (67.31)	10,809
Banj oak (*Q. leucotrichophora*) (19,00 m)	N	2999 (37.93)	89 (1.13)	4819 (60.95)	7907
	P	136 (46.42)	5 (1.71)	152 (51.88)	293
	K	912 (35.45)	14 (0.54)	1647 (64.01)	2573
	Ca	4488 (47.95)	113 (1.21)	4758 (50.84)	9359

Rianj oak (*Q. lanata*)-Site I (2100 m)	N	2339 (29.37)	105 (1.32)	5520 (69.31)	7964
	P	116 (38.93)	3 (1.01)	179 (60.07)	298
	K	552 (23.39)	14 (0.59)	1794 (76.02)	2360
	Ca	3309 (40.77)	116 (1.43)	4692 (57.80)	8117
Rianj oak (*Q. lanata*)-Site II (2150 m)	N	4105 (40.38)	64 (0.63)	5996 (58.99)	10,165
	P	317 (60.96)	2 (0.38)	201 (38.65)	520
	K	2862 (92.26)	16 (0.52)	224 (7.22)	3102
	Ca	3915 (39.71)	70 (0.71)	5874 (59.58)	9859
Tilonj (*Q. floribunda*) oak- Site I (2000 m)	N	4562 (38.77)	87 (0.74)	7119 (60.49)	11,768
	P	153 (39.95)	3 (0.78)	227 (59.27)	383
	K	1404 (49.16)	7 (0.25)	1445 (50.60)	2856
	Ca	5595 (49.33)	77 (0.68)	5671 (50.00)	11,343
Tilonj (*Q. floribunda*) oak—Site II (2200 m)	N	5598 (50.67)	60 (0.54)	5390 (48.79)	11,048
	P	462 (72.07)	3 (0.47)	176 (27.46)	641
	K	3986 (95.56)	23 (0.55)	162 (3.88)	4171
	Ca	5901 (51.45%)	84 (0.73)	5484 (47.82)	11,469
Horse-chestnut (*Aesculus indica*) (2300 m)	N	2773 (31.62%)	58.6 (0.67)	5937.4 (67.71)	8769
	P	99 (25.28)	1.6 (0.41)	291.0 (74.31)	391.6
	K	1267.1 (30.37)	15.4 (0.37)	2890.0 (69.26)	4172.5
	Ca	2911.1 (36.69)	66.4 (0.84)	4957.3 (62.48)	7934.8
Pindrow fir (*A. pindrow*) (2500 m)	N	2786.3 (32.03)	78.2 (0.90)	5834.2 (67.07)	8698.7
	P	147.4 (41.16)	3.6 (1.0)	2,07.1 (57.83)	358.1
	K	1270.4 (43.37)	8.8 (0.30)	1650 (56.32)	2929.2
	Ca	3221.9 (43.03)	69.0 (0.92)	4196.7 (56.05)	7487.6
Kharsu oak (*Q. semecarpifolia*) (2650 m)	N	4102.5 (37.15)	99.5 (0.90)	6840 (61.94)	11,042.6
	P	245.8 (45.57)	5.2 (0.96)	288.4 (53.47)	539.4
	K	1532.6 (39.84)	15.4 (0.40)	2298 (59.75)	3846

(continued)

Table 6.9 (continued)

Forest types with elevation in parentheses	Nutrient	Nutrient (kg ha^{-1}) storage in ecosystems components (% of total nutrient storage in parentheses)			
		Vegetation (trees)	Litter	Soil	Total forest
	Ca	5078.9 (46.72)	100.1 (0.92)	5691.8 (52.35)	10,870.8
Maple (*Acer* spp.) (2750 m)	N	1555 (21.96)	104 (1.47)	5423 (76.57)	7082
	P	130.3 (41.76%)	5.1 (1.63)	176.6 (56.60)	312
	K	660.9 (21.39)	32.8 (1.06%)	2396 (77.55)	3089.7
Birch (*B. utilis*) (3150 m)	N	730.7 (7.45)	74 (0.75)	8998 (91.79)	9802.7
	P	70.5 (15.82)	5.1 (1.14)	370 (83.03)	445.6
	K	335.2 (9.94)	27.9 (0.83)	3008.4 (89.23)	3371.5
R. campanulatum (3300 m)	N	173.1 (2.69)	62 (0.96)	6190 (96.34)	6425.1
	P	13.2 (7.08)	7.1 (3.81)	166.1 (89.11)	186.4
	K	94.1 (4.81)	22.4 (1.14)	1841.7 (94.05)	1958.2

Nutrient amounts in the components are as kg ha^{-1} and values in parentheses are the percentage of total nutrient storage of the ecosystem in the components (Singh and Singh 1987, 1992; Chaturvedi 1983; Khanna 1986; Rawat 1983; Garkoti and Singh 1994; and Adhikari et al. 1995) were generally less disturbed than forests of lower elevations (from the foothills to 1700–1800 m), and they were old-growth or/and in mature stage. They seem to represent the potential storage of biomass and nutrients for the region.

In much of the ICH forests, vegetation and soil were the important nutrient pools, and the litter on forest floor was only a minor in size, though functionally so vital. In several oak and sal forests the amount of N stock in vegetation was between 4124 and 4550 kg ha^{-1}. Nitrogen stock in vegetation of sal forest of Nepal (5738.2 kg) was found to be still greater. The percentage of total nitrogen mass in a forest ecosystem that occurs in vegetation decreases rapidly above 2600 m, from about 34.5% in kharsu oak forest to 7.6% in treeline birch forest and 2.71% in krummholz of *R. campanulatum*. Nevertheless, total N mass in ecosystem in the high elevations remains fairly high, 6400–9700 kg ha^{-1}, because of a large soil N pool that contains 91%–96% of total ecosystem N stock. Among the lower elevation forests, the proportion of nitrogen mass in soil is high in chir pine and the forests having sizeable chir pine trees (~80%). In sal and oak forests, standing vegetation accounts for about 30.0%–52.3% of total ecosystem N pool. Forest floor litter generally accumulated less than 1% of ecosystem N, which, however, could be a little underestimation.

Phosphorus

Phosphorus (P) mass in vegetation ranged about 35-fold from 13.2 kg ha^{-1} in the treeline forest of *R. campanulatum* to 462 kg ha^{-1} in tilonj oak-dominated forest (Table 6.9). The second highest P content in vegetation (here vegetation and biomass have been used interchangeably) was in sal old-growth forest (403 kg ha^{-1}). In sal forest of eastern Nepal (SEN) P stock (537.7 kg ha^{-1}) exceeded that of ICH sal forest (in fact any of the forests). Along the elevation gradient, the standing P mass in vegetation remained high (>120 kg ha^{-1} in maple forest) up to 2750 m from where it declined sharply in treeline forests of birch and rhododendron (*R. campanulatum*). The values of P storage in vegetation in some of the Himalayan forests are higher than those reported for tropical moist forests (42–164 kg ha^{-1} yr^{-1}, Vitousek and Sanford 1986) with similar forest biomasses. Of the total amount of P in ecosystem, generally 21.4–56.4%, occurred in vegetation. As expected within vegetation, most of P mass was in trees (generally >95%). Phosphorus in forest floor litter mass was small 2.5–13 kg ha^{-1}, but more than that of the tropical moist forests of Brazilian Atlantic regions (2.6–5.4 kg ha^{-1}). In Central Himalayas, soil (0–30 cm depth) P mass ranged between 106 kg ha^{-1} (rhododendron forest) and 370 kg ha-1 (birch forest). Along the elevation gradient it remained fairly high up to treelines (370 kg ha-1 birch forest). The proportional allocation to soil increased considerably at the higher end of elevation gradient, from 53.7% in kharsu oak forest to 82.9 and 89.1% in treeline forests of birch and rhododendron, respectively. The total P mass in ecosystem ranged from 184.7 to 645.1 kg ha^{-1} in rhododendron and tilonj oak-dominated forests, respectively. Values were relatively higher in oak forests, generally from 250–645 kg ha^{-1}.

Potassium

The elevation pattern of standing state of K was similar to that of N and P. The standing state of K in vegetation was high up to about 2650 m (1532.6 kg ha^{-1}), and then declined to 94.1 kg ha^{-1} in the treeline (Table 6.9). Between foothills and 2200 m elevation (the lower transect) it ranged between 628 kg ha^{-1} and 3986 kg ha^{-1} (Table 6.9). Values in a similar range (664–3020 kg ha^{-1}) have been reported for the tropical moist forests with similar standing forest biomass values (Vitousek and Sanford 1986). In soil it ranged from 110 kg ha^{-1} in chir pine forest to 3008.4 kg ha^{-1} in birch forest, and increased with elevation from 1650 kg ha^{-1} in silver fir forest to birch forest.

The total ecosystem K was between 649 kg ha^{-1} in mixed oak-pine forest and 4192.5 kg ha^{-1} in banj oak forest. The proportional allocation to soil increased sharply with elevation, and in the treeline forests soil accounted for 89.3%–94% of total ecosystem K. In contrast, in the lower part of the elevation gradient much of the ecosystem K was in vegetation (generally 58%–92%). The K mass in the forest floor litter was invariably small, ranging from 12 to 54.6 kg ha^{-1}.

Calcium

The standing Ca mass in vegetation ranged from 854 kg ha^{-1} in chir pine forest to 5901 kg ha^{-1} in tilonj-dominated mixed oak forest (Table 6.9). The Ca mass in vegetation was higher in oak, silver fir, and sal forests than in others. In tropical moist forests with similar biomass the Ca storage in vegetation ranges from 432 to 3900 kg ha^{-1} (Vitousek and Sanford 1986). Total Ca in forest ecosystem ranged from 5180 kg ha^{-1} in chir pine forest to 12,224 kg ha^{-1} in sal seedling-growth forest (Table 6.9). High (above 10,000 kg and more) Ca mass in ecosystem occurs also in all the oak forests (Table 6.9). These values of Ca are comparable with those of N; however, Ca in soil pool was proportionately larger than N soil pool, particularly in forests having sal, banj oak, and chir pine (66.2%–87.6%). In other forests, soil pool was relatively small (45.5% in kharsu oak forest to 63.0% in horse-chestnut forest). The forest floor litter pool of Ca ranged from ~45 kg ha^{-1} in pine-mixed broad-leaved forest to 150 kg ha^{-1} in chir pine forest.

Sodium

The vegetation pool of Na ranged from 38.9 kg ha^{-1} in chir pine forest to 318.9 kg ha^{-1} in sal old-growth forest. The total Na in ecosystem ranged from 100.4 kg ha^{-1} in pine mixed forest to 772 kg ha^{-1} in kharsu oak forest. In Pinder valley forests, generally soil Na pool was markedly larger than vegetation Na pool.

Forest Floor Litter: A Relatively Small Component

In the monsoon-dominated Himalayan regions (outer ranges), the storage of nutrients in forest floor litter pool is small and remains so even in subalpine forests. Except in a chir pine forest during the pre-monsoon period, forest floor litter remains an inconspicuous feature in these forests. A comparison with similar forests in Japan (plantations in temperate belt of *Cryptomenia japonica*, called Japanese cedar, *Chamaecyparis obtusa* called Japanese cypress, broad-leaved deciduous species—*Fagus crenata*, *Betula* spp., *Quercus* spp., broad-leaved evergreen (*Castanopsis, Lithocarpus*), and subalpine forests of fir (*Abies* spp.) and spruce (*Picea* spp.) as an example, indicates that in general, forest floor nutrient pool was distinctly larger in subalpine forests of Japan than in ICH counterparts. For example in the ICH forests N in forest floor litter ranges from 54–110 kg ha^{-1}, while in Japan it is between 261 and 533 kg ha^{-1} in subalpine forests and 89–123 kg ha^{-1} in plantations and natural forests of temperate belt (Takahashi 2021). The litterfall during pre-monsoon (March–April) in Himalayas is immediately followed by the torrential monsoon rain (July–September) when litter decomposition is the fastest in the year. Herbs mostly deposit litter after monsoon when conditions are relatively dry and cold. However, because of higher nutrient concentrations and low lignin content, they decompose rapidly.

A comparison with forest ecosystems of central Yunnan Plateau, China having similar forests (*Castanopsis carlesii, Lithocarpus mairei, P. yunnanensis, P. armendi, Q. semecarpifolia,* and *Keteleeria evelyniana*) may help in generalizing the pattern of standing state of nutrients of the forests of Himalayas and adjacent monsoonal areas. Like ICH, precipitation in Yunnan is strongly seasonal, 80% annual precipitation occurring during monsoon months (May to October), though the amount is lower. In Yunnan Plateau ecosystems, N storage ranges from 4470 to 8910 kg ha^{-1}, with an average allocation of 86.8% in soil, 10.27% in vegetation, and 2.58% in litter. In the central Himalayan transect the total ecosystem N storage was quite similar, 5200 kg ha^{-1}–11,700 kg ha^{-1}; however, the mean allocation of nutrient storage in vegetation was higher (32%), and that in soil (67.1%) and forest floor litter mass (0.93%) lower.

In brief, several ICH forests, namely sal, tilonj, and kharsu oak forests are sites of large amounts of nutrient storage. Forest biomass accounts for over 40% of total ecosystem nutrient stock up to 2500 m elevations or so. Soil becomes an increasingly predominant nutrient pool toward the treeline. Regardless of elevation, forest floor litter is a minor soil storage component, making these forests closer to those of tropical forests.

Standing State of Nutrients by Tree Components

On an average across the ICH forests from sal at foothills to birch at treeline stem wood storage of nutrients was 1265.5 kg ha^{-1} N, 97.54 kg ha^{-1} P, 715.2 kg ha^{-1} K, 1469.2 kg ha^{-1} Ca, 66.1 kg ha^{-1} Na, and 131.4 kg ha^{-1} Mg. These values of nutrient storage in stem wood are high and comparable with those of tropical forests of Borneo, Congo, French Guiana, and Panama studied by Bauters et al. (2022). The average nutrient uptake values by the forest vegetation are 20.21 kg ha^{-1} yr^{-1} N, 1.68 kg ha^{-1} yr^{-1} P, 11.5 kg ha^{-1} yr^{-1} K, 21.19 kg ha^{-1} yr^{-1} Ca, and 1.12 kg ha^{-1} yr^{-1} Na. These values of nutrient uptake are comparable with those of the above tropical moist forests reported by Bauters et al. (2022): 19 kg ha^{-1} yr^{-1} for Ca and 9 kg ha^{-1} yr^{-1} for K. The nutrient uptake has high carbon cost for trees, as they are required to bear the cost of fine root production and ectomycorrhizal association (Usman et al. 1997; Pande et al. 2007).

Wood Storage Nutrient Conservation Strategies

It should be seen in view of the fact that wood has much less a physiological role than short-lived leaves and reproduction structures that are short-lived but have high demand, unlike that of heartwood, where nutrients may be locked for decades and centuries, unless they move internally to short-lived tissues. Recently Roy et al. (2022) indicated the translocation of K and Mg to near bark region from near the pith region. Bauters et al. (2022) think that internal changes in wood nutrient concentration is poorly understood, and wonder whether wood storage of nutrients is a nutrient conservation strategy. In Himalayan ecosystems are highly vulnerable to nutrient losses due to erosion and leaching, particularly during peak of the monsoon rainfall (July–August) (Singh and Singh 1992). Furthermore, a species might accumulate nutrients in wood to outcompete the neighboring species. Prevalence of monodominant forests is a common feature of Himalayas (Singh and Thadani 2013).

We have limited understanding about the resorption of nutrients of heartwood and sapwood (see Box 6.1 for general features of sapwood and heartwood), though it has been assumed for long that some elements are resorbed from sapwood during the heartwood formation (Bamber and Fukazawa 1985). In fact, nutrient resorption when heartwood is formed has been equated with leaf senescence that enables nutrient conservation and reutilization (Attiwill 1980). But it is hardly known whether nutrients are retranslocated from heartwood to sapwood and reused, though the study of Roy et al. (2022) gives some evidences for this. Heineman et al. (2016) reported a decrease in nutrients on bark-to-pith gradient. This may be partly because of nutrient resorption while transitioning from sapwood to heartwood (Bauters et al. 2022). In brief, the large nutrient stocks and annual uptake in wood and wide interspecies variations emphasize that wood nutrients should be prominently included in forest biogeochemistry studies (Bauters et al. 2022).

Box 6.1 Sapwood and Heartwood

Sapwood is the living outer layer of wood (having living parenchyma), while heartwood is inner layer mostly of dead tissues, generally vessels in it are blocked with tyloses. Sapwood is light and heartwood dark in color. Secondary compounds accumulate in heartwood, for example resins, phenols, and turpentine, while storage products such as starch, soluble sugars, and amino acids occur in sapwood. Minerals might be removed from senescing sapwood.

All wood begins as sapwood that is formed by cambium just under bark. The cambium produces bark to outside and wood cells to the inside. The sapwood occupies a large area in growing trees, but as trees age, heartwood increases in area, and typically occupies a higher proportion of wood in a stem cross section than sapwood.

Generally, sapwood is more moist, has higher concentrations of nutrients N, P, and K, and decomposes rapidly than heartwood (Meerts 2002). Meerts (2002) showed that sapwood had higher nutrient concentrations than heartwood in 42 out of 56 species for N, 59 out of 64 species for P, 60 out of 80 species for K, 49 out of 92 species for Ca, and 38 out of 76 species for Mg. Wood nutrient concentrations are higher in angiosperms than gymnosperms.

Tree species and individuals of a species vary widely in the proportion of sapwood and heartwood and is modified by environment. For example the wood of sal (*S. robusta*) and sheesham (*Dalberia sissoo*) are largely heartwood, while *Terminalia bellerica* and *Fraxinus micrantha* have more of heartwood (Gamble 1985). In oaks, pines, deodar, and teak both sapwood and heartwood are present prominently.

On an average across all forests and study nutrients, stem wood (bole) accounted for 37.39 ± 2.77% (P) to 42.99 ± 2.24% (Na) of total nutrient storages in trees. The percentage of nutrients in stem wood was significantly higher in canopy species than undercanopy species (for N $p < 0.01$), and for P, K, and Ca $p < 0.05$. Across the nutrients, the mean percentage of nutrient storage in stem wood (of total in trees) was between 39.30 ± 1.99% (Mg) and 45.14 ± 1.74% (Ca) for canopy species and between 24.11 ± 2.99% (N) and 40.02 ± 4.90% (Na) for undercanopy species (Table 6.10). Relatively lower allocation of biomass to stem might slow down height growth and result in increased allocation to branches. For example the understory saplings of *A. amabilis* are reported to grow slowly in height because of overall low rates of biomass production and low allocation of biomass to stem and high allocation to branches (King 1997). On the other hand, undercanopy forest species had significantly higher nutrient allocation to branches and twigs than canopy tree species. For example undercanopy species, *Lyonia ovalifolia* had from 21.6% (K) to 31% (N) nutrient allocation to branches, compared to the average range of allocation across all tree species from 18.11% (P) to 18.7% for K.

Table 6.10 Amount of nutrient mass (kg ha^{-1}) in stem wood of trees

Forest	N	P	K	Ca	Na	Mg
Canopy species						
Sal old-growth	2537.85 (56.2)	241.02 (59.1)	1472.28 (58.0)	1547.1 (48.5)	194.64 (60.1)	207.89 (42.6)
Sal seedling-growth	1471.99 (49.2)	136.04 (51.4)	862.78 (51.2)	934.89 (42.4)	109.97 (54.0)	121.85 (39.7)
Chir pine-mixed broad-leaved	537.36 (44.2)	55.57 (46.3)	237.67 (33.7)	538.2 (43.5)	19.06 (41.2)	54.27 (41.2)
Chir pine	649.14 (54.9)	77.67 (58.2)	227.38 (35.4)	464.51 (53.2)	21.79 (54.6)	66 (48.1)
Banj oak-chir pine	1080.1 (38.9)	111.32 (41.4)	402.88 (32.4)	1383.09 (40.0)	35.71 (37.3)	98.54 (35.9)
Rianj oak	1433.96 (34.4)	94.83 (29.5)	1152.45 (39.8)	1505.73 (38.2)	55.5 (33.7)	117.35 (32.8)
Tilonj oak	1901.17 (33.6)	148.45 (31.9)	1537.59 (38.4)	2175.8 (36.6)	66.71 (29.9)	177.05 (34.9)
Kharsu oak	2044.5 (49.9)	108.2 (44.0)	705.2 (46.0)	2463.9 (48.6)	61.2 (43.6)	–
Aesculus indica	835.2 (30.2)	26.2 (26.5)	511.7 (40.5)	1173.5 (40.4)	54 (46.9)	–
A. pindrow	1145.4 (41.1)	62.3 (42.3)	601.7 (47.4)	1574.3 (48.9)	61.7 (38.9)	–
Q. semecarpifolia	2021.6 (50.7)	107.4 (44.3)	694.9 (46.4)	3664.2 (56.0)	60 (43.8)	–
Q. floribunda	42.61 (38.2)	1.08 (34.1)	15.26 (38.9)	72.5 (45.5)	1.4 (43.8)	–
Maple (*Acer caesium*)	564.3 (37.2)	41.8 (32.8)	264.7 (41.6)	–	–	–
Birch forest	257.2 (36.6)	25 (36.3)	128.9 (40.1)	–	–	–
Average ± SE	**1180.17 ± 200.53 (42.52 ± 2.23)**	**88.53 ± 16.58 (41.29 ± 2.72)**	**629.67 ± 128.71 (42.13 ± 1.88)**	**1458.14 ± 282.45 (45.14 ± 1.74)**	**61.81 ± 14.53 (43.98 ± 2.56)**	**120.42 ± 21.09 (39.30 ± 1.99)**
Undercanopy species						
R. arboreum	17 (20.0)	0.6 (27.5)	7.4 (29.6)	43.2 (35.4)	0.74 (31.0)	–
Symplocos chinensis	2.9 (21.6)	0.13 (36.1)	1.43 (34.6)	8.1 (40.1)	0.29 (48.3)	–

Lyonia ovalifolia	1.56 (21.0)	0.04 (19.4)	0.73 (26.2)	2.96 (29.8)	0.08 (32.1)	–
Ilex dipyrena	1.48 (36.0)	0.07 (37.2)	0.76 (42.2)	2.17 (39.1)	0.072 (48.6)	–
R. campanulatum	34.9 (21.9)	1.4 (12.0)	24.9 (29.1)	–	–	–
Average ± SE	**11.57 ± 6.52 (24.11 ± 2.99)**	**0.45 ± 0.26 (26.45 ± 4.85)**	**7.04 ± 4.64 (32.32 ± 2.81)**	**14.11 ± 9.79 (36.12 ± 2.32)**	**0.30 ± 0.16 (40.02 ± 4.90)**	–

Source: Khanna (1986), Adhikari (1992), Garkoti and Singh (1994), and Adhikari et al. (1995)
Values given in parentheses are the percentage of nutrient mass of trees in stem wood (bole)

Between evergreen broad-leaved species and deciduous broad-leaved species groups, the average nutrient percentage in stem wood was significantly higher for evergreen species for N, P, and Ca ($p < 0.05$ in each). In deciduous species, in *Aesculus indica*, as an example, stem wood accounted for 26.7% N, 22.1% P, and 18.2% K compared to the average of 41.7%, 40.7%, and 40.9% for N, P, and K respectively for all canopy species.

The next in importance to stem wood were either coarse roots (containing from 17.47% for N to 20.75% for Na of total nutrients in trees) or branches (accounting for 13.47% for Mg to 20.55% for Ca). Though high in nutrient concentrations, twigs and leaves were the smallest contributors to tree nutrient standing state, forming between 6.75% (for Ca) and 10.46% (for N) and between 5.4% (Na) and 10.14% (Mg) of total mass, respectively.

Distribution of nutrient stoichiometry in different life forms and tree components differ (Xing et al. 2022).

Nutrient Uptake

Of the plant nutrients, the uptake values are the highest either for N or Ca, K is the next in magnitude, and P and Na were much lower. Calcium (Ca) uptake, particularly net uptake (nutrient uptake in net production, adjusted for nutrient resorption from senescing leaves and other parts) is higher than N uptake generally in oaks, partly because resorption of Ca from senescing leaves is negligible or absent, while N resorption from senescing leaves is generally high (Aerts 1996). Between the oak and pine forests, net uptake is generally distinctly higher in oak forests for all nutrients than in chir pine forest, because of relatively higher leaf nutrient concentration and lower resorption from senescing oak leaves (Ralhan and Singh 1987). For example N resorption from the senescing chir pine is 53%–55%, compared to less than 30% for oaks (Ralhan and Singh 1987). The nutrient uptake declines sharply above 2600–2700 m, by 78%–93% from that in kharsu oak and silver fir forests to treeline forests largely because of declining forest productivity. Over the entire elevational range of about 3000 m, the net N uptake varied between 253.4 kg ha^{-1} yr^{-1} (in kharsu oak forest) and 33.2 kg ha^{-1} yr^{-1} (in *R. campanulatum* forest of treeline) (Table 6.11). Thus, the decline in the uptake was 7.6-fold for N, 4.4-fold for P, and 7.2-fold for K. For Ca, for which data are available only up to 2650 m, the range of annual net uptake was from 278.2 (kharsu oak forest) to 88 kg ha^{-1} (gross) (chir pine forest) (Table 6.11). In each case the highest uptake of nutrients was for one or other oak forest and the lowest in the treeline forests.

Table 6.11 Gross (nutrient accumulation in net production) and net (nutrient accumulation in net production after adjustment for nutrient resorption) uptake of nutrients (for trees only) across the Indian Central Himalayan (ICH) forests

	Nutrient uptake		Trees only (kg ha^{-1})				
Source	Forest	Gross/net	N	P	K	Ca	Na
Khanna (1986)	Sal (*S. robusta*) old-growth	Gross	155	13	7	119	–
		Net	121	10	6	113	–
	Sal seedling new-growth	Gross	178	15	85	135	–
		Net	148	12	74	129	–
	Chir pine (*P. roxburghii*)-Baldiyakhan	Gross	164	15	79	95	–
		Net	114	13	67	88	–
Chaturvedi (1983)	Chir pine	Gross	144	14	–	–	–
		Net	91	9	–	–	–
Khanna (1986)	Banj oak (*Q. leucotrichophora*)-pine	Gross	151	14	66	148	–
		Net	117	11	52	135	–
Rawat (1983)	*Banj oak* (*Q. leucotrichophora*)-Kailakhan	Gross	165	9	44	191	–
		Net	133	8	39	171	–
Khanna (1986)	Rianj oak (*Q. lanata*)–Kilbury	Gross	179	12	100	153	–
		Net	142	9	83	135	–
Rawat (1983)	Rianj oak (*Q. lanata*)—Site II	Gross	179	9	52	223	–
		Net	168	8	43	214	–
Khanna (1986)	Tilonj (*Q. floribunda*) oak-Kilbury	Gross	257	19	143	232	–
		Net	213	16	111	160	–
Rawat (1983)	Tilonj (*Q. floribunda*) oak—Site II	Gross	209	9	59	222	–
		Net	196	8	44	197	–
Adhikari et al. (1995)	Horse-chestnut (*Aesculus indica*)	Gross	–	–	–	–	–
		Net	147.6	5.4	52.6	125.2	3.4
	Pindrow fir (*A. pindrow*)	Gross	148.0	–	–	–	–
		Net	131.4	6.2	50.3	120.5	4.1
	Kharsu oak (*Q. semecarpifolia*)	Gross	283.6	–	–	–	–
		Net	253.4	12.8	82.6	278.2	6.7
Garkoti and Singh (1994)	Maple (*Acer* spp.)	Gross	120.2	10.5	51.3	–	–
		Net	89.3	8.0	35.5	–	–
	Birch (*B. utilis*)	Gross	63.1	5.9	24.8	–	–
		Net	45.3	4.5	18.9	–	–
	R. campanulatum	Gross	44.8	3.6	20.3	–	–
		Net	33.2	2.8	15.4	–	–

Box 6.2 Nutrient Cycling in Sal Forest of Eastern Nepal

Sal forms extensive forests in the moist plains of eastern Nepal. The eastern Nepal has a monsoon pattern of rainfall (~2000 mm, annual rainfall with 79% during monsoon months) and warm temperature (10.9–25.3 °C minimum, and 22.1–32.2 °C maximum). The common associates of Sal in that forest are *Haldina cordifolia, Lagerstroemia parviflora, and Terminalia tomentosa.*

In sal forest of Eastern Nepal (SEN) studied by Gautam and Mandal (2016), the values of nutrient uptake, return to soil, and turnover rate resembled those of sal old-growth forest of central Himalayas. However, generally the values of K in SEN were relatively higher. The net uptake by vegetation (kg ha^{-1} y^{-1}) in SEN was 2077 ± 17 N, 19.4 ± 1.2 P, and 164.2 ± 9.4 K, which are on the higher side of the range of values across the central Himalayan forests. In fact, the net uptake of K in SEN exceeded them. The amounts of nutrient return (kg ha^{-1} y^{-1}) in litterfall in SEN, 99.7 ± 8.0, 8.3 ± 0.07, and 64.4 ± 4.1 kg, respectively. For N, P, and K are also on the higher side of range of values for central Himalayan forests. The higher values in SEN was partly because of a better estimate of fine root herb contributions. Leaf litter was the dominant component of litter nutrient, accounting for ~75% N, ~70% P, and 87% K. The total nutrient return (both above ground and belowground, root, and combined) in SEN accounted for 69%, 60%, and 60% of net uptakes of N, P, and K, respectively, which are similar to those of central Himalayan sal forest. As expected for a sal forest the turnover time for nutrients was short, 1.45 yr, 1.47 yr, and 1.53 yr for N, P, and K, respectively. Like central Himalayan forests, in SEN also the role of herbs was greater than that of their standing crop indicated.

Effect of disturbance

The study of Gautam and Mandal (2016) of SEN sheds light on the effect of forest disturbance on nutrient cycling (branches and foliage of trees are frequently cut, resulting in many trees with denuded trunks). The disturbed stand of SEN had 45%–47% less of nutrient stock than that of undisturbed stand; however it had more than 50% net nutrient uptake. The disturbed vegetation resorbed relatively smaller proportion of nutrients from senescing leaves, resulting in relatively larger percentage of litterfall input, relative to nutrient stock.

Share of Nutrient Uptake by Shrubs and Herbs

The difference between nutrient uptake of trees and total forest vegetation, which represents the combined nutrient uptake of shrubs and herbs, varied considerably across the forests and nutrients. While total nutrient uptake (tree + shrubs + herbs) decreased rapidly in higher elevations toward the treeline, from kharsu oak forest to birch and *R. campanulatum* forest, the combined share of herbs and shrubs increased sharply from generally less than 30% to about 40%–70% along this elevation gradient.

Compared to nutrient uptake of trees alone, the nutrient uptake by entire vegetation (trees + shrubs + herbs) was 14% (N) to 40.8% (K) greater in sal old-growth forest. In chir pine forest the trees are more predominant; hence the uptake of entire forest vegetation was only 10.1%–16.5% higher than that of trees alone. In comparison the difference in nutrient uptake values between entire forest vegetation, and trees alone was much higher, 18.7% (Ca)–51.7% (K) in banj oak forests and 20.8% (P)–27.2% (K) in rianj oak-dominated forest. The difference in uptake of nutrients between total vegetation and trees alone increased dramatically toward treeline with the opening of tree and canopy consequent increase in the growth of herbs and shrubs, ranging from 38.7% (P) to 62.8% (K) in maple forest, from 51.1% (P) to 72.4% (N) in birch forest and from 55.7% (N) to 73.4% (K) in *R. campanulatum* forest.

Generally, herbs accounted for a significantly higher percentage of nutrient uptake than shrubs, largely because of the greater production and nutrient concentration of herbs than shrubs. In the three major forests of lower elevations sal, chir pine, and banj oak forests, trees accounted for 68.8%–91.2% of nutrient uptake (N, P, K, and Ca). In these forests shrubs accounted for from 0.4% of total Ca uptake in chir pine forest to 14.4% K in banj oak forest. On an average, in these three forests of the total forest uptake shrubs accounted for 4.37% N, 5.53% P, 6.40% K, 4.33% Ca, 5.67% Na, and 1.30% Mg, and herbs accounted for 14.10% N, 10.93% P, 15.87% K, 12.97% Ca, 14.37% Na, and 31.3% Mg (Table 6.12). Generally, the influence of herbs on the cycling of nutrients is disproportionately high relative to their biomass (Gilliam 2007). For Mg the contribution of herbs was particularly high, from 29.3% to 33% across the ICH forests. Compared to tree foliage, the herb foliage in general has 30% more N and P concentration, two-fold Mg concentration, and three-fold K concentrations (Muller and Bormann 1976; Gilliam 2007).

Table 6.12 Percentage of total nutrient uptake of vegetation by growth forms in three principal lower elevation forests of Central Himalayas

Percent uptake	Growth form	N	P	K	Ca	Na	Mg
Sal	Trees	79.00	86.40	80.80	78.80	66.20	68.00
	Shrubs	4.00	3.40	4.10	4.50	3.60	2.30
	Herbs	17.00	10.10	15.00	16.70	10.30	29.60
Chir pine	Trees	85.60	91.20	84.10	86.60	85.10	66.70
	Shrubs	0.70	1.10	0.70	0.40	1.10	0.30
	Herbs	13.70	7.70	15.20	13.00	13.80	33.00
Oak	Trees	80.40	72.90	68.20	82.70	68.80	–
	Shrubs	8.40	12.10	14.40	8.10	12.30	–
	Herbs	11.60	15.00	17.40	9.20	19.00	–
Average	Trees	**81.67**	**83.50**	**77.70**	**82.70**	**73.37**	**67.35**
	Shrubs	**4.37**	**5.53**	**6.40**	**4.33**	**5.67**	**1.30**
	Herbs	**14.10**	**10.93**	**15.87**	**12.97**	**14.37**	**31.30**

Source: Khanna (1986), Singh and Singh (1992)

Table 6.13 Percentage of total nutrient uptake of vegetation by growth forms in high elevation forests of Central Himalayas

Forest type	Growth form	Percent uptake				
		N	P	K	Ca	Na
Horse-chestnut (2300 m)	Trees	89.00	86.01	81.50	77.7	81.5
	Shrubs	3.10	5.90	4.90	3.5	5.7
	Herbs	8.90	18.00	13.60	8.8	12.8
Silver fir (2500 m)	Trees	91.00	83.20	86.20	88.9	86.0
	Shrubs	0.60	0.80	0.80	0.7	8.0
	Herbs	8.40	16.00	13.00	10.4	13.2
Kharsu oak (2650 m)	Trees	92.70	88.30	88.20	94.8	90.0
	Shrubs	1.60	2.60	2.20	1.7	2.3
	Herbs	5.70	9.10	9.60	5.5	7.5
Maple (2750 m)	Trees	70.20	72.07	61.14	–	–
	Shrubs	4.70	2.70	7.09	–	–
	Herbs	25.07	25.20	31.47	–	–
Birch (3150 m)	Trees	58.0	66.17	60.0	–	–
	Shrubs	12.03	8.82	22.85	–	–
	Herbs	29.96	25.00	17.10	–	–
Rhododendron (3300 m)	Trees	64.21	58.33	57.67	–	–
	Shrubs	0.96	0.625	0.086	–	–
	Herbs	34.80	39.58	41.57	–	–

Source: Garkoti and Singh (1994), Adhikari et al. (1995)
Values in parentheses are elevations in meter

As canopy of trees opened up with increasing elevation and forest floor became more lighted, the cover of herbs and shrubs increased. Among all the forests, the shrubs attained their highest expression in the birch forest and the herbs in *R. campanulatum* krummholz community forest (Table 6.13).

Globally, herbs with less than 1% of forest biomass, and about 4% of NPP account for up to 12%–16% annual nutrient return through litter fall in forests. Moreover, the concentration of N and P in foliage is 30% higher in herbs than trees, and concentration of Mg is nearly two-fold and of K three-fold higher (Muller and Bormann 1976; Muller 2003; and Welch et al. 2007).

Nutrient Return Through Litterfall

Nutrient return to soil through aboveground litterfall is a vital step in the forest nutrient cycling, accounting for bulk of total return of nutrients to soil, the other sources being precipitation, and root mortality. The senesced leaves that fall on to the forest floor greatly influence the soil properties (Viera and Schumacher 2009). There are species that withdraw most of the leaf nutrients before leaf drop, such as *Fraxinus micrantha*, while the oaks may deposit most of the leaf nutrients they contain in mature leaves to soil. A study on the effect of senesced tree leaves on forest soil in Poland showed that the exotic oak, *Q. rubra* significantly modified soil properties of native forests of *Q. robur, Fagus sylvatica*, and *Carpinus betulus*. The low-quality senesced leaves of *Q. rubra* (0.61% N compared to 0.87% in native species) decomposed slowly and generated a physical barrier limiting seed germination and seedling-growth of several understory species (Stanek et al. 2020).

Leaf litter fall accounted for 72.9%–95.0% of total nutrient return in litter fall (except 63.0% for Ca in mixed oak rianj-dominated forest) of ICH forests. Though the percentages were similar across the forests and nutrients, some variations occurred. In chir pine forest the contribution of leaf litter for N, P, and K return was relatively higher (~90–95%) than in other forests (72.9%–91.8%). The other components of litter, namely coarse and fine wood, bark, and reproductive parts, seem to be lower in chir pine forest.

In high elevation forests, on an average leaf litterfall accounted for 68.4% of N, 75% of P, and 74.2% of K in maple forest (Garkoti and Singh 1994). In treeline forests, the contribution of woody litter was relatively higher because of the large-scale damages of branches due to snowfall, nevertheless leaf litter's contribution was around 60%.

If we exclude the values of treeline forests and some exceptionally high values of old-growth oak forests, the N return through litterfall ranges between 56.1 and 83.3 kg ha^{-1} yr^{-1}. These values are distinctly lower than those of the moist tropical forests (124–224 kg ha^{-1} yr^{-1}) of moderately fertile soils; but are comparable with those of the tropical montane forests 37–90 kg ha^{-1} yr^{-1} (Vitousek and Sanford 1986). But distinctly higher N return values through litter fall are reported for tropical moist forests of Brazilian Atlantic. In these Brazilian forests N return as an example is 294.2 kg ha^{-1} yr^{-1} in a semi-deciduous primary forest and between 123.2 and 217.8 kg ha^{-1} yr^{-1} in secondary forests (Viera et al. 2016). In treeline areas, the litterfall N return dropped to 34.1 and 25.6 kg ha^{-1} yr^{-1} in birch and *R. campanulatum* forest, respectively. Among other forests, litterfall N was relatively lower 46.6–69.0 kg ha^{-1} yr^{-1} in chir pine forests (Mehra et al. 1985). In sal forest including that of the eastern Nepal annual N return occurred within a narrow range of 69.9–74.3 kg ha^{-1} yr^{-1}, indicating that the sal forest maintains a high degree of consistency in litterfall N amount across a wide range of moisture conditions. The amount of nutrient return through litterfall in the Himalayan forests would be greater, if coarse woody debris were also included.

The total N return amounts to 38.5%–81.5% (generally between 49.0% and 65.8%) of annual net uptake, indicating that at least in some forests the total nutrient return may match the uptake.

Across the ICH forests, P return in tree litterfall ranges between 2 and 7 kg ha^{-1} yr^{-1} (average 3.66 kg ha^{-1} yr^{-1}). As usual, it declines toward the treeline rapidly to 3.2 and 2.0 kg ha^{-1} yr^{-1} in birch and *R. campanulatum* krummholz, respectively (Table 6.14). The values were relatively higher in old-growth sal, and oak forests, particularly kharsu oak forest of Pindar valley.

The amount of Ca in tree litterfall generally varied between 72.5 and 91.7 kg ha^{-1} yr^{-1}. Exceptionally high values of Ca deposition; however, were recorded in the two oak forests, kharsu oak (140 kg ha^{-1} yr^{-1}) and rianj oak forests (129.9 kg ha^{-1} yr^{-1}).

Table 6.14 Amount of nutrient return (kg ha^{-1} yr^{-1}) through tree litterfall in forests of Central Himalayan high-altitude forests (2600–3300 m) of Pindar Valley (Central Himalayas)

Return through tree litter fall	Horse-chestnut (*Aesculus indica*)[a]	Pindrow fir (*A. pindrow*)[a]	Kharsu oak (*Q. semecarpifolia*)[a]	Maple (*Acer* spp.)[b]	Birch (*B. utilis*)[b]	*R. campanulatum*[b]
N	107.2	78.2	163	56.1	34.1	25.6
P	3.7	3.6	7	4.8	3.2	2
K	32.5	8.8	48.29	23.3	12.2	10
Ca	80.2	69	140.8	–	–	–
Na	2.1	0.6	3.5	–	–	–

Based on Garkoti and Singh (1994)[a], Adhikari et al. (1995)[b]

Partioning of Precipitation and Nutrients Along with It

Apart from litter fall, incident precipitation (water that falls on the forest directly, also referred to as gross fall) is the main flux through which nutrients move from vegetation to soil (Fig. 6.5). Though, the incident precipitation is intercepted by canopy to an extent, much of the precipitation (generally >70%) reaches ground as throughfall (precipitation that passes through the canopy and falls to the ground) and stem flow (the water running down the branches and stems and deposited at the base of the tree) (Parker 1983). Throughfall, that is the rain water dripping from branches and leaves and reaching ground and stem flow, and rain water flowing down the stems of trees are the main fluxes through which nutrients move from the vegetation to soil (Herborn and Congdon 1998). The sum of throughfall and stem flow is referred to as net rainfall or effective rainfall (Parker 1983). Generally, throughfall is proportionally less in conifer forests than in broad-leaved forests (Neary and Gizyn 1994; Zhu and Li 2013). Conversely canopy interception is higher in conifer forests (e.g. 34.1% in *Cedrus deodara* forest) than in broad-leaved evergreen forests (e.g. 24.8% in *Q. leucotrichophora* forest, Jana et al. 2021; for details see Table 6.15 in Box 6.3.

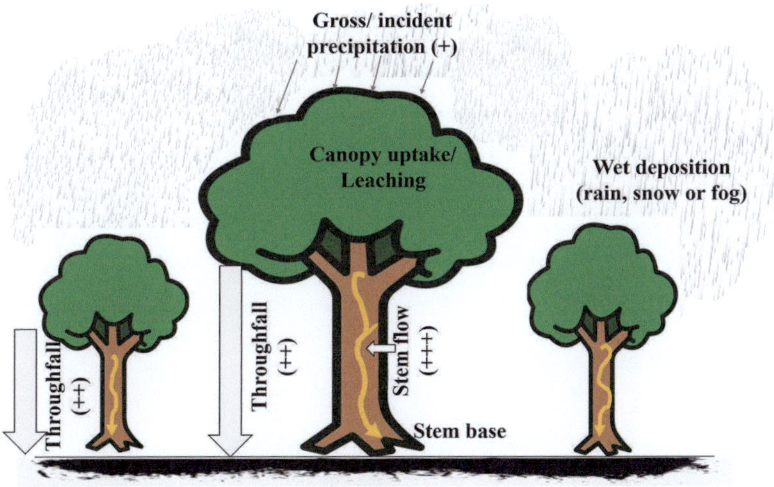

Fig. 6.5 Schematic representation of hydrologic nutrient flow paths in forest ecosystems. Gross precipitation (incident rainfall), throughfall and stem flow carry nutrients, the concentration of which is mostly the highest in stem flow (+++), followed by through fall (++) and bulk precipitation (+), while pH follows the opposite trend. But the total amount of nutrients is far greater in throughfall than in stem flow because of its bulk. Furthermore, throughfall-nutrients are distributed throughout the forest floor, while stem flow is concentrated to stem bases. The higher nutrient concentration in stem flow could be attributed to longer residence time on tree and larger flow area (Yang et al. 2022)

Box 6.3

Table 6.15 Apportionment of precipitation in central Himalayan forests (Pathak et al. 1985)

Forest	Cross rainfall	Apportionment of bulk precipitation (mm)			Apportionment of throughfall (mm)		
		Through fall	Stem flow	Canopy interception	Litter interception	Over last flow	Ground vegetation inception + infiltration
Sal	1153	950 (82.4)	10 (0.86)	193 (16.7)	113 (11.9)	8 (0.84)	839 (88.3)
Chir pine-mixed broad-leaved	1179	1079 (91.5)	5 (0.46)	95 (8.05)	85 (7.91)	15	984 (91.2)
Chir pine	1234	922 (74.7)	3 (0.24)	308 (33.4)	89	10	727 (78.8)
Oak-chir pine	915	158 (82.3)	3 (0.33)	154 (16.81)	81 (10.7)	5	675 (75.8)
Tilonj oak	1364	1155 (84.6)	5 (0.37)	204 (14.9)	128 (9.4)	7	1026 (88.3)
Rianj oak	1240	1002	11 (0.89)	227	114	2	897

Because of climate change, throughfall may increase globally and lead to increase in the soil CO_2 flux in temperate forests (Xu 2023). The study predicts that an increase of 10% in annual precipitation due to climate change would result in 6.9% increase in annual CO_2 emission in forest ecosystems on a global scale because of increased throughfall.

Recent researches indicate that plant surface character affects throughfall apportionment and throughfall types (see Box 6.4), and in this respect broad-leaved and conifer forests differ.

Box 6.4 Throughfall Types

Rainfall interception by forests is the rain caught, stored, and evaporated from surfaces and litter. It is affected by throughfall types that are (i) free throughfall, (ii) throughfall splash, and (iii) canopy drip (Levia et al. 2019). Drop sizes of open rainfall and throughfall are measured using a disdrometer (Nanko et al. 2006) that consists of a laser transmitter and a receiver.

In a study in terms of number, canopy drips accounted for 2%–16%, splash throughfall 42%–66%, and free throughfall 17%–45% across conifer broad-leaved with leaf, and broad-leaved without leaf. But in drop volume, canopy drips are most important as indicated below (Table 6.16).

In brief, conifer trees generated lower canopy drop volume percentage (51%) than broad-leaved trees (69%), and smaller diameter splash throughfall drops (0.76 mm) than broad-leaved trees (0.85 mm). Canopy drips have a larger kinetic energy and erosivity potential than free or splash throughfall. But a lot of these properties depend on plant surface characteristics that alter the proportion of throughfall type and volume (Levia et al. 2019). Future research should focus on interactions of plant surface characteristics and properties of water droplets.

Table 6.16 Distribution of throughfall types in number and volume between conifer and broad-leaved forests

	Conifer trees		Broad-leaved trees	
Number of drops	% number of drops	Proportion of drop volume	% number of drops	Proportion of drop volume
% canopy drip	2	51	7	69
% splash throughfall	53	18	52	11
% free throughfall	45	31	41	21

From Levia et al. (2019)

Concentration of nutrients in throughfall and stem flow are higher than in incident rainfall (Mehra et al. 1985; Singh and Singh 1992; and Oziegbe et al. 2011). These nutrients are in a more easily available form. Between throughfall and stem flows, nutrient concentrations are higher in stem flow, because of a longer residence time of water along branches and stem. For example N concentrations (mg g^{-1}) in incident rainfall, throughfall, and stem flow were 5.3, 8.6, and 11.52, respectively, in sal forest, 6.0, 10.7, and 14.9, respectively, in chir pine forest and 6.0, 11.0, and 17.0, respectively, in tilonj forest. Changes from rain water, throughfall, and stem flow were more apparent for other nutrients, as exemplified by the following: K in sal forest, 10.6, 15.1, and 70 mg g^{-1}, respectively, in rainfall, throughfall, and stem flow; Ca in mixed oak-pine forests 15.4, 19.8, and 62.2 mg g^{-1}, respectively; P in rianj oak-dominated forests 1.0, 1.5, and 7.6 mg g^{-1}, respectively; and Mg in tilonj oak-dominated forest 10.0, 15.1, and 22.9 mg g^{-1}, respectively. In a lowland tropical rainforest, K concentration (mg l^{-1}) was found to be 1.34 in incident rainfall, 30.8 in throughfall, and 38.22 in stem flow (Oziegbe et al. 2011).

The annual net precipitation input (throughfall + stem flow) of nutrient in the Nainital forests varied as follows for different nutrients:

N = 7.99 (mixed oak-pine forest) to 13.79 (tilonj-dominated oak forest) kg ha^{-1}
P = 1.47 (mixed oak-pine forest) to 2.17 (pine forest) kg ha^{-1}
K = 14.42 (sal forest) to 30.27 (tilonj-dominated oak forest) kg ha^{-1}
Ca = 19.20 (mixed oak-pine forest) to 31.70 (rianj-dominated oak forest) kg ha^{-1}
Organic carbon = 11.73 (mixed oak-pine forest) to 18.15 (tilonj-dominated oak forest) kg ha^{-1}

In some forests, nutrient net precipitation (throughfall + stem flow) could be a larger source of nutrient return than litter. For example in the Nigerian lowland typical rainforest, net precipitation accounted for 80% of K, 73.2% of Na, and 63.8% of P inputs. However, litterfall was the major source of inputs for N (70%), Ca (90.2%), and several more other elements (Oziegbe et al. 2011).

In the ICH forests, the nutrient input through precipitation is in the following order Ca > K > Mg > N > P (Mehra et al. 1985). Jana et al. (2021), however, found that the highest throughfall gain was for K followed by Ca. Among Nainital forests of ICH, nutrient input through precipitation was the highest in tilonj oak forest for N, Ca, Mg, P, and organic carbon and its average input across all nutrients was the highest (12.4 kg ha^{-1} yr^{-1}), followed by rianj oak forest (10.4 kg ha^{-1} yr^{-1}), pine-mixed broad-leaved forest, pine forest, sal forest, and mixed-oak forest.

The annual contribution of precipitation to the total nutrient return to forest floor across the Nainital forests was 11%–17% for N, 24%–29% for P, and 17%–26% for Ca (Mehra et al. 1985). Bulk precipitation (incident rainfall), throughfall, and stem flow are important sources of nutrients, particularly for K in forests of humid areas (Su et al. 2019). Generally, the K-limitation to forest productivity is probably more widespread than thought and the external inputs of K, such as atmospheric depositions are of critical importance to deal with K deficiency (Cornut et al. 2022). Chemical composition of throughfall and stemflow is determined by inputs of atmospheric deposition and exchange with vegetation (Su et al. 2019). In the USA atmospheric K deposition is considered to be a major component of K nutrient-cycling (Mikhailova et al. 2019). For some nutrients, throughfall is the larger source of nutrient return to soil than litterfall.

In a study on litter and throughfall nutrient inputs under planted trees of 7 forest species (alder, *Alnusglutinosa*; oak, *Q. robur*; birch, *B. pendula*; willow, *Salix caprea*; and rowan, *Sorbus pendula*), Carnol and Bazgir (2013) found that the nutrient inputs from litterfall and throughfall varied considerably across the planted forests, as given below.

- Leaf litterfall as percentage of total litter fall was 85% for beech, spruce, and oak forests; 70% for alder and willow forests; and 55% for birch and rowan forests.
- Throughfall formed 20% (rowan) to 60% (oak) of total Ca return.
- Mg return in throughfall ranged from 30% (rowan) to 76% (beech) of the total return. Much of the remaining nutrients returned to ground were from litter.

Root mortality (the belowground litter) generally accounted for 6%–10% of total nutrient return to soil across the ICH forests. Only a few values were greater than 10%–11.2% Ca in rianj oak forest, 12% Ca in tilonj oak forests, 14.5% K in tilonj oak forest, and 11.3% N in chir pine forest. As referred to earlier, root mortality is generally underestimated, as the samplings are not adequate to consider spatial and temporal variations.

To summarize, aboveground litterfall is the predominant source of nutrient return to soil, generally accounting for 55%–90% of nutrient return across ICH forests. Nutrient movement, through root mortality can be substantial, but its rough estimates only are available. However, for K, precipitation was also an important source. In some forest studies elsewhere precipitation is a greater source of nutrient return than litterfall (Oziegbe et al. 2011). In ICH forests, on an average, litterfall accounted for 68.1% of K, compared to the averages of over 80% for the other nutrients. Litter from herbs was a major source of nutrient return, far more than its NPP indicates largely because of higher nutrient concentrations in them. Herbs are particularly important contributors to Mg return to soil. Because of the decrease in grazing and fodder and litter collections, the litter nutrient inputs are likely to increase in future in the Himalayan forests, which might enhance the nutrient turnover rate as herbs decompose much more rapidly than woody plants litter. So, in future, while more herb litter may result in an enhanced nutrient cycling, the increase in the accumulation of coarse woody debris on forest floor may affect nutrient cycling differently.

The total return of nutrients as the percentage of net uptake ranged from 38.5% for N in tilonj oak forest to 94% for P in sal old-growth forest, percentages were generally between 60% and 75% of net uptake.

Turnover Time (T) at Forest Floor Litter

The nutrient turnover time in central Himalayan forests is short, closer to that of tropical forests than temperate forests. It ranges between 1.51 and 3.13 years for N, 1.30–4.35 years for P, 1.22–3.63 years for K, 1.56–3.03 years for Ca, and 1.40–3.03 for Na (Table 6.17). It was relatively longer in forests above 2600 m, and in forests having chir pine (*P. roxburghii*) among the forests below 2600 m elevation.

Compared to these, the mean residence time (years) for nutrients in forest floor for some other biomes are as follows: N, 2 for tropical rain forest, 5.5 for temperate deciduous forest, 17.9 for coniferous forest, and 27.1 for boreal forest, P, 1.6 for tropical rain forest, 5.8 temperate deciduous forest, 15.3 for temperate coniferous forest, and 15.2 for boreal forest, and K, 0.7 for tropical rain forest, 1.3 for temperate deciduous forest and 2.2 temperate coniferous forest, and 10.0 for boreal forest (Zobel and Singh 1997). In an experimental site of Okinawa Island, organic matter turnover time was found to be 6–12 years for *P. luchuensis* and pine-mixed broad-leaved forest (Xu and Hirata 2002).

Of all the study nutrients across the forests of Nainital transect, total nutrient return as percentage of net uptake was mostly more than 50%, and in several instances, it exceeded 70%. In high elevation forests up to those of treeline, nutrient return through litterfall accounted for from 51% (K in birch forest) to 72.5% (N in *R. campanulatum* stand) of net nutrient uptake.

Table 6.17 Turnover time (t in yr) of forest floor in central Himalayan forest from foothills to treelines

	Turnover time (t in yr)					
Forest	N	P	K	Ca	Na	Source
Sal (*S. robusta*) old-growth	1.75	1.49	1.45	1.67	1.85	Khanna (1986), Singh and Singh (1992)
Sal seedling new-growth	1.82	1.54	1.47	1.64	1.85	
Chir pine (*P. roxburghii*) mixed broad-leaved	1.96	1.56	1.56	1.89	1.69	
Chir pine	2.43	2.50	1.89	2.17	1.56	Chaturvedi (1983)
Chir pine (*P. roxburghii*)-baldiyakhan	2.94	2.78	2.63	3.03	3.03	Khanna (1986), Singh and Singh (1992)
Mixed Banj oak (*Q. leucotrichophora*)-pine	2.44	2.00	2.04	2.27	2.50	
Mixed banj oak	1.51	1.30	1.22	1.56	2.19	Rawat (1983)
Mixed oak Rianj-dominated	1.89	1.47	1.49	1.79	1.92	Khanna (1986)
Mixed oak Tilonj (*Q. floribunda*)-dominated	1.85	1.61	1.72	1.92	2.38	
Kharsu oak, silver fir, and horse-chestnut	1.83	1.68	1.39	1.89	1.40	Adhikari et al. (1995)
Maple (*Acer* spp.)	2.63	1.85	2.33	–	–	Garkoti and Singh (1994)
Birch (*B. utilis*)	2.70	2.56	3.03	–	–	
Rhododendron forest	3.13	4.35	3.03	–	–	

Periodicity of Nutrient Movements

In Himalayan forests occurring in outer ranges, hence directly exposed to monsoon thrust, the timings of nutrient return through litterfall and precipitation components are clearly separated. Decline in leaf nutrient concentration generally begins in winter (December–February) in many broad-leaved evergreen species, but the duration varies considerably, being 2–3 months (December–February) in *S. robusta* at foothills (mean annual temperature of 21.9 °C) to 4–5 months for oaks occurring over 1500 m, with mean annual temperatures below 16.4 °C (Negi et al. 2020). Much of the litterfall occurs during pre-monsoon (particularly March–April), while nutrient inputs through precipitation is the phenomenon of monsoon months (typically July–August). Monsoon is also the time of rapid litter decomposition and nutrient release. Litter from herbs fall on ground largely during winters. As for nutrient uptake, it peaks during the monsoon, while most of the nutrient resorption (internal cycling) takes place during pre-monsoon in E1 evergreen species (species with concentrated leaf drop and simultaneous leafing during pre-monsoon months) and in several pre-monsoon deciduous species. However, in winter deciduous species, such as maple, birch, and horse-chestnuts nutrient resorption occurs during autumn–winter months. Thus, monsoon and pre-monsoon are periods of principal nutrient fluxes (Fig. 6.6). During monsoon the nutrient storage in net production protects nutrients against losses through leaching, erosion, and landslides. Pre-monsoon is the time when intraplant nutrient cycling (nutrient resorption from old senescing leaves as well as nutrient transfer to new shoots) and nutrient return through litterfall peak. Herbs and some winter deciduous woody species; however, largely return nutrients from vegetation to soil during winter months.

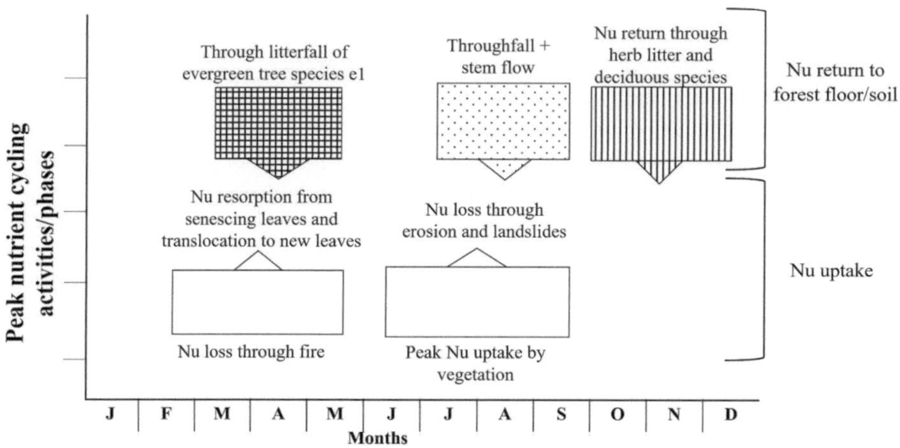

Fig. 6.6 Periodicity of major nutrient (Nu) cycling steps. Nutrient return refers to nutrient movement from vegetation to soil. Nutrient resorption in senescing and other tissues and retranslocation to new tissues/storage influences Nu return as well as Nu uptake. Nu losses from ecosystems generally involve erosion/landslides during monsoon and pre-monsoon. Pre-monsoon and monsoon are the periods of high activities and winter is the period of low nutrient fluxes except in some deciduous trees that shed leaves in winters

Overland Flow and Sediment Loss

The average nutrient loss (kg ha^{-1} yr^{-1}) through overland flow was the highest in pine-mixed broad-leaved forests (0.18), followed by sal (0.11), chir pine (0.102). In the remaining forests the average loss was from 0.04 to 0.07 kg ha^{-1} yr^{-1} (Singh and Singh 1992).

The nutrient loss through sediment output was the highest for sal forest (0.54 kg ha$^{-;1}$ yr^{-1}), followed by chir pine-mixed broad-leaved forest (0.41 kg ha^{-1} yr^{-1}), and chir pine forest (0.29 kg ha^{-1} yr^{-1}). In the remaining forests, the loss was much lower (0.107–0.1916 kg ha^{-1} yr^{-1}).

The average nutrient loss across all forests through overland flow in ICH was in the order: Ca (0.183 kg ha^{-1} yr^{-1}) > N (0.103 kg ha^{-1} yr^{-1}) > Mg (0.102 kg ha^{-1} yr^{-1}) > K (0.09 kg ha^{-1} yr^{-1}) > P (0.25 kg ha^{-1} yr^{-1}), and through sediment output it was Ca (0.52) > Mg (0.45) > K (0.29) > N (0.068) > P (0.015). The N loss through overland flow was relatively high but sediment output was low (Pathak et al. 1984).

Nutrient Resorption from Senescing Leaves

Nutrient resorption (NuR) in plants involves the reallocation of nutrients from senescing tissues to surviving tissues (Killingbeck 1996) and new-growth (sink), which reduces their dependence on current nutrient uptake from soil and other sources (Aerts and Chapin 1999; Killingbeck 2004). As leaves senesce, breakdown products are released in simple metabolites that could be loaded into the phloem and reused, thus contributing to plant fitness (Aerts 1996). For example proteolysis is the principal mechanism for N mobilization, while P mobilization involves diverse catabolic pathways (Estiarte et al. 2023). Furthermore, NuR from senescing leaves has important implications for nutrient cycling in an ecosystem as it affects the nutrients left in litter and hence litter decomposition, and their release (high litter nutrient concentration) or immobilization (low nutrient concentration). Reproductive efforts impose an additional demand with potential to deplete nutrients from leaves (Tully et al. 2013).

In plant ecophysiological studies, the term nutrient resorption efficiency (NuRE) is used as the fraction of nutrients present at peak nutrient levels in mature leaves that are withdrawn before leaves are shed or dead (Estiarte et al. 2023). Aerts (1996) found that (i) globally nutrient resorption efficiency of 50% for N ($n = 287$), and 52% for P ($n = 226$); and (ii) it (N) was higher for deciduous woody species (54%) than for evergreen woody species (47%). More N is allocated to cell wall in evergreen species than deciduous species (Takashima et al. 2004), which could account for the lower NuR in evergreen species. Subsequently, using 86 studies from 31 countries, and 1000 data value points, Vergutz et al. (2012) found significantly higher resorption values than those of Aerts (1996), averages being 62.1%, 64.9%, and 70.1% for N, P, and K, respectively. Among others, the average resorption was 10.9% for Ca, 23.2% for C, and 28.6% for Mg. Between evergreen woody species and deciduous woody species, evergreen woody species had lower average resorption for N (56.1% vs 61.2%), while they were similar for P resorption (56.4% for evergreen and 56.5% for deciduous). As for conifers, on an average they resorbed 62.7% N and 77.0% of P (Vergutz et al. 2012). Graminoids emerged as the most prominent with regard to nutrient resorptions, the percentages being 74.6% for N and 82.1% for P and 84.9% for K.

The average resorption of N in the Himalayan species is 50.3% based on 71 species, largely from ICH forests (compiled from Ralhan and Singh 1987; Negi and Singh 1993; Garkoti and Singh 1994; Athokpam and Garkoti 2015; Negi and Mehta (Final Technical Report, IHTRP 2021; and personal communication). In a study of Himalayas at each site/area species, N resorption was significantly higher in deciduous species 46.7%–62.8% than evergreen species (31.9%–50.9%) (Table 6.18). As for comparison between evergreen (11 species mostly broad-leaved) and deciduous species (15 species) of Nainital transect, the deciduous species have relatively higher average N concentration in mature leaves (3.18% vs 2.45% in evergreen species), higher proportional dry mass loss during leaf senescence (32.6% vs 26.7% in evergreen species), and higher fraction of N resorption (58% vs 42%) (Negi and Singh 1993). The resorption of N in some of the winter deciduous species is particularly high: 75.3% in *Fraxinus micrantha*, 69.9% in *Tectona grandis* (planted), *Sapium sebiferum* (66.7%), and *Populus ciliata* (63.9%). The average N resorption across all species in each area was close to 50%, except in Kumaun forests of lower elevations (300–2200 m) studied by Ralhan and Singh (1987) in which it was distinctly low (36.4%), but similar to that of China's temperate forests (35.6%) studied by Xu et al. (2017).

As for P resorption, the gap between evergreen (36.1%) and deciduous species (54.3%) was still larger. With regard to K resorption, in lower elevation forests of Central Himalayas the difference between evergreen (28.8%) and deciduous species (42.8%) was again large, but the values were almost the same in higher elevation forests (~54.4% for evergreen species and 54.3% for deciduous species).

In the Himalayan tree species, the average resorption of Ca was low (average 13.06%) and close to the global average, 10.9% (Vergutz et al. 2012). Being a structural element of cell wall, Ca is resorbed less than other elements (Kazakou et al. 2007). Both for Ca and Na (average 22.55%) the difference between evergreen (lower) and deciduous (higher) species persisted as described above.

In general, the Himalayan forest trees including those of moist tropical forests of Assam have relatively lower percentages of nutrient resorptions, which may indicate that the forests have relatively higher soil fertility. Across the species of Nainital elevation transect (26 species), N resorption was found positively correlated with foliage N concentration (Negi and Singh 1993).

Recently Singh and Negi (2023) analyzed N, P, and K retranslocation in relation to elevation (300–2200 m). In this study of the years 2014–16, they found relatively higher N resorption (48.7%–76.8%) than in earlier studies, and N resorption was correlated, but weakly, with green leaf N concentration ($R^2 = 0.46$, $p < 0.05$). The mean N resorption in this study (60%) was close to that of global average estimated by Vergutz et al. (2012). The green leaf N and K concentrations increased with

Table 6.18 Nutrient resorption from senescing leaves (%)

Area/site	Type	N concentration in green leaves	% nutrient resorption	Source
Nainital transect (300–2200 m),[a] measured in 1985–86—N	Evergreen ($n = 9$)	1.58 ± 0.11	31.9 ± 3.3	Ralhan and Singh (1987)
	Deciduous ($n = 4$)	1.98 ± 0.14	46.7 ± 4.8	
	Total average ($n = 13$)	**36.4**		
Central HimalayanNainital elevational transect (300–2200 m)[b]—N	Evergreen ($n = 7$)	2.46 ± 0.17	42.0 ± 2.32	Negi and Singh (1993)
	Deciduous ($n = 15$)	3.18 ± 0.16	57.1 ± 2.90	
High altitude area—Pindar Valley (2700–3200 m)[c]—N	Evergreen ($n = 4$)	–	40.7	Garkoti and Singh (1994)
	Deciduous ($n = 1$)	–	62.8	
	Total average ($n = 5$)	**51.75**		
Treeline, tungnath areas[d] (average of 2 years) (300–3300 m)—N	Evergreen (E1) ($n = 3$)	–	70.9	Negi and Mehta (personal communication) and Final Technical Report, IHTRP (2021)
	Evergreen with multiyear species ($n = 1$)	–	44.2	
	Deciduous ($n = 1$)	–	52.4	
	Average of all species ($n = 5$)		**61.92**	
Nepal evergreen warm temperate and cold temperate zoneTemperate forests[e]—N	Evergreen broad-leaved ($n = 5$)	–	23.6	Poudyal et al. (2012)
Moist tropical forest, Assam[f]—N	Evergreen ($n = 7$)	1.58 ± 0.06	50.96 ± 1.88	Athokpam and Garkoti (2015)
	Deciduous ($n = 15$)	1.98	56.10 ± 3.02	
	Average ($n = 22$)		**52.6**	

Nainital transect, measurement in 2014–16 (species range and average in parentheses)	–	–	N 48.7 (*M. esculenta*), 76.8 (*M. philippensis*) 59.7,	Singh and Negi (2023)
	–	–	P 45.4 (*Q. leucotrichophora*), 56.0 (*M. philippensis*), 52.85	
	–	–	K 42.4 (*Q. floribunda*), and 65.2 (*Shorea robusta*) 59.66	

A comparison between evergreen (mostly broad-leaved species with about 1 year leaf life span-E1) and deciduous species (both winter deciduous and pre-monsoon broad-leaved deciduous species) in Indian Central Himalayas (ICH). Values are averages for the group, and *n* is the species number

[a]Included *S. robusta, P. roxburghii, and Q.* species (3 spp., *Q. leucotrichophora, Q. floribunda*, and *Q. lanata*), that are major canopy species; the deciduous species included *Lyonia ovalifolia* and *Symplocos* among others

[b]Included most of the above species, plus several deciduous species, such as *Fraxinus micrantha, Populus ciliata, Pyrus pashia, Prunus cerasoides, and Aesculus indica*

[c]Included *R. arboreum, R. campanulatum* (evergreen), and *Acer cappadocium,* and *B. utilis* (deciduous)

[d]Included rhododendrons as above, *Q. semecarpifolia* (E1 evergreen), *Abies spectabilis* (multiyear evergreen) and *B. utilis* (deciduous)

[e]Included evergreen broad-leaved species including oaks, and *Castanopsis indica* and *Schima wallichii*

[f]Included tropical moist forest species of *Dipterocarpus, Syzygium, Toona ciliata, Terminalia bellerica*

elevation, while P concentration was invariant. Singh and Negi (2023) found that resorption of N and P tended to decrease with elevation, which is contrary to the high N resorption values reported for species growing in high ranges toward treeline.

Nutrient resorption was consistently higher on South aspect (average 61%) than on North aspect for all the three nutrients (Singh and Negi 2023). Though data were not collected aspect-wise, generally south aspect is drier and more eroded than North aspect in the Himalayas (Tyagi et al. 2023).

High NuRE in Treeline Species

Nitrogen resorption values were relatively higher for the Tungnath treeline (Negi and Mehta (personal communication) (Final Technical Report, IHTRP 2021). It was 76.2% for *R. campanulatum* which is exceeded only by *Mallotus philippensis* among the Himalayan species studied. The oak, *Q. semecarpifolia* resorbed 66.8% N that was more than twice as much as generally found for the Himalayan evergreen oaks of the lower elevations. *R. arboreum* in which resorption in lower elevation was 53.6% resorbed 69.8% near the treeline. This tends to be consistent with the findings of Vergutz et al. (2012) based on a global analysis, who showed that nutrient resorption is negatively correlated with mean annual temperature and positively correlated with latitudes. In high altitude forests the release of nutrients from organic matter is slow because of low temperature (He et al.2015); hence, nutrient availability is low to plants. Treeline species are required to have high resorption percentage to reduce their dependence on nutrient uptake as soil may remain too cold to supply N during the short growing period (Körner 2012). High resorption efficiency is expected to enable species to store nutrients for the following growing season (Du et al. 2021). The average N resorption (61.92%) in treeline species was distinctly higher than those for the lower Himalayan sites (for N resorption, 62.1%) (Vergutz et al. 2012).

Oaks vs Chir Pine Nutrient Resorption

Among the main forest forming species of ICH, nutrient resorption in the three evergreen oaks species was notably low, the average being 23.3% for N, 24.6% for P, and 25.8% for K (Ralhan and Singh 1987). Lower percentages of N absorption for these evergreen oaks (20%–22%) and *Castanopsis indica* (another member of *Fagaceae*), and *Schima wallichii* are also reported from Nepal (Poudyal et al. 2012). Compared with oaks, chir pine reabsorbs distinctly more N from its senescing needles (e.g. 53.5%–55% N). Given that resorption reduces dependence on nutrient uptake, plants are expected to maximize nutrient resorption. However, to resorb nutrients from senescing leaves a species might need a sink and an adequately long period of senescence, which has cost in terms of the reduction in net assimilation. In the senescing red leaves of *Liquidambar formosana* net assimilation rate is reported to be 34%–58% lower than that of green leaves (Luo et al. 2017). Then, in the trees in which leafing induces leaf drop such as the Himalayan oaks and pines, the presence of sink in a form of new-growth is important for nutrient resorption. For example, in banj oak in the year of seed masting in Nainital the drain on trees was so severe that they withheld new leaf production that took place only in the following year, resulting in almost doubling of nitrogen resorption (Negi and Singh 1993). Reproductive efforts are known to increase nutrient resorption from leaves (Brant and Chen 2015). The enhanced leaf production after a gap of a year is also expected to increase N resorption by providing a larger sink (Singh et al. 1990). Year-to-year and site-to-site variations in nutrient resorption are common among the oaks (Du et al. 2021). There is a need to conduct a long-term study on nutrient resorption in E1 type of evergreen species.

The advantage of the strategy of keeping NuR low in the evergreen oaks could be the maximization of the duration and rate of net assimilation. Afterall, senescing leaves can fix carbon to a sizeable extent (Luo et al. 2017). In contrast, the deciduous oaks in China (*Q. aliena, Q. glandulifera,* and *Q. variabilis*) are reported to have high NuRE from senescing leaves: N, 57.7%–78.0%; P, 52.8%–86.5%, and K, 67.3%–78.2% (Du et al. 2021).

Oak and Pine Response to Increased Nutrient Availability

The nutritional control on nutrient resorption efficiency is unclear (Aerts 1996). Experimental studies show that in majority of cases (63% for N and 57% for P) there was no response to increase in soil fertility. In the remaining species, the response was the decrease in nutrient resorption (Aerts 1996). However, within a species NuR decreases with increase in soil nutrients. Singh and Bisht (1992) found that increase in soil nutrient concentration brought about drastic reductions in N, P, and K resorption in the seedlings of both chir pine and banj oak (Fig. 6.7). Consistent with natural forests, chir pine had higher nutrient resorption from senescing leaves than banj oak in this experiment. At the highest soil fertility level P resorption was even below 5%. Both NuRE and nutrient uptake from soil by roots have relative costs. As soil nutrient concentration increases the relative cost of nutrient uptake by roots from soil decreases, but the energy cost of NuR from senescing leaves remains invariant (Estiarte et al. 2023). Consequently, species decreases its NuRE as soil fertility increases.

In each species four phenophases could be identified with regard to leaf dry mass and nutrient mass: (i) leaf expansion with increasing leaf dry mass; (ii) increasing leaf dry mass without leaf expansion; (iii) steady-state (constant leaf dry mass and nutrient mass); and (iv) declining leaf dry mass and declining nutrients in senescing leaves. In deciduous species the steady state with constant leaf mass and nutrients are distinctly shorter than evergreen species; however, they too undergo these phases within an annual cycle. Both banj oak and chir pine are similar with regard to the four phases. The steady state in them for N lasts for 3–4 months. Their leaves begin to lose dry mass during later half of winters even before apparent leaf senescence (Table 6.19).

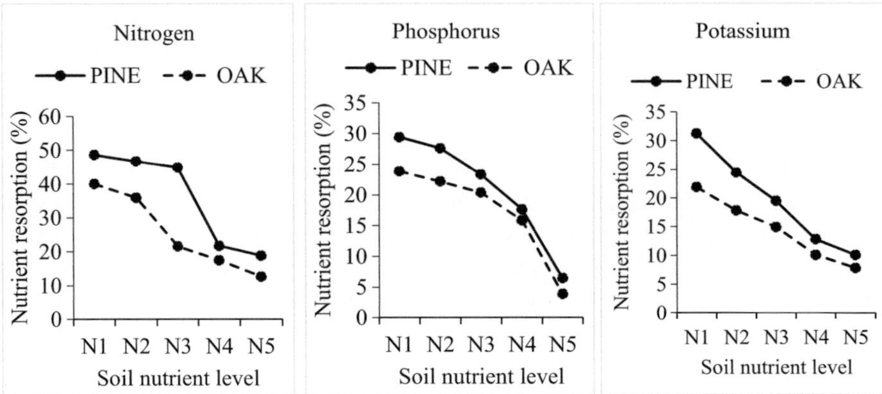

Fig. 6.7 Nutrient resorption in chir pine and banj oak decrease with the increasing nutrient level. Soil nutrients increase from N1 to N5

Table 6.19 Comparison between evergreen and deciduous species of Central Himalayan forests for nutrient (N, P, K, Ca, and Na) resorption (% nutrient mass of mature leaves) from senescing leaves

Forest	Type	N	P	K	Ca	Na
Lower elevation transect (300–2200 m), Nainital (Ralhan and Singh 1987)	Evergreen	31.9 ± 3.3	33.1 ± 3.0	28.8 ± 2.2	9.4 ± 1.0	19.2 ± 2.0
	Deciduous	46.7 ± 4.8	58.6 ± 4.7	42.9 ± 5.9	21.3 ± 5.0	30.1 ± 2.4
	Average	36.45	40.9	33.1	13.06	22.59
	Oaks[a]	23.3	24.6	25.8	7.7	15.9
	Pine[b]	53.5	49.2	44.3	9.7	19.2
Higher elevation forests (2750–3300 m), Pindar Valley (Garkoti and Singh 1997)	Evergreen	40.7	39.5	54.4	–	–
	Deciduous	62.8	50.0	54.3	–	–
	Average	51.75	44.75	54.35	–	–
	Grand average of all species	40.04	41.8	38.1	–	–

[a]Oaks (*Q. leucotrichophora, Q. floribunda,* and *Q. lanata*)
[b]Pine (*P. roxburghii*)

Nitrogen Enrichment in Forests of Himalayas in View of the Alkaline Air

The Himalayan forests are bounded to south by the Indo–Gangetic Plain (IGP), one of the largest regions of the continuous and intense farming. Manure, slurry, and urea-based fertilizers (Sigurdarson et al. 2018) have resulted in very high atmospheric urea (NH_3) concentration in the IGP, possibly more than other regions of the world (Ellis et al. 2022). The nitrogen-rich air mass that NH_3 forms/alkaline air is pushed toward the Himalayan forests on south aspects (Wang et al. 2020). The excessively N-rich air is a pollutant with adverse impact on biodiversity, and ecosystem function (Bobbink et al. 2003). Using lichens' extraordinary sensitivity to atmospheric N pollution, Ellis et al. (2022) showed the Himalayan forests are exposed to high atmospheric N concentrations. The main features of their findings were as follows: (i) the mean value of NH_3 critical level in atmosphere was 1.44 μ gm^{-3}, and (ii) the critical total N deposition load was 8.26 kg ha^{-1} yr^{-1}, (iii) the most forest areas in Himalaya are under the impact of excess atmospheric N, and the impact increases toward western and central Himalayas, and (iv) overall 73%–79% of Himalayan forests exceed the critical NH_3 level (as per the best-case scenario, Ellis et al. 2022). How forest ecosystems are responding to what Ellis et al. (2022) call "nitrogen risk" is hardly known. The gaseous NH_3 is thought to have a stronger effect on lichens than N. The problem of alkaline air (NH_3-rich) arose because of the successful control of acidifying SO_2 and NO_x emissions, which left NH_3 in excess in IGP and other similar agricultural regions. The use of urea-based fertilizers and large populations of cattle and buffalo are the source of NH_3 in IGP. The low SO_2 and NO_x emissions and high temperature result in limited conversion of NH_3 to NH_4, and it is the key reason for high NH_3 loading and IGP (Wang et al. 2020). The current status of nutrient concentration, stock, and cycling can be summarized as follows:

In brief, the data on Himalayan forests indicate that they are the large nutrient storage sites, mature/old-growth stands of oaks and sal with biomass exceeding 500 t ha^{-1}, store 10–12 t N ha^{-1} in ecosystem with >30% in biomass. The thin mass of nutrient accumulation on floor litter and short turnover time even up to treeline are the characteristic features of these forests, which emphasize their unmistakable tropical linkage. Throughout the outer ranges, exposed to the monsoon-thrust, the forest floor nutrient pool is thin. Somehow, nutrients in bole wood (and large branches) have not been given importance in forest nutrient cycling. However, the Himalayan forest data and those of some tropical rain forests indicate that stem wood accounts for considerable storage and uptake of nutrients. There is a need to analyze the strategy of accumulating nutrients in stem wood that is physiologically much less active than leaves.

Nutrient ratios, such as N:P and N:K, on an average, across the species are fairly similar in tree components, and across forests. Litterfall is the main source of nutrient return from vegetation to soil, but the role of precipitation as nutrient source is poorly assessed because of data scarcity. Since, the periodicity of these nutrient inputs is different, they meet a forest's nutrient requirement adequately. Both nutrient storage and nutrient flux were high up to 2600–2700 m, whereafter they sharply declined, being the lowest in the treeline forest. However, soil storage of nutrients diminished slowly with elevation, resulting in proportionately higher representation of soil nutrient pool in high elevation forests. Nutrient movements were characterized by strong seasonality, in which the pre-monsoon (March–May) and monsoon (July–September) were of special importance. Among the lower elevation forests, chir pine is different in several ways from other forests. Chir pine has markedly more proportional nutrient storage in soil, low foliage nutrient concentration, and slow litter decomposition release of nutrients and nutrient turnover. On an average, nutrient resorptions are higher for deciduous tree species than evergreen species. Species also reabsorb Ca, but to a limited extent. Evergreen oaks have relatively lower resorption values of all nutrients. How nutrient resorption from leaf senescence affects forest floor, litter nutrient was and release of nutrients and its availability to plants is poorly known.

References

Adhikari BS (1992) Biomass, productivity and nutrient cycling of kharsu oak and silver fir forests in central Himalaya. Dissertation. Kumaun University, Naini Tal, India

Adhikari BS, Rawat YS, Singh SP (1995) Structure and function of high-altitude forests of Central Himalaya II. Nutrient dynamics. Ann Bot 75(3):249–258

Aerts R (1996) Nutrient resorption from senescing leaves of perennials: are there general patterns? J Ecol:597–608

Aerts R, Chapin FS III (1999) The mineral nutrition of wild plants revisited: a re-evaluation of processes and patterns. In: Advances in ecological research, vol 30. Academic, pp 1–67

Athokpam FD, Garkoti SC (2015) Dynamics of foliar nitrogen of evergreen and deciduous plant species in a wet tropical forest, South Assam, India. Plant Ecol 216:1117–1135

Attiwill PM (1980) Nutrient cycling in a Eucalyptus obliqua (L'Herit.) forest. IV. Nutrient uptake and nutrient return. Aust J Bot 28(3):199–222. https://doi.org/10.1071/BT9800199

Attiwill PM, Adams MA (1993) Nutrient cycling in forests. New Phytol 124(4):561–582. https://doi.org/10.1111/j.1469-8137.1993.tb03847.x

Bamber RK, Fukazawa K (1985) Sapwood and heartwood: a review. For Abstr 46:567–580

Bauters M, Grau O, Doetterl S, Heineman KD et al (2022) Tropical wood stores substantial amounts of nutrients, but we have limited understanding why. Biotropica 54(3):596–606

Becker GS, Braun D, Gliniars R, Dalitz H (2012) Relations between wood variables and how they relate to tree size variables of tropical African tree species. Trees 26:1101–1112

Bobbink R, Ashmore M, Braun S, Flückiger W, Van den Wyngaert IJ (2003) Empirical nitrogen critical loads for natural and semi-natural ecosystems: 2002 update. cLempN:43–170

Brant AN, Chen HY (2015) Patterns and mechanisms of nutrient resorption in plants. Crit Rev Plant Sci 34(5):471–486

Carnol M, Bazgir M (2013) Nutrient return to the forest floor through litter and throughfall under 7 forest species after conversion from Norway spruce. For Ecol Manag 309:66–75

Chaturvedi OP (1983) Biomass structure, productivity and nutrient cycling in *Pinus roxburghii* forest. Ph.D. Thesis. Kumaun University, Nainital, p 347

Chen Y, Han W, Tang L, Tang Z, Fang J (2013) Leaf nitrogen and phosphorus concentrations of woody plants differ in responses to climate, soil and plant growth form. Ecography 36(2): 178–184

Chen X, Wang M, Li M, Sun J, Lyu M, Zhong Q, Cheng D (2020) Convergent nitrogen–phosphorus scaling relationships in different plant organs along an elevational gradient. AoB Plants 12(3):plaa021

Cornut I, Delpierre N, Laclau JP, Guillemot J, Nouvellon Y, Campoe O, Stape JL, Fernanda Santos V, le Maire G (2022) Potassium-limitation of forest productivity, part 1: a mechanistic model simulating the effects of potassium availability on canopy carbon and water fluxes in tropical eucalypt stands. EGUsphere 1-37. https://doi.org/10.5194/egusphere-2022-883

Deng Q, Hui D, Luo Y et al (2015) Down-regulation of tissue N:P ratios in terrestrial plants by elevated CO_2. Ecology 96(12):3354–3362

Du B, Ji H, Liu S, Kang H, Yin S, Liu C (2021) Nutrient resorption strategies of three oak tree species in response to interannual climate variability. Forest Ecosys 8(1):1–11

Ellis CJ, Steadman CE, Vieno M et al (2022) Estimating nitrogen risk to Himalayan forests using thresholds for lichen bioindicators. Biol Conserv 265:109401

Estiarte M, Campioli M, Mayol M, Penuelas J (2023) Variability and limits of nitrogen and phosphorus resorption during foliar senescence. Plant Commun 4(2):100503

Fortunel C, Fine PV, Baraloto C (2012) Leaf, stem and root tissue strategies across 758 N eotropical tree species. Funct Ecol 26(5):1153–1161

Gamble JS (1985) A manual of Indian timbers: an account of the growth, distribution, and uses of the trees and shrubs of India and Ceylon, with descriptions of their wood-structure, 2nd edn. International Book Distributors, Dehra Dun

Garkoti SC, Singh SP (1994) Nutrient cycling in the three central Himalayan forests ranging from close canopied to open canopied treeline forests. India Arct Alp Res 26(4):339–348. https://doi.org/10.1080/00040851.1994.12003078

Garkoti SC, Singh SP (1997) Structure and function of herbaceous vegetation in high mountains of central Himalaya. Trop Ecol 38(1):153–156

Gautam TP, Mandal TN (2016) Effect of disturbance on biomass, production and carbon dynamics in moist tropical forest of eastern Nepal. For Ecosyst 3:11. https://doi.org/10.1186/s40663-016-0070-y

Gilliam FS (2007) The ecological significance of the herbaceous layer in temperate forest ecosystems. BioScience 57(10):845–858. https://doi.org/10.1641/B571007

Güsewell S (2004) N:P ratios in terrestrial plants: variation and functional significance. New Phytol 164(2):243–266. https://doi.org/10.1111/j.1469-8137.2004.01192.x

Güsewell S, Koerselman W (2002) Variation in nitrogen and phosphorus concentrations of wetland plants. Perspect Pl Ecol Evol Syst 5:37–61

Han W, Fang J, Guo D, Zhang Y (2005) Leaf nitrogen and phosphorus stoichiometry across 753 terrestrial plant species in China. New Phytol 168(2):377–385

Han WX, Fang JY, Reich PB, Ian Woodward F, Wang Z (2011) Biogeography and variability of eleven mineral elements in plant leaves across gradients of climate, soil and plant functional type in China. Ecol Lett 14(8):788–796

He YT, Zhang WJ, Xu MG, Tong XG, Sun FX, Wang JZ, He XH (2015) Long-term combined chemical and manure fertilizations increase soil organic carbon and total nitrogen in aggregate fractions at three typical cropland soils in China. Sci Total Environ 532:635–644

Heineman KD, Turner BL, Dalling JW (2016) Variation in wood nutrients along a tropical soil fertility gradient. New Phytol 211(2):440–454

Herborn JL, Congdon RA (1998) Ecosystem dynamics at disturbed and undisturbed sites in North Queensland wet tropical rainforest. lll. Nutrient returns to the forest floor through litter fall. J Ecol 14:219–229

Jana P, Dasgupta S, Todaria NP (2021) Throughfall and stemflow nutrient flux in deodar and oak forests, Garhwal Himalaya, India. Water Supply 21(4):1649–1656. https://doi.org/10.2166/ws.2021.009

Johnson DW, Turner J (2019) Nutrient cycling in forests: a historical look and newer developments. For Ecol Manag 444:344–373. https://doi.org/10.1016/j.foreco.2019.04.052

Jones JM, Heineman KD, Dalling JW (2019) Soil and species effects on bark nutrient storage in a premontane tropical forest. Plant Soil 438:347–360

Kazakou E, Garnier E, Navas ML, Roumet C, Collin C, Laurent G (2007) Components of nutrient residence time and the leaf economics spectrum in species from Mediterranean old-fields differing in successional status. Funct Ecol 21(2):235–245

Khanna RK (1986) Standing state of nutrients in different forest ecosystems of Kumaun Himalaya. Ph.D. Thesis. Kumaun University, Nainital, p 400

Killingbeck KT (1996) Nutrients in senesced leaves: keys to the search for potential resorption and resorption proficiency. Ecology 77(6):1716–1727

Killingbeck KT (2004) Nutrient resorption. In: Plant cell death processes. Academic, pp 215–226

King DA (1997) Branch growth and biomass allocation in *Abies amabilis* saplings in contrasting light environments. Tree Physiol 17(4):251–258

Kleyer M, Minden V (2015) Why functional ecology should consider all plant organs: an allocation-based perspective. Basic Appl Ecol 16(1):1–9

Körner C (2012) Treelines will be understood once the functional difference between a tree and a shrub is. Ambio 41:197–206

Kraft NJ, Valencia R, Ackerly DD (2008) Functional traits and niche-based tree community assembly in an Amazonian forest. Science 322(5901):580–582

Kumar A, Kumar M, Pandey R, ZhiGuo Y, Cabral-Pinto M (2021) Forest soil nutrient stocks along altitudinal range of Uttarakhand Himalayas: an aid to nature based climate solutions. Catena 207:105667

Levia DF, Nanko K, Amasaki H et al (2019) Throughfall partitioning by trees. Hydrol Process 33(12):1698–1708

Luo Z, Guan H, Zhang X, Liu N (2017) Photosynthetic capacity of senescent leaves for a subtropical broadleaf deciduous tree species *Liquidambar formosana* Hance. Sci Rep 7(1):6323

Martin AR, Erickson DL, Kress WJ, Thomas SC (2014) Wood nitrogen concentrations in tropical trees: phylogenetic patterns and ecological correlates. New Phytol 204:484–495

Meerts P (2002) Mineral nutrient concentrations in sapwood and heartwood: a literature review. Ann For Sci 59(7):713–722

Mehra MS, Pathak PC, Singh JS (1985) Nutrient movement in litter fall and precipitation components for Central Himalayan forests. Ann Bot 55(2):153–170

Mikhailova EA, Post GC, Cope MP, Post CJ, Schlautman MA, Zhang L (2019) Quantifying and mapping atmospheric potassium deposition for soil ecosystem services assessment in the United States. Front Environ Sci 7:74

Muller RN (2003) Nutrient relations of the herbaceous layer in deciduous forest ecosystems. In: Gilliam FS, Roberts MR (eds) The herbaceous layer in forests of eastern North America. Oxford University Press, New York, pp 15–37

Muller RN, Bormann FH (1976) Role of *Erythronium americanum* Ker. in energy flow and nutrient dynamics of a northern hardwood forest ecosystem. Science 193(4258):1126–1128

Nanko K, Hotta N, Suzuki M (2006) Evaluating the influence of canopy species and meteorological factors on throughfall drop size distribution. J Hydrol 329(3–4):422–431

Neary AJ, Gizyn WI (1994) Throughfall and stemflow chemistry under deciduous and coniferous forest canopies in south-central Ontario. Can J For Res 24(6):1089–1100. https://doi.org/10.1139/x94-145

Negi GCS, Singh SP (1993) Leaf nitrogen dynamics with particular reference to retranslocation in evergreen and deciduous tree species of Kumaun Himalaya. Can J For Res 23(3):349–357

Negi GCS et al (2020) Long term ecological monitoring in western Himalaya and knowledge generation for decision making. Final Technical Report, GBPNIHESD, Kosi-Katarmal, Almora, pp 90

Oziegbe MB, Muoghalu JI, Oke SO (2011) Litterfall, precipitation and nutrient fluxes in a secondary lowland rain forest in Ile-Ife, Nigeria. Acta Bot Bras 25:664–671

Pan JW, Guo QQ, Sun XG, Gao C (2020) Contents and stoichiometric characteristics of C, N, P and K under different stand ages of *Pinus massoniana* plantations. J Plant Nutr Fertil 26(4):746–756

Pande V, Palni VT, Singh SP (2007) Effect of ectomycorrhizal fungal species on the connections outcome of two major forest species of Central Himalaya. Curr Sci 92:82–84

Parker GG (1983) Throughfall and stemflow in the forest mineral cycle. Adv Ecol Res 13:57–133

Pathak PC, Pandey AN, Singh JS (1984) Overland flow, sediment output and nutrient loss from certain forested sites in Central Himalaya, India. J Hydrol 71:239–251

Pathak PC, Pandey AN, Singh JS (1985) Apportionment of rainfall in central Himalayan forests (India). J Hydrol 76(3–4):319–332

Phoenix GK, Booth RE, Leake JR, Read DJ, Grime JP, Lee JA (2003) Simulated pollutant nitrogen deposition increases P demand and enhances root-surface phosphatase activities of three plant functional types in a calcareous grassland. New Phytol 161:279–289

Pietsch KA, Ogle K, Cornelissen JHC, Cornwell WK, Bönisch G, Craine JM et al (2014) Global relationship of wood and leaf litter decomposability: the role of functional traits within and across plant organs. Glob Ecol Biogeogr 23:1046–1057

Poudyal K, Jha PK, Zobel DB (2012) Role of wood water properties and leaf dynamics in phenology and response to drought in evergreen Himalayan tree species. Ecoprint: An Int J Ecol 19:71–84

Ralhan PK, Singh SP (1987) Dynamics of nutrients and leaf mass in central Himalayan forests trees and shrubs. Ecology 68:1974–1983

Rawal R (2022) Assessment of floristic diversity and rarity patterns along the altitudinal gradient in the forested landscapes of west Himalaya. Ph.D. Thesis,. Kumaun University, Naini Tal

Rawat YS (1983) Plant biomass, net primary production and nutrient cycling in oak forests. Ph.D. Thesis,. Kumaun University, Naini Tal

Reich PB, Oleksyn J (2004) Global patterns of plant leaf N and P in relation to temperature and latitude. PNAS 101:11001–11006

Rodríguez-Soalleiro R, Eimil-Fraga C, Gómez-García E et al (2018) Exploring the factors affecting carbon and nutrient concentrations in tree biomass components in natural forests, forest plantations and short rotation forestry. For Ecosyst 5:35. https://doi.org/10.1186/s40663-018-0154-y

Roy S, Leban JM, Zeller B, van der Heijden G et al (2022) Removing harvest residues from hardwood stands affects tree growth, wood density and stem wood nutrient concentration in European beech (Fagus sylvatica) and oak (Quercus spp.). For Ecosyst 9:100014

Sardans J, Janssens IA, Alonso R, Veresoglou SD, Rillig MC, Sanders TG et al (2015) Foliar elemental composition of European forest tree species associated with evolutionary traits and present environmental and competitive conditions. Glob Ecol Biogeogr 24(2):240–255

Sardans J, Grau O, Chen HY, Janssens IA, Ciais P, Piao S, Peñuelas J (2017) Changes in nutrient concentrations of leaves and roots in response to global change factors. Glob Change Biol 23(9): 3849–3856

Sharma CM, Baduni NP, Gairola S, Ghildiyal SK, Suyal S (2010) Effects of slope aspects on forest compositions, community structures and soil properties in natural temperate forests of Garhwal Himalaya. J For Res 21(3):331–337

Sigurdarson JJ, Svane S, Karring H (2018) The molecular processes of urea hydrolysis in relation to ammonia emissions from agriculture. Rev Environ Sci Biotechnol 17(2):241–258

Singh SP, Bisht K (1992) Nutrient utilization in *Quercus leucotrichophora* and *Pinus roxburghii* seedlings at five soil fertility levels. J Veg Sci 3(5):573–578

Singh P, Negi GCS (2023) Altitudinal patterns of foliar nutrients in dominant tree species: are central Himalayan forests nutrient stressed. World J For Res 2(2):92–109

Singh JS, Singh SP (1987) Forest vegetation of Himalaya. Bot Rev 53:80–193

Singh JS, Singh SP (1992) Forests of Himalaya: structure, functioning and impact of man. Gyanodaya Prakashan, Nainital, p 294

Singh SP, Thadani R (2013) Valuing ecosystem services flowing from the Indian Himalayan states for incorporation into national accounting. In: Lowman M, Devy S, Ganesh T (eds) Treetops at risk: challenges of global canopy ecology and conservation. Springer, New York, pp 423–434

Singh SP, Rawat YS, Rana BS, Negi GCS (1990) Effect of unusually large seed crop on litter fall and nitrogen retranslocation in Himalayan oaks. For Ecol Manag 32:79–86

Stanek M, Piechnik Ł, Stefanowicz AM (2020) Invasive red oak (Quercus rubra L.) modifies soil physicochemical properties and forest understory vegetation. For Ecol Manag 472:118253

Su L, Zhao C, Xu W, Xie Z (2019) Hydrochemical fluxes in bulk precipitation, throughfall, and stemflow in a mixed evergreen and deciduous broadleaved forest. Forests 10(6):507

Switzer GL, Nelson LE (1972) Nutrient accumulation and cycling in loblolly pine (*Pinus taeda*) plantation ecosystems: the first 20 years. Soil Sci Soc Am Proc 36:143–147. https://doi.org/10.2136/sssaj1972.03615995003600010033x

Takahashi M (2021) Nutrient storage and stoichiometry of the forest floor organic matter in Japanese forests. Soil Syst 5(3):51. https://doi.org/10.3390/soilsystems5030051

Takashima T, Hikosaka K, Hirose T (2004) Photosynthesis or persistence: nitrogen allocation in leaves of evergreen and deciduous *Quercus* species. Plant Cell Environ 27(8):1047–1054

Tang Z, Xu W, Zhou G, Bai Y, Li J, Tang X et al (2018) Patterns of plant carbon, nitrogen, and phosphorus concentration in relation to productivity in China's terrestrial ecosystems. PNAS 115(16):4033–4038

Tian D, Yan Z, Niklas KJ, Han W, Kattge J, Reich PB et al (2018) Global leaf nitrogen and phosphorus stoichiometry and their scaling exponent. Natl Sci Rev 5(5):728–739

Tully KL, Wood TE, Schwantes AM, Lawrence D (2013) Soil nutrient availability and reproductive effort drive patterns in nutrient resorption in Pentaclethra macroloba. Ecology 94(4):930–940

Tyagi V, Singh SP, Singh RD, Gumber S, Thadani R, Pandey R (2023) Influence of slope position and aspect on the vegetation attributes and tree-water relations in forests of the central Himalayas. J Mt Sci 20(9):2592–2602

Usman S, Rawat YS, Singh SP, Garkoti SC (1997) Fine root biomass production and turnover in evergreen forests of central Himalaya. India Oecol Mont 6(1–2):4–8

Vergutz L, Manzoni S, Porporato A, Novais RF, Jackson RB (2012) Global resorption efficiencies and concentrations of carbon and nutrients in leaves of terrestrial plants. Ecol Monogr 82:205–220

Verhoeven JTA, Koerselman W, Meuleman AFM (1996) Nitrogen-or phosphorus-limited growth in herbaceous, wet vegetation: relations with atmospheric inputs and management regimes. Trends Ecol Evol 11(12):494–497

Viera M, Schumacher MV (2009) Nutrients concentration and retranslocation in the Pinus taeda L. needles. Ciência Florestal 29:375–382

Viera M, Caldeira MVW, Rovani FFM, Castro KC (2016) Ecological and environmental aspects of nutrient cycling in the Atlantic forest, Brazil. In: Beanco JA, Chang S, Lo Y (eds) Tropical forests. InTech CBS Publishers. https://doi.org/10.5772/64188

Vitousek PM, Sanford RL Jr (1986) Nutrient cycling in moist tropical forest. Ann Rev Ecol Syst 17(1):137–167

Wang T, Song Y, Xu Z, Liu M, Xu T, Liao W et al (2020) Why is the Indo-Gangetic Plain the region with the largest NH3 column in the globe during pre-monsoon and monsoon seasons? Atmos Chem Phys 20(14):8727–8736

Welch NT, Belmont JM, Randolph JC (2007) Summer ground layer biomass and nutrient contribution to above-ground litter in an Indiana temperate deciduous forest. Am Midl Nat 157(1): 11–26

Xing S, Cheng X, Kang F, Wang J, Yan J, Han H (2022) The patterns of N/P/K stoichiometry in the *Quercus wutaishanica* community among different life forms and organs and their responses to environmental factors in northern China. Ecol Indic 137:108783

Xu X (2023) Effect of changes in throughfall on soil respiration in global forest ecosystems: a meta-analysis. Forests 14(5):1037. https://doi.org/10.3390/f14051037

Xu XN, Hirata E (2002) Forest floor mass and litterfall in Pinus luchuensis plantations with and without broad-leaved trees. For Ecol Manag 157(1–3):165–173

Xu S, Zhou G, Tang X et al (2017) Different spatial patterns of nitrogen and phosphorus resorption efficiencies in China's forests. Sci Rep 7(1):10584

Yan Z, Li P, Chen Y, Han W, Fang J (2016) Nutrient allocation strategies of woody plants: an approach from the scaling of nitrogen and phosphorus between twig stems and leaves. Sci Rep 6:20099

Yang X, Tang Z, Ji C, Liu H et al (2014) Scaling of nitrogen and phosphorus across plant organs in shrubland biomes across Northern China. Sci Rep 4:5448

Yang T, Li Y, Ouyang X, Wang B, Ge X, Tang L (2022) Rainfall physical partitioning and chemical characteristics in evergreen coniferous and deciduous broadleaved forest stands in a high nitrogen deposition region. China For 13(10):1644. https://doi.org/10.3390/f13101644

Zanne AE, Oberle B, Dunham KM, Milo AM, Walton ML, Young DF (2015) A deteriorating state of affairs: how endogenous and exogenous factors determine plant decay rates. J Ecol 103(6): 1421–1431

Zhao N, Yu G, He N, Wang Q et al (2016) Coordinated pattern of multi-element variability in leaves and roots across Chinese forest biomes. Glob Ecol Biogeogr 25(3):359–367
Zhu F, Li H (2013) Difference in canopy interception of coniferous forest and broad-leaf forest in reservoir watershed. Int J Appl Environ Sci 8(9):1027–1032
Zobel DB, Singh SP (1997) Himalayan forests and ecological generalizations. BioScience 47(11): 735–745. https://doi.org/10.2307/1313096

Biomass, Carbon, and Productivity

7

Ashish Tewari, Ikramjeet Maan, Arif Ansari, Yogesh Chandra Tripathi, and Shruti Shah

Abstract

The Himalayan forests represent significant carbon reservoirs that play a critical role in global climate regulation. This chapter synthesizes current knowledge on forest biomass, carbon sequestration, and productivity across the Indian Himalayan Region (IHR). In undisturbed forests, tree biomass typically ranges from 200 t ha^{-1} in *Pinus roxburghii* to 400 t ha^{-1} in *Quercus* species, with old-growth forests sometimes exceeding 500–700 t ha^{-1}. Net primary productivity in oak forests varies between 5.68–13.2 t ha^{-1} yr^{-1} in the tree layer, with allocation patterns showing highest productivity in foliage, followed by bole, roots, twigs, and branches. Soil organic carbon increases with elevation, with maximum stocks (138.37 t ha^{-1}) observed above 2500 m. Annual litterfall contributes 5–9 t ha^{-1} yr^{-1} to forest floors, with seasonal patterns showing maximum leaf drop during spring. Root research reveals that approximately 50% of fine roots occur in the top 30 cm soil layer, but species vary in rooting depth (90–180 cm) and response to seasonal drought. Banj oak forests demonstrate 56.2% more fine root biomass than chir pine forests, with annual production rates of 3.62 and 2.50 t ha^{-1} yr^{-1}, respectively. Carbon sequestration rates in community-managed forests range from 2–7 t ha^{-1} yr^{-1} depending on forest condition. These forests serve as vital carbon sinks that require conservation. The chapter provides comprehensive allometric equations for biomass estimation across multiple Himalayan tree species and documents carbon stocks in different forest types, emphasizing their importance in climate change mitigation strategies.

A. Tewari (✉) · I. Maan · A. Ansari · Y. C. Tripathi · S. Shah
Department of Forestry, DSB Campus, Kumaun University, Nainital, Uttarakhand, India
e-mail: atewari69@gmail.com

S. P. Singh, R. Thadani (eds.), *Himalayan Forests in a Changing World*,
https://doi.org/10.1007/978-981-95-0978-2_7

Keywords

Net production · Forest carbon · Root biomass · Litter decomposition

Introduction

Biomass refers to living organic matter, expressed in dry mass (t ha^{-1}). Forest biomass dynamics encompass three primary processes: biomass gain through plant recruitment, biomass growth through surviving plants, and biomass loss through mortality. These processes collectively determine net changes in forest aboveground biomass (Brienen et al. 2015). In forest ecosystems, carbon is stored in living floral and faunal biomass and dead organic matter. The largest carbon pools in most forests are aboveground biomass and soil organic matter, with roots and surface detritus in smaller but significant amounts.

Several factors beyond growth rates determine biomass levels in forests. Carbon accumulation capacity is notably influenced by disturbance regimes (Pregitzer and Euskirchen 2004; Keeling and Phillips 2007). Biomass and productivity share strong interrelationships, with biomass functioning as a direct outcome of productivity, both influenced by ecological factors such as temperature, moisture, and nutrient availability (Keeling and Phillips 2007).

In ecological terms, a carbon pool refers to a reservoir capable of storing or releasing carbon. Key forest carbon pools include biomass, soil, and the atmosphere. Among these, aboveground live biomass represents the most rapidly changing pool, which can be estimated accurately through allometric approaches (Jenkins et al. 2003). However, quantifying changes in other carbon pools remains equally critical for a comprehensive understanding.

To better comprehend forest productivity, it is essential to unravel the underlying processes in net biomass change. Net ecosystem productivity (NEP) signifies the rate of carbon accumulation in an ecosystem, representing the sum of variations across carbon pools. Biologically, NEP is the difference between net primary productivity (NPP, the annual net carbon fixation by plant photosynthesis) and heterotrophic respiration (CO_2 emission by non-photosynthetic organisms). A highly sensitive relationship exists between NEP, carbon pool sizes, and forest management activities.

Forests and Carbon in the Himalayan Region

Forests constitute a major component of the world's carbon reserves, significantly influencing all living organisms (Whittaker and Likens 1975). They account for 70%–90% of total terrestrial biomass (Houghton et al. 2009) and store more than 80% of all terrestrial aboveground carbon and more than 70% of all soil organic carbon (SOC) (Ameray et al. 2021). As a result, changes in forest carbon stocks directly impact global climate change patterns.

The carbon storage in forest ecosystems is strongly affected by climate, forest type, stand age, disturbance regimes, and edaphic conditions (Pregitzer and Euskirchen 2004). Climate change can modify tree species composition and migration patterns, which further influence forest composition and affect carbon sequestration rates (Bunker et al. 2005). Forest carbon therefore serves as a valuable measure for assessing changes in forest structure and composition.

Carbon Stock in the Indian Himalayan Region (IHR)

According to the 2020 Global Forest Resource Assessment by FAO, the total global forest carbon stock is approximately 662 Gigatonnes (Gt). Within this, India accounts for approximately 7.2 Gt (about 1.09% of the world's total) as per the India State of Forest Report (FSI 2021).

Among Indian Himalayan states, there are reported to be significant increases in carbon stock and biomass, ranging from about 3% to 60% between 2011 and 2021 as per India State of Forest Reports (FSI 2011; FSI 2021), but these are possibly due to methodological changes in data collection rather than real increases in Forest Carbon.

In the tropical forests of the Indian Himalayan region, biomass and total carbon density are highest in the Western Himalayan region, followed by the Central Himalayas, and lowest in the Northeast region. However, in subtropical and temperate forests, the maximum biomass and carbon are observed in the Central Himalayan region, followed by Northeast India, with the Western Himalayan region showing the lowest values (Table 7.1).

Allometric Equations for Himalayan Tree Species

Allometric equations relating biomass of different tree components to circumference at breast height (1.37 m) are essential tools for forest carbon estimation. Table 7.2 presents allometric relationships for important tree species of the Central Himalayas.

Table 7.1 Total carbon in different forest types and Indian Himalayan regions

Forests type	Western Himalayas (Himachal Pradesh)	Central Himalayas (Uttarakhand)	Northeast India
Tropical Forest	177	119	112
Sub-tropical Forest	93	272	113
Temperate Forest	148	291	135

Source: Sahoo et al. (2021), Kaushal and Baishya (2021), Singh and Verma (2018), Sharma et al. (2010a)

Note: Carbon values are in t ha^{-1} in each forest type

Table 7.2 Allometric relationships between biomass of tree components (Y, kg per tree) and circumference at breast height (X, cm at 1.37 m height)

Species name	Component		Intercept (a)	Slope (b)
Dalbergia sissoo	AGB	Bole	−0.367	1.3457
		Branch	−1.4581	0.7708
		Twig	−0.2932	0.1461
		Leaf	−0.4501	0.283
		Stump root	−0.4017	0.2744
	BGB	Lateral root	−0.2723	0.1923
		Fine root	−0.0948	0.0664
Shorea robusta	AGB	Bole	−2.832	1.976
		Branch	−2.037	1.501
		Twig	−2.688	1.463
		Leaf	−1.736	1.175
Mallotus phillippensis	AGB	Bole	−2.1425	1.398
		Branch	−2.282	1.215
		Twig	−2.3285	0.81
		Leaf	−3.8605	1.066
Interspecies of Shorea robusta forest	AGB	Bole	−5.0299	2.3333
		Branch	−5.2096	2.081
		Twig	−4.633	1.683
		Leaf	−4.9646	1.679
Acacia catechu	AGB	Bole	−1.9418	1.3604
		Branch	−0.2451	0.1958
		Twig	−0.2262	0.2519
		Leaf	0.1098	0.2194
		Stump root	0.081	0.1234
	BGB	Lateral root	−0.4909	0.2123
		Fine root	−0.0322	0.0147
Albizia lebbeck	AGB	Bole	−0.2976	0.4172
		Branch	−0.1617	0.1561
		Twig	−0.1579	0.1411
		Leaf	−0.1247	0.1853
		Stump root	−0.0162	0.0707
	BGB	Lateral root	−0.0614	0.061
		Fine root	−0.0007	0.0007
Pinus roxburghii	AGB	Bole	−6.418	2.598
		1st order branch	−9.833	2.978
		Other branch	−9.338	2.63
		Foliage	−6.111	1.872
		Stump root	−7.22	2.448
		Lateral root	−9.161	2.593
		Fine root	−9.102	2.069
Quercus leucotrichophora	AGB	Bole	−0·523	1·367
		Branch	−0·718	1·302

(continued)

Table 7.2 (continued)

Species name	Component		Intercept (a)	Slope (b)
		Twig	0·065	0·895
		Foliage	−0·976	0·854
		Stump root	0·982	0·904
	BGB	Lateral root	−0·312	0·809
		Fine root	−1·326	0·504
Interspecies for Banj oak-Chir pine	AGB	Bole	−0.861	1.425
		Branch	−0.908	1.327
		Twig	−0.506	1.028
		Foliage	−1.106	1.042
	BGB	Stump root	0.098	0.948
		Lateral root	−2.346	0.997
		Fine root	−2.874	0.197
Aesculus indica	AGB	Bole	1.9322	0.9809
		Bole bark	1.8096	0.6836
		Branch	0.7989	1.0233
		Twig	0.3294	0.8008
		Foliage	−0.0689	0.7308
	BGB	Stump root	2.3831	0.6438
		Lateral root	1.0028	0.8118
		Fine root	0.1808	0.7429
Juglans regia	AGB	Bole	1.9626	0.9002
		Bole bark	−0.993	1.1049
		Branch	0.0445	0.873
		Twig	0.5687	0.5707
		Foliage	−2.6355	0.9973
	BGB	Stump root	0·9494	0·8010
		Lateral root	0·2858	0·7140
		Fine root	−3·2823	1·1510
Litsea umbrosa	AGB	Bole	−0·0598	0·9413
		Bole bark	−3·2379	0·9836
		Branch	−1·5706	1·0297
		Twig	−2·0961	0·8336
		Foliage	−2·1761	0·7369
	BGB	Stump root	0·2594	0·6362
		Lateral root	−0·2290	0·5692
		Fine root	−2·5342	0·7400
Symplocos chinensis	AGB	Bole	−0·2204	0·8635
		Bole bark	−2·0041	0·6066
		Branch	−1·8699	1·0151
		Twig	−2·4490	0·9425
		Foliage	−0·6680	0·2999
	BGB	Stump root	−4·0061	1·4903
		Lateral root	−0·4620	0·1756

(continued)

Table 7.2 (continued)

Species name	Component		Intercept (a)	Slope (b)
		Fine root	−0·4337	0·1425
Ilex dipyrena	AGB	Bole	0·8036	0·7985
		Bole bark	−3·0025	1·0073
		Branch	−1·9720	1·2225
		Twig	−2·0474	0·8015
		Foliage	−1·6810	0·6184
	BGB	Stump root	−1·7826	0·1056
		Lateral root	−2·6710	1·1322
		Fine root	−2·7198	0·9505
Lyonia ovalifolia	AGB	Bole	−0·6884	1·0865
		Bole bark	−2·6248	0·9400
		Branch	−0·9054	1·0530
		Twig	−1·5042	1·0496
		Foliage	−1·6954	0·8085
	BGB	Stump root	−0·1944	0·7920
		Lateral root	−2·8928	0·8537
		Fine root	−1·6587	0·4573
Rhododendron arboreum	AGB	Bole	0·2070	0·9638
		Bole bark	−3·5917	1·0207
		Branch	−0·1113	0·8288
		Twig	−0·4734	0·8790
		Foliage	−1·5638	0·8044
	BGB	Stump root	0·9364	0·5874
		Lateral root	−0·4078	0·6924
		Fine root	−0·9856	0·3754
Cupressus torulosa	AGB	Bole	−5.2566	2.3762
		Bole bark	−4.8379	1.5999
		Branch	−1.5634	1.3825
		Twig	0.7641	0.5677
		Foliage	−0.0433	0.7322
	BGB	Stump root	−2.8713	1.5145
		Lateral root	1.662	0.3638
		Fine root	−1.5051	0.7495
Cedrus deodara	AGB	Bole	−5.5628	2.3452
		Bole bark	−7.945	2.2078
		Branch	−2.059	1.2736
		Twig	0.0963	0.6103
		Foliage	−0.0191	0.5572
	BGB	Stump root	−2.252	1.257
		Lateral root	−0.3328	0.6661
		Fine root	−1.8519	0.5475
Quercus floribunda	AGB	Bole	2·0815	0·8858
		Bole bark	−0·3287	0·8789

(continued)

Table 7.2 (continued)

Species name	Component		Intercept (a)	Slope (b)
		Branch	1·4970	0·8350
		Twig	1·6624	0·4244
		Foliage	1·3092	0·4997
	BGB	Stump root	2·4764	0·4804
		Lateral root	2·2269	0·3882
		Fine root	1·1272	0·3727
Interspecies of oak forest	AGB	Bole	−0·861	1·425
		Branch	−0·908	1·327
		Twig	−0·506	1·028
		Foliage	−1·106	1·042
		Stump root	0·098	0·948
	BGB	Lateral root	−2·346	0·997
		Fine root	−2·874	0·529
Quercus semecarpifolia	AGB	Bole	1.1230	1·1217
		Bole bark	−1·2514	1·1100
		Branch	−0·2312	1·1231
		Twig	−0·8548	0·8673
		Foliage	−0·9425	0·8416
	BGB	Stump root	0·5886	0·9549
		Lateral root	−0·4084	0·9896
		Fine root	−0·8265	0·8770
Abies pindrow	AGB	Bole	1·5380	1·0088
		Bole bark	−0·1066	0·8876
		Branch	0·0356	0·9977
		Twig	−0·0146	0·8374
		Foliage	0·2464	0·6429
	BGB	Stump root	−0·4874	1·0909
		Lateral root	−0·6510	0·9947
		Fine root	1·0137	0·4604

Source*:* Brahma et al. (2021), Singh et al. (2011), Adhikari et al. (1995), Singh and Singh (1992), Rawat and Singh (1988), Chaturvedi and Singh (1987), Rana (1985), Rawat (1983)
The equation form is: $\ln Y = a + b \ln X$, where ln is natural logarithm, a is the intercept, and b is the slope or regression coefficient
Note: *AGB* aboveground biomass, *BGB* belowground biomass

Biomass and Net Production in Himalayan Forests

Values for forest biomass and carbon stock in forests from foothills (280 m) to high elevation (3300 m) in the Uttarakhand Himalayan region vary considerably with species composition, forest type, and altitudinal zone (Table 7.3).

In the Central Himalayas, *Shorea robusta* forest biomass (dry matter) is 577.16 t ha^{-1}, with a carbon sequestration rate of 4.63 t ha^{-1} yr^{-1} and a net primary

Table 7.3 Total biomass, carbon stocks (above and below ground) and net primary productivity (NPP) of different forest types of Uttarakhand

Forest type	Biomass (t ha^{-1})	Carbon stocks (t ha^{-1})	Carbon sequestration (t ha^{-1} yr^{-1})	NPP (t ha^{-1} yr^{-1})	Region	Source
Shorea robusta forest	396–669	186–314	6.6–8.8	12.9–18.8	Central Himalayas	Rana et al. (1989)
Shorea robusta forest	577	288	4.63	9.75 ± 1.27	Central Himalayas	Joshi et al. (2021a, b)
Shorea robusta forest	346	159	–	–	Garhwal Himalayas	Sharma et al. (2010b)
Pinus roxburghii forest	113–283	53–134	3.6–8.8	7.6–18.7	Central Himalayas	Gosain et al. (2015)
Pinus roxburghii forest	285	142	4.8	10.17 ± 0.43	Central Himalayas	Joshi et al. (2021a, b)
Pinus roxburghii forest[a]*	48–102	24–51	–	–	Tehri Garhwal	Kumar and Sharma (2015)
Pinus roxburghii forest	298.04	137.1	–	–	Garhwal Himalayas	Sharma et al. (2010b)
Pinus roxburghii forest*	119.1	56.6	2.64	9.04	Kumaun Himalayas	Gosain et al. (2015)
Mixed Banj oak-Chir pine forest	426	202	7.1	15	Central Himalayas	Rana et al. (1989)
Quercus leucotrichophora forest	230.1	115.1	3.9	8.4 ± 2.8	Central Himalayas	Joshi et al. (2021a, b)
Quercus leucotrichophora forest	387	183	6.2	13.2	Central Himalayas	Rawat and Singh (1988)
Quercus leucotrichophora forest	189.8	91.1	3.9	5.68	Kumaun Himalayas	Gosain et al. (2015)
Quercus leucotrichophora forest*	46–63	23–31	–	–	Tehri Garhwal	Kumar and Sharma (2015)
Quercus lanata forest	285–557	135–264	7.3–8.4	15.5–17.8	Central Himalayas	Rawat and Singh (1988)
	533.28	134.51	–	–	Garhwal Himalayas	

Quercus floribunda forest						Sharma et al. (2010b)
Quercus floribunda forest	458–782-	217–371	7.8–11.9	16.6–25.1	Central Himalayas	Rana et al. (1989)
Quercus semecarpifolia forest	590.2	295.1	–	–	High-altitude Central Himalayas	Adhikari et al. (1995)
Quercus semecarpifolia forest	388–615	184–292	–	–	Central Himalayas	Verma et al. 2012
Quercus semecarpifolia forest	279.3 ± 26.3	128.4 ± 12.1	–	–	Upper West Himalayas	Sharma et al. (2010b)
Mixed oak forest	227.23	113.62	4.47	9.42 ± 2.50	Central Himalayas	Joshi et al. (2021a, b)
Cedrus deodara forest	464.2	232.1	–	–	Uttarkashi District	Sharma et al. (2016)
Abies spectabilis forest	460.9	230.45	–	–	Uttarkashi District	Sharma et al. (2016)
Abies pindrow forest	377.6 ± 38.6	173.7 ± 17.7	–	–	Upper West Himalayas	Sharma et al. (2010b)
Abies pindrow forest	296	148	–	–	Chamoli Garhwal	Gairola et al. (2011)
Mixed broadleaved conifer forest	298.6	149.3	–	–	Chamoli Garhwal	Gairola et al. (2011)

In the degraded and young forests of *Q. leucotrichophora* and *P. roxburghii* lower biomass ranging from 40–100 t ha^{-1} can be observed in the Himalayan forests

[a](Editors note) Values with asterisk(*) likely pertain to degraded forests

productivity (NPP) of 9.75 ± 1.27 t ha^{-1} yr^{-1} (Joshi et al. 2021b). Studies in the Garhwal Himalayas report biomass of 346.48 tonnes/ha for *Shorea robusta*, highlighting the variability within regions and differences in forest age (Sharma et al. 2010b).

The average biomass of *P. roxburghii* is approximately 200 t ha^{-1} in the Uttarakhand Himalayan region, while oak forests in relatively undisturbed areas typically reach about 400 t ha^{-1} (Singh and Singh 1992). The biomass accumulation ratio of Oak forests is higher compared to *Pinus roxburghii*. Among forest types, evergreen broad-leaved forests of the warm temperate zone are highly productive, while deciduous broad-leaved forests in cold temperate zones show the lowest productivity (Satoo 1971; Rawat and Singh 1988).

The oak forests of Uttarakhand have a total net production in the tree layer ranging from 5.68 to 13.2 t ha^{-1} yr^{-1}. The average shrub production is around 1.65 t ha^{-1} yr^{-1}, and herbaceous net production ranges between 1.2 and 1.5 t ha^{-1} yr^{-1}. In temperate forests, the aboveground herbaceous production ranges from 0.1 to 1.0 t ha^{-1} yr^{-1}. In mid-altitude oak forests and pine forests, net production ranges from 7.6–25.1 t ha^{-1} yr^{-1} (Chaturvedi and Singh 1987). The total (tree, shrub, and herb) net primary productivity in oak forests is generally between 13.2–16.6 t ha^{-1} yr^{-1} (Fig. 7.1).

Sharma et al. (2016) reported the biomass of *Cedrus deodara* and *Abies spectabilis* at 464.2 and 460.9 t ha^{-1}, respectively, in Uttarkashi district of Uttarakhand. In the high-altitude central Himalayas region, Adhikari et al. (1995) reported the biomass of *Quercus semecarpifolia* at 590.2 t ha^{-1}. These figures suggest that the temperate belt in Uttarakhand Himalayas still contains old-growth forests with biomass around 500 t ha^{-1}.

In the Central Himalayan forests, NPP of trees can vary from 5.6 to 25.1 t ha^{-1} yr^{-1} (Singh and Singh 1992; see also Table 7.3). The wide range of NPP can be attributed to the condition of the forests and its age. If we remove the unusually low and high values, the range of the net primary production would be 8-18 t $ha^{-1}Yr^{-1}$ which can appropriately represent the NPP rates of the forests of their region. However, unusually low values in disturbed forests of chir pine and banj oak have been reported from the Lamgara block of Almora (Karky et al. 2007). Such degraded or young forests of banj or chir had biomass values ranging from 40–100 t ha^{-1}.

Biomass Allocation in Different Forest Types

In trees, the largest fraction of net production is allocated to foliage, followed by bole, roots, twigs, and branches. In shrubs, the allocation pattern differs slightly, with maximum allocation to foliage, followed by roots and stem.

At higher altitudes in Uttarakhand, horse chestnut shows higher aboveground biomass (AGB) net production compared to other deciduous forests globally. The higher-altitude brown oak (*Quercus semecarpifolia*) has the highest AGB among all the Oak forests of the region (459.7 t ha^{-1}). The total root biomass in higher-altitude

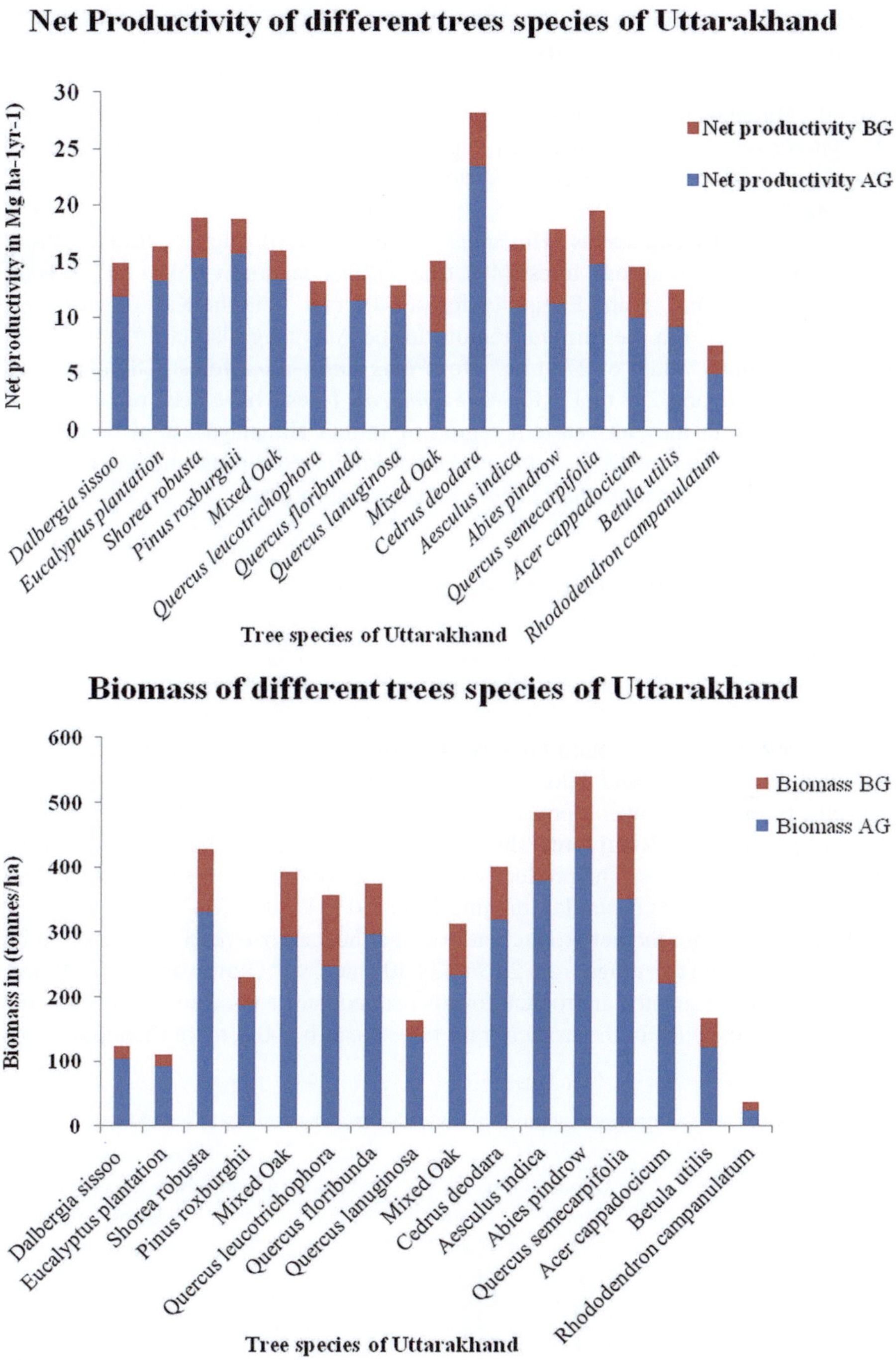

Fig. 7.1 The biomass and net productivity of major deciduous and evergreen tree species of Uttarakhand (Lodhiyal et al. 2014; Rawat and Singh 1989; Singh and Singh 1992; Singh et al. 1994)

forests (Spruce) generally ranges between 106 and 131 t ha^{-1}, which exceeds that of mid-altitude Oak and Pine forests.

The contribution of aboveground biomass varies between 79.4% and 85.6% in Central Himalayan tree species, being maximum in *Q. floribunda* and *Q. lanuginosa*. Chir pine also has high aboveground allocation. *In contrast, Q. leucotrichophora* has a deep root system with higher allocation to belowground components.

In the Western Himalayas (Himachal Pradesh region), higher altitudes show greater biomass in Banj oak forest (492 t ha^{-1}), Deodar forest (304 t ha^{-1}), Blue pine forest (361 t ha^{-1}), and Fir/spruce forest (407 t ha^{-1}) in the Kullu region, and spruce (335 t ha^{-1}) in the Sirmour region. In the Anantnag district of Jammu and Kashmir, biomass values of 272 t ha^{-1} for *Pinus wallichiana* forest, 284 t ha^{-1} for *Cedrus deodara*, and 295 t ha^{-1} for *Abies pindrow* forests have been reported.

In Manipur, in the Northeastern region of Indian Himalayas, total biomass is 179 t ha^{-1} in *Schima wallichii* and *Ficus cunia*-dominated forests, and 246 t ha^{-1} in *Phoebe hainesiana* and *Gmelina arborea*-dominated forests.

Biomass and Productivity of Herbs and Shrubs

The understory components also contribute significantly to forest biomass and productivity. Table 7.4 presents data for different forest types in the Central Himalayas.

Sal forest has the maximum biomass of shrubs (5 t ha^{-1}) and herbs (2 t ha^{-1}), followed by Rianj oak and Tilonj oak forests. Among Central Himalayan forests, chir pine forests have lower biomass of shrubs (0.8 t ha^{-1}) and herbs (1.2 t ha^{-1}). In terms of productivity, Rianj forest shows the maximum values: 2.5 t ha^{-1} for shrubs and 2.3 t ha^{-1} for herbs. The productivity of shrubs and herbs in sal, chir pine, and mixed banj oak forests is similar, ranging from 2.0 to 2.3 t ha^{-1}.

Across the world, the net production (tons per hectare per year) is reported to be highly variable and can range from 2.07 to 29.40 t ha^{-1} y^{-1} (Rawat and Singh 1988). Terrestrial NPP is greatest in tropical forests near equator and decreases towards the poles. The forest NPP in the tropical rain forests can be 60% more than that in the

Table 7.4 Biomass and net primary productivity of herbs and shrubs in different forest types of Central Himalayas

	Biomass (t ha^{-1})		Net primary productivity (t ha^{-1} yr^{-1})	
Forest types	Shrubs	Herbs	Shrubs	Herbs
Sal forest	5.0	2.0	1.5	2.1
Chir pine	0.8	1.2	1.0	2.2
Rianj oak	5.0	1.5	2.5	2.3
Tilonj oak	4.4	1.5	2.2	1.9
Mixed banj oak-chir pine	3.5	2.0	1.1	2.0
Banj oak	2.3	1.6	0.7	1.8

temperate forests (Huston and Wolverton 2009). The NPP in the tropical rain forests of Brazil, Thailand, and Malaysia were 16.8, 28.6, and 19.22 t ha^{-1} y^{-1}, respectively. The NPP of Indian Oaks ranges between 15.86–20.57 t ha^{-1} y^{-1} which is comparatively higher than the NPP of Oaks across the world 6.8–15.16 t ha^{-1} y^{-1} (Rawat and Singh 1988).

Old-Growth Forests

Tree biomass values ranging between 500–700 tons have been reported for old-growth forests in the Central Himalayas across elevations from 300–3000 m. At the foothills, old-growth Dipterocarp Sal (*Shorea robusta*) forests can have biomass close to 710 t ha^{-1} (Rana et al. 1989), while at higher altitudes, *Quercus floribunda* and *Quercus semecarpifolia* forests (old-growth) have biomass above 500 t ha^{-1} (Rodin and Bazilevich 1967). Tree biomass values of 545.8 to 571.1 t ha^{-1} have been reported for certain mixed oak forests between 2270–2390 m in Nepal (Yoda 1968).

These studies indicate that biomass values above 500 t $ha^{-1}$are common for mature, undisturbed oak forests of the Himalayan region. The percentage allocation in bole biomass of *Shorea robusta* (sal), *Pinus roxburghii* (Chir pine), and *Cedrus deodara* can vary between 70% and 81%, compared to 50%–60% in *Quercus* species. Branch allocation in Oaks approaches 30%, whereas in Sal, chir pine, and Deodar, it ranges between 12% and 17%. The percent root biomass of total biomass across Central Himalayan species ranges from 14.72% to 28.3%, reaching maximum in *Q. leucotrichophora* and minimum in *Quercus lanuginosa*. In an old-growth *Pinus kesiya* forest in Northeastern India, the total biomass pool is 460.5 t ha^{-1}, of which 91.20% is aboveground biomass and 8.8% belowground biomass (Baishya and Barik 2011).

For comparison, in boreal old-growth forests, maximum biomass approaches 400 t ha^{-1}, while in temperate old-growth coniferous forests of the Pacific Northwest, values range between 800–1000 t ha^{-1}, with a maximum of 2500 t ha^{-1} reported in coastal redwood forests. In the Himalayan region, values close to 700 t ha^{-1} have been reported for old-growth sal forests (Singh and Singh 1992).

Agroforestry Systems (AFS)

The stored carbon and biomass in agroforestry systems is greatly influenced by the management of the system and its species composition (Albrecht and Kandji 2003; Rajput et al. 2015). These systems have significant potential to remove atmospheric CO_2 and work as carbon sinks through growth of trees and understory vegetation (Yadav et al. 2017). According to IPCC, the worldwide average amount of stored carbon (aboveground) in agroforestry systems ranges between 40–150 t C ha^{-1} (IPCC 2006). In India, the average carbon sequestration potential of agroforestry systems is estimated ranging between 0.29 and 15.21 t ha^{-1} yr^{-1} (Nair et al. 2009;

Yadav et al. 2017). It is predicted that in the next 50 years, agroforestry systems will sequester up to 1.1–2.2 Gt C (Bhardwaj et al. 2023).

In the foothill regions of Uttarakhand, *Eucalyptus* and Popular plantations are being raised extensively in agroforestry models. The biomass and productivity of *Eucalyptus tereticornis* and *Populas deltoids* clones in the Tarai range from 112–171 t ha^{-1} in different aged plantations and the NPP generally varies 13–21 t ha^{-1} yr^{-1} (Singh 1998). A significant amount of biomass and carbon has been estimated in the pure Poplar agroforestry plantation in the Tarai region of central Himalayas, which acts as an additional carbon sink in the region. In an 8-year-old *P. deltoides* plantation (short rotation), a total of 96 t C ha^{-1} was found to be stored and, on an average, 0.192 t C is stored in a single tree (Singh and Lodhiyal 2009). Of the total carbon, 78.68% carbon is stored in the aboveground components, whereas 21.32% carbon in the belowground components of the trees (Singh and Lodhiyal 2009). Another study by Lodhiyal 2014 indicates 66.83 t ha^{-1} total biomass in 5-year-old *E. tereticornis* plantation in Tarai region of Uttarakhand and NPP range from 17–21 t ha^{-1} yr^{-1} (Bargali et al. 1992). Five-year-old plantations of the fast-growing eucalypts had carbon sequestration potential of 0.03 t C per tree (Lodhiyal 2014). Plantations of short rotation are considered as 'energy forests' and used for biomass production, the biomass produced annually could then be as high as 20 t ha^{-1} yr^{-1} if the condition of soil and moisture are favourable (Mosiej et al. 2012; Lodhiyal 2014).

Carbon and Biomass in Community-Managed Forests

Van Panchayat Forests in Uttarakhand

Van Panchayat (VP) represents the oldest participatory Forest Management Institutions in India and were notified under the VP rule of the Indian Forest Act (IFA) 1927. Legally, these forests have the status of village forests under Section-28 of IFA. Currently, VPs are governed by the Uttarakhand Panchayati Forest Rules (UPFR 2005). According to the VP atlas of 2006 for Uttarakhand, there are 12,089 VPs covering an area of 544,965 ha. Chir pine and banj oak are the most common trees in these forests.

Data on carbon sequestration collected from three VPs in Uttarakhand over a three-year period indicates that sequestration rates vary between 2–4 t ha^{-1} yr^{-1} depending on forest condition. Studies of VP forests in the Lamgarah block of Almora found that total biomass in even-aged banj oak forests and mixed banj oak forests ranged from 358–528 t ha^{-1}. However, values were significantly lower in degraded and very young forests.

Rawat (2013) reported that the carbon stock of 308.3 t ha^{-1} in Van Panchayat forests at 1500-1800 m elevation, with the tree layer contributing the most, followed by herbs and shrubs. Litter biomass in these forests seasonally varies between 5.4 t ha^{-1} (rainy season) and 6.6 t ha^{-1} (winter season). Carbon sequestration rates varied from 2.03 to 5.68 t ha^{-1} yr^{-1}.

Community Forests in Nepal and Northeast India

In Nepal, Community forests (CF) account for about 20% of the total carbon stock of all forests (Pokharel and Byrne 2009). The carbon sequestrated in government-managed forests in Nepal is 596 million tons, while CF contributes 183 million tons (Pokharel and Byrne 2009; Neupane and Sharma 2014).

The forests of Ilam (located in the low hills) in Nepal are dominated by bamboo, *Lannea grandis*, and *Schima wallichii*; Lamatar forests (in the mid-hills) are dominated by temperate broad-leaved species, particularly *Schima-Castanopsis*; and Manang (in high hills) are characterized by temperate conifer forest dominated by *Pinus wallichiana*. In Nepal's community forests, tree biomass in Ilam and Lamatar exceeds 100 t ha^{-1}, whereas Manang's value is approximately half of this amount (Table 7.5).

The tree biomass reported from Nepal (up to 300 t ha^{-1}) and Northeast India (up to 150 t ha^{-1}) was considerably lower than those reported from the Uttarakhand Himalayas (up to 500 t ha^{-1}). This is possibly because the forests studied in Nepal were either younger (as in Lamatar) or more sparse (as in Manang). It's worth noting that the figures for Nepal only account for aboveground biomass of trees >5 cm dbh

Table 7.5 Total biomass and carbon in Van Panchayat (VPs) forests in the Himalayan region

Region	Total biomass (t ha^{-1})	Total carbon (t ha^{-1})
Dhaili VP forest, Almora, Uttarakhand		
Even-aged banj oak forest	351	167
Dense mixed banj oak forest	519	247
Young banj oak with pine	42	20
Toli VP forest, Almora, Uttarakhand		
Young banj oak with chir pine forest	322	153
Chir pine with bushy banj oak	124	59
Young pure chir pine forest	147	70
Dhaili VP forest, Almora, Uttarakhand		
Young pure chir pine forest	24	12
Mixed banj oak forest	312	148
Chapanala, Assam, north east		
Sal plantations	156	74
Nepal		
Lannea grandis forests (Ilam)	364	173
Schima-Castanopsis forests (Lamatar)	105	50
Pinus wallichiana forests (Manang)	64	30
Dolakha district, Nepal		
Pine forest (Simsungure)	93	44
Pine forest (Mahankal)	74	35
Mixed forest (Mathani)	67	32
Pine forest (Sitakunda)	94	45
Mixed forest (Barkhe Dadapari)	40	19
Mixed forest (Chyansi Bhagwati)	56	27

Source: Karky et al. (2007), Chaudhury et al. (2022)

and exclude biomass in herbs/grass, litter, and trees <5 cm dbh, thereby underestimating the trust biomass. Belowground biomass is calculated using a default value of 12.5% of the aboveground biomass.

Soil Carbon Pool in Himalayan Forests

The soil carbon pool represents the largest carbon reservoir in many forests, though detecting changes in soil carbon pool size is challenging due to high spatial variability. Soil carbon pools exhibit complex responses to land use changes such as afforestation and deforestation (McKinley et al. 2011), with the magnitude and direction of response dependent on vegetation, soil characteristics, and climate (Guo and Gifford 2002).

Organic carbon accumulation in soil is favoured by temperate climate, higher elevation, and low temperatures, which restrict microbial activities and slows decomposition processes, resulting in high litter accumulation. Forest soil carbon data for the Uttarakhand Himalayas are primarily based on topsoil (30 cm depth), which does not account for most soil carbon. On average, the percentage of carbon in the top 20 cm relative to that in the first metre of soil is 50%, and the amount in the next 2 metres is 56% of that in the first metre (Fig. 7.2). Therefore, forest soils likely contain about three times as much soil carbon as reported values based on the top 30 cm of soil.

The projected carbon in the forest soils of Uttarakhand up to 150 cm depth is 263.58 million tons. Across different altitudinal ranges and land uses in Uttarakhand, the maximum SOC stock (138.37 t ha^{-1}) is observed above 2500 m altitude, followed by 2001–2500 m (115 t ha^{-1}) (Table 7.6).

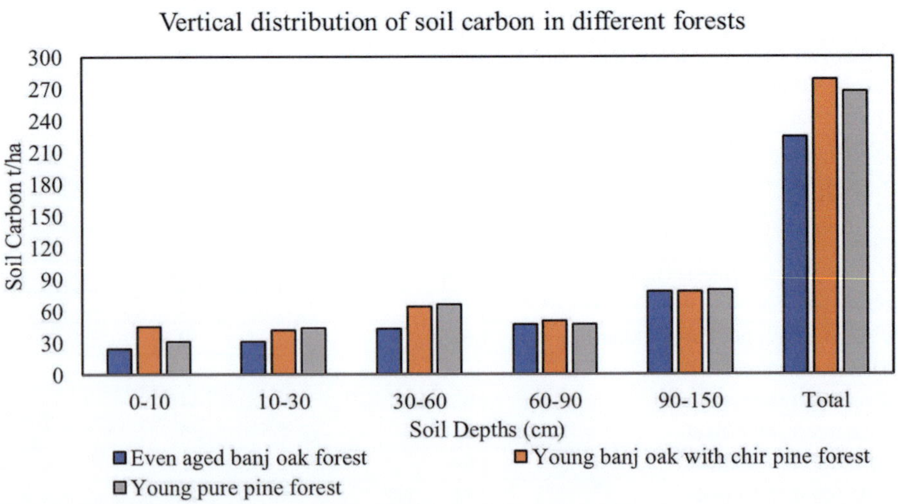

Fig. 7.2 Vertical distribution of soil carbon (up to 150 cm depth, in t ha^{-1}) in even-aged banj oak forests, Young banj oak with chir pine forest and Young pure pine forests in the Central Himalayan region (Tewari and Karky 2007)

Table 7.6 Soil organic carbon (up to 30 cm depth, in t ha^{-1}) under different land uses in relation to the altitude range in the Himalayan region (Murthy et al. 2012; Gupta et al. 2018)

		Uttarakhand			Himachal Pradesh		
S no.	Altitude range	Natural forests	Plantation	Grassland	Farm forestry	Community land	Restoration forests
1	<500 m	52.52	38.46	–	–	–	–
2	501–1000 m	60.08	47.31	37.09	13.39	22.20	7.40
3	1001–1500 m	57.5	62.98	41.15	49.88	31.36	38.19
4	1501–2000 m	84.95	52.18	97.8	–	45.01	56.48
5	2001–2500 m	130.71	76.36	105.28	–	–	–
6	>2501 m	133.45	–	142.14	–	–	–

Elevation significantly influences carbon storage, with soil organic carbon stock relatively low below 2000 m. Generally, SOC increases with increasing elevation. Above 2000 m, soil organic carbon reaches its maximum in each vegetation type: natural forests (~132 t ha^{-1}), plantations (~76 t ha^{-1}), and grasslands (~120 t ha^{-1}).

Natural forests and grasslands accumulate more SOC at higher elevations compared to plantations or horticultural land uses (Singh et al. 2019). The SOC pool is highest in forest lands, followed by grasslands, orchards, and plantation areas (Gupta and Sharma 2011). Total nitrogen, total phosphorous, and SOC contents in forest soil increase with altitude, and soil carbon sequestration potential follows the order: *Abies pindrow* > *Cedrus deodara* > *Quercus leucotrichophora* > *Pinus roxburghii* (Kumar et al. 2021a; b).

Litterfall and Decomposition

Litter production and decomposition serve as reliable indicators of forest health and viability (Rai et al. 2021) and play a central role in nutrient cycling, which is crucial for sustainable forest production. Annual litterfall typically varies between 5.0 and 9.0 t ha^{-1} yr^{-1} in forest stands, though higher values are not uncommon (Rai et al. 2021) and forest floor litter biomass can be as much as 12 t ha^{-1} in some forests. In Himalayan oak forests, typical litterfall is 5–6 t ha^{-1} yr^{-1}. In undisturbed forests of chir pine forest, floor litter may be higher, though this is not the case in most chir forests where fires burn away the litter (Lodhiyal 2014; Rawat and Singh 1989; Singh and Singh 1992).

Litter production is indicative of forest productivity—higher litter production and subsequent decomposition release more nutrients, resulting in higher productivity. It also serves as the primary source of organic matter for forest soils, especially in fast-growing tree plantations like *Eucalyptus globulus* (Yang Wan Qin et al. 2006). Forest floor components are integral to maintaining soil nutrient pools and facilitating nutrient cycling, with an intricate exchange of nutrients occurring through litter decomposition (Rawat et al. 2010).

In the outer Himalayan region and in Central and Western Himalayas, evergreen tree species typically have a leaf life span of approximately 12 months (Krishna and Garkoti 2022; see Chap. 3 for a description of these e1 type species). Leaf drop is a concentrated event generally occurring with the onset of spring (March–April). This concentrated leaf drop can reduce canopy coverage by 25–35% in oaks, 50%–60% in Chir pine, and up to 80% in *Shorea robusta*. Through litterfall and decomposition, nutrients return to the soil. The litter layer, which can be relatively deep in Oak forests, protects the soil, influences water runoff, and affects seedling establishment and development, playing a significant role in community development. Impacts of leaf litter removal, especially from an oak forest are discussed in Chap. 9 (Chronic disturbance).

In Himalayan forests, vegetation (trees, shrubs, and herbs) contributes between 5.0 and 9.0 t litter biomass ha^{-1} yr^{-1} to the forest floor (Table 7.7). These values slightly exceed those reported for warm temperate forests (Singh and Singh 1992).

Table 7.7 Annual litter fall (t dry wt ha^{-1} yr^{-1}) and seasonal pattern of litter fall (percentage) in different forests of Central Himalayas (Rawat and Singh 1989; Singh and Singh 1992)

		Seasonal pattern of litter fall (in %)		
Forest types	Total litter*	Pre-monsoon	Rainy	Winter
Sal	7.6	75.1	18.9	5.9
Chir pine mixed broadleaf	5.0	69.8	18.8	11.5
Chir pine	7.0	60	32.1	7.9
Rianj oak	9.0	69.3	18.1	12.6
Tilonj oak	7.0	58.1	25.1	14.8
Mixed banj oak-chir pine	7.2	70.9	10.7	18.9
Banj oak	7.0	81.9	12.5	5.5
Kharsu oak	5.6	60.4	23.2	16.4

Note*—Total litter includes sum of leaf litter, wood litter, miscellaneous litter, and herb litter

Exceptionally high values of 9 t ha^{-1} yr^{-1} have been reported for an ombrophilous broadleaf forest in the Eastern Himalayas (Collinson 1988).

Kharsu Oak forest, typically found between 2600 and 3200 m, has low litter biomass (4.6–5.6 t ha^{-1} yr^{-1}), as does *Cupressus torulosa* forest (5.0 t ha^{-1} yr^{-1}). Sal, pure chir pine, rianj oak, mixed banj oak forests, and tilonj oak generally produce litter in the range of 5.0–9.0 t ha^{-1} yr^{-1}. Compared to pine forests globally, annual litterfall is higher in the case of chir (*Pinus roxburghii).*

Tree leaves contribute 70%–80% of the total litterfall from woody vegetation in Central Himalayan forests. Woody litter (twigs, branches, etc.) accounts for 14%–31% of the total. Herbaceous vegetation makes a smaller contribution, generally varying from 0.6 t ha^{-1} in chir pine forests to 1.2 t ha^{-1} in rianj oak forests.

The maximum litterfall from woody vegetation (>60%) occurs during spring and early summer, except in *Cupressus*, where it peaks during the rainy season. In all deciduous species of the Himalayan region, concentrated litterfall generally occurs during autumn (Oct-Nov), occurring somewhat earlier in high-altitude species like *Betula utilis* and *Carpanus vimnea*. The seasonal trend for Central Himalayan species shows maximum litterfall in summer, followed by rainy and winter seasons.

Litter Decomposition

In an experiment conducted by Upadhyay et al. (1985), *Q. leucotrichophora* leaf litter placed at several forest sites along an elevation gradient (300–2150 m) decomposed fastest at low elevations in Sal-dominated forests and slowest in chir pine forests. The acidic leaf litter of *P. roxburghii* reduces decomposer biota, thereby suppressing decomposition of *Q. leucotrichophora* litter.

Weight loss in *Q. leucotrichophora* leaf litter placed in different forest sites of the Central Himalayas (Fig. 7.3) varied from 75% to 99% over 1 year, with the lowest rate in chir pine forests (75%) and the highest in Sal forests (99%) (Nautiyal 2015).

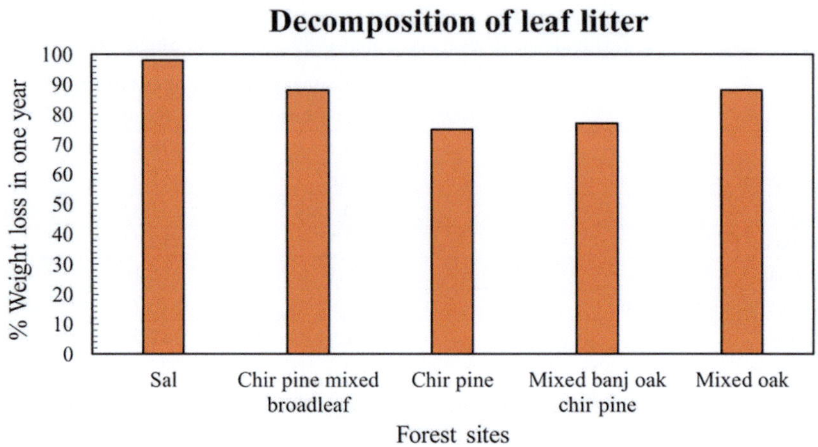

Fig. 7.3 The percent weight loss (%) in common leaf litter in different types of forests in Central Himalayas (Singh and Singh 1992)

A similar study by Bohara et al. (2020) in Nepal Himalayas examining multiple tree species along an elevation gradient from 1300–2500 m found that leaf mass decrease correlates inversely with elevation: highest at the lowest elevation (63.65%), followed by middle elevation (46.59%), and lowest at the highest elevation (44.48%).

The impact of global warming and climate change on litter decomposition in Himalayan forests requires further research. It can be hypothesized that increased moisture and higher temperatures might accelerate litter decomposition. However, if sites become drier due to increased evapotranspiration, even higher temperatures may not significantly influence decomposition rates. The faster decomposition observed during the rainy season (which is generally warm and humid) compared to other seasons in the Himalayan region confirms the favourable effects of moisture and temperature in promoting decomposition.

Root Biomass and Productivity[1]

Across ten Himalayan tree species, about 50% of roots occur below 30 cm. Species vary in rooting depth (from 90 cm to 180 cm) and generally species respond to winter drought by extending root depths. Strategies can however be divergent. For example, banj oak responses to winter droughts by increasing root depth, while chir pine roots become shallower and sparser during winters. Banj oak roots, which are richer in nitrogen concentration than those of chir pine, decompose relatively more rapidly.

Largely confined to upper 100 cm soil depth, roots play a far more important role in ecosystem processes than is generally perceived. The root functions include water and nutrient uptake, anchorage of plants, storage of excesses carbohydrates,

[1] Section on roots contributed by S. P. Singh and Ripu Daman Singh.

transportation of nutrients, water and hormones through plants, production of chemicals and resultant alteration of surrounding soils, and providing home for fungi and microbes. Soil root properties are important to determine changes in soil carbon, which accounts for 45% global carbon (FAO 2020). In terrestrial ecosystems, fine root productivity accounts for 22%–40% net primary productivity (McCormack et al. 2015).

Root diameter influences root traits more than other characters across species, growth forms, and biomes. Plants have evolved thinner roots since they first occurred in terrestrial ecosystem (Ma et al. 2018), which resulted in an enhanced root efficiency to explore soil per unit carbon invested. It enabled plants to reduce their dependence on mycorrhizal fungi, which can exert a considerable energy cost.

For a number of forest types, roots are poorly investigated, and this particularly applies to Himalayan forests. Here, we largely dwell upon (i) depth distribution of fine roots of major forest tree species from foothills to 2200 m, and root biomass and production in a few principal Central Himalayan forest types.

Vertical Distribution of Roots

We have now some ideas about root distribution, both fine (<3 mm diameter) and coarse roots (≥ 3 mm diameter) of some major forests of Kumaun Himalayas (Table 7.8). The data are for September (post monsoon), when shoot and root development was maximum for most species, and of winter to analyse the root responses to the beginning of soil drying and cold condition.

Vertical Root Distribution During Monsoon Season

On average across the ten forest types, the upper 30 cm soil accounts for 56.7% of total fine roots of the soil column, with the range from 46.5% in *Rhododendron* to 62.0% in *Fraxinus* (Table 7.9). This average percentage was 41.1% for coarse roots and 51.53% for total roots (Table 7.9). The average percentage of fine roots in top 30 cm soil in Central Himalayan forests is similar to that of temperate coniferous forest (50%, Jackson et al. 1996), and central Amazonian forests (53.9%, Cordeiro et al. 2020).

The rooting depth (the highest soil depth beyond which roots were not found) of fine roots (<3 mm diameter) in the nine study species ranges from 90 cm (in *Carpinus* and *Fraxinus*) to 150 cm (*Cornus*, *Rhododendron*, *Machilus*, and *Q. floribunda*). The percentage of roots (all diameter classes) in upper 30 cm soil profile varies from 37.7% in *Rhododendron arboreum* to 60.3% in *Fraxinus micrantha*. Though the root density decreases with depth rapidly, on average 20.6 ± 2.8% roots occur below 60 cm depth, the limit which is generally considered in water relation studies.

Table 7.8 General distribution of the Himalayan tree species investigated for vertical root distribution

Tree species (family)	Altitudinal range (m asl)	Distribution/habitat
S. robusta Gaertn.[e] (Dipterocarpaceae)	300–1200	Widely distributed in the foothills of Himalayas from Himachal Pradesh in the west to eastern Himalayas
P. roxburghii Sarg.[e] (Pinaceae)	1000–2000	Forms extensive forests in outer ranges; common on ridges of the Siwalik Hills flanking the Himalayas, generally between 1000–2000 m
Q. leucotrichophora A. Camus[e] (Fagaceae)	1200–2400	Widely distributed in western and Central Himalayas, between 1000–2200 m
Machilus duthiei King[e] (Lauraceae)	1200–2400	Distributed in western Himalayas, in ravines and shady places as undercanopy of oak forests
Rhododendron arboreum Smith[ee] (Ericaceae)	1500–2400	Occurs as undercanopy in oak forests in western Himalayas; but also extends to eastern Himalayas
Fraxinus micrantha Lingelsh.[a] (Oleaceae)	1800–2100	Occurs in the Himalayan region from Sikkim westwards, distributed in patches in mixed temperate forests
Carpinus viminea Lindley[d] (Betulaceae)	1800–2400	Occurs in isolated patches mainly on the bouldery/ old landslides and moist shady places of oak forests
Cupressus torulosa D. Don[ee] (Cupressaceae)	1800–2700	Distributed throughout the in outer and middle ranges of the Himalayas in isolated small pockets
Cornus macrophylla Wall.[a] (Cornaceae)	1800–3000	Limited to moist ravines, temperate and lower temperate mixed forests
Q. floribunda Lindley ex A. Camus[e] (Fagaceae)	2000–2400	Distributed in western Himalayas from Nepal westwards; common in Kumaun Himalayas, between 2000–2400 m

Source: Troup (1921), Champion and Seth (1968), and Sahni (1990), Singh et al. (2006)
[a]= deciduous; [e] = evergreen with about 1 year leaf life span; [ee] = multi-year evergreen

In the Kumaun trees, the percentage of roots (all diameter classes) in the top 90 cm soil ranges from ~87% in *Cornus* to 100% in *Fraxinus* and *Carpinus*. For the two oaks, it was 92.2% for *Q. leucotrichophora* and 93.1% in *Q. floribunda*. Globally, 95% of all fine roots (<5 mm diameter) occur within 58 cm soil depth in boreal forests, 104–121 cm in temperate forests, and 91–94 cm in tropical forests (Schenk and Jackson 2002). Thus, with regard to vertical distribution of roots, the Himalayan forest trees are closer to temperate forests. Across all biomes, 50% roots are reported from 12–23 cm. In comparison, in top 30 cm soil of Kumaun forests, close to 50% roots occur. On average across all samples of all species, fine roots account for 68.3% of total roots, the rest being coarse roots (up to diameter of 25 mm).

Table 7.9 Percentage of roots (density) in top 30 cm of soil column in monsoon (warm and wet) and winter (dry and cold) seasons of Kumaun Himalayan forests

	Fine roots (≤3 mm)		Coarse roots (>3 mm)		Total roots	
	Monsoon	Winter	Monsoon	Winter	Monsoon	Winter
Average (%)	56.72	49.79	41.13	40.89	51.53	45.58
Minimum (%)	46.52 (*Rhododendron*[a])	30.48 (*Cornus*[b])	27.00 (*Rhododendron*[a])	27.77 (*Cornus*[b])	37.74 (*Rhododendron*[a])	29.36 (*Cornus*[b])
Maximum (%)	62.00 (*Fraxinus*[c])	60.96 (*Fraxinus*[c])	53.31 (*Fraxinus*[c])	66.23 (*Pinus*[d])	60.29 (*Fraxinus*[c])	64.86 (*Pinus*[d])

Derived from Pathak et al. (2021)
Values are averages of all species along with minimum and maximum values across the species
[a]*Rhododendron arboreum,* [b]*Cornus macrophylla,* [c]*Fraxinus micrantha,* and [d]*Pinus roxburghii*

Table 7.10 Comparison in average root density (across all species) across the Kumaun forests between monsoon and winter season

	Total roots	Fine roots	Coarse roots
Monsoon	41.18	28.18 (68.3%)	13.00 (31.7%)
Winter	27.91	14.47 (52%)	13.40 (48%)

From Pathak et al. (2021)
Values in parentheses are percentage of total roots

Response of Roots to Dry Winters

From monsoon to winter when soil becomes drier and colder, in Kumaun Himalayan trees, the following changes occur in root's vertical distribution: (i) the rooting depth increases in all species, except in *Pinus* and *Machilus* in which it decreases; (ii) the contribution of upper 30 cm soil layer to the roots in entire soil column decreased, the winter average being 49.8% for fine roots, 40.9% for coarse roots, and 45.6% for total roots; (iii) however, the coarse roots show only a negligible change in density from monsoon to winter (13.94 root tips 100 cm^{-2} in the soil column in winters and 13.0 root tips 100 cm^{-2} during monsoon), while the fine root density in soil column is almost halved from monsoon (28.18 roots per 100 cm^2) to winter (14.47 roots per 100 cm^2); (iv) on average across all the study species, the percentage of fine roots in total roots decreases from 68.3% in monsoon season to 52% in winter season (Table 7.10).

Under drought conditions, fine roots are reduced (however, very fine roots <0.5 mm) may increase (Olmo et al. 2014). Rooting depth, however, may increase in response to seasonal drought in forest ecosystem (Brunner et al. 2015). In Kumaun forests, both increases and decreases were observed after monsoon as soil lost moisture (Pathak et al. 2021).

Species Wise Changes in Kumaun Himalayan Forests

Shorea robusta Rooting depth increases from monsoon (90 cm) to winter (120 cm). Of the total roots, fine root density accounts for 51.8% in winter compared to 80.3% during monsoon. The proportion of all roots in upper 30 cm is similar in the two seasons, 51.6% in monsoon at 53.7% in winter.

Pinus roxburghii Fine root density decreases drastically from monsoon to winter; rooting depth decreases from 120 cm in monsoon to 90 cm in winter; the proportion of roots in top 30 cm soil increases from ~51% to ~77%.

Quercus leucotrichophora In contrast to the pine, in this oak, rooting depth increases from 120 cm in monsoon to 180 cm in winter, and fine root density shows only a limited decline in winters. Of the total roots in soil, the top 30 cm soil accounts for ~56% during monsoon and 42.6% during winter.

Q. floribunda It resembled *Q. leucotrichophora*: the rooting depth increases from monsoon (150 cm) to winter (180 cm); fine root density decreases only slightly and the top 30 cm soil accounts for ~41% of total fine roots in winter compared to ~56% in monsoon.

Machilus duthei The rooting depth decreases from 150 cm during monsoon to 130 cm during winter. During this, the density of fine roots declines by about 57% in monsoon and the percentage of fine roots in total roots declines from 54.5% (monsoon) to 46.7% (winter).

Rhododendron The rooting depth increases from 150 cm to 160 cm, and the fine root density is similar in monsoon (28.56) and winter (26.62). Among all species, *Rhododendron* has the lowest percentage of total roots in top 30 cm soil (37.7% in monsoon and 36.4% in winter).

Carpinus *Carpinus* responds to winter drought by extending its rooting depth from 90 cm (in monsoon) to 130 cm depth, and by declining in the proportion of fine roots from 76.3% of total roots to 53.1% in total roots.

Fraxinus In *Fraxinus,* rooting depth is slightly more in winter (100 cm) than monsoon (90 cm). Like *Carpinus,* there is a marked drop in the proportion of fine roots from monsoon (80.3%) to winter (64.2%). However, the percentage of roots in upper 30 cm soil remains about 60% in both seasons.

Cornus The rooting depth in *Cornus* in monsoon is 150 cm, and it does not change in winter. However, in deeper soils (120–150 cm), there were more roots in winter than in monsoon. Roots are more evenly distributed across the depths, particularly in winter than in the other two deciduous species.

The two oaks and *Rhododendron* show a marked increase in rooting depth during winters and maintain a relatively stable fine root density across the seasons. The winter has a marked negative effect on roots in chir pine and *Machilus*.

The Himalayan species seems to differ considerably in root density. It calls for collecting more precise data. Total fine root density in soil column (sum of densities of all depths) ranges from 16.7 roots 100 cm^{-2} in *Q. florinbunda* to 38.56 roots 100 cm^{-2} in *Machilus* (Table 7.11). Among the remaining species it is high in *Carpinus* and *Cornus* (both deciduous) and *Shorea* (nearby deciduous) and low for the two dominant forest species *Q. leucotrichophora* (22.8) and *P. roxburghii* (17.3) (Table 7.11). The root density is less related to elevation; however, data are available only for a limited elevation range. In upper 30 cm soil, fine root density was the highest in *Machilus* (23 roots 100 cm^{-2}) and the lowest in chir pine and *Q. floribunda*

Table 7.11 Changes in root distribution from monsoon to winter in soil column of Kumaun forests

	Density (root tips 10 cm^{-2})					
	Fine roots (<3 mm)		Coarse roots (3 mm to 20 mm)		All roots (fine + coarse roots)	
Species	Monsoon	Winter	Monsoon	Winter	Monsoon	Winter
Pinus	17.3 (10.4)	2.4 (1.7)	10.6	9.1	27.98	11.5
Carpinus	38.0 (20.9)	11.9 (6.7)	11.79	10.5	49.8	22.4
Fraxinus	25.0 (15.5)	12.6 (7.9)	6.1	7.1	31.1	19.7
Cornus	33.34 (18.3)	19.8 (6.2)	16.3	13.5	49.63	33.3
Shorea	33.36 (18.3)	12.0 (7.0)	14.0	11.2	47.30	23.2
Machilus	38.56 (23)	16.7 (8.8)	13.59	13.8	52.15	30.5
Q. leucotrichophora	22.8 (12.7)	14.9 (7.2)	11.96	15.7	34.76	30.6
Q. floribunda	16.7 (10.2)	13.2 (6.3)	9.34	14.5	26.04	27.7
Rhododendron arboreum	28.56 (13.29)	26.62 (9.7)	23.37	25.7	51.93	52.3

Derived from Pathak et al. (2021)
Root density in the top 30 cm is given in parentheses

(10.4 roots 100 cm^{-2}). Banj oak and chir pine have among the lowest root densities. The coarse roots density ranges from 9.34 roots 100 cm^{-2} in *Q. floribunda* to 23.4 roots 100 cm^{-2} in *Rhododendron*. The total root density (roots of all diameter classes) in the entire soil column is higher (47.3–52% roots 100 cm^{-2}) in *Machilus*, *Rhododendron*, *Carpinus*, and *Shorea* and then in the rest of species (26.04–31.1% roots 100 cm^{-2}).

Generally, the decline in fine root density from monsoon to winter season was more in upper 30 cm soil than deeper soil profile. As for coarse root density, it is generally stable in oaks and *R. arboreum* from monsoon to winter season, largely because of increase in deeper soil layers.

Data on *Cupressus torulosa*—a conifer which forms sporadic monodominant stands, particularly on steep slopes, rich in limestone (Troup 1921) are available only for winter season. Its root density, both fine (25.9/ 100 cm^2) and coarse (12.5/ 100 cm^2), is much higher than of chir pine, the other conifer. Occurring up to 140 cm depth, the roots of *Cupressus* in top 30 cm account for about 47% of total roots in soil. The top 90 cm soil accounted for 97.6% of the total roots. Fine roots accounted for 67.5% of total roots of soil column.

Among the three deciduous species which occur in temperate forests, *Cornus* was far more deep rooted (150 cm in monsoon) than the other two (90 cm in monsoon).

The fine root density in *Cornus* (33.3/100 cm^2) and *Carpinus* (~38/100 cm^2) was higher than in *Fraxinus* (25/100 cm^2). In all of them, root density decreases from monsoon to winter, but root depth either increases or does not change.

Root Biomass and Productivity

In the Himalayas, our knowledge about root biomass and productivity is largely based on chir pine and banj oak.

Compared to chir pine, banj oak had 56.2% more fine root biomass and 32.1% more net fine root primary productivity as reported by Usman et al. (2000) (Table 7.12). Chand et al. (2022) who sampled roots in upper 40 cm soil, found that fine root biomass in banj oak was 18% greater than that of chir pine. Data on root density sampled from the entire soil column having roots, indicated 32.6% more value for banj oak than chir pine (Pathak et al. 2021) (Table 7.12). Sal (*Shorea robusta*) had far greater root density than both chir pine and banj oak. Compared to chir pine, the root density in sal is 94% greater when entire soil column is considered and 74.3% greater when upper 30 cm soil is considered.

Annual production of live roots ($\leq$ 1 mm diameter) in banj oak forest (3629 kg h^{-1} yr^{-1}) is 1.45 times higher than in chir pine forest (2508 kg h^{-1} yr^{-1}). Dead fine root accumulates annually at the rate of 1527 kg h^{-1} yr^{-1} in banj oak forest and 1197 kg h^{-1}

Table 7.12 A comparative account of root parameters of three important Central Himalayan tree species

	Chir pine (*Pinus roxburghii*)	Banj oak (*Quercus leucotrichophora*)	Sal (*Shorea robusta*)	Reference
Biomass (kg ha^{-1})	2072 2150	3236 (56.2%) 3120	–	Usman et al. (2000) Verma et al. (2021)
Soil N concentration (%) top 20 cm soil 40–60 cm soil	0.64–0.68 0.47–0.48	0.71–0.76 0.59–0.62	–	Usman et al. (2000)
Net productivity (kg ha^{-1} yr^{-1})	2459	3428	–	
Biomass (top 40 cm soil) (kg ha^{-1})	610	720	1100	Chand et al. (2022)
Root density in entire soil column[a] (roots 100 cm^{-2})	17.2	22.8 (32.6%)	33.3 (94.2%)	Pathak et al. (2021)
Root density top 30 cm soil	10.5	12.7 (21.1%)	18.3 (74.3%)	Pathak et al. (2021)

Percentages by which values of banj oak roots were greater than those of chir pine are indicated in parentheses

[a]Soil root column was of 90 cm in chir pine, 150 cm in banj oak, and 120 cm in sal. Usman et al. (2000) considered only 1 mm $\leq$ roots, Chand et al. (2022) 2 mm, and Pathak et al. (2021) <3 mm

Table 7.13 Fine root (≤1 mm) production (kg/ha^{-1} yr^{-1}, both live and dead) in banj oak and chir pine forests using two methods, ingrowth and turnover in parentheses

	Banj oak		Chir pine	
Soil depth (cm)	Ingrowth core method (live + dead)	Maximum and minimum method	Ingrowth core method (live + dead)	Maximum and minimum method
0–20	2972.0	1960 (0.66)	2078.0	1524 (0.730)
20–40	1410.4	1249 (0.749)	1146.6	767 (0.700)
40–60	654.8	219 (0.826)	358.3	168 (0.705)
Total	**5036.0 (3629 + 1527)**	**3428**	**3582.9 (2508 + 1197)**	**2459**

From Usman et al. (2000)

Table 7.14 Seasonal changes in fine root biomass (kg^{-1} ha^{-1}), soil depths (0–20, 20–40, and 40–60 cm) of banj oak (*Quercus leucotrichophora*) and chir pine (*Pinus roxburghii*)

	Banj oak			Chir pine		
Season	Live	Dead	Total	Live	Dead	Total
Monsoon	1686	3337	5023	1239	2192	3431
Winter	475	1000	1475	326	736	1062
Pre-monsoon (summer)	1055	2176	3231	635	1197	1832

Values are sum of all the three soil depths (Usman et al. 2000)

yr^{-1} in chir pine forest (the former being 1.27 times higher than the latter) (Table 7.13). In both the forests, production declines with soil depth, the uppermost 20 cm accounting for 63% of the total root production (live + dead) in banj oak and 53% in chir pine forest and the bottom soil layer (40–60 cm) accounted for less than 10% biomass in both (Table 7.13). The proportion of dead roots was 29% in the oak and 33% in chir pine (Usman et al. 2000).

The fine root production both in banj oak and chir pine increases throughout pre-monsoon and monsoon seasons to peak in October, whereafter it declines during winter (Table 7.14). The seasonal patterns of root production are the same for live and dead roots in both forests; however, the monsoon time increase was sharper in banj oak.

Which method measures fine root production more accurately is uncertain. Studies in Uttarakhand found that the ingrowth core method measures significantly higher fine root production than maximum minimum difference method. However, generally the productivity estimates elsewhere are lower with ingrowth core method.

A study on banj oak-chir pine mixed forest (based on forest sites) of Kumaun showed that mean fine root mass occur between 2.15 Mg ha^{-1} and 3.12 Mg ha^{-1}. During this monthly mean, necromass varies from 0.42 to 0.95 Mg ha^{-1} (Verma et al. 2021). The nitrogen concentration in dead fine roots (0.69%–0.76%) is higher than in live fine roots (0.67%–0.73%) (Verma et al. 2021), indicating the lack of resorption in fine roots before their mortality and their association with soil biota or/and mycorrhizal fungi (Verma et al. 2021).

Globally, mean fine root production (all plants i.e., trees + understorey in kg ha^{-1} yr^{-1}) is estimated as 311 ± 259 in boreal ($n = 39$), 428 ± 315 in temperate ($n = 71$), and 596 ± 478 in tropical forests ($n = 32$), respectively (Finér et al. 2011). The corresponding annual fine roots turnover was estimated as 0.77 ± 0.70, 1.21 ± 1.04, and 1.44 ± 0.76, respectively. In Uttarakhand, the total fine root production (live + dead) in banj oak (5.3 t ha^{-1} yr^{-1}) and chir pine (3.7 t ha^{-1} yr^{-1}) lie within the range 1.4–11.5 t ha^{-1} yr^{-1} for various temperate forests (compiled by Usman et al. 2000). The mean global fine root production for all roots is estimated at 493 ± 434 g m^{-2} (Finér et al. 2011), when only trees were considered the fine root production decreased by 17%–21%, compared to when entire vegetation is considered.

Because of higher productivity and tissue nitrogen concentration, nitrogen uptake in banj oak is ~1.8 times higher than in chir pine. In both, the nitrogen uptake sharply decreases with depth, at 40–60 cm depth the uptake values being only ~9% in banj oak and ~7% in chir pine of those at 0–20 cm soil depth (Usman et al. 1999).

Root Decomposition of Chir Pine and Banj Oak

The rate of root decomposition was 0.083% per day in banj oak and 0.071% per day in chir pine. Compared to leaf litter decomposition, it is 0.32 times in banj oak (leaf litter decomposition 0.253% per day), and 0.56 times in chir pine (leaf litter decomposition 0.126% per day) (Usman et al. 2000). The rate of root decomposition increases with decreasing root diameter and accompanying increase in nitrogen concentration. Thus, the nutrient turnover is expected to be more in species with finer roots.

In brief, the Himalayan forests are grossly understudied for roots, so more data are required to characterize them. Comparison of data with those of the other studies indicates that trench method of root distribution is fairly reliable.

Due to high forest cover, the Himalayan forests are large reservoir of carbon. An undisturbed oak forest can easily have up to 400 t ha^{-1} biomass. Similarly, biomass values are also found in some conifers like *C. deodara* and *A. pindrow* forests. Information about roots is scarce. Though carbon sequestration in forest biomass and soils can be estimated accurately, information on carbon sequestration rates of soils is extremely limited. The forest soils of Himalayan region have a large soil carbon pool (263.58 million ton SOC up to 150 cm depth), for which it is essential that they should be conserved and protected. It is worth considering that new plantations or naturally raised forests would require 80–100 years to create a similar stock of carbon. Relentless degradation results in loss of soil carbon which holds up to two times more carbon than the vegetation biomass. Both biomass and root primary productivity are potentially high up to 2600–2800 m altitude. In the changing climatic conditions and warming environment, estimation of the biomass productivity and the carbon capture potential of the forests are essential. Good-quality forests have diminished over the past century and are localized in certain patches which need to be conserved. The old growth forests which can have more than 500 t per hectare carbon are becoming difficult to locate and estimates of soil carbon and biodiversity that they support needs to be examined.

References

Adhikari BS, Rawat YS, Singh SP (1995) Structure and function of high-altitude forests of central Himalaya I. Dry matter dynamics. Ann Bot 75:237–248. https://doi.org/10.1006/anbo.1995.1017

Albrecht A, Kandji ST (2003) Carbon sequestration in tropical agroforestry systems. Agric Ecosyst Environ 99:15–27. https://doi.org/10.1016/S0167-8809(03)00138-5

Ameray A, Bergeron Y, Valeria O et al (2021) Forest carbon management: a review of silvicultural practices and management strategies across boreal, temperate and tropical forests. Curr For Rep 7:1–22. https://doi.org/10.1007/s40725-021-00151-w

Baishya R, Barik SK (2011) Estimation of tree biomass, carbon pool and net primary production of an old-growth Pinus kesiya Royle ex. Gordon forest in North-Eastern India. Ann For Sci 68: 727–736. https://doi.org/10.1007/s13595-011-0089-8

Bargali SS, Singh SP, Singh RP (1992) Structure and function of an age series of eucalypt plantations in central Himalaya: I. Dry matter dynamics. Ann Bot 69:405–411. https://doi.org/10.1093/oxfordjournals.aob.a088361

Bhardwaj DR, Salve A, Kumar J et al (2023) Biomass production and carbon storage potential of agroforestry land use systems in high hills of north-western Himalaya: an approach towards natural based climatic solution. Biomass Convers Biorefinery 14:18079–18092. https://doi.org/10.1007/s13399-022-02964-8

Bohara M, Acharya K, Perveen S et al (2020) In situ litter decomposition and nutrient release from forest trees along an elevation gradient in Central Himalaya. Catena 194:104698. https://doi.org/10.1016/j.catena.2020.104698

Brahma B, Nath AJ, Deb C et al (2021) A critical review of forest biomass estimation equations in India trees for people. 5:100098. https://doi.org/10.1016/j.tfp.2021.100098

Brienen RJW, Phillips OL, Feldpausch TR et al (2015) Long-term decline of the Amazon carbon sink. Nature 519:344–348. https://doi.org/10.1038/nature14283

Brunner I, Herzog C, Dawes MA, Arend M, Sperisen C (2015) How tree roots respond to drought. Front Plant Sci 6:547. https://doi.org/10.3389/fpls.2015.00547

Bunker DE, DeClerck F, Bradford JC et al (2005) Species loss and aboveground carbon storage in a tropical forest. Science 310:1029–1031. https://doi.org/10.1126/science.1117682

Champion HG, Seth SK (1968) A revised survey of forest types of India. Govt. of India Press, New Delhi, p 404

Chand HB, Joshi G, Bhatta RP, Singh S, Kumar A, Joshi NR, Bohara R (2022) Fine root biomass differs significantly across different forest types and soil depth in Central Himalaya. India Curr Sci 123(2):194–199

Chaturvedi OP, Singh JS (1987) The structure and function of pine forest in central Himalaya I. Dry matter dynamics. Ann Bot 60:237–252. https://doi.org/10.1093/oxfordjournals.aob.a087442

Chaudhury G, Darji CB, Basumatari M et al (2022) Stand structure, biomass, and carbon stock along disturbance gradients in differently managed tropical forests of Assam, Northeast India trees for people. 9:100296. https://doi.org/10.1016/j.tfp.2022.100296

Collinson AS (1988) Introduction to world vegetation, 2nd edn. Springer, Dordrecht. https://doi.org/10.1007/978-94-015-3935-7

Cordeiro AL, Norby RJ, Andersen KM et al (2020) Fine-root dynamics vary with soil depth and precipitation in a low-nutrient tropical forest in the Central Amazonia. Plant Environ Interact 1(1):3–16. https://doi.org/10.1002/pei3.10010

FAO (2020) Global forest resources assessment 2020—key findings. Rome. https://doi.org/10.4060/ca8753en
Finér L, Ohashi M, Noguchi K, Hirano Y (2011) Fine root production and turnover in forest ecosystems in relation to stand and environmental characteristics. For Ecol Manag 262(11): 2008–2023. https://doi.org/10.1016/j.foreco.2011.08.042
Forest Survey of India (2011) India state of forest report. Forest Survey of India, Ministry of Environment and Forests, Government of India, Dehradun
Forest Survey of India (2021) India state of forest report. Forest Survey of India, Ministry of Environment and Forests, Government of India, Dehradun
Gairola S, Sharma CM, Ghildiyal SK et al (2011) Live tree biomass and carbon variation along an altitudinal gradient in moist temperate valley slopes of the Garhwal Himalaya (India). Curr Sci 100:1862–1870
Gosain BG, Negi GCS, Dhyani PP et al (2015) Ecosystem services of forests: carbon stock in vegetation and soil components in a watershed of Kumaun Himalaya, India. Int J Ecol Environ Sci 41:177–188
Guo LB, Gifford RM (2002) Soil carbon stocks and land use change: a meta-analysis. Glob Chang Biol 8:345–360. https://doi.org/10.1046/j.1354-1013.2002.00486.x
Gupta MK, Sharma SD (2011) Sequestrated carbon: organic carbon pool in the soils under different forest covers and land uses in Garhwal Himalayan region of India. Int J Agric For 1:14–20
Gupta B, Sarvade S, Singh M (2018) Species composition, biomass production, and carbon storage potential of agroforestry systems in Himachal Pradesh. In: Gupta SK, Panwar P, Kaushal R (eds) Agroforestry for increased production and livelihood security. New India Publishing Agency, New Delhi, pp 245–269. https://doi.org/10.56093/ijas.v93i6.123589
Houghton RA, Hall F, Goetz SJ (2009) Importance of biomass in the global carbon cycle. J Geophys Res Biogeosci 114. https://doi.org/10.1029/2009JG000935
Huston MA, Wolverton S (2009) The global distribution of net primary production: resolving the paradox. Ecol Monogr 79:343–377
IPCC (2006) In: Eggelston S, Buendia L, Miwa K, Ngara T, Tanabe K (eds) Guidelines for national greenhouse gas inventories, vol 4. Institute for Global Environmental Strategies (IGES) for the IPCC, pp 4.11–4.13
Jackson RB, Canadell J, Ehleringer JR, Mooney HA, Sala OE, Schulze ED (1996) A global analysis of root distributions for terrestrial biomes. Oecologia 108(3):389–411. https://doi.org/10.1007/BF00333714
Jenkins JC, Chojnacky DC, Heath LS et al (2003) National-scale biomass estimators for United States tree species. For Sci 49:12–35. https://doi.org/10.1093/forestscience/49.1.12
Joshi R, Singh H, Chhetri R et al (2021a) Carbon sequestration potential of community forests: a comparative analysis of soil organic carbon stock in community managed forests of Far-Western Nepal. Eurasian J Soil Sci 10:96–104. https://doi.org/10.18393/ejss.825066
Joshi VC, Negi VS, Bisht D et al (2021b) Tree biomass and carbon stock assessment of subtropical and temperate forests in the Central Himalaya, India. Trees For People 6:100147. https://doi.org/10.1016/j.tfp.2021.100147
Karky BS, Banskota K, Skutsch M (2007) Reducing carbon emissions through community managed forests in the Himalaya. ICIMOD, Kathmandu
Kaushal S, Baishya R (2021) Stand structure and species diversity regulate biomass carbon stock under major Central Himalayan forest types of India. Ecol Process 10:1–18. https://doi.org/10.1186/s13717-021-00283-8
Keeling HC, Phillips OL (2007) The global relationship between forest productivity and biomass. Glob Ecol Biogeogr 16:618–631. https://doi.org/10.1111/j.1466-8238.2007.00314.x
Krishna M, Garkoti SC (2022) Evergreen and deciduous tree species show distinct strategies to synchronize with seasonality in mid-elevational forests of central Himalaya. For Ecol Manag 526:120567. https://doi.org/10.1016/j.foreco.2022.120567

Kumar A, Sharma MP (2015) Estimation of carbon stocks of Balganga reserved forest, Uttarakhand, India. For Sci Technol 11:177–181. https://doi.org/10.1080/21580103.2014.990060

Kumar A, Kumar M, Pandey R et al (2021a) Forest soil nutrient stocks along with an altitudinal range of Uttarakhand Himalayas: an aid to nature-based climate solutions. Catena 207:105667. https://doi.org/10.1016/j.catena.2021.105667

Kumar M, Kumar A, Kumar R et al (2021b) Carbon stock potential in Pinus roxburghii forests of Indian Himalayan regions. Environ Dev Sustain 23:12463–12478. https://doi.org/10.1007/s10668-020-01178-y

Lodhiyal LS (2014) Biomass productivity, carbon sequestration and nutrient cycling of eucalypt plantations. Eucalypts in India, pp 244–266

Lodhiyal LS, Lodhiyal N, Bhakuni N (2014) Standing tree biomass and carbon content in natural forests of Kumaun in central Himalaya. J Plant Dev Sci 6:167–174

Ma Z, Guo D, Xu X, Lu M, Bardgett RD, Eissenstat DM et al (2018) Evolutionary history resolves global organization of root functional traits. Nature 555(7694):94–97. https://doi.org/10.1038/nature25783

McCormack ML, Dickie IA, Eissenstat DM et al (2015) Redefining fine roots improves understanding of below-ground contributions to terrestrial biosphere processes. New Phytol 207:505–518. https://doi.org/10.1111/nph.13363

McKinley DC, Ryan MG, Birdsey RA et al (2011) A synthesis of current knowledge on forests and carbon storage in the United States. Ecol Appl 21:1902–1924. https://doi.org/10.1890/10-0697.1

Mosiej J, Karczmarczyk A, Wyporska K, Rodzkin A (2012) Biomass production in energy forests – short rotation plantations. In: Management of organic waste. IntechOpen, pp 131–152. https://doi.org/10.5772/33843

Murthy IK, Alipuria AK, Ravindranath NH (2012) Potential for increasing carbon sink in Himachal Pradesh, India. Trop Ecol 53:357–369

Nair PK, Mohan Kumar B, Nair VD (2009) Agroforestry as a strategy for carbon sequestration. J Plant Nutr Soil Sci 172:10–23. https://doi.org/10.1002/jpln.200800030

Nautiyal A (2015) Is Chir Pine displacing Banj Oak in the central Himalaya? Socioeconomic implications for local people and the conservation of oak forest biodiversity. PhD Thesis,. University of Arkansas

Neupane B, Sharma RP (2014) An assessment of the effect of vegetation size and type, and altitude on aboveground plant biomass and carbon. J Agric Crop Res 2:44–50

Olmo M, Lopez-Iglesias B, Villar R (2014) Drought changes the structure and elemental composition of very fine roots in seedlings of ten woody tree species. Implications for a drier climate. Plant Soil 384(1):113–129. https://doi.org/10.1007/s11104-014-2178-6

Pathak GC, Joshi H, Singh RD, Tewari A, Pandey R, Singh SP (2021) Vertical root distribution in Himalayan trees: about half of roots occur below 30 cm, the generally sampled depth. Trop Ecol 62(3):479–491. https://doi.org/10.1007/s42965-021-00167-0

Pokharel BK, Byrne S (2009) Climate change mitigation and adaptation strategies in Nepal's forest sector: how can rural communities benefit? vol Discussion Paper 7. Nepal Swiss Community Forestry Project (NSCFP), Kathmandu. http://www.forestrynepal.org/publications/nscfp/4428

Pregitzer KS, Euskirchen ES (2004) Carbon cycling and storage in world forests: biome patterns related to forest age. Glob Chang Biol 10:2052–2077. https://doi.org/10.1111/j.1365-2486.2004.00866.x

Qin YW, Yuan WK, Kellomaki S, Zhang J (2006) Annual and monthly variations in litter macronutrients of three subalpine forests in Western China. Pedosphere 16:788–798. https://doi.org/10.1016/S1002-0160(06)60116-6

Rai YK, Bhutia R, Joshi R (2021) Leaf litter production and decomposition dynamics of selected tree species in Sikkim Himalaya. J Adv Plant Sci 11(1):45–59. G.B. Pant National Institute of Himalayan Environment and Sustainable Development

Rajput BS, Bhardwaj DR, Pala NA (2015) Carbon dioxide mitigation potential and carbon density of different land use systems along an altitudinal gradient in North-Western Himalayas. Agrofor Syst 89:525–536. https://doi.org/10.1007/s10457-015-9788-8

Rana BS (1985) Biomass and net primary productivity in different forest ecosystems along an altitudinal gradient in Kumaun Himalaya. PhD Thesis. Kumaun University, Nainital, p 323

Rana BS, Singh SP, Singh RP (1989) Biomass and net primary productivity in Central Himalayan forests along an altitudinal gradient. For Ecol Manag 27:211–232. https://doi.org/10.1016/0378-1127(89)90107-2

Rawat YS (1983) Plant biomass, net primary production and nutrient cycling in oak forests. PhD Thesis. Kumaun University, Nainital, p 355

Rawat VS (2013) Carbon sequestration rates in van panchayat forests and their benefits under REDD. J Asian Sci Res 3:396

Rawat YS, Singh JS (1988) Structure and function of oak forests in central Himalaya I. Dry matter dynamics. Ann Bot 62:397–411. https://doi.org/10.1093/oxfordjournals.aob.a087673

Rawat YS, Singh JS (1989) Forest floor biomass, litter fall and nutrient return in central Himalayan oak forests. Vegetatio 82:113–125

Rawat N, Nautiyal BP, Nautiyal MC (2010) Litter decomposition rate and nutrient release from different litter forms in a Himalayan alpine ecosystem. Environment 30:279–288. https://doi.org/10.1007/s10669-010-9275-8

Rodin LE, Bazilevich NI (1967) Production and mineral cycling in terrestrial vegetation. Oliver and Boyd, London

Sahni KC (1990) Gymnosperms of India and adjacent countries. Bishen Singh Mahendra Pal Singh, Dehra Dun, p 169

Sahoo UK, Tripathi OP, Nath AJ et al (2021) Quantifying tree diversity, carbon stocks, and sequestration potential for diverse land uses in Northeast India. Front Environ Sci 9:724950. https://doi.org/10.3389/fenvs.2021.724950

Satoo T (1971) Primary production relations in plantations of Norway spruce in Japan–Materials for the studies of growth in stands. Bull Tokyo Univ For 65:125–142

Schenk HJ, Jackson RB (2002) The global biogeography of roots. Ecol Monogr 72(3):311–328. https://doi.org/10.1890/0012-9615(2002)072[0311:TGBOR]2.0.CO;2

Sharma CM, Gairola S, Ghildiyal SK, Suyal S (2010a) Physical properties of soils in relation to forest composition in moist temperate valley slopes of the Central Western Himalaya. J For Environ Sci 26:117–129. https://doi.org/10.7747/JFS.2010.26.2.117

Sharma CM, Baduni NP, Gairola S et al (2010b) Tree diversity and carbon stocks of some major forest types of Garhwal Himalaya, India. For Ecol Manag 261:1862–1870. https://doi.org/10.1016/j.foreco.2010.09.014

Sharma CM, Mishra AK, Krishan R et al (2016) Variation in vegetation composition, biomass production, and carbon storage in ridge-top forests of high mountains of Garhwal Himalaya. J Sustain For 35:119–132. https://doi.org/10.1080/10549811.2015.1118387

Singh B (1998) Biomass production and nutrient dynamics in three clones of Populus deltoides planted on Indogangetic plains. Plant Soil 203:15–26

Singh P, Lodhiyal LS (2009) Biomass and carbon allocation in 8-year-old poplar (Populus deltoides Marsh) plantation in Tarai agroforestry systems of central Himalaya, India. New York Sci J 2:49–53

Singh JS, Singh SP (1992) Forests of Himalaya: structure, functioning and impact of man. Gyanodaya Prakashan, Nainital

Singh S, Verma AK (2018) Biomass and carbon stocks in different forest types of Western Himalaya. Trop Ecol 59:647–658

Singh SP, Adhikari BS, Zobel DB (1994) Biomass, productivity, leaf longevity, and forest structure in the central Himalaya. Ecol Monogr 64:401–421

Singh SP, Zobel DB, Garkoti SC, Tewari A, Negi CMS (2006) Patterns of water relations of central Himalayan trees. Trop Ecol 47:159–182

Singh V, Tewari A, Kushwaha SPS, Dadhwal VK (2011) Formulating allometric equations for estimating biomass and carbon stock in small diameter trees. For Ecol Manag 261:1945–1949. https://doi.org/10.1016/j.foreco.2011.02.019

Singh R, Bhardwaj DR, Pala NA, Rajput BS (2019) Biomass production and carbon stock potential of natural vegetation, agroforestry, and cultivated land use systems along altitudinal gradient in north-western Himalaya. Range Manag Agrofor 40:94–103

Tewari A, Karky BS (2007) Carbon measurement methodology and results. In: Reducing carbon emissions through community-managed forests. International Centre for Integrated Mountain Development, Kathmandu, pp 39–45

Troup RS (1921) The silviculture of Indian trees, vol III. Oxford University Press, Lauraceae to Coniferae

Upadhyay VP, Pandey U, Singh JS (1985) Effect of habitat on decomposition of standard leaf-litter species. Biol Fertil Soils 1:201–207. https://doi.org/10.1007/BF00257638

Usman S, Singh SP, Rawat YS (1999) Fine rootproductivity and turnover in two evergreen central Himalayan forests. Ann Bot 84(1):87–94

Usman S, Singh SP, Rawat YS (2000) Fine root productivity and turnover in two evergreen central Himalayan forests. Ann Bot 84:87–94. https://doi.org/10.1006/anbo.1999.0894

Uttarakhand Panchayati Forest Rules (2005) Published by the Government of Uttarakhand

Verma A, Tewari A, Shah S (2012) Carbon storage capacity of high-altitude Quercus semecarpifolia forests of Central Himalayan region. Scand J For Res 27:609–618. https://doi.org/10.1080/02827581.2012.689003

Verma AK, Garkoti SC, Singh S, Kumar S, Kumar M (2021) Fine root production and nutrient dynamics in relation to stand characteristics of chir pine mixed banj oak forests in central Himalaya. Flora 279:151808. https://doi.org/10.1016/j.flora.2021.151808

Whittaker RH, Likens GE (1975) The biosphere and man. In: Lieth H, Whittaker RH (eds) Primary productivity of the biosphere, Ecol Stud, vol 14. Springer, Berlin, pp 305–328. https://doi.org/10.1007/978-3-642-80913-2_15

Yadav RP, Bisht JK, Bhatt JC (2017) Biomass, carbon stock under different production systems in the mid hills of Indian Himalaya. Trop Ecol 58:15–21

Yoda K (1968) A preliminary survey of the forest of eastern Nepal. II. Plant biomass in the sample plots chosen from different vegetation zones. J Coll Arts Sci Chiba Univ:277–302

Net Ecosystem Productivity and Carbon Sequestration: Methodological Concepts and Initial Learning from the Himalayan Forests

8

Priyanka Lohani, Pramit Kumar Deb Burman, Sumit Pundir, and Sandipan Mukherjee

Abstract

In the last few decades, both ecological indices-based biometric method and flux-based eddy covariance technique have been used extensively for quantification of the ecosystem productivity and carbon sequestration potential. This study presents a narrative review of past and current methods used for the quantification of ecosystem productivity and carbon sequestration, and provides distinction in terminologies used by the field and index-based method and flux-based methods. In addition, efforts are also made to compare the seasonal variation of net ecosystem carbon exchanges, gross primary productivity, and ecosystem respiration between three forest types of Himalayas, i.e., a needle leaf Pine forest, a broad leaf Oak forest of central Himalaya, and a mixed deciduous forest of eastern Himalaya. This comparative assessment is made based on data obtained from the flux-based eddy covariance technique which simultaneously measures

P. Lohani
Centre for Land and Water Resource Management, GB Pant National Institute of Himalayan Environment, Kosi-Katarmal, Almora, India

Graphic Era Deemed To Be University, Dehradun, India

P. K. Deb Burman
Centre for Climate Change Research, Indian Institute of Tropical Meteorology, Ministry of Earth Sciences, Pune, India

Department of Atmospheric and Space Sciences, Savitribai Phule Pune University, Pune, India

Department of Environmental Science, Savitribai Phule Pune University, Pune, India

S. Pundir
Graphic Era Deemed To Be University, Dehradun, India

S. Mukherjee (✉)
Ladakh Regional Centre, GB Pant National Institute of Himalayan Environment, Leh, India
e-mail: sandipan@gbpihed.nic.in

S. P. Singh, R. Thadani (eds.), *Himalayan Forests in a Changing World*,
https://doi.org/10.1007/978-981-95-0978-2_8

selected micrometeorological parameters. One of the major findings of this study indicates that the accelerated carbon uptake in the needle leaf Pine forest compared to broad leaf Oak and deciduous forest of eastern Himalaya is linked to lesser ecosystem respiration, particularly during the monsoon season. Moreover, it is noted that the needle leaf Pine forest has higher responses to changes in the environmental conditions compared to Oak and deciduous forests of Himalaya.

Keywords

Ecosystem exchange of carbon · Eddy covariance · Ecosystem productivity

Abbreviations

DBH	Diameter at breast height
EC	Eddy covariance measurements
GPP	Gross primary productivity
IHR	Indian Himalayan region
NEE	Net ecosystem exchange
NEP	Net ecosystem productivity
NPP	Net primary productivity
Reco	Ecosystem respiration
SMI	Soil moisture index
VPD	Vapor pressure deficit

Introduction

Traditionally, carbon exchange within an ecosystem is the conversion of carbon between inorganic and organic forms (Chapin III et al. 2006). With recent advancements, atmospheric scientists and geophysicists have been measuring carbon flux or combinations of C-fluxes through instruments and using geo-spatial techniques conceptually based on the general ecosystem C-cycle, whereas the ecologists use field and indices-based biometric method to quantify ecosystem productivity and carbon sequestration potential of any ecosystem as photosynthesis and respiration (Baldocchi 2003, 2014; Campioli et al. 2016). Since most ecological phenomena are the results of interactions of a group of relevant ecophysiological factors, identifications of the most informative factors governing these phenomena using quantitative approaches provide deeper insights into the underlying mechanisms of ecological dynamics. Moreover, continuously measured values of carbon and water fluxes of terrestrial ecosystems could be used as a general indicator of forest ecosystem functioning. Generally, ecologists and atmospheric scientists use two approaches to quantify ecosystem C-balance: the eddy-covariance (EC) technique and ecological field and indices-based biometric (BM) method.

The BM method uses field-based and empirical approach to assess plant growth which is further used to estimate ecosystem productivity (Campioli et al. 2016). On the contrary, EC approach estimates the ecosystem fluxes by the concurrent measurements of vertical wind speed and gas concentration at a certain height above the canopy that provides more objective and accurate information about the emission and consumption rate of various gases. The footprint of EC measurements spans from a few hundred meters to kilometers (Burba and Anderson 2009; Baldocchi 2003). Therefore, EC-based approach for measuring carbon fluxes has higher spatial extent.

Conceptually, BM and EC methods are capable of quantifying the ecosystem productivity but have differences in their measurement approaches and terminologies. The balance between carbon inputs through gross primary production (GPP) and carbon losses through respiration govern the carbon balance of forests which can be termed as the net ecosystem productivity (NEP) wherein the terrestrial ecosystem respiration (Reco) includes both autotrophic (Ra) and heterotrophic respirations (Rh) (Clark et al. 2001). Traditionally, NEP is measured by (i) estimating carbon stocks in vegetation and soil or (ii) by integrating CO_2 fluxes in and out of vegetation, i.e., through EC-based approach, termed as the net ecosystem exchange (NEE). In the EC measurement, the negative of this NEP is termed as NEE. As indicated by Curtis et al. (2002), conceptually, NEE and NEP are equivalent as both constitute the difference between GPP and terrestrial ecosystem respiration (Reco = Ra + Rh). Similarly, as described by Lovett et al. (2006), NEE can be equal to the NEP with sources and sinks for CO_2 that do not involve conversion to or from organic carbon. Therefore, as per the convention of meteorologists, the NEE is of opposite in sign, so that the fluxes into the ecosystem are negative, and can be represented as follows:

$$-\text{NEE} = \text{NEP} + \text{Inorganic sinks for } CO_2 - \text{Inorganic sources of } CO_2 \quad (8.1)$$

The NEP can further be divided in to above ground net primary productivity (NPP) and below ground net primary productivity (NPP_b), wherein NPP is the difference between GPP and Ra. Moreover, NPP represents the rate of production of living organic matter, and primary productivity of ecosystem was related to energy flow in the early 1970s by E. Odum (Odum 1971). Similarly, the Ra and Rh can be sub-divided for above and below ground value, and a pictorial representation of the NEP partitioning is provided in Fig. 8.1.

According to the Forest Survey of India (FSI), 2021, approximately 21.72% of the total geographic area of India falls under forest. The ecosystem productivities of some of these forests are being quantified using both BM- and EC-based approaches. Although, a plethora of literature has been available globally on both approaches, there is limited scientific consensus on the terminologies used in these methods. Consequently, it is being pointed out that estimates of forest NPP are based on incomplete, and sometimes inappropriate field measurements, and a variety of techniques are used to make corrections for the estimation of annual NEE. Here, we present a comprehensive review of the past and current methods used for the

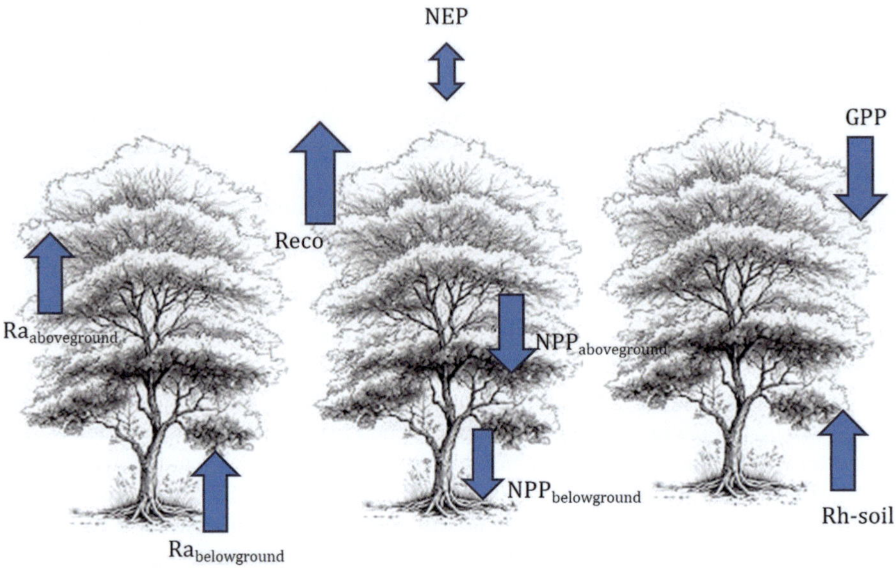

Fig. 8.1 Pictorial representation of the different components of ecosystem productivity. NEP is net ecosystem productivity, NPP is net primary productivity, GPP is gross primary productivity, Ra is autotrophic respiration and Rh is heterotrophic respiration

quantification of ecosystem productivity and carbon sequestration, and provide distinction in terminologies used by each method so that a scientific understanding of the applicability of these terminologies can be made. Moreover, efforts are also made to enhance understanding of Himalayan forests by presenting a comparative assessment of NEE, GPP, and RE of the already established sites. In addition, we have used net ecosystem exchange and micrometeorological data obtained using flux-based instrumentation to derive seasonal and annual carbon exchanges and its relationships with environmental conditions.

General Approaches of Carbon Budget Estimation

Ecological Field and Indices-Based Biometric C-Cycling Concepts

Forests are accountable for around 75% of terrestrial gross primary production (Beer et al. 2010). Terrestrial ecosystem works as either source or sink of CO_2 and defining C-cycle terminologies unambiguously is important. Since the field-based direct measurements of NPP and GPP are not possible (Waring and Schlesinger 1985), and field-based measurement of Reco leads to substantial uncertainty (Lavigne et al. 1997), indirect methods are adopted for field-based estimation of NPP (Clark et al. 2001). This indirect method of NPP estimation includes measurement of the amount of new organic matters retained by plants within a time interval, and the amount of organic matter produced and lost by the plants within the same time interval (Clark et al. 2001). In this context the Intergovernmental Panel for Climate Change (IPCC) protocol for greenhouse gas inventory indicates that carbon stock change measurement to be made from five carbon pools, i.e., above ground biomass, below ground biomass, dead wood, litter, and soil organic matter; and the annual change in the biomass to be computed by two approaches, Gain-Loss method and Stock-Difference method (IPCC 2006). However, this is to be noted that field measurements of biomass are carried out under many assumptions, and in most cases a sum of above ground biomass and litter is reported as above ground NPP.

The NPP of natural forest communities having individual trees of different ages and weights can be estimated by considering the biomass of the trees by felling the trees of a plot, and measuring different size classes and then measuring DBH (diameter at breast height) and height. This is a destructive method. Additionally, the dry weight of the tree parts such as stem, branch, twigs, leaves, stump roots, lateral roots, flowers, and fruits also need to be determined for each tree. Similarly, wood and miscellaneous litter fall is also added to the biomass of twigs/branches. The total data obtained as mentioned above is further used for regression analysis to relate the dry weight of each component with the DBH of the tree (Woodwell and Whittaker 1968; Kira and Shidei 1967). Finally, the allometric regression equation is used to estimate biomass is:

$$Y = a + b \log X \tag{8.2}$$

where Y is dry weight of the component (kg tree^{-1}), $X =$ DBH (cm^2 tree^{-1}), a and b are the intercept, and slope or regression coefficient. These equations can be developed for each component of the tree or for the tree as a whole, where in each case Y will be the dry weight. Once the regression equations for different species are developed, biomass can finally be estimated from the equations for different species, for which no further felling of trees is required. Moreover, the biomass of different components is also calculated from the DBH or by using d^2H (DBH squared multiplied by the height) as using height with DBH was also suggested by Kira and Shidei (1967). For calculating net primary productivity (NPP), the biomass of the respective forest is measured twice, first initially at the start of the study and the next time usually after one or 2 years. The value is calculated as follows:

$$\text{NPP} = Y_2 - Y_1 \tag{8.3}$$

where Y_1 is the initial biomass and Y_2 is the biomass recorded later. Although, carbon stock in a forest is assumed to be half of the total biomass per year, whereas carbon sequestration is assumed to be half of the NPP due to heterotrophic respiration (Magnussen and Reed 2004). Some data on biomass and ecosystem productivity in the Indian Himalayan region could be found in Singh et al. (1985), Tiwari et al. (1985), Singh and Singh (1987), and Tewari et al. (1989) in which the dynamics of forests over IHR and ecosystem productivity were estimated by analyzing structural composition of the trees and with standardized allometric equations. Nowadays, these equations are used for the estimation of biomass in which the mean DBH value for each species and a girth class is used to get an estimate of mean tree biomass for the respective girth class.

Using diameter-based allometric equations are not a very good way of estimating biomass as tree biomass not only depends on diameter and species but also a lot on the site. As an alternative to allometric equation-based method, the site index can be used to quantify the potential productivity of a forest stand based on height, age, and site quality (potential productivity) in even-aged single-species stands. They are used for predicting stand height development, and for assessing site quality. In general, a tree growing on a good fertile site can have twice the height of one with the same diameter growing on a poor quality site, hence the biomass would also differ a lot. However, if we measure a tree at a particular site after X number of years and use the difference in biomass (Eq. 8.3), this may also give a better indication of growth—though allometric equations, even if correctly done, are done for particular 'average' site indices.

Eddy Covariance-Based Carbon Cycling Concepts

Eddy covariance (EC) is one of the most direct and appropriate ways to measure exchanges of gas, energy, and momentum between ecosystems and the atmosphere. The eddy covariance system estimates the ecosystem fluxes by the concurrent measurement of vertical wind speed and gas concentration at a certain height above the canopy. However, the sensor position depends on the canopy height and area of interest (Burba and Anderson 2009). It provides accurate information about the emission and consumption rate of various gases and water fluxes over areas ranging from a few hundred meters to kilometers. The ecosystem flux exchange measurements using EC system is generally carried out under the assumption that airflow is primarily horizontal with each eddy parcel moving in the upward and downward direction, and each eddy has 3-D wind components. The vertical flux of a scalar (S) could be computed at a point once the number of particles moving upward and downward within a known volume of air is measured and simultaneous wind information (W) is available (Fig. 8.2).

Fig. 8.2 Conceptual framework of atmospheric eddy transport. Half circles with arrows represent turbulent eddies. (Adapted from Burba and Anderson (2009))

Box 8.1

In an EC system, data is measured continuously, which represent the net ecosystem exchange for the gases of interest, particularly CO_2, and this net flux exchange can be partitioned into individual components such as GPP and Reco. Theoretically, GPP is always non-negative during the day and is zero at night (no sunlight means no photosynthesis, GPP = 0), and Reco is always positive and in general, is non-zero for both night-time and day-time. NEE can either be positive (mainly at night when GPP = 0 but Reco >0) or negative (mainly during the day when GPP >0 and Reco >0). There are several approaches to partition NEE into GPP and Reco. The method by Lasslop et al. (2010) is commonly used for partitioning of the NEE to GPP and Reco that includes vapor pressure deficiency (VPD) dependency. In this approach, the response of NEE measurements to temperature is assumed to be independent of systematic measurement errors during periods of poorly developed turbulence during nighttime. Moreover, the magnitude of the respiration is also quantified by using only daytime data, thus independent of such biases.

Swinbank (1951) had first time presented the practical implementation of vertical transfer of heat and water vapor by eddy movement. The EC method was fully designed in 1980s with complete instrumentation and theoretical framework for flux measurements for short periods. Subsequently, some additional advances were made in the late 1980s and early 1990s providing ground to the researchers for accurate flux measurements for longer time scale.

Consequently, the EC measurement has grown worldwide, and at present, more than 500 EC flux towers across five continents are operating for long-term monitoring of the carbon and water fluxes through many network observation programs such as FLUXNET (World), OzFLUX (Australia), AsiaFLUX (Japan, China, and Korea), etc.

Comparison Between BM and EC Approaches for Measuring Productivity

Both the BM and EC approaches have frequently captured different components of NEP and NEE, and can therefore be compared across scales only by carefully specifying the fluxes included in the measurements. The EC measurement provides efficient results on land–atmosphere interactions. However, EC measurements face challenges with standardized inventory-based assessments because of storage, loss terms, and lateral fluxes such as fluxes that are not related to respiration. Moreover, the EC method is based on underlying assumption that surface sources are homogeneously distributed and atmospheric flows are one-dimensional to the measurement sensors. Though, such assumptions are non-trivial to address for real-world scenarios, EC approaches require a significant correction of actual observed fluxes.

Similarly, BM measurements are also labor-intensive and area-specific, and mostly used for above ground NPP estimation. In this context, Curtis et al. (2002) have shown that the estimated carbon pool more closely matches with NEE rather than NEP. Similarly, Baldocchi (2003) have also shown the most of the studies are in agreement with eddy covariance-based measurement in computation of biomass approach. A recent study by Campioli et al. (2016) has also shown confidence in EC-based measurement of ecosystem productivity. Since EC measurement is non-destructive, have larger spatial extent for productivity measurement, together with geo-spatial technology, EC-based approach of ecosystem productivity assessment have better prospect for estimating the carbon budget of forest ecosystems. However, these measurements are site-specific and safety measures are needed to be taken care of before site selection and instrument installation. This is also to note that EC system requires horizontally homogeneous distribution of source areas (i.e., forests and grasslands), else requires footprint estimation and further data correction. Conclusively, EC-based approaches are gaining interest in India, and the following section highlights examples of successful implementation of EC approaches for measuring productivity for various forests of the Indian Himalayan region.

Eddy Covariance-Based Measurements for Forests of Indian Himalayan Region

Current Status

Although EC measurements over the Indian Himalayan region are very sparse yet give an idea of some NEE values and their seasonal patterns. The first such measurement over Himalaya was carried out by Chatterjee et al. (2018) for the evergreen coniferous Japanese Cedar (*Cryptomeria japonica,* exotic species) Darjeeling in which biosphere–atmosphere exchange was reported for the period of two months, i.e., March–April of the year 2015. The observed results indicated that the high-altitude Himalayan coniferous forest in eastern Himalaya acts as a net sink of CO_2, with a net ecosystem exchange of -2.90 gC m^{-2} day^{-1} during the spring season. Similarly, Mukherjee et al. (2018) have studied biosphere–atmosphere interactions and quantified the impact of monsoon rainfall on ecosystem fluxes of Pine (*Pinus roxburghii*)-dominated vegetation of Indian western Himalaya using wavelet analysis in which EC-based measurements were used to analyze scale interaction. Results of the analysis indicated that *Pinus roxburghii* (Pine) dominated ecosystem was a sink of CO_2 (-3.21 g.C.$m^{-2}day^{-1}$), and *Pinus roxburghii*-dominated forest productivity of Western Himalaya, India, are expected to increase with the increment of heavy rainfall events in the near future. A similar study from central Himalaya by Watham et al. (2020) have also presented a comparative assessment of eco-physiological responses of two forest types (Sal-dominated forest and mixed dry deciduous forest) in the foothills of the Indian western Himalaya generated from the eddy covariance towers. The study highlighted that both the ecosystems (deciduous forest and young mixed plantation) sequester −1.97 and −1.28 g.C.m^{-2}day^{-1}, respectively. The relationship between Reco and NEE with the GPP was also established providing an insight into the mechanism of carbon exchange and may be useful in the development of biophysical models. In addition to the above, Deb Burman et al. (2021) have presented a thorough comparative assessment of ecosystem fluxes of Kaziranga National Park, India, dominated by deciduous broad leaf trees, particularly, *Gmelina arborea* Roxb., *Mallotus repandas* (Willd) Müll. Arg., and *Tetrameles nudiflora* R. Br with a central Himalayan Pine-dominated ecosystem at Kosi-Katarmal, Almora, India. Results of the study indicated Pine-dominated ecosystem as a strong sink of CO_2 while Kaziranga National Park remained a moderate source to weak sink of CO_2. This study also reported the first-time assessment of the carbon and water coupling mechanisms for the selected forests using Big-Leaf Framework (Knauer et al. 2018) which quantify physical and physiological ecosystem properties from eddy covariance data. The insights from this study could be helpful for improved understanding of carbon and water budgets of sub-tropical forests and their controls in this region, and directly support development and parameterization of plant functional type descriptions in the earth system models. A similar comparison was made by Lohani et al. (2023b) between Himalayan Chir-Pine and Banj-Oak-dominated ecosystems, and quantified their responses with rainfall pattern. Both the ecosystems were found to be a sink of CO_2, Chir-Pine-dominated ecosystem was found to sequester approx. 1.8 times higher carbon than Banj-Oak-dominated ecosystem. A network theory approach-based analysis was also carried out by Khadke et al. (Khadke et al. 2024a, b) for Pine and Oak-dominated ecosystems to identify hydrometerological factors affecting the ecosystem growth. The hydrometeorological factors such as net radiation, relative humidity, air temperature, and sensible heat flux were found to be consistently affecting the NEE and Reco of Pine-dominated ecosystem. Moreover, moisture-stressed dry periods have more casual links with leading NEE. Conclusively, EC-based measurement is noted to be pivotal for prescribing plantation policy cues of forests within the Indian region. In order to show the distribution of EC-based ecosystem exchanges over Himalaya, Fig. 8.3 is prepared, and based on Fig. 8.3, ratios of GPP and Reco were estimated in order to identify the forest ecosystem as a potential sink or source of CO_2. To provide a summary of GPP and Reco of the forests of Himalaya, as measured using EC-based methods, a comparative chart is prepared (Table 8.1). It can be noted from the table that the GPP–Reco ratio when >1 indicates shift of net production from positive to negative whereas GPP-Reco ratio when <1 indicates vice versa (Bott and Newbold 2023). It can be noted from Table 8.1 that Himalayan forests are potential sinks of CO_2 and it can be inferred from the study that the Pine ecosystem has higher sink capacity than other forest ecosystems. Moreover, it can further be noted from Table 8.1 that broad leaf deciduous forest at Kaziranga National Park was having lower GPP–Reco ratio value due to the marginal difference in GPP and Reco, while Pine-dominated ecosystem with higher GPP, found to have higher GPP–Reco ratio value due to higher sequestration rate. One of the major findings of this study indicates that the accelerated carbon uptake in the needle leaf Pine forest compared

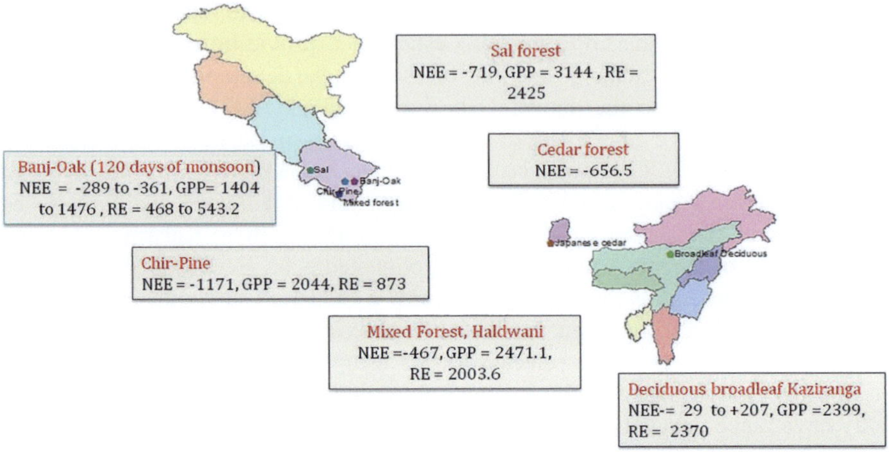

Fig. 8.3 Selected carbon flux measurements over the Indian Himalayan region using EC methods. The unit of NEE, GPP, and Reco was gC m^{-2} year^{-1}

Table 8.1 Productivity and respiration ratio as described in various studies available over IHR are presented

Forest ecosystem	Period of study	Altitude (m)	Age (years)	Canopy height (m)	GPP–Reco ratio
Pine-dominated ecosystem, Almora	Yearly	1217	25–28	12–15	2.70
Pine-dominated ecosystem, Almora	Monsoon	1217	25–28	12–15	6.75
Oak-dominated ecosystem, Pithorgarh	Monsoon	1650	20–30	8	2.73
Broad leaf Deciduous, Kaziranga National Park	Yearly	80	Mature forest	20	1.01
Sal Forest, Barkot	Yearly	415	Mature forest	32	1.29
Mixed forests, Haldwani	Yearly	280	15	16	1.23

to broad leaf Oak and deciduous forest of eastern Himalaya is linked to lesser ecosystem respiration, particularly during the monsoon season.

However, it is noted that during the compilation of the EC-based forest productivity assessment over Himalaya that no concerted national effort (such as ChinaFLUX for China, JapanFLUX for Japan, and KoFLUX for Korea) is yet made for deciphering the carbon dynamics of most of the Himalayan ecosystems. In this context, this is also to highlight that inner western Himalayan forests, i.e., forests dominated by Blue Pines, Rhododendrons, Betula spps, are barely studied in terms of land–atmosphere interactions and carbon cycling. Similarly, forests of the north east Himalayan regions, i.e., wet temperate, temperate coniferous, Fir and sub-alpine Rhododendrons are not yet explored. Since EC-based approaches of measuring forest productivity has the advantage of simultaneous measurement of microclimatic parameters at multiple temporal scales, as currently shown by Khadke et al. (2024a, b), new mathematical modeling frameworks such as network theory, artificial intelligence, and machine learning could be used for deciphering intricate relationships between forest productivity and its linkages to hydrological and meteorological parameters. With the current context of climate change, such studies would become absolutely necessary for framing and adapting new forest plantation and conservation policies.

The following section highlights such insights of land–atmosphere interactions as observed using EC-based methods. Therefore, the following sections could be considered as case studies wherein EC-based fluxes are reported for three forest types of Himalaya (Pine and Oak forests of central Himalaya and a deciduous forest of eastern Himalaya) and their daily scale interactions with air temperature (T_{air}), vapor pressure deficit (VPD), and soil moisture index (SMI) are compared.

Comparative Assessment of Pine, Oak, and Deciduous Forests in the Indian Himalayan Region

Site Description and Data Collection

Quantification of forest growth in terms of carbon balance involves calculating the assimilation of carbon and its distribution at different levels in the system, such as above and below ground components. To show the general comparative assessment, Eddy Covariance-based NEE, GPP, and Reco and their relationships with other environmental parameters, data from the already established three sites (Pine, Oak, and Kaziranga, abbreviated as KNP) were used. The basic land–atmosphere interactions were already assessed for these sites and could be accessed through Deb Burman et al. (2021) and Lohani et al. (2023b). The first selected site was Pine site, particularly *Pinus roxbughii*-dominated ecosystem with an average canopy of 12–15 m height (this Pine can easily attain heights 25–35 m), situated in the campus of G. B. Pant National Institute of Himalayan Environment, Kosi-Katarmal, Almora, at an elevation of 1217 masl. Details of the vegetation diversity and slope can be found in Mukherjee et al. (2018), Mukherjee et al. (2021), and Lohani et al. (2023a, b). The area around the flux tower was 27 Ha out of which 6.75 Ha was dominated by *Pinus roxburghii* (893 ind/ha) and the remaining 20.2 Ha had mixed plantation such as *Quercus leocotrichophora, Populus ciliate, Celtis australis, Prunus cerasoides, Melia azedarach, Toona ciliate, Grevillea robusta, Morus alba, Pyrus pashia, and Quercus glauca*. High species richness was recorded in the Pine site as the site is within the protected area of GBPNIHE, Almora, campus. The 10 Hz data were collected from 3-D sonic anemometer (CSAT3, Campbell Scientific) and CO_2/H_2O (EC150, Campbell Scientific) gas analyzer installed at 30 m height along with slow sensors. The EC measurement for this site was available since 2014. The second site i.e. Oak site was particularly *Quercus leucotrichophora* (Banj -Oak) dominated site, with an average canopy height of 8 m (it grows 25–30 m tall) height, for which details could be found in Lohani et al. (2023a) and Lohani et al. (2023b). The dominant species surrounding the flux tower were *Quercus leucotrichophora* (660 ind/ha) followed by *Prunus cerasoides and Cedrus Deodara*. The site was situated at Gangolihat block, district Pithoragarh, at an elevation of 1650 masl. A combined 3-D sonic anemometer (CSAT3, Campbell Scientific) and CO_2/H_2O gas analyzer was installed at 10 m height along with slow sensors. This is to highlight here that these measurements are comparable as all the measurements were carried out within the surface boundary layer and above the roughness sub-layer. Moreover, corrections were made in order to compensate slope in the underlying terrain at all three sites and the actual measurement height was computed by removing displacement height (d) from measurement height ($z = z_m - d$) as per Foken and Nappo (2008), i.e., $d = 2/3h$, where h is the canopy height. Furthermore, this is to be noted here that these forest ecosystems are not limited to only one species as the mountains mostly have mixed vegetation, except for few monoculture plantations.

The third site is a semi-evergreen moist deciduous forest at Kaziranga National Park, referred to as "KNP site", with an average canopy height of 20 m. The flux tower at the KNP site was 50 m tall in which a sonic anemometer Windmaster Pro by Gill Instruments, UK and CO_2 gas analyzer (LI-7200 enclosed pathCO_2/H_2O analyzer, Licor inc.) were installed at 37 m height, whereas net radiometer was installed at the height of 20 m. More details of the site and instrumentation could be found in Deb Burman et al. (2017). As provided in Sarma et al. (2022), the KNP has different floristic composition, namely eastern wet alluvial grasslands, Assam alluvial plains semi-evergreen forests, tropical moist mixed deciduous forests, Eastern Dillenia swamp forests, wetlands and sandy "chars". Although, the flux tower was installed around the semi-evergreen deciduous forest and the major dominant species around the flux tower are *Gmelina arborea* Roxb., *Mallotus repandas* (Willd) Müll. Arg., and *Tetrameles nudiflora* R. Br.

For Pine and Oak sites, data used and analyzed in this study are for two seasons, particularly monsoon (1 June, 2017 to 30 Sep, 2017) and winter (1 Nov 2017 to 31 Dec, 2017) seasons of 2017, whereas 2018 data (for monsoon, 1 June, 2018 to 30 Sep, 2018 and for winter, 1 Nov, 2018 to 31 Dec, 2018) were used for KNP. The basic flux data processing was carried out using the EddyPro software (v6.0, LiCor-Bioscience, USA) and following similar approaches presented in Mukherjee et al. (2018, 2020), Lohani and Mukherjee et al. (2021), Deb Burman et al. (2021), and Lohani et al. (2023b). The 30-min runs obtained from EddyPro were further used for partitioning of NEE into GPP and Reco, which was done using Lasslop et al. (2010). The 30-min runs were further checked for quality by removing outliers. Finally, daily values were calculated from 30-min runs. Soil moisture index (SMI) data were extracted for Pine, Oak, and KNP sites through Google earth engine using Sentinel-1 (https://sentinels.copernicus.eu/) at 10 m resolution and on 4–5 days interval. In order to produce relationships between soil moisture and NEE, only those data were used when both SMI and NEE data were available for all the sites.

Results

Seasonal Variation of NEE, GPP, and Reco

The monthly variation of NEE, GPP, and Reco for three sites of IHR is presented in Fig. 8.4. Both Pine and Oak forest sites were found to be a sink of CO_2 for selected seasons with an average NEE of −4.54 (±0.17) g.C.m^{-2}.d^{-1} and – 3.01 (±0.15) g. C.m^{-2}.day^{-1} during the monsoon season (Lohani et al. 2023b), which changed to – 2.03 (±0.01) g.C.m^{-2}.day^{-1} and −2.67 (±0.01) g.C.m^{-2}.day^{-1} during the winter season. A marginally higher sequestration at the Oak forest site during the winter season could be attributed to higher photosynthesis during the season. When the comparison of NEEs of Pine and Oak sites were made with the KNP site, the deciduous forest of KNP was found to be a weak sink during monsoon with an average value of −0.19 (±0.12) g.C.m^{-2}.day^{-1}; however, the winter season had slight shift of NEE from sink to source 0.07(±0.01) due to the deciduous nature of the ecosystem which subsides the growth rate during the winter season. Moreover, the reason behind such responses of the KNP site could be seasonality as continuous flushing events during the spring season could increase the growth rate and sequestration potential of the ecosystem even during the monsoon season. When GPP was compared between Pine and Oak ecosystems, we found similar GPP values (11.9 ± 0.001 and 11.7 ± 0.0017 g.C.m^{-2}.day^{-1}, respectively) during the monsoon season which changed to 5.27 (±0.11) and 4.29 (±0.07) g.C.m^{-2}.day^{-1} during the winter season. The similar value of GPP was also observed by Watham et al. (2020) for Sal forest (11.5 g.C.m^{-2}.day^{-1}) and Mixed forest (10.07 g.C.m^{-2}.day^{-1}) of western Himalaya during the monsoon season (July-Sep). The higher GPP during the monsoon season could be attributed to higher vegetative growth. However, during the monsoon season, the KNP site had a lower value of GPP of 8.92 (±0.01) g.C.m^{-2}.day^{-1} than the Pine and Oak ecosystems while during the winter

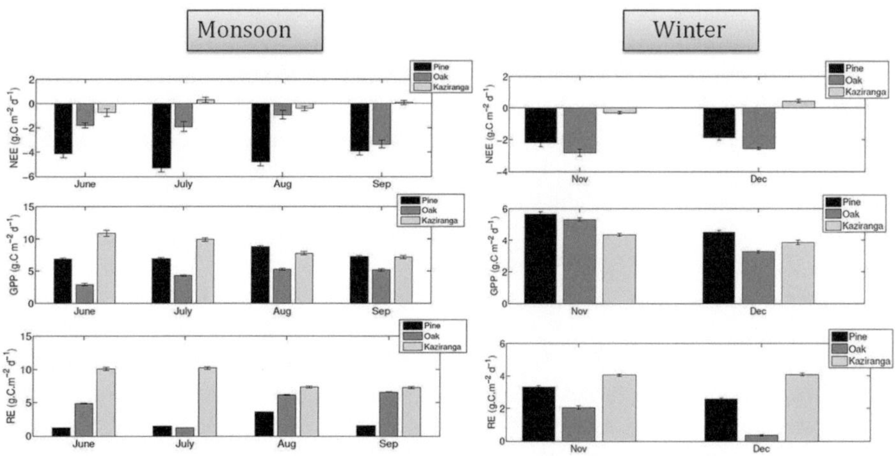

Fig. 8.4 Monthly variation of NEE, GPP, and Reco for monsoon and winter seasons of Pine, Oak, and KNP sites. The bars in the figure indicate standard errors

season, the KNP site was having higher GPP 4.19 (±0.01) g.C.m^{-2}.day^{-1}. Ecosystem respirations during winter had contrasting results to that of the monsoon season between Pine and Oak sites. During the monsoon season, the Pine ecosystem had lower Reco (1.88 ± 0.17 gC.m^{-2}.day^{-1}) than the Oak ecosystem (4.53 ± 0.04 gC. m^{-2}.day^{-1}), while during the winter season, the Pine ecosystem had higher Reco (2.94 ± 0.02 g.C.m^{-2}.day^{-1}) than Oak (1.18 ± 0.01 g.C.m^{-2}.day^{-1}). The reason for such higher ecosystem respiration for the Pine site during winter could be linked with occurrences of forest fire (around 175 events) during the year 2017 as reported by Forest Department, Uttarakhand. The Reco for the KNP site was found to be 6.08 (±0.19) and 4.41 (±0.12), during the monsoon and winter season, respectively. The higher Reco during the winter season of the KNP site could be linked to higher decomposition rate due to leaf shading.

Table 8.2 shows the seasonal variation of NEE of different forest ecosystems of the Indian Himalayan region based on available literature. It can be easily observed from Table 8.2 that there was a strong seasonality impact on the carbon uptake process of the studied forest ecosystems. Moreover, the Pine site was found to have highest sequestration during monsoon whereas the Oak-dominated ecosystem and KNP's deciduous forest were having higher carbon uptake during pre-monsoon. The reason for such higher sequestration during pre-monsoon could be the accelerated ecosystem growth after the winter season. During the pre-monsoon season, as indicated by Tyagi et al. (2023), Pine is sensitive to drought-like conditions and closes stomata leading to lower pre-monsoon NEE compared to monsoon. However, in KNP's deciduous forest, leaf flushing events can be corroborated to enhanced growth rate during pre-monsoon and subsequently, expected to increase the carbon sequestration rate. On the other hand, due to leaf drop during winter, the growth rate decreased and a sudden shift from sink to source was observed. Similarly, Oak forest where the deep roots allow for relatively good photosynthetic activity even during the dry pre-monsoon season, the NEE is actually highest in the pre-monsoon although microbial activity and photo respiration are lower at this time. On contrary, Sal forest at Barkot (BFS) and Mixed forest at Haldwani (HFS) had highest sequestration during the post-monsoon season whereas lowest was recorded during monsoon and winter at BFS and HFS, respectively. Such behavior of Sal forest could be linked to temperature variation. In Sal forests, high temperatures and drought in the pre-monsoon limit photosynthesis activity. As temperature improves in the post-monsoon, NEE improved gradually.

Diurnal Variation of NEE, GPP, and Reco

To further elaborate intra-seasonal variability of NEE, GPP, and Reco, mean diurnal changes in the NEE, GPP, and Reco of the selected ecosystems, i.e., Pine, Oak, and KNP are presented in Fig. 8.5. Irrespective of the seasons and ecosystems, highest NEE (negative implying larger net carbon uptake), due to enhanced carbon assimilation, was noted around mid-day (11:00 to 13:00 IST) time. Moreover, Pine had higher carbon uptake during both monsoon (−17.83 μmol.m^{-2}.s^{-1}) and winter (−11.35 μmol.m^{-2}.s^{-1}) seasons in comparison to Oak and KNP. The Oak and KNP had almost a similar pattern of NEE during the day time, although due to higher carbon release during nocturnal hours by the KNP's deciduous ecosystem,

Table 8.2 Seasonal variation of Net ecosystem exchanges of study sites available over IHR

Forest ecosystem	Net ecosystem exchange (g.C.m^{-2}.day^{-1})					
	Data period	Pre-monsoon	Monsoon	Post-monsoon	Winter	Source
Pine-dominated ecosystem	2014–2016	−3.58 (March–April)	−4.00 (July–September)	–	−2.56 (November–January)	Mukherjee et al. (2018)
Oak-dominated ecosystem	2020	−2.72 (March–May)	−1.96 (June–September)	−1.95 (October–November)	–	Lohani et al. (2023a)
KNP deciduous forest	2018	−0.87 (March–May)	−0.19 (June–September)	–	0.07 (November and December)	Deb Burman et al. (2021)
Sal forest, Barkot (BFS)	2016–2018	−1.30 (April–June)	−1.06 (July–September)	−2.94 (October–December)	−2.54 (January–March)	Watham et al. (2020)
Mixed forest, Haldwani (HFS)	2016–2018	−1.17 (April–June)	−1.26 (July–September)	−2.01 (October–December)	−0.65 (January–March)	Watham et al. (2020)

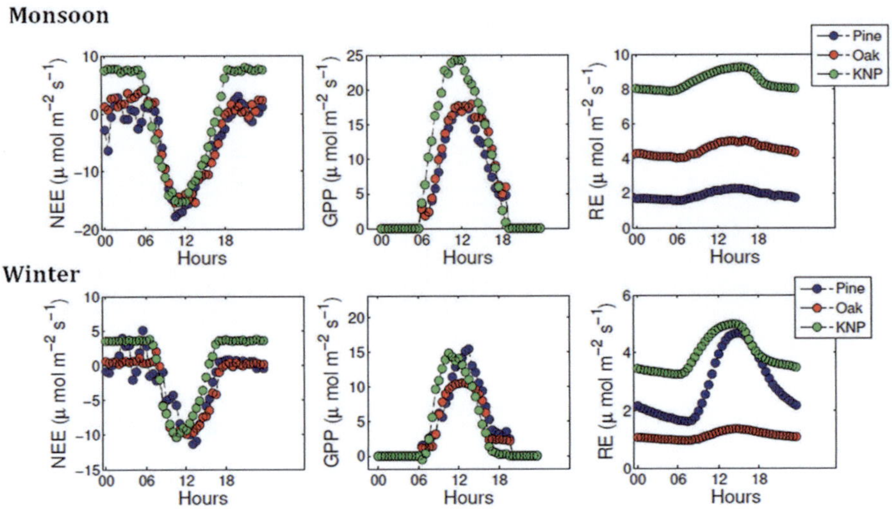

Fig. 8.5 Diurnal variation of NEE, GPP, and RE at Pine, Oak, and Kaziranga sites

significant difference could be observed in the carbon assimilation rate of Oak and KNP's deciduous ecosystem. It can further be noted from Fig. 8.5a, d that all the three sites had the maximum carbon sequestration observed around 12:00 IST during the monsoon season whereas during the winter season, it was around 11:00 IST for Oak and KNP sites and 13:00 IST for the Pine site. A similar pattern was also observed for GPP and Reco. Similar to NEE, GPP (Fig. 8.5b, e) peaked during mid-day (11:00 to 13:00 IST) time; however, a slight shift in GPP diurnal values was observed at the Pine site. As already indicated in Deb Burman et al. (2021), irrespective of the season, KNP's deciduous ecosystem was observed to have a comparatively higher Reco (Fig. 8.5c, f) than Pine and Oak ecosystems. However, during monsoon, Pine was observed to have lowest ecosystem respiration amongst the three forest types due to low microbial activities whereas during the winter season, Pine had higher Reco than Oak for which GPP was also observed to be high. The reason for such variation could be related to substantial changes in the autotrophic and heterotrophic respiration.

Seasonal Variation of NEE with VPD, T_{air}, and SMI

In this section, attempts are made to quantify the impact of T_{air}, VPD, and SMI on NEEs of Pine, Oak, and KNP ecosystems. In general, T_{air} increased gradually from the pre-monsoon season till mid of monsoon season resulting in higher carbon sequestration for all the sites (Fig. 8.6). When a linear regression model was fitted to NEE $-$ T_{air} relationship for all the sites, statistically significant relationships were observed for Pine ($R^2 = 0.58$) and KNP sites ($R^2 = 0.93$) during monsoon. During the winter season (Fig. 8.6), except for Pine site, Oak and KNP sites had statistically significant relationships between NEE and T_{air} at $R^2 = 0.55$ and $R^2 = 0.91$, respectively. The overall inference could be made from the analysis that temperature could

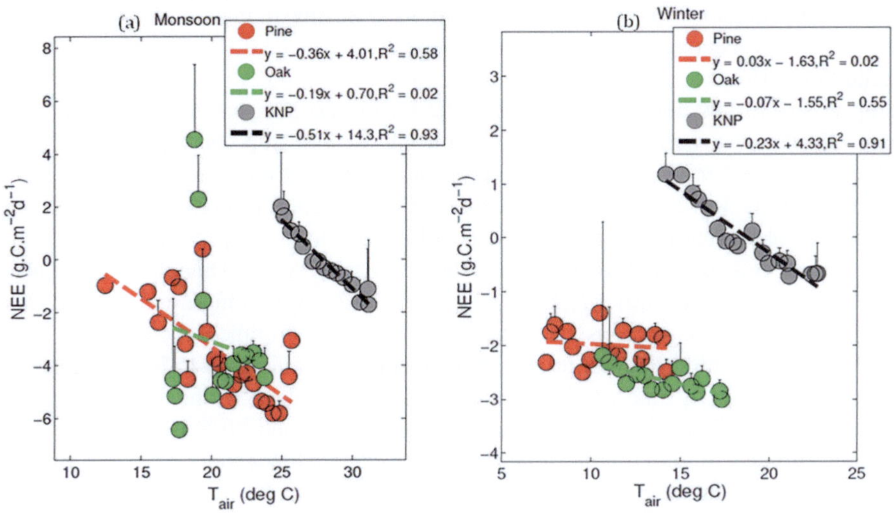

Fig. 8.6 Variation of NEE with respect to air-temperature for Pine, Oak, and KNP sites

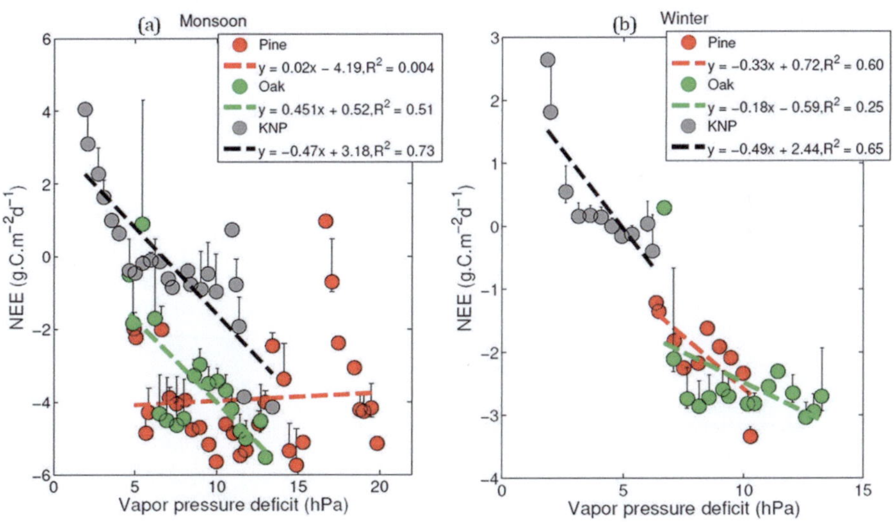

Fig. 8.7 Variation of NEE with respect to VPD for Pine, Oak, and KNP sites

be one of the controls of carbon uptake of the Pine ecosystem during monsoon, Oak ecosystem during winter, and KNP ecosystem during monsoon and winter, respectively.

As mentioned by Grossiord et al. (2020), plants are highly responsive to VPD variation and higher VPD conditions could lead to stomata closure. In this study, during the monsoon season, NEE of Pine had no statistically significant relationship with VPD as higher vapor pressure deficit led to reduction in the carbon sequestration potential (Fig. 8.7). On the contrary, Oak and KNP sites were having statistically significant relationships between NEE and VPD during monsoon ($R^2 = 0.51$ and $R^2 = 0.73$, respectively) signifying low dependency of broad leaf forests on VPD conditions. However, during the winter season, no statistically significant relationship was observed between NEE and VPD for the Oak ecosystem, although, Pine and KNP sites had statistically significant relationship between NEE-VPD ($R^2 = 0.60$, and $R^2 = 0.65$, respectively). In addition, at the Pine site, irrespective of the seasons, lower VPD (<10 hPa) conditions resulted in higher NEE. Another important observation was reduced NEE of Pine due to limited ecosystem functioning, unlike Oak and KNP ecosystems, with increasing VPD. This observation also corroborates with the earlier report by Tyagi et al. (2023) that highlights higher Oak mortality due to non-responsive changes of Oaks to environmental conditions.

It may be pointed out that longer spells of rainfall could lead to lowering of net carbon exchange of Pine and Oak ecosystems of Himalaya (Lohani et al. 2023b). However, no relationship between soil moisture index and NEE is established for Indian Himalayan ecosystems. Therefore, in order to understand the role of soil moisture in modulating CO_2 uptake rate by the Pine, Oak, and KNP ecosystems, linear regression models were fitted between soil moisture index and net ecosystem exchange (Fig. 8.8). It can be clearly observed from Fig. 8.8a–c that during the monsoon season, NEE increased with increasing soil moisture index for the Pine ecosystem ($R^2 = 0.73$) whereas contrasting impact was observed for Oak

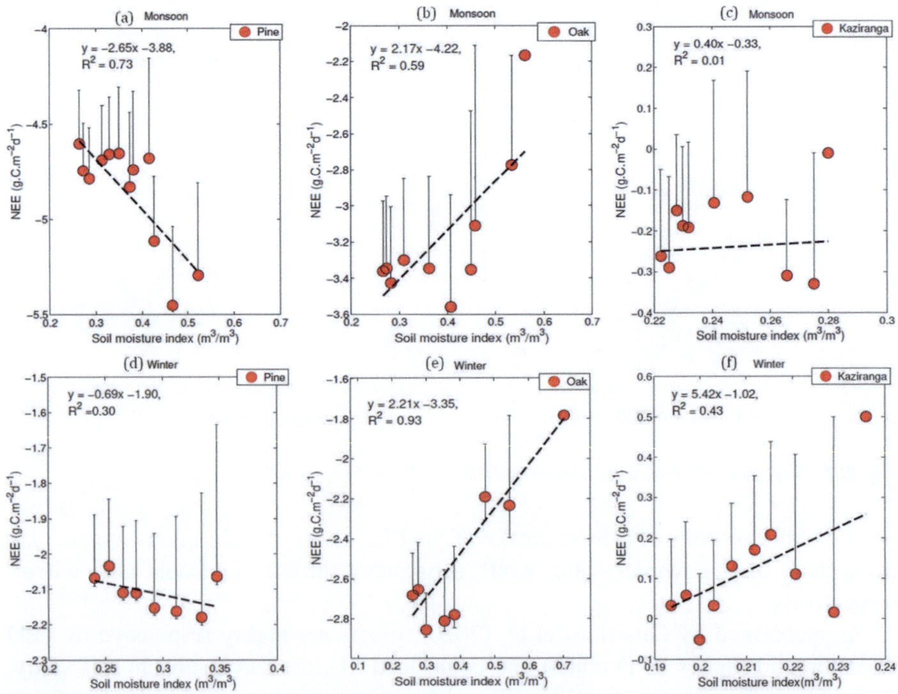

Fig. 8.8 Variation of NEE with respect to soil moisture index for Pine, Oak, and KNP sites

($R^2 = 0.59$) and Kaziranga ecosystems ($R^2 = 0.01$). Similar to monsoon, during the winter season (Fig. 8.8d–f, increasing soil moisture resulted in increasing NEE for the Pine ecosystem ($R^2 = 0.59$); however, the Oak and KNP ecosystems were having lesser NEE with increasing SMI ($R^2 = 0.93$ and 0.43, respectively). These SMI – NEE relationships for Pine, Oak, and KNP could be attributed to the fact that Oak and KNP ecosystems, contrary to the Pine ecosystems, were already having sufficient water availability throughout the year. Hence, additional soil moisture through rainfall prohibited further ecosystem functioning and lowering of the carbon sequestration.

Summary and Way Forward

Terrestrial ecosystems gain carbon through photosynthesis and lose it through respiration in autotrophs (plants and photosynthetic bacteria) and heterotrophs (fungi, animals, and some bacteria). Depending on the composition and distribution of plant biodiversity, forest ecosystems could be of many types, consequently, carbon and water fluxes and their annual budgets are expected to vary. This study presents a narrative review of the past and current methods used for the quantification of ecosystem productivity and carbon sequestration, and provide a distinction in terminologies used by each method so that a scientific understanding of the applicability of these terminologies can be made. The review suggests that NEE is a better estimate of the carbon budget and by integrating geo-spatial technology, the eddy covariance-based approach would provide a better estimate of the carbon budget of Indian forests in comparison to conventional ecology-based methods as this method is laborious and area-specific. The benefit of eddy-based method for measuring NEE is the additional measurement of micrometeorological parameters within the same setup. Consequently, data obtained through this approach can be used to derive relationships amongst NEE and other micrometeorological parameters. In order to present the efficacy of EC-based method, this study further presents seasonal variation of NEE of three forests (i.e., needle leaf Pine, broad leaf Oak, and deciduous broad leaf) of the Indian Himalayan region. The comparison of GPP and Reco measured through the EC-based method shows Pine ecosystem is a major carbon sink, however, the higher carbon sequestration is attributed to lesser Reco. Moreover, air temperature and soil moisture were found to be major drivers of net ecosystem exchange of Pine.

For the Indian context, the horizon of carbon budget research has the scope for expansion when process-based modeling of carbon cycle would be applied at a large scale, as these models do not rely on the remote-sensing data, hence can be used to study the pre-satellite era period as well as could be used to develop future scenarios. It is also to be noted that the current effort of ecosystem productivity assessment for forests of Indian Himalaya is insufficient; hence, expansion of a country-wide robust network of eddy covariance-based observations can lead to a better understanding of the land atmosphere interactions and national carbon budget.

Acknowledgments The National Mission on Himalayan Studies (NMHS), Ministry of Environment Forest and Climate Change, Govt. of India (Grant no: NMHS-2019-20/MG/64) is acknowledged for financial support to conduct this study. The Oak site was established through NMSHE, TF-3 (Phase 1), DST, Govt. of India, and the data analyses of Oak site were carried out through support from NMSHE, Phase 2, DST, Govt. of India. Ms. Lohani acknowledges a research grant from the Council of Scientific and Industrial Research, India (RA fellowship no: 09/0560 (19454)/2024—EMR—I), for supporting this work. The Director of G. B. Pant National Institute of Himalayan Environment, Uttarakhand, India, is acknowledged for providing the computational facilities. IITM is fully financially supported by the Ministry of Earth Sciences (MoES), the Government of India, New Delhi. The flux tower facilities at KNP is funded by the MoES under the MetFlux-India project. We acknowledge the Director of the Indian Institute of Tropical Meteorology (IITM), Pune, and MetFlux-India project members at IITM and Tezpur University for maintaining and providing us access to this data for research. We also thank the State Forest Department of Assam for granting us the necessary permission and support to continue the measurements.

References

Baldocchi DD (2003) Assessing the eddy covariance technique for evaluating carbon dioxide exchange rates of ecosystems: past, present and future. Glob Change Biol 9:479–492

Baldocchi DD (2014) Measuring fluxes of trace gases and energy between ecosystems and the atmosphere-the state and future of the eddy covariance method. Glob Change Biol 20:3600–3609

Beer C, Reichstein M, Tomelleri E, Ciais P, Jung M, Carvalhais N, Rödenbeck C, Arain MA, Baldocchi D, Bonan GB, Bondeau A (2010) Terrestrial gross carbon dioxide uptake: global distribution and covariation with climate. Science 329(5993):834–838

Bott TL, Newbold JD (2023) A multi-year analysis of factors affecting ecosystem metabolism in forested and meadow reaches of a Piedmont stream. Hydrobiologia 850:4243–4269. https://doi.org/10.1007/s10750-023-05296-2

Burba G, Anderson D (2009) A brief practical guide to eddy covariance flux measurements: principles and workflow examples for scientific and industrial applications, LI-COR Biosciences (Version 1.0.1)

Campioli M, Malhi Y, Vicca S, Luyssaert S, Papale D, Peñuelas J, Reichstein M, Migliavacca M, Arain MA, Janssens IA (2016) Evaluating the convergence between eddy-covariance and biometric methods for assessing carbon budgets of forests. Nat Commun 7:1371

Chapin FS III et al (2006) Reconciling carbon-cycle concepts, terminology, and methods. Ecosystems 9:1041–1050

Chatterjee A, Roy A, Chakraborty S, Karipot AK, Sarkar C, Singh S, Ghosh SK, Mitra A, Raha S (2018) Biosphere atmosphere exchange of CO_2, H_2O vapor and energy during spring over a high altitude Himalayan forest in Eastern India. Aerosol Air Qual Res 18:2704–2719

Clark DA, Brown S, Kicklighter DW, Chambers JQ, Thomlinson JR, Ni J (2001) Measuring net primary production in forests: concepts and field methods. Ecol Appl 11:356–370

Curtis PS, Hanson PJ, Bolstad P, Barford C, Randolph JC, Schmid HP, Wilson KB (2002) Biometric and eddy-covariance based estimates of annual carbon storage in five eastern North American deciduous forests. Agri For Meteorol 113(1–4):3–19. https://doi.org/10.1016/S0168-1923(02)00099-0

Deb Burman PK, Sarma D, Williams M, Karipot AK, Chakraborty S (2017) Estimating gross primary productivity of a tropical forest ecosystem over north-east India using LAI and meteorological variables. Journal of Earth System Science 126(7) https://doi.org/10.1007/s12040-017-0874-3

Deb Burman PK, Launiainen S, Mukherjee S, Chakraborty S, Gogoi N, Murkute C, Lohani P, Sarma D, Kumar K (2021) Ecosystem-atmosphere carbon and water exchanges of subtropical evergreen and deciduous forests in India. For Ecol Manag 495:119371

Foken T, Nappo C (2008). Micrometeorology. Springer, Berlin

FSI (2021) India State of Forest Report. Forest Survey of India (Ministry of Environment & Forests and Climate Change, Govt. of India), Kaulagarh Road, Dehradun

Grossiord C, Buckley TN, Cernusak LA, Novick K, Poulter B, Seigwolf R, Sperry J, McDowell N (2020) Plant responses to rising vapour pressure deficit. New Phytol 226(6):1550–1566

Intergovernmental Panel for Climate Change (IPCC) (2006) Guidelines for national greenhouse gas, Chapter 4: Forest Land. https://www.ipcc-nggip.iges.or.jp/public/2006gl/pdf/4_Volume4/V4_04_Ch4_Forest_Land.pdf

Khadke L, Mukherjee S, Kumar K, Ghosh S (2024a) Hydrometeorological factors affecting the carbon exchange of the Himalayan pine-dominated ecosystem. Ecol Inform 79:102446

Khadke L, Mukherjee S, Ghosh S (2024b) Micrometeorological drivers affecting the variability of CO2 uptake in the Himalayan Oak and Pine dominated ecosystems: an assessment of causal relationships. J Hydrol, 647:132333

Kira T, Shidei T (1967) Primary production and turnover of organic matter in different forest ecosystems of Western pacific. Japan J Ecol 17:70–87

Knauer J, El-Madany T, Zaehle S et al (2018) Bigleaf – an R package for the calculation of physical and physiological ecosystem properties from eddy covariance data. PLoS One 13(8):e0201114

Lasslop G, Reichstein M, Papale D, Richardson A, Arneth A, Barr A, Stoy P, Wohlfahrt G (2010) Separation of net ecosystem exchange into assimilation and respiration using a light response curve approach: critical issues and global evaluation. Glob Change Biol 16(1):187–208. https://doi.org/10.1111/j.1365-2486.2009.02041.x

Lavigne MB et al (1997) Comparing nocturnal eddy covariance measurements to estimates of ecosystem respiration made by scaling chamber measurements at six coniferous boreal sites. J Geophys Res 102:28977–28985

Lohani P, Mukherjee S, Sekar KC (2023a) Seasonal variation of ecosystem fluxes of a Himalayan Banj-Oak-dominated vegetation. In: Sharma S, Kuniyal JC, Chand P, Singh P (eds) Climate change adaptation, risk management and sustainable practices in the Himalaya. Springer, Cham. https://doi.org/10.1007/978-3-031-24659-3_13

Lohani P, Mukherjee S, Sekar KC, Mehta P, Kumar K, Dimri AP (2023b) Impact of monsoon season rainfall spells on the ecosystem carbon exchanges of Himalayan Chir-Pine and Banj-Oak dominated forests: a comparative assessment. Environ Monit Asses 195:827

Lovett GM, Cole JJ, Pace ML (2006) Is net ecosystem production equal to ecosystem carbon accumulation? Ecosystems 9(1):152–155. https://doi.org/10.1007/s10021-005-0036-3

Magnussen S, Reed D (2004) Modelling for estimation and monitoring. Knowledge reference for national forest assessments, Food and Agriculture Organisation (FAO-IUFRO), 111

Mukherjee S, Sekar KC, Lohani P, Kumar K, Patra PK, Ishijima K (2018) Investigation of scale interaction between rainfall and ecosystem carbon exchange of Western Himalayan Pine dominated vegetation. Biogeosci Discuss. https://doi.org/10.5194/bg-2018-299

Mukherjee S, Lohani P, Kumar K, Chowdhuri S, Prabhakaran T, Karipot A (2020) Assessment of new alternative scaling properties of the convective boundary layer: application to velocity and temperature spectra. Boundary-Layer Meteorology,176,271–289. https://doi.org/10.1007/s10546-020-00525-w

Mukherjee S, Lohani P, Tiwari A, Sturman AP (2021) Characteristics of convective surface layer turbulence over central Himalaya with an emphasis to monin-obukhov similarity theory. J Atmos Solar Terres Phys 225:105748. https://doi.org/10.1016/j.jastp.2021.105748

Odum E (1971) Fundamentals of ecology. Saunders, Philadelphia

Sarma D et al (2022) Quantifying the net ecosystem exchange at a semi-deciduous forest in Northeast India from intra-seasonal to the seasonal time scale. Agric For Meteorol 314: 108786. https://doi.org/10.1016/j.agrformet.2021.108786

Singh JS, Singh SP (1987) Forest vegetation of the Himalaya. Bot Rev 53(1):80–192

Singh SP, Khanna RK, Singh JS (1985) Accumulation in wood: a nutrient conserving strategy of tropical forests. Environ Conserv 12:170–173

Swinbank WC (1951) The measurement of vertical transfer of heat and water vapor by eddies in the lower atmosphere. J Meteorol 8:3

Tewari JC, Rikhari HC, Singh SP (1989) Compositional and structural features of certain tree stands along an elevational gradient in Central Himalaya. Vegetatio 85:117–120. https://doi.org/10.1007/BF00042261

Tiwari AK, Saxena AK, Singh JS (1985) Inventory of forest biomass for Indian Central Himalaya. In: Singh JS (ed) Environmental regeneration in the Himalaya: concepts and strategies. Central Himalayan Environment Association, Nainital, pp 236–247

Tyagi V, Singh SP, Singh RD, Gumber S (2023) Chir Pine and Banj Oak responses to pre-monsoon drought across slope aspects and positions in central Himalaya. Environ Monit Assess 195:258. https://doi.org/10.1007/s10661-022-10860-9

Waring RH, Schlesinger WH (1985) Forest ecosystems: concepts and management. Academic, New York

Watham T, Srinet R, Nandy SK, Padalia H, Sinha SK, Patel NR, Chauhan P (2020) Environmental control on carbon exchange of natural and planted forest in Western Himalaya foothills of Himalaya. Biogeochemistry 151:291–311

Woodwell GM, Whittaker RH (1968) Primary production in terrestrial ecosystems. Zoologist 8:19–30

9 Understanding Photosynthetic Characteristics of Himalayan Forests Through a Remote Sensing Lens

A. Kurian, A. Raj, P. Tripathi, and M. D. Behera

Abstract

Terrestrial ecosystems such as those in the Himalayas, which are characterized by dense vegetation and high biodiversity, appear to have the potential to accommodate a considerable part of the CO_2 emitted into the atmosphere by various sources. In order to understand the actual capacity of Himalayan ecosystems to consume CO_2, accurate measurement of the terrestrial vegetation primary productivity and its understanding are mandatory. This study used the satellite remote sensing-derived (MODIS) land products for trend analysis of Gross Primary Productivity (GPP), Leaf Area Index (LAI), and Absorbed Photosynthetically Active Radiation (APAR). The monthly average GPP value in eastern and western Himalaya is 0.03 and 0.018 kgC/m^2 respectively, as estimated by the MOD17A2H data product, whereas the mean value of FPAR in the eastern Himalaya is 0.648 and 0.581 in the western Himalaya as per the MODIS data analyzed during 2017 to 2023. The average LAI in the eastern Himalaya was observed as 4.31 in low altitude and 4.49 in mid altitude, whereas in western Himalaya, it was 2.74, 3.11, and 1.28 in low-, mid-, and high-elevation belts, respectively, as per the MODIS data. This indicates an underestimation of LAI by MODIS data as reported by many researchers. The western Himalaya with varying climatic conditions, including colder temperatures and lower precipitation compared to eastern Himalaya, influences the seasonal dynamics of FPAR, particularly during the growing season of vegetation as reflected in the Gross Primary Productivity.

A. Kurian · A. Raj · M. D. Behera (✉)
CORAL, IIT, Kharagpur, WB, India
e-mail: mdbehera@coral.iitkgp.ac.in

P. Tripathi
ICIMOD, Kathmandu, Nepal

S. P. Singh, R. Thadani (eds.), *Himalayan Forests in a Changing World*,
https://doi.org/10.1007/978-981-95-0978-2_9

Keywords

Gross primary productivity · Leaf area index · Absorbed photosynthetically active radiation · Google earth engine

Introduction

The Himalayas stand out for their remarkable temperature and moisture fluctuations across short distances. Within just 100 km, ecosystems can shift from tropical to subalpine, encompassing the intermediate subtropical and montane temperate zones. Similarly, there is a considerable contrast in precipitation amounts from the southern Himalayan ranges to the rain shadow areas in the inner ranges (Singh and Thadani 2023). The intensity of monsoons is highest in the eastern regions of the North-western Himalaya and gradually diminishes toward the west. The amount of precipitation is less in the inner regions of the North-western Himalaya, with these areas typically receiving more snowfall during winter. Additionally, the geographical orientation of the mountain range plays a role in influencing the distribution of rainfall (Prasher et al. 2015). The distribution of forests in the Himalayas is primarily dictated by elevation and precipitation. Altitudinal variation is an umbrella term for numerous climatic and physical variables that changes along with it, directly influencing and shaping the composition and structure of the forest, along with other factors, such as anthropogenic activities, historical effects, species-area effects, isolation, and so on (Singh et al. 2017).

Forest ecosystems being one of the essential carbon pools that incessantly exchange carbon dioxide (CO_2) with the atmosphere (Ghosh and Behera 2018), their continuous monitoring is important in order to assess carbon stocks, spatial and temporal variation in vegetation, and to aid climate change mitigation programs (Vijith and Dodge-Wan 2020; Dash et al. 2020). It will be crucial to comprehend how vegetation reacts to climate variations to establish effective adaptation strategies, particularly in regions of the Indian tropics with unpredictable rainfall patterns (Gautam Govinda et al. 2013). Since vegetation tends to respond gradually to environmental shifts, there can be a delay between climate changes and alterations in vegetation, potentially leading to shifts in biome composition before any irreversible damage becomes evident at the ecosystem level (Quillet et al. 2010). The Himalayan region faces formidable conservation challenges due to the limited availability of long-term data sets and documentations elucidating the ongoing consequences of environmental changes and ecosystem processes (Chandra Sekar et al. 2017). Proper documentation of these responses and their likely consequences is necessary to provide a basis for future comparisons (Negi et al. 2019). Therefore, the continuous monitoring of these sensitive and fragile ecosystems is paramount, given their substantial contribution to global biodiversity.

The emergence of remote sensing technology has marked a new era in ecophysiological studies in mountainous areas concerning climate change. Leveraging this technology can significantly enhance our understanding of vegetation responses to a changing climate (Jeganathan et al. 2024). The advantages of Remote Sensing (RS) data, including its synoptic view, extensive area coverage, and multitemporal capabilities, make it a highly valuable tool, especially over inaccessible terrain (Behera 1999). Environmental decision-makers rely on a nuanced comprehension of both seasonal and interannual vegetation dynamics. This work attempted to quantify the biophysical and climatic variability of eastern and western Himalayas separately in three different altitudinal zones using remotely sensed data, by analyzing the seasonal and annual variations of leaf area index, FPAR, and GPP. The significance of this research lies in its potential to unravel the complexities of the ecosystems in the Himalayan region and offer a comprehensive picture of the overall carbon dynamics and productivity of forests.

Materials and Methodology

Study Area

The western (Jammu & Kashmir, Ladakh, Himachal Pradesh, and Uttarakhand) and eastern (Sikkim and Arunachal Pradesh) Indian Himalayan range have been selected for this study (Fig. 9.1). The altitudinal gradient of the study area ranges from approximately 80 to 8500 m and has been divided into three zones (low, <1200 m; mid, 1200–3000 m, and high, >3000 m) for analysis. The western and eastern regions have distinct topographic, climatic, and ecological conditions (Das

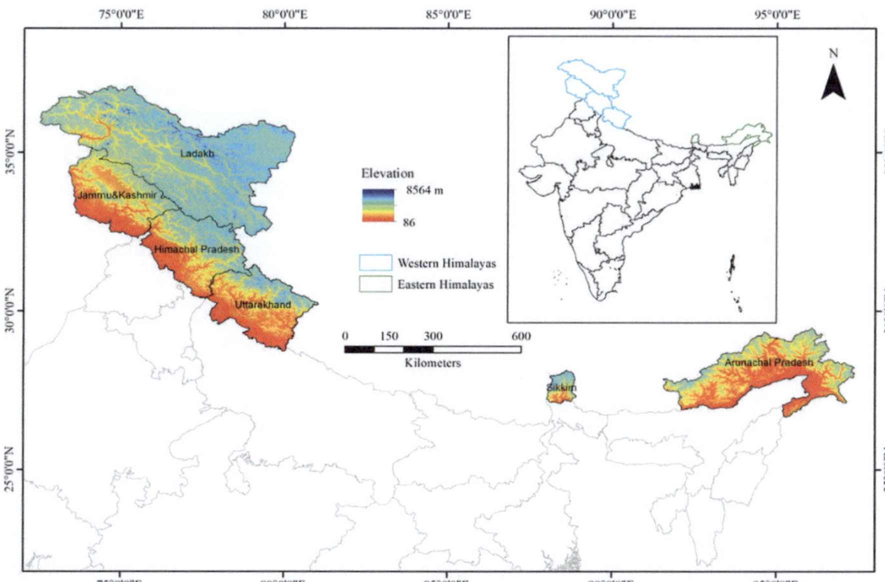

Fig. 9.1 Study area map of the eastern and western Indian Himalayan range

and Meher 2019). The ombrophilous broadleaf forests situated in the submontane zone (below 1000 m) of the eastern region resemble typical tropical rainforests (Behera et al. 2005). Conversely, at elevations exceeding 3000 m, the communities bear similarities to subalpine and alpine types. Moving from more favorable to less favorable environments, either due to decreasing moisture from east to west or declining temperatures from low to high elevations, results in a gradual shift in forest characteristics (Chitale et al. 2014). The forests progressively become more open, shorter in stature, and simpler in structure, with minimal vertical stratification (Singh and Singh 1987). The western region receives >75% of the total annual rainfall (100–150 cm) during the summer monsoon (June–September) (Upadhyaya et al. 2023), whereas the eastern region receives >60% of the total annual rainfall (200–250 cm) (Goswami and Prasad 2023).

Datasets

This study used the following MODIS land products for interannual and seasonal trend analysis: Gross Primary Productivity (GPP), Leaf Area Index (LAI), and Absorbed Photosynthetically Active Radiation (APAR).

Gross Primary Productivity

GPP holds a central role in the global carbon balance and influences nearly all ecosystem processes (Gilmanov et al. 2003). GPP is defined as the amount of carbon produced by photosynthesis, which is directly related to the absorbed photosynthetically active radiation (APAR) (Behera et al. 2017). By utilizing the MOD17A2H product, GPP can be equated to the concept of light use efficiency (LUE), which helps in estimating the exchange of carbon between vegetation and the atmosphere within an ecosystem. The MODIS GPP algorithm is based on a simple light use efficiency model (MOD17) and it requires daily inputs of incoming photosynthetically active radiation (PAR), minimum temperature over the 24 h period, and daytime average vapor pressure deficit. These meteorological data are provided by the NASA Data Assimilation Office (DAO) based on a general circulation model that is continuously assimilating observations from space and ground stations. Additional MOD17 inputs include FPAR (multiplied by PAR to get absorbed PAR) and LAI (used to estimate biomass for the purposes of estimating Ra), which are both standard MODIS products. The LUE of photosynthetically active vegetation is defined as the quantity of fixed carbon per unit of APAR (Chitale et al. 2012). The LUE approach is one of the most frequently used methods for estimating GPP (Varghese and Behera 2019).

Leaf Area Index

LAI is defined as the one-sided green leaf area per unit ground surface area for broadleaf canopies, and the maximum projected green leaf area per unit ground surface area in needleleaf canopies (Paramanik et al. 2022). LAI is a dimensionless quantity, and its estimated retrieval uncertainty is at the second level of processing of the 3-D canopy Radiative Transfer (RT) model (Myneni et al. 1997). This metric measures the quantity of foliage and regulates the rate at which water and carbon are exchanged with the atmosphere. In simpler terms, the LAI is the most critical metric for describing the state of a region's vegetation (Mudi et al. 2022). LAI values commonly range from 0 to 6, with 0 indicating bare land, 1 denoting a single layer of leaves covering the ground completely, and values exceeding 1 indicating a layered canopy with multiple leaf layers per unit ground surface area.

The MODIS LAI/FAPAR algorithms consist of a main Look-up-Table (LUT)-based procedure that exploits the spectral information content of the MODIS red (648 nm) and near-infrared (NIR, 858 nm) surface reflectance, and the back-up algorithm that uses empirical relationships between NDVI and canopy LAI and FAPAR. The LUT was generated using a 3D radiative transfer equation with the inputs such as (i) vegetation structural type, (ii) sun-sensor geometry, (iii) BRFs at red (648 nm) and near-infrared (NIR, 858 nm) spectral bands, and (iv) their uncertainties. LAI is the light interceptive surface area of a plant with a higher LAI value indicating more photosynthesis and green biomass.

Fraction of Photosynthetically Active Radiation (FPAR)

Photosynthetically active radiation (PAR) encompasses radiation within the spectral range of 400–700 nm and is utilized by plants for photosynthesis. The fraction of absorbed photosynthetically active radiation (FPAR) denotes the ratio of PAR absorbed by vegetation to the total PAR within an integrated plant canopy. FPAR is intricately associated with canopy processes such as photosynthesis, carbon assimilation, and evapotranspiration rates. It serves as a crucial biophysical parameter for estimating vegetation productivity, monitoring growth conditions, and analyzing climate change impacts (Mahanand et al. 2021). The determination of FPAR involves the integration of spectral absorptance across the wavelength interval of 0.4–0.7 μm (Mahanand et al. 2022). The FPAR data from MCD15A3 is used here. The MODIS LAI/FPAR algorithm selects the most optimal pixel from all Terra and Aqua sensor acquisitions within a 4-day composite.

Methodology

The current study focused on the spatiotemporal analysis of forest productivity and the associated biophysical variables with data acquired from MODIS biophysical data products from 2017 to 2023. The entire study area is divided into three altitudinal zones, low altitude (<1200 m), middle altitude (1200–3000 m), and high altitude (>3000 m) for analyzing the influence of elevation on vegetation productivity, LAI, and FPAR. The nonvegetated areas, water bodies, urban and built-up lands, and areas with permanent snow and ice are masked using the MODIS annual plant functional type-based land cover map (MCD12Q1). The time series analysis for all variables was carried out on the Google Earth Engine platform after quality control and preprocessing at the pixel level, generating seasonal and monthly averages of variables for eastern and western regions separately. The Mann-Kendall test and Sen's slope estimator are used to analyze the changes that occurred on a seasonal scale, and spatial maps are generated using ArcGIS. The seasonal average raster images (JJAS, DJF, and MAM) of each season are stacked separately for each variable, and Sen's slope estimator was applied after performing the Mann-Kendall test to identify the magnitude of change of each pixel during the study period. The correlation analysis among the ecosystem photosynthetic variables was carried out to understand the most important factors that determine the vegetation productivity of the Indian Himalaya.

Results

Spatiotemporal Analysis of the Biophysical Variables in the Eastern and Western Himalaya

This study analyzed the temporal variability and seasonal change of biophysical parameters such as LAI and FPAR, which are in−/directly linked to GPP, in three different elevation zones, between January 2017 and December 2023.

Gross Primary Productivity

The monthly mean GPP exhibits a clear and consistent annual variation pattern in all three elevation belts, with the highest GPP in the low altitude, followed by middle (1200–3000 m) and high (>3000 m). The highest daily GPP is observed during the monsoon season (JJAS). Gradually, it decreases during the postmonsoon period, reaching minima in winter (DJF), and then the trend reverses, with an increase in GPP during the premonsoon period (MAM). This well-defined annual variation in GPP reflects the influence of seasonal changes on plant photosynthetic activity, with maximum productivity coinciding with the favorable conditions of the monsoon. In other words, carbon fixation is most likely at its maxima during the monsoon and premonsoon periods (Varghese and Behera 2019).

The GPP in the Eastern Himalayas (EH) is highest in the low-altitude region (1200–3000 m), where evergreen and deciduous broadleaf forest types dominate. The GPP value (kgC/m^2) ranges from 0.02 kgC/m^2 (January 2019) to 0.08 kgC/m^2 (May 2019) in low altitude, 0.01 kgC/m^2 (January 2022) to 0.07 kgC/m^2 (July 2017) in mid altitude, and 0.001 kgC/m^2 (February 2022) to 0.03 kgC/m^2 (August 2022) in high-altitude zone as per the MOD17A2H data product (Fig. 9.2). The lowest GPP (kgC/m^2) measured in WH during the study period is 0.006 kgC/m^2 (April 2022) at low altitude, 0.014 kgC/m^2 (January 2019) at middle altitude and 0.0003 kgC/m^2 (January 2020) at high-altitude region. The highest values are 0.06 kgC/m^2 (August 2021), 0.07 kgC/m^2 (August 2022), and 0.04 kgC/m^2 (July 2022) at low, middle, and high altitudes respectively as estimated by the MOD17A2H data product (Fig. 9.3). In WH, GPP peaks during the summer monsoon and a secondary peak is observed during the winter precipitation that may be attributed to westerlies (December–February).

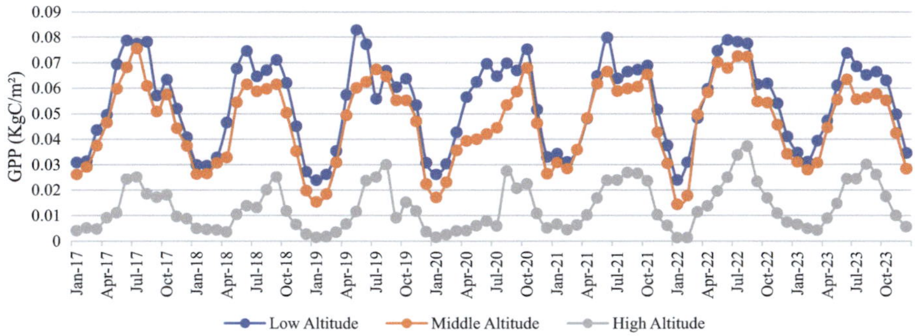

Fig. 9.2 Temporal change of monthly mean GPP over the eastern Himalaya (2017–2023)

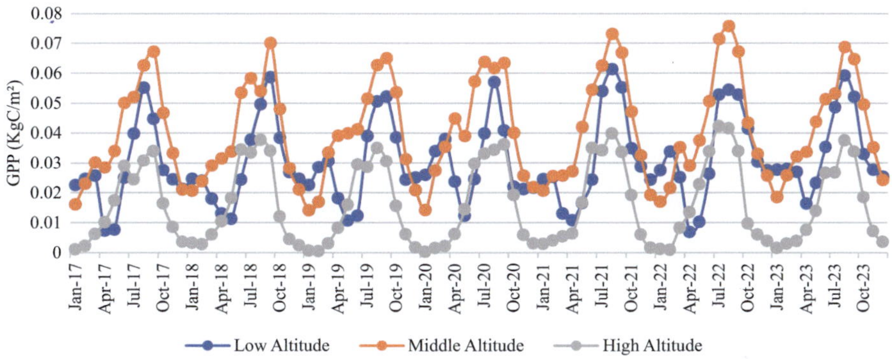

Fig. 9.3 Temporal change of monthly mean GPP over the western Himalaya (2017–2023)

The MODIS-based GPP in both regions drops to its lowest during the winter and reaches its peak during the monsoon. The seasonal fluctuations in vegetation productivity depend upon the vegetation type, edaphic, climatic, and other environmental factors (Kharkwal et al. 2005). The average GPP value in eastern and western Himalaya is 0.03 and 0.018 kgC/m^2 respectively, as estimated by the MOD17A2H data product.

Seasonal Variation in Gross Primary Productivity Over the IHR

The seasonal GPP at the pixel level reveals distinct trends across different regions and time frames. In EH, there is an overall positive trend in GPP observed during the premonsoon and summer monsoon (JJAS) periods, except for southern Sikkim (Fig. 9.4a). Conversely, during the winter monsoon period (Fig. 9.4b), most parts of EH exhibit a decline in GPP. In contrast, the WH exhibits an upward trend in GPP during the winter season. However, during the summer (premonsoon) months (MAM), vegetation productivity in WH exhibits a downward trend in middle altitude regions, while low-altitude regions display a positive trend (Fig. 9.4c). The magnitude of change in GPP varies seasonally, with the highest positive (0.08

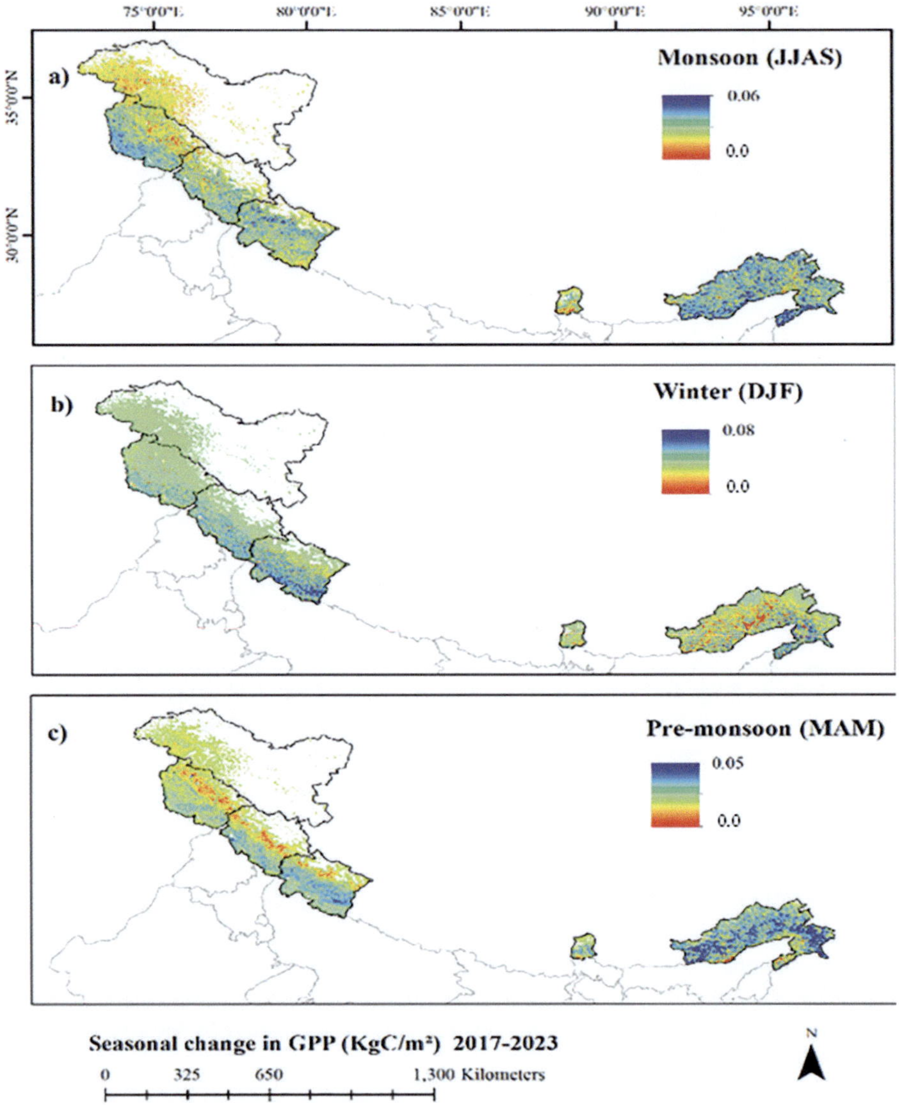

Fig. 9.4 Seasonal change in GPP over the eastern and western Himalaya (2017–2023)

kgC/m^2) change observed during the winter season as per the MOD17A2H data products. During JJAS, the values range up to 0.06 and 0.05 kgC/m^2 during the premonsoon period.

LUE stands out as a crucial factor in the assessment of GPP, with its dynamics influenced by various climatic factors, including rainfall, temperature, moisture/humidity, Photosynthetically Active Radiation (PAR), as well as the age of leaves and the plant itself. However, among these factors, PAR, temperature, and precipitation emerge as the dominant factors governing LUE. These factors play crucial roles in shaping the growth, development, and overall health of vegetation within forest ecosystems (Gwal et al. 2020; Koju et al. 2020). The significant difference in GPP between the EH and WH can be attributed to these parameters. The variation in GPP reveals that broad-leaved forests have higher productivity (Koju et al. 2020). The Eastern Himalayas experience increased precipitation and a shorter dry season, leading to significant effects on forest vegetation. The majority of wet forest types are predominantly located in the Indian Eastern Himalayas, with dense broadleaf evergreen forests until an elevation of about 900 m (Champion et al. 1968; Prasher et al. 2015). High LAI and FPAR values throughout the year in the tropical evergreen forests of EH compared to WH can be one reason for the noticeable variation in annual average GPP, along with other factors such as optimum LST and high-precipitation accumulation in EH during the summer monsoon period.

(a) Leaf Area Index

LAI is primarily directly proportional to transpiration and GPP, while being inversely proportional to canopy transmissivity (Asner et al. 2003). LAI is impacted by factors such as elevation (or temperature), water availability, soil fertility, and topography. Leaf characteristics play a crucial role in shaping both the composition and adaptation features of a community, directly and indirectly impacting the physiognomy of vegetation. These traits are instrumental in defining the overall characteristics of individual plants (Reich et al. 1992). Throughout the study period, LAI exhibited peaks during the monsoon season (JJAS) and a decline in winter (DJF), as shown in Figs. 9.5 and 9.6. Additionally, there were discernible ascending trends during the premonsoon period and descending trends in the postmonsoon phase as per the MODIS data.

In EH, the peak LAI occurs during the monsoon (August–September) and premonsoon period (March–April), and the decreasing pattern during the winter months. In the lower elevation forests of the Eastern Himalaya, trees exhibit

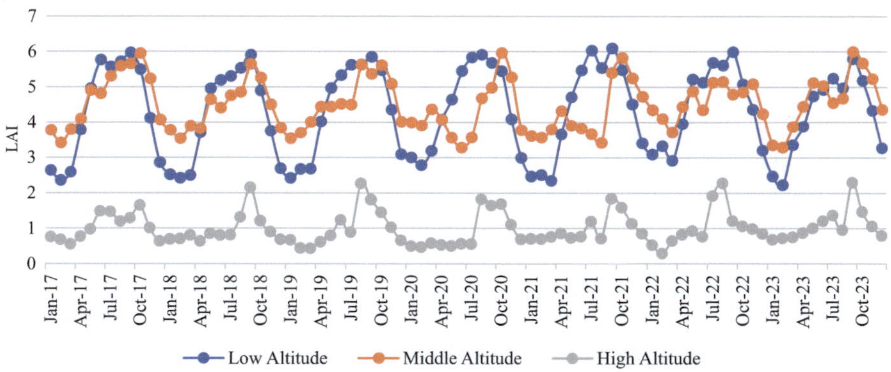

Fig. 9.5 Monthly mean LAI over the eastern Himalaya (2017–2023)

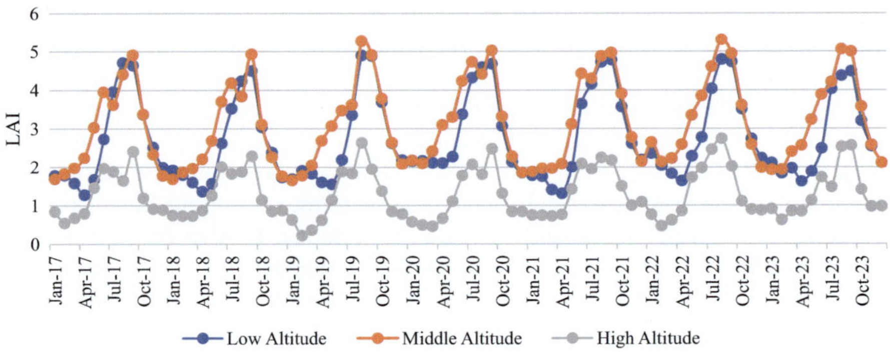

Fig. 9.6 Temporal change of monthly mean LAI over the western Himalaya (2017–2023)

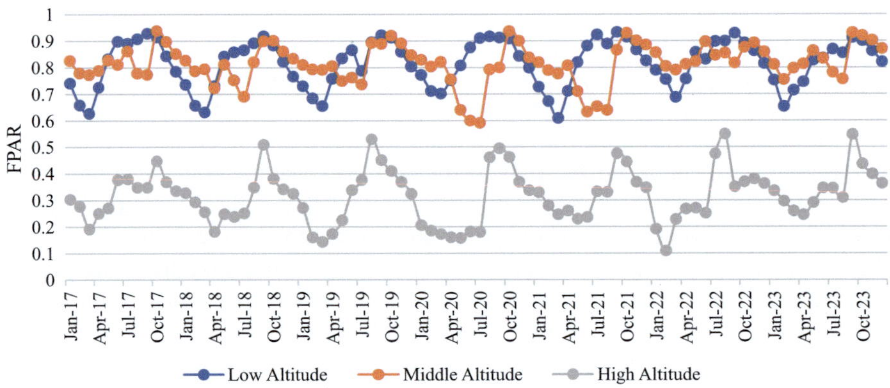

Fig. 9.7 Monthly mean FPAR over the eastern Himalaya (2017–2023)

evergreen characteristics and display multiple flushes throughout the year. Notably, the leaf life span of these species is typically less than 1 year (Devi and Garkoti 2013). Underestimation of LAI by MODIS pixels is also reported by several authors (Behera et al. 2024).

In the WH, during the summer monsoon months, the LAI peaks, reflecting the maximum foliage during the active growth phase. The patterns described by the monthly mean LAI values closely align with the onset and retreat of the southwest monsoon (Dash et al. 2010). In tropical and subtropical evergreen forests of the EH, a bimodal curve is evident, showing two peaks, one during the postwinter months and the other during the monsoon. The average LAI in EH was observed as 4.31 in low altitude and 4.49 in mid altitude, whereas in WH, it was 2.74, 3.11, and 1.28 in low-, mid-, and high-elevation belts, respectively as per the MODIS data.

(b) Fraction of Absorbed Photosynthetically Active Radiation

The seasonal fluctuation of FPAR in EH and WH is shown in Figs. 9.7 and 9.8, with varying magnitudes of absorbed PAR. Owing to the high LAI values in EH, the

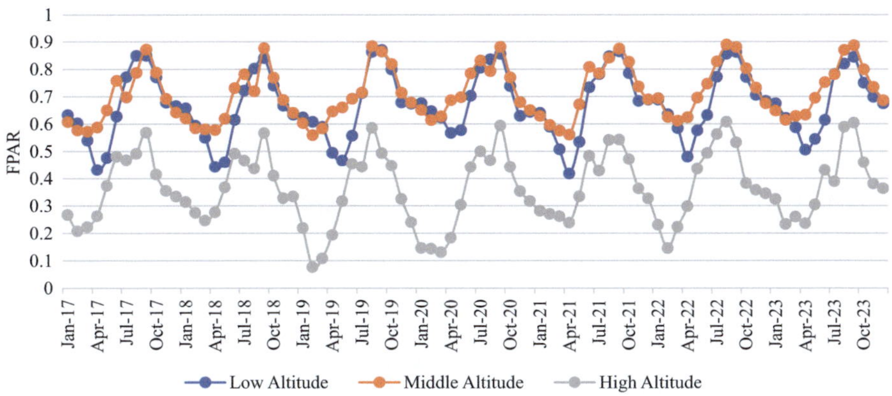

Fig. 9.8 Monthly mean FPAR over the western Himalaya (2017–2023)

fraction of PAR absorbed by the vegetation will be relatively higher compared to the WH. The mean value of FPAR during the study period in the EH is 0.648 and 0.581 in the WH as per the MODIS data. In the EH, the dense evergreen broadleaf forests contribute to a constant layer of foliage all year (Seyoum et al. 2012), resulting in higher FPAR values, with peaks occurring during the premonsoon and summer monsoon months. In EH and WH, the FPAR values are the highest during the monsoon, slowly declining through the prewinter months, reaching their lowest during the winter.

The seasonal fluctuation in LAI represents the principal biophysical expression of vegetation phenology. Phenological changes in LAI, coupled with the seasonal variation in solar geometry, play a pivotal role in driving concurrent variation in the fraction of photosynthetically active radiation absorbed by vegetation (Baret and Guyot 1991).

(c) Seasonal Change in Leaf Area Index Over the IHR

The pixel-level analysis of seasonal variations in LAI across the study area is depicted in (Fig. 9.6). Throughout all three seasons, the EH region notably displays a negative trend in LAI, particularly pronounced during the winter and premonsoon periods. During the premonsoon season, there is a significant upward trend in LAI observed in the southern parts of Uttarakhand and the southwest region of Himachal Pradesh, dominated by deciduous and evergreen broadleaf forests. Conversely, the rest of the WH region exhibits a declining trend in LAI during this period. During the winter monsoon months, with the exception of sparsely vegetated areas in Ladakh, most parts of WH exhibit a positive trend in LAI. In the northern parts of Uttarakhand, dominated by oak forests, a notable upward trend in LAI is evident during the summer monsoon, whereas other areas in WH either exhibit a downward trend or a weak positive trend (Table 9.1).

Table 9.1 Correlation of GPP with other variables over the Indian Himalayan range

	LAI	FPAR
GPP (MAM)	0.78	0.59
GPP (JJAS)	0.54	0.61
GPP (DJF)	0.68	0.89

Interlinkage of Productivity-Dependent Variables Over the Indian Himalayan Range

The productivity of an ecosystem is influenced by a complex interplay of factors. However, in the Indian Himalaya, these interactions vary significantly in the Eastern and Western regions due to distinct climate, vegetation type, and topography. Understanding these interlinkages will be essential for better explaining the ecosystem productivity and for predicting the impacts of climate change on these sensitive and diverse ecosystems. During the premonsoon period, GPP exhibits the highest positive correlation with LAI (0.78) and FPAR (0.59).

Discussion

The satellite-derived time-series data on vegetation greenness to analyze seasonal trends in vegetation across the Uttarakhand Himalayas identified seasonal trends that are closely linked to changes in LULC and/or shifts in climatic drivers, such as alterations in precipitation and temperature regimes (Mishra and Chaudhuri 2015). Limited knowledge exists regarding the structure and function of the high-altitude forests in the Himalayas. These ecosystems thrive in an environment that is distinct from the well-documented subalpine and alpine communities. Establishing baseline information on the structural and functional aspects of these high-altitude forests is crucial. This baseline data will facilitate the assessment of future changes in ecosystem characteristics, especially in light of climate change. Given that these forests serve as significant carbon sinks and experience relatively lower human impacts, understanding their dynamics becomes even more essential (Garkoti 2008).

The LAI values in the WH show a single peak during the monsoon season of every year, whereas in the eastern Himalaya, a bimodal curve exists in monthly mean LAI values, with one peak in monsoon and the other during summer, which is consistent with the findings of Ralhan and Singh (1987). The peak LAI occurs in September/October, reflecting the climax of the growing season, then declines during November and December due to leaf fall, reaching its lowest values. This cyclical pattern highlights the seasonal fluctuations in vegetation cover, capturing the dynamic nature of leaf growth and senescence (Chhabra and Panigrahy 2011). In subtropical evergreen forests, leaf initiation starts in February, peaks in May before premonsoon showers, and reaches high LAI values during September and October. It is attributed to vigorous vegetation growth following the monsoon (Chhabra and Panigrahy 2011). The MODIS LAI values and other meteorological variables used for the assessment of GPP in the tropical forest ecosystem of North-East India (Deb Burman et al. 2017) revealed similar LAI patterns to those observed in the present study over EH. The WH with varying climatic conditions, including colder temperatures and lower precipitation compared to EH, influences the seasonal dynamics of FPAR, particularly during the growing season of vegetation. The dense evergreen forests of the EH contribute to relatively higher LAI and FPAR values compared to the Western Himalayan states.

Conclusions

Understanding the forest ecosystem dynamics is crucial for maintaining the Earth's ecological system and its function. This study examined the Indian Himalayan stretch from east to west. It stands out as a highly under-explored system in remote sensing studies. This vast region is marked by extensive ecological complexities and a rapidly changing climate and is recognized as a biodiversity hotspot that sustains the livelihoods of millions of people. Therefore, the findings of this study can offer insights into the interlinkage between various biophysical, edaphic, and climatic variables that control vegetation productivity. Seasonal and annual fluctuations in these parameters were analyzed separately for both regions from 2017 to 2023. Factors such as extreme temperatures, differences in forest type, precipitation regimes, and nutrient limitations can increase the energy expended by plants for physiological adaptations and survival, which can be the reason for low net primary productivity in the WH compared to EH. The distinct monthly mean LAI patterns align with the vegetation phenology and the onset and retreat of the southwest monsoon. These variations, coupled with seasonal variations in solar geometry, influence the absorption of photosynthetically active radiation. In subtropical evergreen and deciduous forests, seasonal fluctuations in LAI are one of the key determinants of evapotranspiration rates, alongside edaphic and climatic factors. This study demonstrated the potential of utilizing open-source satellite-derived products for the spatially explicit representation of vegetation dynamics in the highly biodiverse but understudied Himalayan region.

LAI is one of the most essential biophysical variables as it directly regulates the physiological process and contributes to vegetation dynamics models. It plays an essential role in forest cover and primary productivity monitoring and has been identified as one of the climate variables that support climate change studies. Remote sensing technologies have emerged as one of the most effective sensing techniques for measuring LAI in mountain regions and consequently identifying its dynamic characteristics, where LAI is mainly dependent on forest structure. Optical sensors onboard remote sensing satellites measure the canopy greenness via spectral reflectance, and the band-specific reflectance is used in empirical biophysical models for LAI estimation. Optical data, on the other hand, has several constraints for estimating LAI measurements. Acquiring quality data is challenging due to cloud conditions, especially in mountainous terrain. Sentinel-2 sensor has three narrow bands in the red-edge spectrum, and higher spatiotemporal resolution significantly impacts biophysical variables with the highest correlation, including forest fuel. Thus, Sentinel-2 offers prospects for algorithm development and prediction of various climate variables, burned area mapping, species mapping, and other emerging applications (Mudi et al. 2022). Researchers have utilized Landsat and Sentinel data for LAI estimation and reported on the efficacy of these two open sources of satellite datasets.

Primary productivity, either gross primary productivity (GPP) or net primary productivity (NPP), is an essential ecological parameter for evaluating the efficiency of terrestrial ecosystems for carbon transfer from the atmosphere to biomass. On a global scale, estimation of GPP, the total amount of carbon fixed by terrestrial vegetation per unit time via the photosynthetic chopping of CO_2 into organic matter, provides information about the rate at which carbon is assimilated by an ecosystem and the influence of climate change. Tracing large-scale fluctuations in GPP will give insights into the feedback of global ecosystems to climate change and catch shifts in biomes. Estimation of GPP helps evaluate the photosynthetic levels of various ecosystems such as forests and croplands, which in turn help manage these systems. Even though a wide variety of carbon pools exists on earth, the vegetation of terrestrial ecosystems plays a vital role in the intake of atmospheric CO_2. Gross primary productivity (GPP) is a vital ecosystem variable that is used as a proxy to study the functional behavior of a terrestrial ecosystem and its ability to regulate atmospheric CO_2 by working as a carbon pool. The Himalayas, having the potential terrestrial ecosystem dynamics to absorb the atmospheric carbon dioxide to some extent, is one of the least-explored regions in terms of carbon monitoring studies.

References

Asner GP, Scurlock JMO, Hicke JA (2003) Global synthesis of leaf area index observations: Implications for ecological and remote sensing studies. Glob Ecol Biogeogr 12(3):191–205. https://doi.org/10.1046/j.1466-822X.2003.00026.x

Baret F, Guyot G (1991) Potentials and limits of vegetation indices of LAI and APAR assessment. Remote Sens Environ 35:161–173

Behera MD (1999) Remote sensing and environment. Employ News 24(13):1–2

Behera MD, Kushwaha SPS, Roy PS (2005) Rapid assessment of biological richness in a part of Eastern Himalaya: an integrated three-tier approach. For Ecol Manag 207(3):363–384. https://doi.org/10.1016/j.foreco.2004.10.070

Behera SK, Sahu N, Mishra AK, Bargali SS, Behera MD, Tuli R (2017) Aboveground biomass and carbon stock assessment in Indian tropical deciduous forest and relationship with stand structural attributes. Ecol Eng 99:513–524. https://doi.org/10.1016/j.ecoleng.2016.11.046

Behera MD, Krishna JSR, Paramanik S, Kumar S, Behera SK, Anto S, Singh SN, Verma AK, Barik SK, Mohanta MR, Sahu SC, Jeganathan C, Srivastava PK, Pradhan B (2024) Digital hemispherical photographs and Sentinel-2 multi-spectral imagery for mapping leaf area index at regional scale over a tropical deciduous forest. Trop Ecol 65(2):258–270. https://doi.org/10.1007/s42965-024-00327-y

Champion HG, Seth SK et al (1968) A revised survey of the forest types of India. Govt. of India Press, New Delhi

Chandra Sekar K, Rawal RS, Chaudhery A, Pandey A, Rawat G, Bajapai O, Joshi B, Bisht K, Mishra BM (2017) First GLORIA site in Indian Himalayan Region: towards addressing issue of long-term data deficiency in the Himalaya. Natl Acad Sci Lett 40(5):355–357. https://doi.org/10.1007/s40009-017-0584-z

Chhabra A, Panigrahy S (2011) Analysis of spatio-temporal patterns of leaf area index in different forest types of India using high temporal remote sensing data. https://isprs-archives.copernicus.org/articles/XXXVIII-8-W20/119/2011/

Chitale VS, Tripathi P, Behera MD, Behera SK, Tuli R (2012) On the relationships among diversity, productivity and climate from an Indian tropical ecosystem: a preliminary investigation. Biodivers Conserv 21(5):1177–1197. https://doi.org/10.1007/s10531-012-0247-9

Chitale VS, Behera MD, Roy PS (2014) Future of endemic flora of biodiversity hotspots in India. PLoS One 9(12):e115264. https://doi.org/10.1371/journal.pone.0115264

Das L, Meher JK (2019) Drivers of climate over the Western Himalayan region of India: a review. Earth-Sci Rev 198:102935. https://doi.org/10.1016/j.earscirev.2019.102935

Dash J, Jeganathan C, Atkinson PM (2010) The use of MERIS terrestrial chlorophyll index to study spatio-temporal variation in vegetation phenology over India. Remote Sens Environ 114(7):1388–1402. https://doi.org/10.1016/j.rse.2010.01.021

Dash J, Behera MD, Jeganathan C, Jha CS, Sharma S, Lucas R, Khuroo AA, Harris A, Atkinson PM, Boyd DS, Singh CP, Kale MP, Kumar P, Behera SK, Chitale VS, Jayakumar S, Sharma LK, Pandey AC, Avishek K et al (2020) India's contribution to mitigating the impacts of climate change through vegetation management. Trop Ecol 61(1):168–171. https://doi.org/10.1007/s42965-020-00075-9

Deb Burman PK, Sarma D, Williams M, Karipot A, Chakraborty S (2017) Estimating gross primary productivity of a tropical forest ecosystem over north-east India using LAI and meteorological variables. J Earth Syst Sci 126(7):99. https://doi.org/10.1007/s12040-017-0874-3

Devi AF, Garkoti SC (2013) Variation in evergreen and deciduous species leaf phenology in Assam, India. Trees – Struct Funct 27(4):985–997. https://doi.org/10.1007/s00468-013-0850-8

Garkoti SC (2008) Estimates of biomass and primary productivity in a high-altitude maple forest of the west central Himalayas. Ecol Res 23(1):41–49. https://doi.org/10.1007/s11284-007-0355-2

Gautam Govinda MR, Timilsina R, Acharya K (2013) An assessment of potential climate change impacts on the Himalayas. http://econ.worldbank.org

Ghosh SM, Behera MD (2018) Aboveground biomass estimation using multi-sensor data synergy and machine learning algorithms in a dense tropical forest. Appl Geogr 96:29–40. https://doi.org/10.1016/j.apgeog.2018.05.011

Gilmanov TG, Verma SB, Sims PL, Meyers TP, Bradford JA, Burba GG, Suyker AE (2003) Gross primary production and light response parameters of four Southern Plains ecosystems estimated using long-term CO2-flux tower measurements. Glob Biogeochem Cycles 17(2). https://doi.org/10.1029/2002GB002023

Goswami G, Prasad RK (2023) Trend analysis of rainfall pattern in Arunachal Pradesh (India). Environ Model Assess 28(6):1093–1125. https://doi.org/10.1007/s10666-023-09903-3

Gwal S, Singh S, Gupta S, Anand S (2020) Understanding forest biomass and net primary productivity in Himalayan ecosystem using geospatial approach. Model Earth Syst Environ 6(4):2517–2534. https://doi.org/10.1007/s40808-020-00844-4

Jeganathan C, Singh B, Singh CP, Behera MD, Srivastava S, Natesha SR, Singh K, Ranjan R, Prakash M, Kumar A, Pandya MR, Bhattacharya BK, Krishna AP, Lala MGN, Rathore VS, Sinha NK, Choudhary K, Bhuyan M, Singh SS, Sardar P (2024) Integrated use of field sensors, PhenoCam, and satellite data for pheno-phase monitoring in a tropical deciduous forest of Dalma Wildlife Sanctuary, Jharkhand, India: initial results from the Indian Phenology Network. Biodivers Conserv 33(12):3539–3565. https://doi.org/10.1007/s10531-024-02889-8

Kharkwal G, Mehrotra P, Rawat YS, Pangtey YPS (2005) Phytodiversity and growth form in relation to altitudinal gradient in the Central Himalayan (Kumaun) region of India. 89(5) https://www.jstor.org/stable/24111035

Koju UA, Zhang J, Maharjan S, Bai Y, Zhang S, Yao F (2020) Analysis of spatiotemporal dynamics of forest net primary productivity of Nepal during 2000–2015. Int J Remote Sens 41(11):4336–4364. https://doi.org/10.1080/01431161.2020.1717667

Mahanand S, Behera, Roy PS, Kumar P, Barik SK and Srivastava PK (2021) Satellite based fraction of absorbed photosynthetically active radiation is congruent with plant diversity in India. Remote Sens 13(2):159. https://doi.org/10.3390/rs13020159

Mahanand S, Behera MD, Roy PS (2022) Rapid assessment of plant diversity using MODIS biophysical proxies. J Environ Manag 311:114778. https://doi.org/10.1016/j.jenvman.2022.114778

Mishra NB, Chaudhuri G (2015) Spatio-temporal analysis of trends in seasonal vegetation productivity across Uttarakhand, Indian Himalayas, 2000–2014. Appl Geogr 56:29–41. https://doi.org/10.1016/j.apgeog.2014.10.007

Mudi S, Paramanik S, Behera MD, Prakash AJ, Deep NR, Kale MP, Kumar S, Sharma N, Pradhan P, Chavan M, Roy PS, Shrestha DG (2022) Moderate resolution LAI prediction using Sentinel-2 satellite data and indirect field measurements in Sikkim Himalaya. Environ Monit Assess 194(12):897. https://doi.org/10.1007/s10661-022-10530-w

Myneni RB, Ramakrishna R, Nemani R, Running SW (1997) Estimation of global leaf area index and absorbed par using radiative transfer models. IEEE Trans Geosci Remote Sens 35(6): 1380–1393. https://doi.org/10.1109/36.649788

Negi VS, Pathak R, Rawal RS, Bhatt ID, Sharma S (2019) Long-term ecological monitoring on forest ecosystems in Indian Himalayan Region: criteria and indicator approach. Ecol Indic 102: 374–381. https://doi.org/10.1016/j.ecolind.2019.02.035

Paramanik S, Behera MD, Dash J (2022) Symbolic regression-based allometric model development of a mangrove forest LAI using structural variables and digital hemispherical photography. Appl Geogr 139:102649. https://doi.org/10.1016/j.apgeog.2022.102649

Prasher IB et al (2015) Wood-rotting non-gilled Agaricomycetes of Himalayas. Springer

Quillet A, Peng C, Garneau M (2010) Toward dynamic global vegetation models for simulating vegetation–climate interactions and feedbacks: recent developments, limitations, and future challenges. Environ Rev 18(NA):333–353. https://doi.org/10.1139/A10-016

Ralhan PK, Singh SP (1987) Dynamics of nutrients and leaf mass in central Himalayan forest trees and shrubs. Ecology 68(6):1974–1983

Reich PB, Walters MB, Ellsworth DS (1992) Leaf life-span in relation to leaf, plant, and stand characteristics among diverse ecosystems. Ecol Monogr 62(3):365–392

Seyoum Y, Fetene M, Strobl S, Beck E (2012) Foliage dynamics, leaf traits, and growth of coexisting evergreen and deciduous trees in a tropical montane forest in Ethiopia. Trees – Structure and Function 26(5):1495–1512. https://doi.org/10.1007/s00468-012-0723-6

Singh JS, Singh SP (1987) Forest vegetation of the Himalaya. Bot Rev 53(1):80–192. https://doi.org/10.1007/BF02858183

Singh SP, Thadani R (2023) A leaf-based classification for Himalayan forests. Int J Ecol Environ Sci 50(1):11–31. https://doi.org/10.55863/ijees.2024.3250

Singh SP, Daman Singh R, Gumber S, Bhatt S (2017) Two principal precipitation regimes in Himalayas and their influence on tree distribution. Trop Ecol 58(4):679–691. www.tropecol.com

Upadhyaya A, Rai AK, Kumar P (2023) Anomalous rainfall trends in the North-Western Indian Himalayan Region (NW-IHR). Theor Appl Climatol 151(1):253–272. https://doi.org/10.1007/s00704-022-04280-5

Varghese R, Behera MD (2019) Annual and seasonal variations in gross primary productivity across the agro-climatic regions in India. Environ Monit Assess 191(10):631. https://doi.org/10.1007/s10661-019-7796-2

Vijith H, Dodge-Wan D (2020) Applicability of MODIS land cover and enhanced vegetation index (EVI) for the assessment of spatial and temporal changes in strength of vegetation in tropical rainforest region of Borneo. Remote Sensing Appl Soc Environ 18:100311. https://doi.org/10.1016/j.rsase.2020.100311

Tree Water Relations of Monsoonal Himalayas

10

S. P. Singh

Abstract

This chapter on tree water relations of the Himalayas is based on the trees of three regions with monsoonal precipitation regime: lower elevation forests and also treeline forests of Kumaun, Uttarakhand; and Kathmandu Valley forests, Nepal. The parameters studied were soil Ψ, tree Ψ, osmotic adjustment, and leaf conductance across the seasons. In Kumaun forests, from monsoon to premonsoon, average soil Ψ across all sites decreased by 5.4 times and average tree species Ψ_{PD} by 2.75 times. In Kumaun forests, all species, but *Shorea robusta,* experienced less than −1 MPa Ψ_{PD} values, in the treeline, 5 out of the 70 instances were lower than −1 MPa, and in Kathmandu, none of the species showed −1 MPa or less Ψ_{PD}. In the monsoon season, Ψ_{PD} was highest and Ψ_{Δ} lowest in Kumaun trees, while in treeline, both Ψ_{PD} and Ψ_{Δ} (Ψ_{PD}—Ψ_{MD}) were the highest. In response to drought, chir pine closed its stomata while Ψ_{PD} was still high (isohydric), while oaks kept stomata open up to much lower Ψ_{PD} (anisohydric). Broad-leaved evergreen species, including oaks, showed high osmotic adjustment across the species, it was 1.11 at Ψ_{sf} and Ψ_{sz} in Kumaun and Ψ_{sf} 1.30 and Ψ_{sf} 1.77 in the treeline. Leaf conductance across all seasons and species was between 60 and 500 m mol m^{-2} s^{-1}. However, extraordinarily high conductance rate (>1000 mmol m^{-2} s^{-1}) occurred during autumn in dry Chitkul sites of the Inner Himalayas. Water stored in a wood plays a critical role in leafing and flowering of some of the species of Kathmandu. In brief, in the monsoonal precipitation regime of the Himalayas, water stress is seasonal and mild, but it is enough to cause the dominance of species with sclerophyllous leaves.

S. P. Singh (✉)
Graphic Era (deemed to be) University, Dehradun, Uttarakhand, India

S. P. Singh, R. Thadani (eds.), *Himalayan Forests in a Changing World*,
https://doi.org/10.1007/978-981-95-0978-2_10

Keywords

Deciduous · Evergreen · Himalayas · Leaf conductance · Osmotic potential · Tree · Water Potential

Introduction

Water relations are key to understanding the ecology of terrestrial plant communities (Prieto et al. 2012). Across much of the Himalayas, precipitation is driven by the Indian Summer Monsoon, which accounts for 70–80% of yearly precipitation, typically from June to September. This warm and wet period is preceded by a long, premonsoon drought from March to May. In monsoon-dominated regions, even winters (December–February) are dry, the percentage of annual precipitation being less than 10% (Singh et al. 2017). Trees in such seasonal climates often experience frequent water deficiencies even when annual precipitation is not low. The warm and dry premonsoon (generally March–May) period is phenologically a critical time for many dominant Himalayan tree species, which put forth new shoots and flowers along with the simultaneous shedding of old leaves during this period (Ralhan et al. 1985; Negi and Singh 1992).

With the revolutionization of the measurements of plant water potential during the 1960s–1980s due to the development of the pressure chamber by Scholander et al. (1965), information on tree water relations progressed rapidly in different parts of the world (Landsberg and Waring 2017).

Because of climate change, trees are expected to be more at drought-induced risk in the future. During droughts, xylem embolism can be lethal; however, species that can refill embolized xylem and recover hydraulic function may survive (Zeppel et al. 2019).

However, the likely controls of embolism recovery, including changes in non-structural carbohydrates (NSCs), stomatal conductance, and hydraulic traits, are unknown (Klein et al. 2018; Zeppel et al. 2019). Evidence suggests that NSCs and xylem recovery are positively correlated (Zeppel et al. 2019).

Research on tree water relations in the Himalayas largely developed during the 1990s, when Prof. Don Zobel of Oregon State University (Corvallis, USA) established collaborative research with some Himalayan universities (see Zobel and Singh 1995).

In this chapter, tree water relations analysis is largely based on studies conducted on forest trees in two regions of the Central Himalayas, namely, Uttarakhand state of India and Kathmandu Valley of Nepal. Most studies were on lower elevation forests (~500–2500 m), and treeline species of Tungnath (>3000 m), Uttarakhand. Both in lower elevation forests and treelines of Uttarakhand and Nepal, premonsoon drought is a common feature. However, because of snow melt, the premonsoon water stress is likely to be less severe in treeline areas than in forest areas.

In this chapter, first we have described the main features of tree (and soil) water relations based on studies on Kumaun, Uttarakhand forests. Then, we have

Table 10.1 A list of tree species investigated for water relations parameters and the instruments used

A. List of tree species
Kumaun (29°23 N 79°27′ E) forest trees (~500–2500 m)
Sal (*Shorea robusta*) e^1 generally in foothills; Chir pine (*Pinus roxburghii*) e^1 between 800 and 2000 m; Oaks e^1 (*Quercus leucotrichophora*, *Q. floribunda*, *Q. lanata*); *Q. semecarpifolia*; *Rhododendron arboreum* e^2; *Machilus duthei* e^2; *Fraxinus micrantha*^{d}; *Cornus macrophylla*^{d}; *Carpinus viminea*^{d}; *Cupressus torulosa* e^{my}
Kathmandu (27°33′N, 85°22′E) forest (1400–2715 m) (Nepal)
Q. lanata e^1, *Q. semecarpifolia* e^1, *Castanopsis indica* e^1, *Schima wallichi* e^1, *R. arboreum* e^2
Treeline, Tungnath (30°49′22″N, 79°21′47″E) (3200–3560 m) (Uttarakhand)
Q. semecarpifolia e^1, *Abies spectabilis* e^{my}, *R. arboreum* e^2, *R. campanulatum* e^2, *Betula utilis*^{d}
B. Parameters investigated along with instruments used (in parentheses)
Tree water potential (Pressure chamber) Soil water potential (Psychrometers) Leaf conductance (Porometer) Hydraulic conductance (High-pressure flow meter) *Seasons of data collection*: monsoon (mid-June to mid-September, wet and warm, also referred to as rainy season), autumn (mid-September to mid-November, dry and cool), early winter (mid-November to mid-January), late winter or spring (mid-January to February), early premonsoon (March to April, dry and warm, also referred to as early summer), and late premonsoon (May to mid-June, also referred to as summer).

e^1 evergreen with leaf life span about 1 year, e^2 evergreen with leaf life span between 1 and 2 years, e^{my} evergreen with leaf life span of 3–4 years, and d deciduous species

compared tree water relations of Nepal forests and treelines of Tungnath (located in Uttarakhand) with those of Kumaun forests (Table 10.1). Kumaun and Nepal both receive about 80% annual rainfall due to the Indian Summer Monsoon, and fairly well represent the tree water relations of the outer Central Himalayan ranges.

Both in Kumaun and Nepal, evergreen tree species with just over one-year leaf longevity (e^1) dominate the forests. Because of a relatively longer monsoon period, seasonal water stress should be moderate in Kathmandu, Nepal, compared to that of Kumaun, India. Treeline formation is driven by heat deficiency, and water is seldom a limiting factor for the growth of trees (Körner 2012). However, as trees of treelines are adapted to moist conditions, even moderate water stress may result in a marked reduction in transpiration (Anfodillo et al. 1998). Furthermore, climatic warming in Alpine areas is known to increase root growth and sap density in trees (Wieser et al. 2014). Sap flow techniques have considerably developed in recent decades, allowing the monitoring of tree transpiration in remote treelines (Brito et al. 2014). The science of water relations is already fairly mature (Landsberg and Waring 2017); however, Himalayas and India-specific knowledge is scarce. Here we have tried to find out the patterns in water relation attributes of Himalayan trees, particularly in relation to seasonal changes. Our focus is on the water stress that the premonsoon (March to mid-June), the warm dry period of the year, imposes on trees. However, trees are also subject to water stress during winters.

While discussing various aspects of tree water relation ecology, we have made comparisons between various tree species groups, such as oak and pine, forest and treeline species, and Kumaun and Nepal species.

Kumaun Forest Trees

Water relations of 11 important species distributed between 500 and 2310 m elevation in the Kumaun Himalayas were analyzed. It included water potential (Ψ) of soil and trees, osmotic potential (Ψ_s) and osmotic adjustment, an index of tissue elasticity (ε), and leaf conductance.

The important forest species of the region which were studied, included the dominant canopy evergreen species, which have a leaf life span of about 1 year: sal (*Shorea robusta*), three species of oak, namely, banj (*Q. leucotrichophora),* moru or tilonj *oak (Q. floribunda),* and rianj oak (*Q. lanata*), two undercanopy broad-leaved evergreen species (*R. arboreum* and *M. duthei*), and a long-needled pine (*P. roxburghii*). A cypress (*C. torulosa*) and also three broad-leaved winter deciduous species (*Fraxinus micrantha*, *Cornus macrophylla,* and *Carpinus viminea*) were also part of the study. Hereafter, in this chapter, we will use their common names or the name of genera with a single species. The oaks were sclerophyllous with specific leaf mass between ~135 and 200 g m^{-2}. Of the two conifers, while chir pine grows widely throughout the western and central Himalayas in outer ranges with direct monsoon exposure, the *Cupressus* occurs sporadically with a highly restricted distribution between 2000 and 2300 m. The three deciduous species generally are part of mixed forests above 1800 m, only occasionally *Carpinus* forms mono-dominant stands. We have considered the timing and severity of water stress in relation to season and phenological traits. While local communities often associate chir pine with excessive water consumption, and drying of soil and extinction of springs, scientific literature regards pines as conservative users of water, as they maintain relatively wide safety margin and close stomata to avoid water loss while water stress is moderate (Brodribb and Cochard 2009; Blackman et al. 2019).

Soil Water Potential

From the end of the monsoon (July–September) to the premonsoon (March–May) of the following year, soil Ψ had a declining trend at all forest sites, except during late winter when it rose somewhat, because of the rainstorms during the study years. The range of average Ψ_{soil} values across the study sites increased seasonally 2.49 times from monsoon (−0.7 MPa) to premonsoon (−1.72 MPa), thus indicating that the sites dried out to different extents after monsoon rains. On average across the sites, premonsoon soil was 5.4 times drier than monsoon soil, while trees on average showed a 2.75 times decline in Ψ_{PD}. Comparison between the chir pine and banj oak forest sites showed that they had similar seasonal soil Ψ patterns and values (Fig. 10.1).

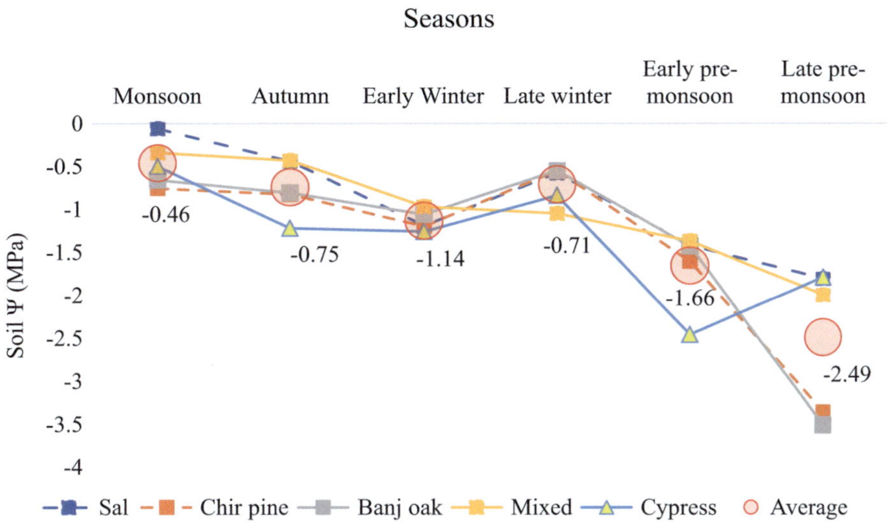

Fig. 10.1 Seasonal pattern of soil water potential ($\Psi_{60\ cm}$) for the study forest types. Average seasonal values are given in the figure. (Derived from Singh et al. 2006)

Tree Water Potential

Tree Ψ_{PD} also decreased from the peak in monsoon to premonsoon, with some small aberrations. The range in Ψ_{PD} across tree species in premonsoon was 2.64 times wider than in monsoon, which corresponds to that of soil Ψ (Fig. 10.2).

The lowest seasonal predawn water potential (Ψ_{PD}) values of the species are moderately severe, compared to trees in other climatic zones. It ranged from −0.84 MPa in foothill sal trees to −2.38 MPa in banj oak (*Q. leucotrichophora*) (Fig. 10.3), and was −2 MPa or less in rianj oak (*Q. lanata*), *Rhododendron,* and *Fraxinus* (Fig. 10.3). In sal, Ψ_{PD} remained above –1 MPa in all seasons and years, while in the rest of the species, the percentage of values ≤–1 MPa ranged from 15% in chir pine to 46% in *Fraxinus*. Among the oaks, this percentage was much higher in *Q. lanata* (45%) and *Q. floribunda* (38%) than in banj oak (18%).

In all the seasons, the highest Ψ_{PD} (least negative) was of *Shorea*, while the species with the lowest Ψ_{PD} varied with seasons: it was of *Q. floribunda* (−0.52 MPa) in monsoon, *Cornus* (−0.86 MPa) in autumn, *Q. lanata* (−1.24 and −1.33 MPa) in early and late winter, and *Carpinus* (−1.12 and −1.36 MPa) in early and late premonsoon.

Tree Ψ_{PD} was the highest in monsoon in all the 11 tree species, while the lowest Ψ_{PD} occurred in late premonsoon (May–June, just before monsoon) in six species, in winter in four species, and in autumn in one species. The lowest winter Ψ_{PD} was recorded for the species occurring toward the higher end of the elevational range, such as *Cornus* and *Q. lanata*.

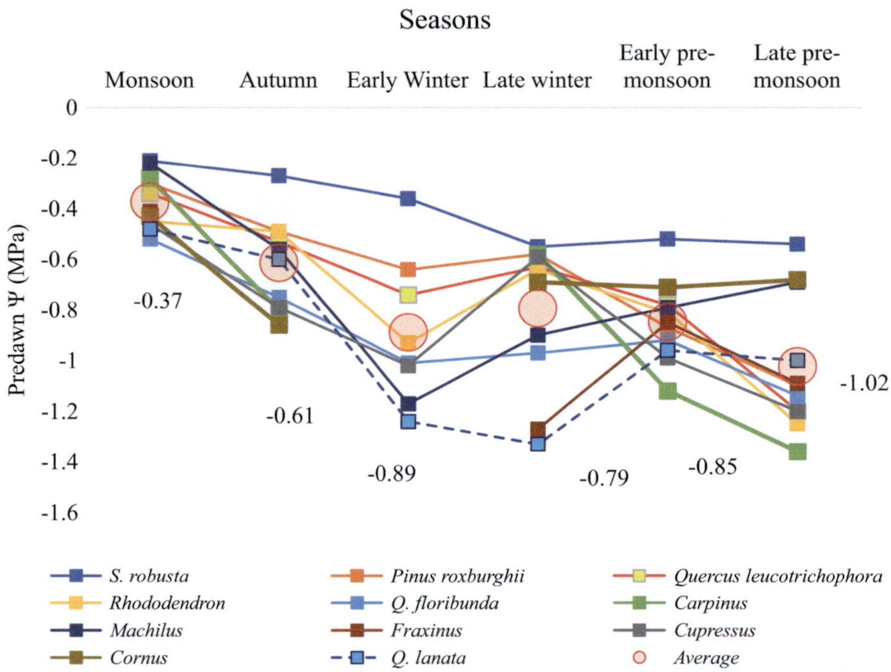

Fig. 10.2 Seasonal pattern of predawn tree water potential for the 11 study species. Values given within the graph are seasonal averages of all species. Their locations are indicated with large circles outlined in red. (Derived from Singh et al. 2006)

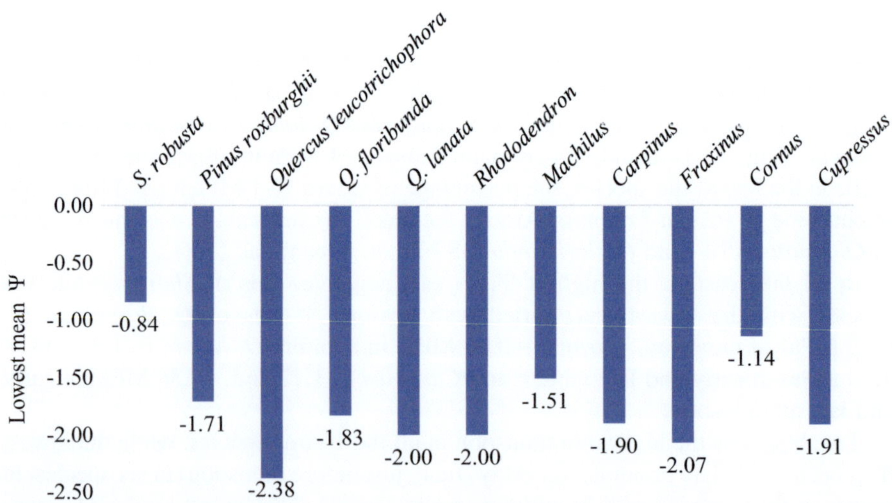

Fig. 10.3 A comparison of 11 Kumaun tree species for seasonally lowest tree Ψ_{PD} (generally in premonsoon). (Derived from Singh et al. 2006)

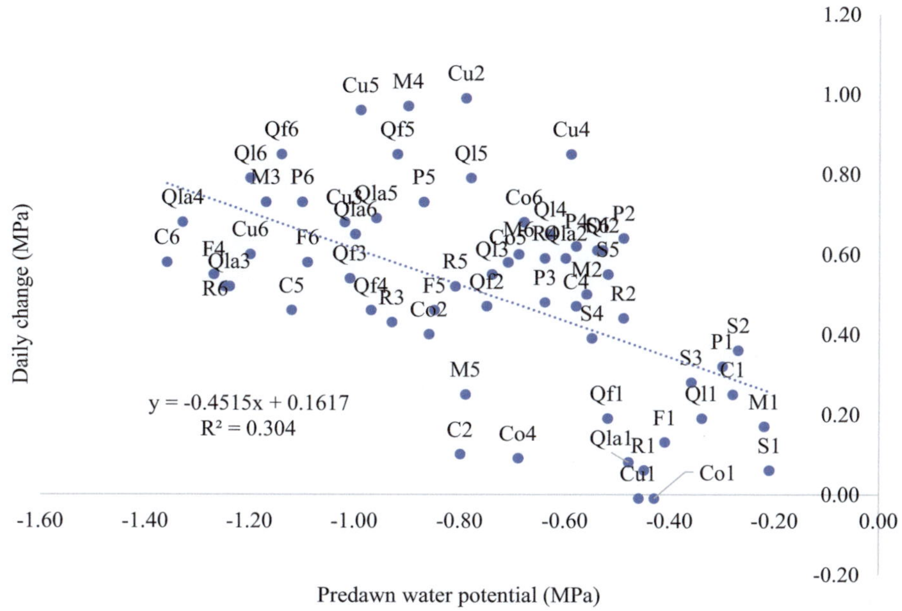

Fig. 10.4 Relationship between predawn water potential and daily change in Ψ (Ψ_{PD}–Ψ_{MD}) of 11 tree species of Kumaun in different seasons. Here, study seasons and species are indicated as following: monsoon/rainy = 1; autumn = 2; early winter = 3; late winter/spring = 4; early summer/premonsoon = 5; and summer/late premonsoon = 6; *Shorea* = S; *Pinus* = P; *Q. leucotrichophora* = Ql; *Rhododendron* = R; *Q. floribunda* = Qf; *Q. lanata* = Qla; *Carpinus* = C; *Machilus* = M; *Fraxinus* = F; *Cupressus* = Cu; and *Cornus* = Co. (Derived from Singh et al. 2006)

The three oaks (all e[1]) were similar with values within a relatively narrow range for several parameters: the lowest Ψ_{PD} from −1.14 MPa (*Q. floribunda*) to −1.33 MPa (*Q. lanata*); the lowest Ψ_{MD} from −1.75 MPa (*Q. lanata*) to −2.01 MPa (*Q. floribunda*); daily change in Ψ from 0.69 MPa (*Q. lanata*) to 0.90 MPa (*Q. leuocotrichophora*); Ψ_{sf} from −2.21 MPa (*Q. lanata*) to −2.61 MPa (*Q. floribunda*); Ψ_{sz} from −2.68 MPa (*Q. lanata*) to −3.04 MPa (*Q. floribunda*).

The difference between Ψ_{PD} and Ψ_{MD}, referred to as daily change (Ψ_{Δ}), is a measure of the water loss through leaf conductance at a given $\Psi_{PD.}$ It considerably varied seasonally. On average across all species, monsoon was characterized by high Ψ_{PD} and low daily Ψ_{Δ}, autumn by low Ψ_{PD} and moderate daily change, and early and late summer (premonsoon) by low Ψ_{PD} and high daily Ψ_{Δ} (Fig. 10.4).

It shows that, on average, Kumaun tree species keep stomata fairly open during premonsoon when water stress is severe. They maintain xylem conductance when water stress is high, indicating that they are cavitation resistant and/or refill embolized tissues. In these species, the daily Ψ_{Δ} was the lowest during the monsoon season when Ψ_{PD} was high; it generally increased in all species subsequently, with decreasing Ψ_{PD} up to about −1 MPa. The highest Ψ_{Δ} was found between Ψ_{PD} −0.8 and −1 MPa for *Cupressus* in autumn and early premonsoon and for *Machilus* in late

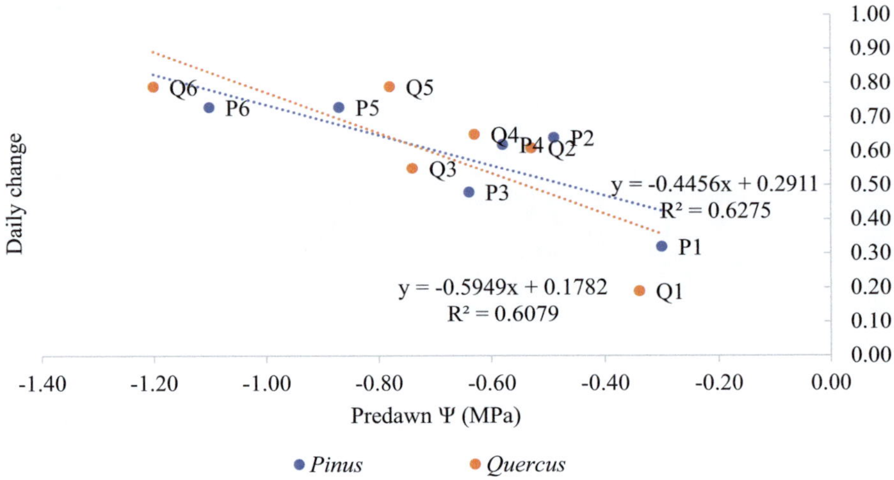

Fig. 10.5 Relationship between Ψ_{PD} and daily Ψ_{Δ} in banj oak and chir pine for different seasons. Here, study seasons and species are indicated as follows: monsoon/rainy = 1; autumn = 2; early winter = 3; late winter/spring = 4; early summer/premonsoon = 5; summer/late premonsoon = 6; chir pine = P; and banj oak = Q. In all seasons, Q is on the left side of P, indicating a lower Ψ_{PD} of oak than pine

winter. The Ψ_{MD} was lower in banj oak than in chir pine in all seasons. Both in chir pine and banj oak, the daily Ψ_{Δ} increased continuously from the rainy season to late premonsoon; however, it was generally more in banj oak than chir pine as Ψ_{PD} declined below −0.8 MPa (Fig. 10.5).

A study on chir pine and banj oak in a mixed forest shows that the oak generally maintained higher daily Ψ_{Δ} than in pine (Fig. 10.5).

In a study conducted on all major forest types of Kumaun below 2600 m, Tyagi (2006) compared the ratio between daily Ψ_{Δ} in relation to Ψ_{PD} during premonsoon, when water stress was the highest or close to that (Fig. 10.6). In sal and pine, the two values were similar or the same, in oaks, Ψ_{Δ} values were distinctly higher than Ψ_{PD}, with a ratio of 1.4:1.6. This comparison highlights the ability of all Kumaun oaks to keep stomata open even, resulting in a large-scale daily reduction in even at severe water stress.

Leaf Conductance

The average leaf conductance across the study species (measured only for broad-leaved species) showed two well-marked peaks, one in autumn and the other in spring. The winter and premonsoon values were relatively much lower (conductance was not measured during the monsoon).

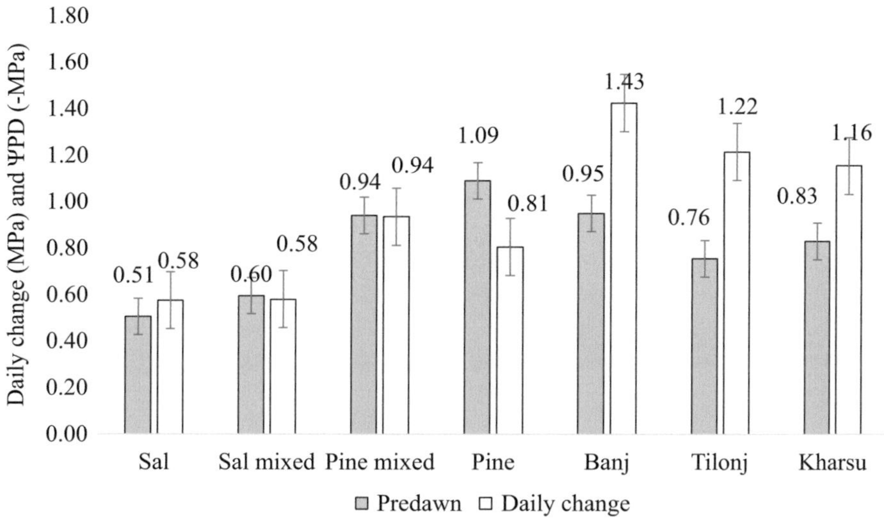

Fig. 10.6 Daily change in tree water potential (Ψ_{PD}–Ψ_{MD}) in comparison to Ψ_{PD} ($-\Psi$ MPa) in important forest tree species during premonsoon when Ψ_{MD} is the lowest (derived from Tyagi 2006). Hatched histogram shows $\Psi_{PD,}$ and the open histogram shows daily Ψ change. Sal (*S. robusta*)—high Ψ_{PD} and low daily Ψ_{Δ}; pine (*P. roxburghii*)—low Ψ_{PD} and moderate daily Ψ_{Δ}; oaks low Ψ_{PD} and high daily Ψ_{Δ}. Banj is *Q. leucotrichophora*, moru *Q. floribunda*, and kharsu *Q. semecarpifolia*

The leaf conductance of sal in autumn (500 m Mol m^{-2} s^{-1}) was much higher than that of any other species in any season. In the rest of the species, the leaf conductance ranged from 44 in *Carpinus* in early premonsoon to 222 m Mol m^{-2} s^{-1} in *R. arboreum* in spring.

Osmotic Potential and Osmotic Adjustment

Osmotic potential varied significantly seasonally and across the species ($p < 0.0001$), both at full (Ψ_{sf}) and zero turgor (Ψ_{sz}), which were closely related as follows: Ψ_{sz} (MPa) 46.8 ± 0.979, Ψ_{sf} (MPa) ($n = 258$, $r^2 = 0.83$, $p < 0.0001$) (Singh et al. 2006). With regard to osmotic potential at full turgor, the species of Angiosperm and Gymnosperm were clearly separable, the average of the former being significantly lower (−1.86 MPa) than that of the latter (−1.69 MPa, $p = 0.038$).

Osmotic adjustment (measured here as the difference between the lowest and highest seasonal values of Ψ_{sf}) was very high for *Cupressus* (1.47 MPa) and *Machilus* (1.43 MPa) (Fig. 10.7). Among others, it was low for the three deciduous species, *Shorea* and *Rhododendron* (<0.6 MPa), and moderate (0.7–0.9 MPa) for the rest of the species.

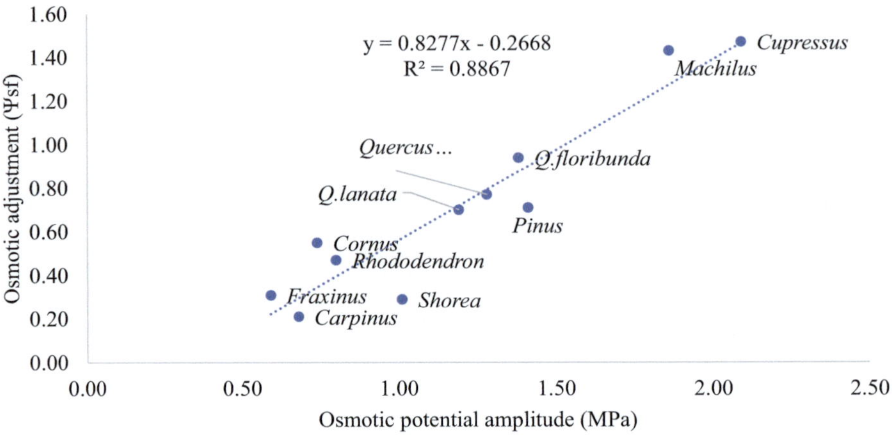

Fig. 10.7 Relationship between osmotic adjustment at full turgor (Ψ_{sf}) and osmotic amplitude (OA), which is the difference between the highest Ψ_{sf} (least negative) and the lowest Ψ_{sz} (most negative) across the 11 study species of Kumaun Himalayas. (Derived from Singh et al. 2006)

Interestingly, the osmotic adjustment was particularly high between the monsoon season and autumn season (Ψ_{sf} decreased in 9 out of 11 species) when Ψ_{PD} was still high (−0.5 to −0.8 MPa) and Ψ_{MD} above the turgor loss point. Days following the wet monsoon are favorable for photosynthesis in this part of the Himalayas: bright sun, still moist soil and warm temperatures, and fairly long day length. Photosynthesis is required to supply compatible solutes for osmotic adjustment (Kramer and Boyer 1995). The pronounced osmotic adjustment during autumn and early winter, in turn, ensures soil water availability for photosynthesis, which peaks in the majority of these species in autumn (Tyagi 2006). This strategy enables the species to gain more carbon than the cost of generating compatible solutes within cells. Measuring leaf nonstructural carbohydrates, photosynthesis, solute, and water potential in *Q. rubra*, Gersony et al. (2020) have shown that transport of nonstructural carbohydrate in daytime occurs at equal or higher rates than at night despite a decrease in Ψ to −1.8 MPa at midday. They estimated that sucrose accumulated over the course of the day contributed ~50% of the 0.4 MPa diurnal osmotic adjustment, necessary for maintaining high midday turgor (Gersony et al. 2020). These findings illustrate the vital role of nonstructural carbohydrate in leaf turgor maintenance and the effective functioning of phloem in the face of diurnal fluctuations of water potential. By prolonging the period of warmer temperatures, climate change is likely to favor species with effective osmotic adjustment when conditions are still not severe. In this regard, oaks are likely to be favored by climate change.

The relationships among tree growth, phenology, and osmotic adjustment differ across the species of tropical dry forest (Eamus and Prior 2001) and temperate forest (Zobel 1996). Within dry tropical forest species, osmotic adjustment is related to the change from wet to dry season (Eamus and Prior 2001). Osmotic adjustment is one of the various strategies that plants employ to deal with water shortage in different situations. For example, between the two major oaks of Uttarakhand, *Q. floribunda* with a relatively lower root density below 60 cm soil depth showed more osmotic adjustment (0.94 MPa) than *Q. leucotrichophora* (0.77 MPa) with dense roots in deeper soils (Pathak et al. 2021). The low osmotic adjustment in the deep-rooted *Shorea* further indicates the trade-off between allocation to roots and generation of compatible solutes for osmotic adjustment.

Osmotic amplitude, calculated as the difference between the highest Ψ_{sf} and lowest Ψ_{sz}, was higher for conifers (average −1.75 MPa, Fig. 10.7) than for different groups of angiosperms: 0.67 MPa for deciduous species and 1.09 MPa for evergreen broad-leaved species. Osmotic adjustment at full turgor was positively correlated with osmotic amplitude, which provides a range within which species can osmotically adjust (Fig. 10.7); the wider the range in osmotic amplitude, more the osmotic adjustment.

How species respond in terms of osmotic adjustment from a moist to a dry habitat? The study of Tyagi (2006) and Tyagi et al. (2023) indicates that osmotic adjustment invariably increases from hill base to hill top and from north-facing slope to south-facing slope in chir pine and banj oak. Between them, it was more in banj oak in all instances. There was a notable consistency in species responses to changes in aspect and position.

Index of Tissue Elasticity (ε)

Plants differ in cell wall elasticity. The opposite of elasticity is ε; the lower the ε, the higher is elasticity. In other words, high ε indicates that cell walls are rigid. Cells with elastic walls can adjust turgor pressure (the pressure water exerts against the cell wall) and turgidity by easily changing cell volume. In contrast, cells with rigid walls reduce Ψ when subjected to drying. The two conifers had significantly higher average tissue elasticity (i.e., low ε, average 4.8 MPa, $p = 0.03$) than the angiosperms ($\varepsilon = 14.7$ MPa). Among the angiosperms, ε was particularly high (rigid walls) for undercanopy evergreen species: *Machilus* (24.2 MPa) and *Rhododendron* (20.2 MPa).

As cells lose water under the influence of drought, elastic walls help to maintain turgor by reducing cell volume. However, when water stresses are extreme, the cell wall may become less elastic. On the other hand, cells with less elastic or stiff walls reduce Ψ faster as they lose water, which in turn results in an enhanced water uptake and the avoidance of solute accumulation to a toxic level (Kramer and Boyer 1995). As expected, cell wall elasticity in trees was highest (ε lowest) when new leaves were expanding during early premonsoon (March–April) (see Tyree and Karamanos 1981). Generally, as soils become progressively dry during autumn and winter, rigid cell walls and osmotic adjustment combine to maintain a sharp gradient for water extraction without a major reduction in RWC (Abrams 1990).

Comparison of Kumaun Himalayan species to trees from similar thermal climates elsewhere, indicates that the Ψ_{PD} in low-elevation *Shorea* forests is higher than those of tropical dry forest (Zobel et al. 2001a). The Kumaun conifers have less severe Ψ than related conifers studied elsewhere, and in evergreen broadleaf forests, Ψ values in oaks and other species are markedly higher than in Mediterranean oaks.

With regard to osmotic potential, the cool temperate species of Kumaun have lower Ψ_{sf} than most North American species, but are less extreme than in Mediterranean climates. In cool temperate species of Kumaun, osmotic adjustment is as great or greater than values reported elsewhere. In lower elevation trees, osmotic adjustment was less than in most dry forests. In warm temperate forest trees of Kumaun, osmotic adjustment was similar to that of Mediterranean forests.

The conifers of Kumaun have higher cell elasticity (low ε) than other published values. In contrast, about half of the angiosperm species of Kumaun have their maximal elasticity less than that of the species studied elsewhere. To what extent conifers as a group have higher cell elasticity than angiosperms, as observed in the Himalayas and elsewhere (Bahari et al. 1985; Anderson and Helms 1994; Zobel 1996; Zobel et al. 2001b), and its implications have not been analyzed.

Deciduous Species: A Heterogenous Group

The three winter deciduous species are at odds with one another in tree water relations, resembling one another only in a few parameters: low leaf conductance, low osmotic adjustment, and limited daily change in Ψ_{PD}. *Cornus*, which grows toward the upper side of the study elevation range, is characterized by high Ψ_{PD} (even the lowest seasonal value being −0.86 MPa), next only to *Shorea*, while the lowest seasonal Ψ_{PD} values of *Carpinus* and *Fraxinus* were on the lower end of the range across the 11 species. The percentage of Ψ_{PD} values below –1 MPa was 46% in *Fraxinus,* which was the highest among the 11 study species, and 18% in *Cornus,* which was toward the lower end of the range. Osmotic amplitude in them ranged from 0.19 MPa in *Cornus* to 0.53 MPa for *Carpinus*. With respect to RWC at zero turgor, *Cornus* was characterized by a wide range 74–89% and *Fraxinus* by a narrow range (0.85–0.87%). The highest leaf conductance in *Cornus* was 119 m Mol m^{-2} s^{-1}, which was 1.65 times higher than that of *Carpinus*. The Huber value (ratio of total xylem cross section to leaf area) in *Cornus* was more than twice as much as in the other deciduous species.

The evergreen broadleaves which dominate forests in the region resemble deciduous tree species in producing new leaves yearly to maximize the advantage of the favorable monsoon period that follows. By retaining leaves throughout the year, they take advantage of mild winters (Sakai and Malla 1981) with fairly long day lengths. These evergreen broad-leaved species undergo a moderate thinning of the canopy during premonsoon when water stress is at peak. However, they restore their full canopy by the time monsoon sets in, as do deciduous species.

Impact of an Extreme Drought on Oaks

An increase in extreme weather events, such as severe droughts, is a worrying prediction associated with climate change. Severe droughts, that lead to deleterious consequences across the forest landscape, have been recorded in Uttarakhand several times in the past century, including in 1916, 1921, and 1999. In 1995, in Nainital town (located at 1950–2200 m in Uttarakhand), rainfall was only 26.5 mm in 8 months preceding the monsoon, compared to the average of 400 mm. That year, the mean April temperature when most trees are in the peak leafing was 31 °C, compared to the average of 26 °C. Both in *Q. leucotrichophora* and *Q. floribunda,* some trees put forth new leaves in March (referred to as population X) in the face of severe drought, while in the other trees, leaf buds either remained dormant or died out (referred to as population Y). Thus, the drought revealed the presence of two populations in both oak species. Individuals of the Y population showed little phenological activity for many weeks with no apparent growth, and shed withering buds and old leaves. They produced new leaves only after several days of rain by May 30, when Ψ was replenished adequately. So, the gap in leafing time of the two populations was about 6 weeks. The average Ψ_{PD} of the Y population during the drought was -4.4 ± 0.047 MPa, nearly three times lower than in normal years, and one and a half times lower than that of the X population in that year. Individuals with Ψ_{PD} −5.5 MPa seemed to be dead, but they revived in June–July only after weeks of monsoon and appeared to quickly start showing normal phenological activity. That year, the osmotic adjustment was 1.7 MPa compared to 1.3 MPa and less in normal years (Table 10.2).

Table 10.2 Summary of the impact of the 1999 premonsoon drought on moru oak (*Q. floribunda*) trees

	Quercus floribunda drought		
	Normal year	Drought year	
Parameters	population	X population	Y population
Ψ_{PD} during (MPa)	−1.14	−3.1	−4.4 (extreme −5.5)
Leaf initiation time	Mid–march	March-end	May-end to June beginning
Osmotic potential at full turgor (Ψsf) (MPa)	−2.6	−1.6	−3.5
at zero turgor (Ψsz) (MPa)	−3.0	−2.5	−4.2
Index of elasticity (MPa)[a]	13.4	4.3	15.0
Leaf population exchange	Most old leaves replaced by new leaves	Most old leaves replaced by new leaves	Remains of a few old leaves, and aborting buds, with no new leaf

Based on Singh et al. (2000)

Average minimal daily Ψ: −1.5 to −4 MPa for tropical rainforests; −1.5 to −2.5 MPa for deciduous temperate forests; −3.5 to −0.5 MPa for sclerophyllous dry region trees; −1.5 to −6.0 MPa for evergreen conifer trees (See in Singh et al. 2000)

[a]The lower the value, the higher the elasticity of the cell wall

We did not find a similar phenological disturbance in chir pine and other associates of oaks. These Himalayan oaks are vulnerable to drought because of the delayed response of their stomata to water stress. In other words, unlike chir pine, their safety margins are narrow. A drought of similar severity occurred widely in 2021 when young oak individuals withheld new leaf production and barely survived through good resprouting.

A meta-analysis in species mortality rate across 475 tree species from 33 studies around the globe has shown that the hydraulic safety margin (HSM) is the only significant predictor (but that too only a modest one) of survival in the face of severe droughts. HSM is the difference between typical minimal Ψ and that causing xylem dysfunction, i.e., the Ψ causing 50% and 88% loss in hydraulic conductivity because of embolism. HSM integrates the safety of xylem to drought-induced embolism response to leaf water potential (i.e., Ψ at which stomata close down).

Gymnosperms have narrow tracheids, but are unable to recover from Ψ_{P50} because of the sparser parenchyma tissue in wood. They are also poor at resprouting. Angiosperms are more susceptible to embolism, but can refill embolized vessels quickly and resprout quickly.

Treelines

Globally, treelines are hardly studied for water relations because they are not considered to be water stressed (Körner 2012). At very low temperatures, evapotranspiration losses are expected to be insignificant. However, studies on tree ring chronology in relation to climate, particularly in the Himalayas, indicate that tree growth is frequently limited by premonsoon water stress (Gaire et al. 2014; Tewari et al. 2018; also see Tewari FTR unpubl.). The treeline tree growth forms, namely evergreen conifers (*Abies spectabilis*), broad-leaved evergreen (e.g., *R. campanulatum*), and broad-leaved deciduous (*Betula utilis*), were the same as in forests of lower elevations of Kumaun described in the earlier section. Since both areas were also dominated by monsoon rainfall, a comparison between them in tree water relations should lead to determining the effect of low temperature and resultant short growth period of treelines. Here, our focus is on differences in treeline water relations between the lower forest tree species described earlier and treeline species.

(i) Consistent with the general understanding, the treeline species were less water stressed, as their average (of 3–4 years) Ψ_{PD} mostly remained higher than −1 MPa in all seasons, while it was below −1 MPa in lower forest trees.
(ii) Both in the treeline and Kumaun forest trees, Ψ was highest during the monsoon months. But the two groups showed contrasting patterns when the daily change in Ψ was compared. It was significantly higher in monsoon in treeline species than in other seasons, while lower forest trees hardly showed daily change in monsoon (Fig. 10.8).

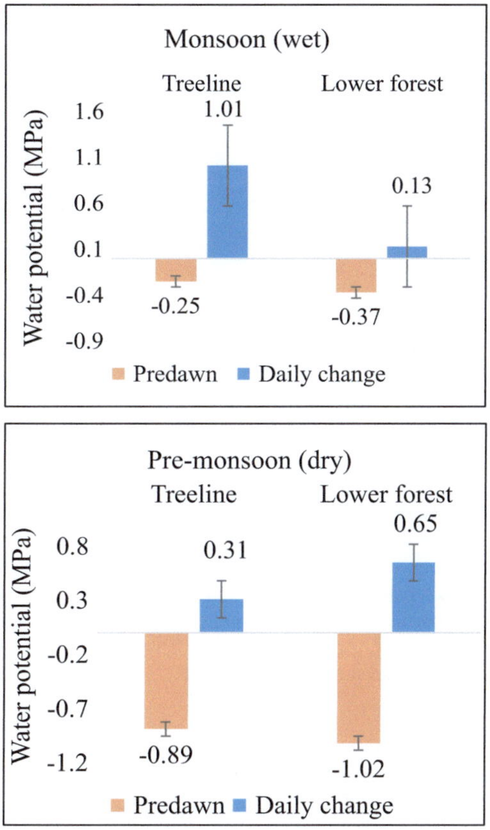

Fig. 10.8 A comparison between the predawn and daily change in tree water potential (difference between predawn and midday on a given day) of treeline and lower forest tree species of Kumaun (also referred to as Kumaun's forests). All are averages of tree species (five treeline and 11 lower forest species). (Source: FTR IHTRP unpubl.)

(iii) Furthermore, the treeline species showed a limited daily change in Ψ during dry premonsoon when tree Ψ_{PD} was most negative. In contrast, lower forest tree species secured the highest average daily Ψ_{Δ} change at the peak of the dry season.

Why do lower forest trees undergo little daily change (difference between Ψ_{PD} and Ψ_{MD}) during monsoon months when treeline species undergo the highest change? During monsoon at Tungnath treeline, Ψ_{PD} was within a narrow range between -0.16 ± 0.30 MPa (*Abies*) and -0.31 ± 0.03 MPa (*R. arboreum*), while daily change was high in all five species, the average being 1.01 MPa (Fig. 10.9).

During the monsoon, forest trees maintain photosynthesis at a considerably high rate involving substantial water loss (Tyagi 2006), so the lack of daily Ψ change in them may indicate the simultaneous replenishment of transpirational loss from soil water. In an open treeline vegetation, trees are expected to get more sunlight, hence more leaf conductance than lower forest trees (Tewari et al. 2018). In the early part of the growth period (premonsoon), treeline species are quite likely to have incompletely developed roots, and thus are incapable of extracting water from cold soils. In

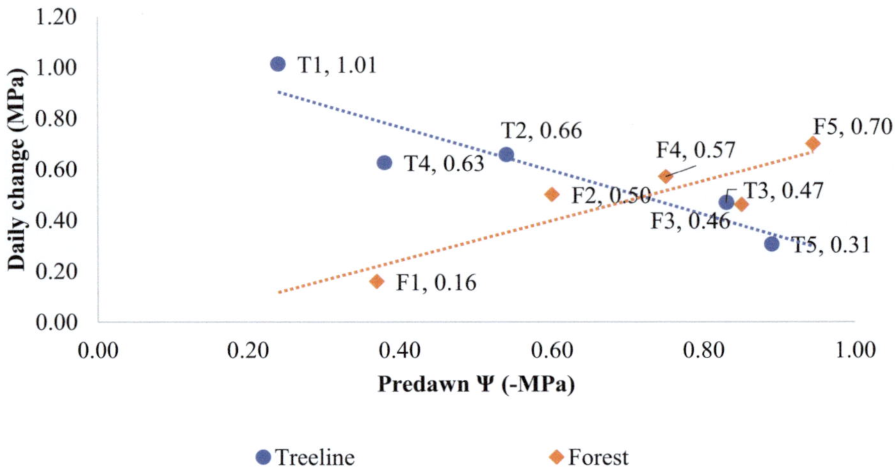

Fig. 10.9 Contrasting pattern of relationship between Ψ_{PD} and daily Ψ_{Δ} (Ψ_{PD}–Ψ_{MD}) of the average of all species across the seasons at treeline and lower forest trees. In the treeline (T), monsoon season is characterized by high Ψ_{PD} and high daily Ψ_{Δ}, while in lower forest trees (F), it is characterized by high Ψ_{PD}, but low daily Ψ_{Δ}, which is highest when Ψ_{PD} is the lowest during premonsoon. The numbers indicate seasons 1 = monsoon; 2 = autumn; 3 = early winter; 4 = late winter (or spring); and 5 = premonsoon (or summer); treeline species are indicated with **T** and forest species are indicated with **F**

treeline species, even May–June may get a cold spell, interrupting cambium growth and vessel and tracheid development. On the other hand, forest trees get a fairly long period before the monsoon to develop new roots and foliage.

Species of both lower forests and treelines secure a considerable amount of osmotic adjustment; however, values were significantly higher in treelines than forests (Table 10.3). This could be partly related to higher nonstructural carbon (NSC) reserves in treeline species, which often lack enough carbon sink to transport NSCs (Körner 2012).

In forest areas, generally leaf water conductance is higher in the morning than in the afternoon, while in treelines, afternoon leaf water conductance is significantly greater (average of all seasons in treelines being 167.12 and 200.65 m Mol m^{-2} s^{-1}, respectively, in morning and afternoon). In *Q. semecarpifolia*, as an example, the afternoon rate (191.55 m Mol m^{-2} s^{-1}) was significantly higher than the morning rate (129.68 m Mol m^{-2} s^{-1}). Only in *Betula,* both morning (275.25 m Mol m^{-2} s^{-1}) and afternoon (281.65 m Mol m^{-2} s^{-1}) leaf conductance values were similar. The other two broad-leaved evergreen species resembled *Q. semecarpifolia* (also a broad-leaved evergreen species). In three of the four broad-leaved species, leaf water conductance and daily Ψ change were the highest during the monsoon season. Leaf conductance was particularly low in premonsoon in *Q. semecarpifolia* and *R. arboreum* (Table 10.3).

Since species of Alpine treeline are adapted to the abundance of soil water, moderate water stress is likely to reduce leaf conductance and transpirational loss

Table 10.3 A comparison between forest and treeline species with respect to Ψ_{PD}, Ψ_D, and leaf conductance

	Lower forests (500–2500 m asl)	Treeline (>3000 m asl)
Ψ_{PD} less than −1 MPa	In all 12 study species, except sal, at least in one season	In only five out of 70 instances (species × seasons × years), at least in a season.
Season of the highest Ψ_{PD}	Monsoon	Monsoon
The season and mean value of the highest daily change in Ψ	Premonsoon (average of all species 0.70 MPa)	Monsoon, average of all species >1.01 MPa
The season of the lowest daily change Ψ	Monsoon (average of all species 0.16 MPa)	Premonsoon (6 out of 15 instances)
Number of Ψ_{MD} <−1.5 MPa across all species, seasons, and years	23 out of 66 instances	9 out of 70 instances
Tree leaf conductance: range across the species and seasons (m Mol m^{-2} s^{-1})	44 in *Carpinus* in early premonsoon to 500 *Shorea* in autumn	26 in *R. Campanulatum* in winter—329 in *Betula* in monsoon
Average osmotic adjustment across the species		
Ψ_{sf} (MPa)	1.11	1.30
Ψ_{sz} (MPa)	1.06	1.77
Average leaf conductance of species across all seasons (m Mol m^{-2} s^{-1})	Higher in the morning than in midday Highest sal 242, followed by *Rhododendron* (146) and banj oak (133)	Lower in morning average, (167) than midday average, (281)
Average leaf conductance (m Mol m^{-2} s^{-1}) of seasons across all species	Highest in autumn (171), followed by late premonsoon—132; monsoon—95; winter—132; early premonsoon—91	Highest in monsoon (241), followed by premonsoon—164; autumn—164; winter—88

(Anfodillo et al. 1998). In the Italian Alpine treeline, where growing season precipitation is much lower (500 mm) than in the Tungnath treeline of the Himalayas, water stress continuously increases with time during the growing period. In Himalayan treelines, the growing season is divisible into an earlier dry period (generally May to mid-June) and a late wet period (mid-June to September) when most of the precipitation occurs. Since the Himalayan treeline species are also not adapted to water stress, they generally undergo a limited daily change (0.31 MPa) during the early growing period (premonsoon) as Ψ_{PD} is generally low (−1.0 to −1.4 MPa). Subsequently, during monsoon, their Ψ_{PD} rises to 0.25 MPa on average, and the daily change in Ψ on average to ~1 MPa. These species thus avoid water stress during premonsoon by minimizing water loss, and take full advantage of monsoon, when their leaves freely conduct water. In the Italian Alps, the Ψ_{PD} is reported to range between −0.29 and −1.0 MPa. The lowest Ψ is relatively higher than in the Himalayas, when it goes down up to −1.40 MPa. However, the most Ψ_{MD} in the Himalayan treeline, −0.86 to −1.78 are higher, compared to the Italian Alps, −0.45 to −1.93 MPa.

High Leaf Conductance at Chitkul: A Monsoon Shadow Site

During winters, tree Ψ declined rapidly after predawn because of frozen soil water supply could not compensate for the loss due to leaf conductance. Thus, the lowest Ψ was reached between 8.30 and 10.30 am (at 12.30 pm for birch). Subsequently, as the day warmed up, soil water supply from the soil was resumed and tree Ψ was revived at Ψ_{PD} level.

At Chitkul, a site located in dry inner valleys, Ψ_{PD} was relatively lower, ranging between −0.94 and −1.1 MPa for trees and between −0.8 and −1.4 MPa for seedlings. At Chitkul in summer, the leaf conductance (m mol m^{-2} s^{-1}) ranged from 165 to 215, but in autumn, it rose to 783.8 in *R. campanulatum* and 667.1 in birch. Strangely, the values were still higher for seedlings, 1147.7 for *R. campanulatum* and 1323.3 for birch. How could trees and seedlings sustain these high rates of leaf conductance in a dry treeline? It could be a temporary phenomenon, hence it needs frequent data collection over a long period.

Kathmandu Valley Trees

Much of the knowledge about tree water relations in Nepal is due to the investigations on five evergreen broad-leaved species (e′) of Phulchowki Mountain, 10 km SE of Kathmandu (27°33'N, 85°22′E) (Poudyal et al. 2004). The study area is at the eastern edge of Central Nepal with annual rainfall of 1882 mm, of which 80% occurs during monsoon months (mid-June to mid-September, of which July alone gets 483 mm). The study period also included the dry year 1999, when only 13 mm of rainfall was received between January and mid-June, and annual rainfall was 1407 mm. At Kathmandu airport (1350 m), the mean temperature ranged from 2.6 to 18.7 °C during winter and 15.8 to 28.2 °C during summer.

Water Potential

Of the five evergreen broad-leaved species of Nepal, *Castanopsis indica* and *Schima wallichii* were sampled at 1400 m (upper subtropical forest, Stainton 1972), *R. arboreum*, *Q. lanata* at 2020 m, and *Q. semecarpifolia* at 2130 m. The Kathmandu (Nepal) study sites were more moist than the Kumaun sites. Though yearly rainfall values were similar to that of Kumaun, in Nepal, the monsoon period is a bit longer, and winters are milder because of lower latitudes.

In Kathmandu forests, none of the species had a mean Ψ less than −1 MPa, while in Kumaun, 10 out of the 11 study species had Ψ_{PD} less than −1 MPa at least in one season, mostly during premonsoon. Seasonal pattern of tree Ψ_{PD} was similar in the two regions, with values generally being highest in monsoon and lowest in premonsoon. However, within premonsoon, the month varied, the lowest Ψ_{PD} being in March in Kathmandu forests, and April–May in Kumaun forests.

For high-elevation species however, it was winter in Kumaun. In Kathmandu forests, the mean Ψ_{PD} and Ψ_{MD} ranged, respectively from −0.40 to −1.18 MPa in *Q. semecarpifolia* to −0.2 and −0.63 MPa in *Schima wallichii* (Table 10.4). In Kumaun, a detailed sampling by Tyagi (2006) based on six sites varying in slope aspect and slope positions, the average Ψ_{PD} and Ψ_{MD} values were respectively −0.41 to −0.76 MPa for sal, −0.60 and −0.99 MPa for chir pine, −0.59 and −1.12 MPa for banj oak, −0.54 and −1.11 MPa for moru oak, and −0.54 and −0.95 MPa for kharsu oak.

As for the daily change in Ψ among the species of Kathmandu, *Q. semecarpifolia* had the highest value, 0.8 MPa; however, it is lower than 0.95 MPa (Tyagi 2006) and 1.16 MPa (Shirish Kumar Singh unpubl.) measured in the China peak area (2550 m) of Nainital and Kunjakharak (2480 m) forest area of Kumaun, respectively. In Nepal, 0.8 MPa of daily Ψ change was also observed in *Castanopsis indica* (Poudyal et al. 2004). A higher daily change in Ψ was recorded (1.43 MPa) in banj oak of Kumaun (Tyagi 2006).

Leaf Water Conductance

In the study of Nepal, generally g_w was observed to decline with a drop in Ψ_{PD}, and in *R. arboreum,* the two (Ψ_{PD} and g_w) were significantly correlated (Poudyal et al. 2004). The average g_w (m Mol m^{-2} s^{-1}) in *Q. semecarpifolia* (172–192 m Mol m^{-2} s^{-1}) at Kathmandu site was higher than the values recorded for Kumaun oaks (132 m Mol m^{-2} s^{-1} for *Q. lanata*, 98 m Mol m^{-2} s^{-1} for *Q. floribunda,* and 133 m Mol m^{-2} s^{-1} for *Q. leucotrichophora* (Singh et al. 2006). However, no species in Kathmandu approached leaf water conductance of sal (average 242 and peak 500 m Mol m^{-2} s^{-1}). In Kathmandu, g_w reached up to 328 m Mol m^{-2} s^{-1}. Higher g_w than Himalayan oaks have been reported for deciduous oaks of other regions, such as *Q. rubra* (400 m Mol m^{-2} s^{-1}, Kubiske and Abrams 1992), *Q. afares* (448 m Mol m^{-2} s^{-1}, Acherar & Rambal 1992), and *Q. douglasii* (>400 m Mol m^{-2} s^{-1}, Xu and Baldocchi 2003). In Nepal, g_w was found to increase ($r = 0.95$) with elevation and Ψ_{PD} to decrease ($r = -0.74$).

Tree species, even those that occur together in Kathmandu valley, differ in the mechanisms they employ to maintain turgor during dry months. *Schima* depends on both osmotic and elastic adjustment. However, in *Castanopsis,* a major associate of *Schima*, osmotic adjustment has little role in turgor maintenance; it largely depends on cell elasticity. Despite low relative water content at zero turgor (RWC_z), *Q. semecarpifolia* exhibited effective lowering of osmotic potential during drought

Table 10.4 Comparison between Kumaun, Indian Himalayan region, and Kathmandu Valley, Nepal, for certain tree water relation parameters

Severity	Kumaun, India	Kathmandu, Nepal
Tree species studied	11 tree species: 2 conifers, 9 angiosperms, of which 6 broad-leaved evergreen and 3 deciduous	Five broad-leaved evergreen species
The lowest soil Ψ across sites and seasons	−3.5 MPa for soil 60 cm (May)	−3.2 MPa in March for soil 30 cm
–Range of mean Ψ_{PD}, of all the study seasons and the species	−0.41 MPa in *S. robusta* to −0.94 in *Q. lanata*	−0.20 MPa in *Schima wallichii* to −0.63 MPa in *Q. emecarpifolia*
Species with mean Ψ_{PD} below −1 MPa	Nine out of 11 species had below −1 MPa at least in one season, mostly late premonsoon	In none of the species
Some of the lowest seasonal Ψ_{PD} among the species	−1.36 MPa in *Carpinus viminea* in late premonsoon (may–June)	1.70 MPa in *Q. semecarpifolia* (in march)
Months of the lowest Ψ_{PD}	April–May, in 7 of 11 species and in 4 late January–February	Mostly march (early premonsoon)
Range in daily change in tree Ψ (Ψ_{PD-MD})- mean of seasons	−0.35 *Cornus macrophylla* to −0.68 *Cupressus torulosa* (mean −0.50)	−0.43 in *S. wallichii* to −0.81 in *C. indica* (mean −0.62)
Mean leaf conductance (m mol m^{-2} s^{-1}) range across the species for seasonal means (morning + afternoon/2)	From 66 (*Carpinus viminea*) to 242 (*S. robusta*)	78 (*C. indica*) to 172 (*Q. semecarpifolia*)
Afternoon		74 (*C. indica*) to 190 (*Q. semecarpifolia*) Average 125
Lowest midday Ψ with season	In eight of 11 species, early premonsoon (March–April), in 2 late winter (January–February), and in 1 early winter (November–December)	Early premonsoon (March–April) in all
Highest daily change, range across the species	From 0.55 MPa, *S. robusta* (early premonsoon) to 0.99 MPa (*Cupressus torulosa*) in fall	From 0.43 MPa *Schima wallichii* to 0.81 MPa *Castanopsis indica*
Highest seasonal daily change average of all species	0.70 MPa premonsoon (March–May)	–

The Kumaun data are based on 11 tree species from tropical to cool temperate forests (500–2300 m), each investigated for six seasons. In Kathmandu, it is based on seasonal values of five species between 1400 and 2020 m. The common species were *R. arboreum, Q. semecarpifolia,* and *Q. lanata*. Ψ_{PD}—predawn water potential, Ψ_{MD}—midday water potential (based on Poudyal et al. 2004; Zobel et al. 2001b; Singh et al. 2006)

Seasons: monsoon: typically mid-June to September, warm and wet; autumn: October to mid-November; winter: mid-November to February (could be divided as early winter from mid-November to December, and late winter: January to February), premonsoon: March to mid-June, generally called summer, warm and dry (March to April, early premonsoon and May to mid-June, late premonsoon)

Table 10.5 Comparison of three tree species common to Kathmandu and Kumaun forests for certain water relation parameters (based on Singh et al. 2006; Poudyal et al. 2012)

	R. arboreum		*Q. lanata*		*Q. semecarpifolia*	
Parameters	Kumaun	Kathmandu	Kumaun	Kathmandu	Kumaun	Kathmandu
RWC_z (%)	86.66–93.4 (91.09)	82–95 (88.6)	67.5–89.5 (81.2)	64–90 (78.7)	56.2–91.4 (77.5)	74–93 (82.9)
Mean Ψ_{sz} and Ψ_{sf} (MPa)	−1.62 and −1.89	−1.86 and −1.53	−1.83 and −2.27	−2.47 and −1.87	−2.16 and −2.70	−2.56 and −2.04
Osmotic adjustment (MPa)	0.27	0.33	0.44	0.61	0.54	0.51
Index of elasticity (ε)[a]	20.45	13.22	10.65	9.2	–	–

Values in parentheses are averages. RWC_z is relative water content at zero turgor; Ψ_{sz} is osmotic potential at zero turgor; Ψ_{sf} is osmotic potential at full turgor
[a]Lower the value, the greater the elasticity

(Table 10.5). The species had the lowest Ψ. As for *Q. lanata*, there was no definite pattern for ε, but it had a wide osmotic amplitude (difference between Ψ_{sf} and Ψ_{sz}). Poudyal et al. (2012) suggest that this irregular pattern could be linked to multiple leaf flushing and heavy leaf insectivory.

Both in Nepal and Kumaun, large-scale lopping of branches can be seen during premonsoon, which results in increased Ψ_{PD} compared to un-lopped trees (Zobel et al. 1995).

Wood Water Content and Leafing

Trees of Nepal are known to have high wood water content (Poudyal et al. 2012) during autumn, which decreases subsequently until the premonsoon season (March–April) when trees put forth new leaves. This phase of phenological activity results in the withdrawal of stored stem water to rehydrate active tissues (Borchert 1994). Poudyal et al. (2012) showed that in *S. wallichi,* the difference in the timing of leaf emergence between the two study years was related to the difference in the time of wood water depletion. In year 2, both leaf emergence and wood water depletion were delayed by a month than in year 1, occurring in March and February, respectively. They (Poudyal et al. 2012) also found a relationship between the flowering of *R. arboreum* and water status, to which water stored in wood tissues contributed significantly. A marked year-to-year difference in flowering is observed in *R. arboreum* (from January to September), partly because of water status and the role of wood water storage in it.

Hydraulic Conductance

Poudyal et al. (2004) have analyzed how the hydraulic conductance of Himalayan trees changes seasonally in response to drought. In all the study species, except *R. arboreum,* hydraulic conductance declined from −0.7 to −1.7 MPa. Their threshold Ψ_{PD} for cavitation values is higher than those reported for *Betula occidentalis* (from −1.42 to −2.2 MPa) (Sperry and Pockman 1993), and *Quercus rubra* (−2 MPa, Cochard and Tyree 1990). In western hemlock (*Tsuga heterophylla*), Kavanagh and Zaerr (1997) found 30% loss in conductivity at Ψ_{PD} less than −2.5 MPa, 50% loss at <−3 MPa, and total loss of conductivity at <−4.0 MPa.

The trees of Nepal are similar to relatively more drought-sensitive species like *Populus balsamifera,* which is reported to cavitate at tree Ψ below 0.7 MPa (Hacke and Sauter 1995). In Nepal, trees revive conductivity as monsoon arrives. Refilling of embolism is a common feature among angiosperms, though the embolism recovery process is not fully understood (Nardini et al. 2017, Klein et al. 2018, Zeppel et al. 2019). The lower vulnerability of *R. arboreum* to cavitation was possibly because of its high stem water content (130%).

G_w and shoot conductivity (K_s) are linearly correlated, indicating that the loss of functional xylem tissues affects leaf conductance (Nardini and Salleo 2000). In the Nepal study, G_w was found to be significantly correlated with shoot conductivity (K_s) in *Q. semecarpifolia* and *C. indica* (Poudyal et al. 2004).

Leaf-specific conductivity, which is a practical measurement of xylem hydraulic supply capacity (Zimmermann 1978), decreased with decreasing water availability in all species of the Nepal study (Poudyal et al. 2004).

KI, which indicates the efficiency of supply of water to leaves, was high for the oaks (4.1×10^{-4} kg s MPa^{-1} m^{-1} and 3.4×10^{-4} kg s MPa^{-1} m^{-1} for *Q. lanata* and *Q. semecarpifolia*, respectively) than other species, indicating that they maintain a favorable water balance despite high evaporative demand. KI values of the Himalayan oaks are higher than those of temperate trees observed by Tyree et al. (1991). However, higher KI values than the oaks of Nepal have been reported for tropical deciduous species *Pseudobombax septenatum* (5.2×10^{-4} kg s MPa^{-1} m^{-1}) and evergreen *Ochroma pyramidale* (11.4×10^{-4} kg s MPa^{-1} m^{-1}) (Machado and Tyree 1994).

To conclude, the tree species of Nepal have developed a diverse combination of hydraulic properties (Poudyal et al. 2004). *Schima* with low H_v and K_s has a low water supply to the leaves. Characterized by high Ψ_{PD} and continuously low G_W during dry periods, *Schima* is adapted to avoid water stress, in which high cell elasticity helps it to maintain turgor when water supply is low. Furthermore, stem water storage helps it to rehydrate tissues. Though xylem is moderate, it is effective in transport. In a way, in avoiding drought, it resembles the chir pine of Kumaun. Like the oaks in Kumaun, the oaks in Nepal are profligate water users (Poudyal et al. 2004). They kept stomata at a considerably low Ψ_{PD}, and they maintained high leaf-specific conductivity.

Water Relations of Kharsu Oak (*Q. semecarpifolia*) in Nepal and Uttarakhand

Occurring generally from 2400 to treeline above 3000 m, *Q. semecarpifolia* has been studied both in lower elevations and treelines, and in Kathmandu, Nepal, as well as Uttarakhand. The average tree Ψ_{PD}, Ψ_{MD}, and daily change in Ψ of *Q. semecarpifolia* across all seasons in Kathmandu were −0.40 MPa, −1.18 MPa, and 0.77 MPa, respectively. In Kumaun, the lowest Ψ_{PD} and Ψ_{MD} were −0.83 and −1.98 MPa (Tyagi 2006), respectively, and the daily change was 1.15 MPa, which was far more than in Nepal. In another study, the Ψ_{PD} and Ψ_{MD} ranged respectively from −0.20 to −1.97 MPa and from −0.56 to −2.57 MPa (Shirish Kumar Singh unpubl.). In this study, the highest Ψ was recorded in monsoon and the lowest in premonsoon. *Q. semecarpifolia* realized marked daily reductions in Ψ even at relatively low Ψ_{PD}. It realized 1.19 MPa daily change at Ψ_{PD} −1.48 MPa (in Autumn), 0.97 MPa at Ψ_{PD} −1.60 MPa, and 0.31 MPa at Ψ_{PD} −1.97 MPa. The 70% drop in daily change from −1.60 to −1.97 MPa Ψ_{PD} is likely to be because of massive xylem embolism. At the treeline site, the highest daily change occurred during the monsoon when Ψ_{PD} was the highest and ranged from 0.76 to 1.14 MPa, with an average of ~1 MPa (Tewari A. and associates, FTR unpubl., IHTRP).

On average across the sites, years, and seasons, the osmotic adjustment was 1.37 MPa at Ψ_{sf} and 1.68sz (Shirish Kumar Singh unpubl.). These values are higher than the values of other species as well as in *Q. semecarpifolia* at other sites.

Saplings and seedlings had more negative Ψ_{PD} and Ψ_{MD} values than trees, and between them, seedlings had more negative values (Shirish Kumar Singh unpubl.). While the highest Ψ was during monsoon, the lowest was in premonsoon for saplings and spring for seedlings (Table 10.6).

Table 10.6 Range in water potential of *Q. semecarpifolia* and daily change in it during two study years, each measured in five seasons at Nainital site (Uttarakhand): monsoon = July–September; autumn = October–November; winter = December–February; spring = March–April; and premonsoon = May–June; PD, predawn; MD, midday

	Trees	Saplings	Seedlings
Ψ_{PD} (−MPa)	0.48 (monsoon)–1.97 (premonsoon)	0.12 (monsoon)–2.57 (premonsoon)	0.12 (monsoon)–3.55 (spring)
Ψ_{PD} (−MPa)	0.56 (monsoon)–2.57 (premonsoon)	0.43 (monsoon)–3.03 (premonsoon)	0.32 (monsoon)–3.73 (spring)
Daily change in Ψ (MPa)	0.08 (monsoon)–1.19 (autumn)	0.06 (monsoon)–1.44 (autumn)	0.9 (monsoon)–1.43 (autumn)

Source: Shirish Kumar Singh unpubl

Throughfall Exclusion Experiment: Effect of Soil Drying

Because of climate change, the uncertainty of precipitation patterns has increased. To simulate a precipitation decline, throughfall exclusion is used, which enables examination of the effect of soil drying on forest ecosystems. In a study in Central Bhutan on oak (*Q. griffithii* and *Q. lanata*, at 2640 m) and mixed conifer-oak (*Tsuga dumosa* and *Q. semecarpifolia*, at 3260 m), throughfall was excluded for three consecutive years during monsoon months, June to September, when 75% precipitation occurs (Wangdi et al. 2023). The main findings were as follows:

In the second year of the experiment, Ψ_{plant} decreased from −2.1 to −1.1 MPa (control) to −2.5 MPa in the oak forest and from −1.9 to −0.9 MPa (control) to −2.0 MPa in the mixed forest. It seems that *Q. griffithii* with very large vessels (252 μ) is more drought sensitive than small-vesseled (~100 μm, Gupta and Gupta 2020) *Q. semecarpifolia*. The drought treatment did not cause tree mortality during the entire three-year study. Stem increment remained unaffected until the second year, but caused growth reduction in the third year, the decline being most in *Tsuga dumosa* (60%). The fine root standing mass was unaffected in all species; however, oak trees increased carbon allocation to the belowground component. In oak trees, root production doubled due to throughfall exclusion. Aboveground litterfall was unaffected until the second year. However, soil CO_2 flux declined in all 3 years of throughfall exclusion. Overall, data shows that both oak and mixed conifer-oak forests of Bhutan were highly resistant to throughfall exclusion during the monsoon period. Perhaps, examining the effect of throughfall exclusion on shoot production during premonsoon could shed more light.

Conclusion

Despite the strong seasonality of rainfall and steep and rugged topography, forest trees are only moderately water stressed, indicating that forests effectively regulate and conserve water in a difficult terrain. The broken and scattered rocks and landslide deposits might contribute to retaining water within the mountains. In forest areas, tree water potential was high in lower elevation areas, particularly in sal despite warm temperatures, and tended to decline with elevation.

The lower forest and treeline species presented contrasting seasonal patterns. While trees had the highest Ψ_{PD} during monsoon months in both, this was the time when lower forest trees exhibited minimal, and the treeline species the maximum daily change in Ψ. Interestingly, the maximum daily change on average across the seasons in the treeline was significantly higher than the average of maximum values in forest trees. It seems that, being adapted to a relatively short growing period, treeline species exploit favorable water conditions for growth more effectively than the forest tree species. The species which occur both in treeline and forest areas, viz., *R. arboreum* and *Q. semecarpifolia,* differ in their seasonal patterns with regard to daily change in Ψ and faithfully follow specific patterns. On the other hand, lower forest trees are more adapted to premonsoon time water deficiency as they generally achieve the highest daily change in Ψ, despite the increasing water stress. These two tree species groups are expected to be differently affected by climate warming directly and through changes in monsoon drought. It is likely that during premonsoon, tissue development in treeline species is still not completed, hence they are not fully equipped for water uptake and leaf conductance.

The major forest species of mid- and low-elevation areas, banj oak and chir pine, are similar to an extent with regard to seasonal Ψ patterns. Both share several intermediate characters; however, chir pine is more conservative and maintains a wider safety margin than banj oak under premonsoon water stress. The deciduous species make a heterogenous group with only a limited number of common features. They generally occur sporadically and look outliers in the region dominated by evergreens. Even in the geological past, they are not reported as a prominent forest-forming tree.

Comparison of Kumaun and Nepal indicates that water stress declines from west to east, which also leads to an increasingly stronger dominance of evergreen broad-leaved species. As for seasonal patterns, they are hardly different between Nepal and Kumaun trees. In the two regions, oaks have several common characteristics, including delayed response of stomata to water stress, rigid cell wall, and a strong osmotic adjustment.

The extraordinarily high leaf conductance in treeline species occurring in the dry inner Himalayas calls for more detailed studies, involving more sites and species, and locations.

Do treeline species confront water stress? Though soil Ψ in treeline is generally less severe than in forest soils, treeline species are not adapted to water stress, possibly because of low temperature and resultant incomplete tissue root development. Tree ring growth analysis indicates that treeline species are vulnerable to premonsoon drought (Singh et al. 2018). Trees of a stationary treeline are likely to benefit with warmer temperatures, provided water stress doesn't intensify, particularly when the treeline becomes densified and trees become larger. Much of the effect of climate change would depend on how the premonsoon and monsoon periods are affected. In the years of severe premonsoon drought, oaks suffer severely, as the observations of 1999 and 2019 indicate, while chir pine survives. Growth phenology of treeline species is tied to monsoon; the species are not as effective as forest species in their adaptation to premonsoon drought. So much of the change would depend on how climate change affects the transition from the dry to the wet period.

Variation across seasons and sites in leaf conductance of adult trees and seedlings in treelines needs to be thoroughly analyzed and modeled. Our data on treelines is too sketchy to make any useful conclusion; in fact, they confuse us. Tree leaf conductance was found in an extraordinary range from less than 150 to over 1000 m Mol m^{-2} s^{-1} on a dry treeline site.

References

Abrams MD (1990) Adaptations and responses to drought in *Quercus* species of North America. Tree Physiol 7:227–238

Acherar M, Rambal S (1992) Comparative water relations of four Mediterranean oak species. Vegetatio 99(100):177–184

Anderson PA, Helms JE (1994) Tissue water relations of Pinus ponderosa and Arctostaphylos patula exposed to various levels of soil moisture depletion. Can J For Res 24:1495–1502

Anfodillo T, Rento S, Carraro V, Furlanetto L, Urbinati C, Carrer M (1998) Tree water relations and climatic variations at the alpine treeline: seasonal changes of sap flux and xylem water potential in *Larix deciduas* Miller, *Picea abies* (L.) Karst. and *Pinus cembra* L. Ann For Sci 55:159–172

Bahari ZA, Pallardy SG, Parker WC (1985) Photosynthesis, water relations and drought adaptation in six woody species of oak-hickory forest in Central Missouri. For Sci 31:557–569

Blackman CJ, Creek D, Maier C et al (2019) Drought response strategies and hydraulic traits contribute to mechanistic understanding of plant dry-down to hydraulic failure. Tree Physiol 39(6):910–924

Borchert R (1994) Soil and stem water storage determine phenology and distribution of tropical dry forest trees. Ecology 75:1437–1449

Brito P, Lorenzo JR, González-Rodríguez AM, Morales D, Wieser G, Jiménez MS (2014) Canopy transpiration of a Pinus canariensis forest at the tree line: implications for its distribution under predicted climate warming. Eur J For Res 133:491–500. https://doi.org/10.1007/s10342-014-0779-5

Brodribb TJ, Cochard H (2009) Hydraulic failure defines the recovery and point of death in water-stressed conifers. Plant Physiol 149(1):575–584

Cochard H, Tyree MT (1990) Xylem dysfunction in Quercus: vessel sizes, tyloses, cavitation and seasonal changes in embolism. Tree Physiol 6:393–407

Eamus D, Prior L (2001) Ecophysiology of trees of seasonally dry tropics: comparisons among phenologies. Adv Ecol Res 32:113–197

Gaire NP, Koirala M, Bhuju DR, Borgaonkar HP (2014) Treeline dynamics with climate change at the Central Nepal Himalaya. Clim Past 10(4):1277–1290

Gersony JT, Hochberg U, Rockwell FE, Park M, Gauthier PP, Holbrook NM (2020) Leaf carbon export and nonstructural carbohydrates in relation to diurnal water dynamics in mature oak trees. Plant Physiol 183(4):1612–1621

Gupta P, Gupta S (2020) Wood anatomy of Indian oaks, with reference to systematic, ecological and evolutionary perspectives. Nord J Bot 38(4)

Hacke U, Sauter JJ (1995) Vulnerability of xylem to embolism in relation to leaf water potential and stomatal conductance in *Fagus sylvatica*, *F. Purpurea* and *Populus balsamifera*. J Exp Bot 46:1177–1183

Kavanagh KL, Zaerr JB (1997) Xylem cavitation and loss of hydraulic conductance in western hemlock following planting. Tree Physiol 17:59–63

Klein T, Zeppel MJ, Anderegg WR et al (2018) Xylem embolism refilling and resilience against drought-induced mortality in woody plants: processes and trade-offs. Ecol Res 33:839–855

Körner C (2012) Alpine treelines: functional ecology of the global high elevation tree limits. Springer, Basel

Kramer PJ, Boyer JS (1995) Water relations of plants and soils. Academic, San Diego

Kubiske ME, Abrams MD (1992) Photosynthesis, water relations and leaf morphology of xeric versus mesic *Quercus rubra* ecotype in Central Pennsylvania in relation to moisture stress. Can J For Res 22:1402–1407

Landsberg J, Waring R (2017) Water relations in tree physiology: where to from here? Tree Physiol 37(1):18–32. https://doi.org/10.1093/treephys/tpw102

Machado JL, Tyree MT (1994) Patterns of hydraulic architecture and water relations of two tropical canopy trees with contrasting leaf phenologies: *Ochroma pyramidale* and *Pseudobombax septenatum*. Tree Physiol 14(3):219–240. https://doi.org/10.1093/treephys/14.3.219. PMID: 14967698

Nardini A, Salleo S (2000) Limitation of stomatal conductance by hydraulic traits: sensing or preventing xylem cavitation? Trees 15:14–24

Nardini A, Savi T, Losso A, Petit G, Pacilè S, Tromba G, Mayr S, Trifilò P, Lo Gullo MA, Salleo S (2017) X-ray microtomography observations of xylem embolism in stems of Laurus nobilis are consistent with hydraulic measurements of percentage loss of conductance. New Phytol 213(3): 1068–1075

Negi GCS, Singh SP (1992) Leaf growth pattern in evergreen and deciduous species of the Central Himalaya, India. Int J Biometeorol 36:233–242

Pathak GC, Joshi H, Singh RD, Tewari A, Pandey R, Singh SP (2021) Vertical root distribution in Himalayan trees: about half of roots occur below 30 cm, the generally sampled depth. Trop Ecol 62(3):479–491

Poudyal K, Jha PK, Zobel DB, Thapa CB (2004) Patterns of leaf conductance and water potential of five Himalayan tree species. Tree Physiol 24:689–699

Poudyal K, Jha PK, Zobel DB (2012) Role of wood water properties and leaf dynamics in phenology and response to drought in evergreen Himalayan tree species. Ecoprint 19:71–84

Prieto I, Armas C, Pugnaire FI (2012) Water release through plant roots: new insights into its consequences at the plant and ecosystem level. New Phytol 193(4):830–841

Ralhan PK, Khanna RK, Singh SP, Singh JS (1985) Phenological characteristics of the tree layer of Kumaun Himalayan forests. Vegetatio 60:91–101

Sakai A, Malla SB (1981) Winter hardiness of tree species at high altitudes in the East Himalaya, Nepal. Ecology 62(5):1288–1298

Scholander PF, Bradstreet ED, Hemmingsen EA, Hammel HT (1965) Sap pressure in vascular plants: negative hydrostatic pressure can be measured in plants. Science 148(3668):339–346

Singh SP, Tewari A, Singh SK, Pathak GC (2000) Significance of phenologically asynchronous populations of the central Himalayan oaks in drought adaptation. Curr Sci 79(3):353–357

Singh SP, Zobel DB, Garkoti SC, Tewari A, Negi CMS (2006) Patterns of water relations of central Himalayan trees. Trop Ecol 47:159–182

Singh SP, Singh RD, Gumber S, Bhatt S (2017) Two principal precipitation regimes in Himalayas and their influence on tree distribution. Trop Ecol 58(4):679–691

Singh U, Phulara M, David B, Ranhotra PS, Shekhar M, Bhattacharyya A, Dhyani R, Joshi R, Pal AK (2018) Static tree line of Himalayan silver fir since last several decades at Tungnath, Western Himalaya. Trop Ecol 59(2):351–363

Sperry JS, Pockman WT (1993) Limitation of transpiration by hydraulic conductance and xylem cavitation in *Betula occidentalis*. Plant Cell Environ 16:279–287

Stainton JDA (1972) Forests of Nepal. Hafner, New York, p 181

Tewari A, Shah S, Singh N, Mittal A (2018) Treeline species in Western Himalaya are not water stressed: a comparison with low elevation species. Trop Ecol 59(2):313–325

Tyagi V (2006) Forest species composition and diversity in relation to water status in an altitudinal transect in Kumaun Himalaya. PhD Thesis. Kumaun University, Nainital

Tyagi V, Singh SP, Singh RD, Gumber S, Thadani R, Pandey R (2023) Influence of slope position and aspect on the vegetation attributes and tree-water relations in forests of the Central Himalayas. J Mt Sci 20(9):2592–2602

Tyree MT, Karamanos AJ (1981) Water stress as an ecological factor. In: Grace JB, Ford ED, Jarvis PJ (eds) Plants and their atmospheric environment. Blackwell Scientific Publications, Oxford, pp 237–261

Tyree MT, Snyderman DA, Wilmot TR, Machado JL (1991) Water relations and hydraulic architecture of a tropical tree (*Schefflera morototoni*): data, models, and a comparison with two temperate species (*Acer saccharum* and *Thuja occidentalis*). Plant Physiol 96(4): 1105–1113. https://doi.org/10.1104/pp.96.4.1105

Wangdi N, Ahmed IU, Mayer M, Nirola MP, Orong K et al (2023) Estimating the response of Himalayan old-growth mountain forests to decreased monsoon precipitation. Agric For Meteorol 336:109471

Wieser G, Gruber A, Oberhuber W (2014) Sap flow characteristics and whole-tree water use of Pinus cembra across the treeline ecotone of the central Tyrolean Alps. Eur J For Res 133:287–295. https://doi.org/10.1007/s10342-013-0760-8

Xu L, Baldocchi DD (2003) Seasonal trends in photosynthetic parameters and stomatal conductance of blue oak (*Quercus douglasii*) under prolonged summer drought and high temperature. Tree Physiol 23:865–877

Zeppel MJ, Anderegg WR, Adams HD et al (2019) Embolism recovery strategies and nocturnal water loss across species influenced by biogeographic origin. Ecol Evol 9(9):5348–5361. https://doi.org/10.1002/ece3.5126

Zimmermann MH (1978) Structural requirements for optimal water conduction in tree stems. In: Tomlinson PB, Zimmermann MH (eds) Tropical trees as living systems. Cambridge University Press, London, pp 517–532

Zobel DB (1996) Variation of water relations parameters with extended rehydration time, leaf form, season, and proportion of leaf. Can J For Res 26(2):175–185

Zobel DB, Singh SP (1995) Tree water relations along the vegetational gradient in the Himalayas. Curr Sci 68(7):742–745

Zobel DB, Ram J, Bargali SS (1995) Structural and physiological changes in *Quercus leucotrichophora* and *Pinus roxburghii* associated with stand disturbance in the Kumaun Himalaya, India. Int J Ecol Environ Sci 21:45–66

Zobel DB, Garkoti SC, Singh SP, Tewari A, Negi CMS (2001a) Patterns of water potential among forest types of the central Himalaya. Curr Sci 80:774–779

Zobel DB, Garkoti SC, Singh SP (2001b) Leaf conductance of primary and mature leaves of Pinus roxburghii:a comparison. J For Res 6:1–5

Community Structure and Ecophysiology in Major Forest Types of Himalayas in Response to Slope Aspects

11

Surabhi Gumber, Ripu Daman Singh, S. P. Singh, and Rajesh Thadani

Abstract

This chapter examines the effect of two contrasting aspects, North and South, and three slope positions (Hill Base, Mid Slope, and Hill Top) on each aspect on vegetation parameters, tree and soil water potential, and net assimilation rate (NAR) in a series of Central Himalayan forests, namely sal (*Shorea robusta*) forest, sal-chir pine (*Pinus roxburghii*) forest, chir pine forest, banj oak (*Quercus leucotrichophora*) forest, tilonj oak (*Q. floribunda*) forest, and kharsu oak (*Q. semecarpifolia*) forest. Water potential (Ψ) and NAR were measured in all seasons, namely winter, pre-monsoon, monsoon, spring, and autumn. The mesic northern slopes and hill bases generally supported higher species richness, diversity, tree density, and basal area compared to southern slopes and hill tops. Ψ_{PD} and Ψ_{soil} were generally higher on moist north slope aspect (−0.78 ± 0.05 MPa and −3.34 ± 0.18 MPa, respectively) than dry south slope aspect (−0.82 ± 0.18 MPa and −3.77 ± 0.18 MPa, respectively). Across six different forests, these values were higher at hill base (−0.71 ± 0.06 MPa and −

S. Gumber
G. B. Pant University of Agriculture and Technology, Pantnagar, Uttarakhand, India

R. D. Singh (✉)
Graphic Era (deemed to be) University, Dehradun, Uttarakhand, India
e-mail: ripuds4777@gmail.com

S. P. Singh
INSA Honorary Scientist, Dehradun, Uttarakhand, India

Centre for Himalayan Studies, Graphic Era (deemed to be) University, Dehradun, Uttarakhand, India

R. Thadani
Centre for Ecology Development and Research (CEDAR), New Delhi, India

Department of Botany, Jai Hind College, Churchgate, Mumbai, India

S. P. Singh, R. Thadani (eds.), *Himalayan Forests in a Changing World*,
https://doi.org/10.1007/978-981-95-0978-2_11

2.77 ± 0.19 MPa, tree predawn water potential and soil water potential, respectively) than other topographical positions. Net assimilation rates (μ mol m^{-2} s^{-1}) followed similar patterns; the average across the four forest sites and five seasons was higher on the north (5.08 ± 0.23) slope aspect than south aspect (4.61 ± 0.22), and among the three slope positions, it was higher on hill base (5.13 ± 0.29) than on mid-slope (4.80 ± 0.27) and hill top (4.61 ± 0.26). The novelty of this research lies in its comprehensive analysis of how topographical factors shape forest composition and functioning across an elevational gradient in the Central Himalayas. By demonstrating the importance of northern slopes and hill bases as potential refugia for species under climate change, this study provides valuable insights for forest conservation and management strategies in the region.

Keywords

Broadleaved evergreen forests · Net Assimilation Rate · Refugia · Slope aspect · Slope positions · Water Potential

Introduction

Topographical factors such as slope aspect and slope position play a crucial role in shaping the local environment, thereby significantly influencing the distribution and characteristics of vegetation (Daws et al. 2002; Moeslund et al. 2013; Sundqvist et al. 2013; Jucker et al. 2018). Slope aspect, in particular, affects solar radiation (see Chap. 2, Box 2.1), leading to varying environments and associated vegetation patterns between equator- and pole-facing slopes (South- and North-facing slopes in the northern Hemisphere). In semi-arid areas, slope aspect significantly influences microclimate variables such as air and soil temperature, evapotranspiration, and wind speed (Cantlon 1953; Bennie et al. 2008; Burnett et al. 2008), as well as soil properties like organic matter content, chemical composition, soil depth, and texture (Geroy et al. 2011; Lozano-García et al. 2016; Rasmussen and Tabor 2007). Topographical factors impact hydrological processes, such as runoff dynamics, hydraulic conductivity, and soil water retention (Casanova et al. 2000; Gutiérrez-Jurado et al. 2007; Broxton et al. 2009; Wang et al. 2011), leading to distinct vegetation types on opposite slopes (Cantlon 1953; Pook and Moore 1966; Holland and Steyn 1975).

Generally, polar-facing slopes (North aspect) are cooler and wetter, with richer organic matter and deeper soils, supporting mesic vegetation than equator-facing slopes (South aspect) (Cantlon 1953; Pook and Moore 1966). Slope degree also plays a crucial role in controlling vegetation size, composition, and species distribution by influencing factors such as wind speed, soil water and nutrient content, solar radiation intensity, and seed dispersal distance (Moeslund et al. 2013). In addition, steeper slopes, being prone to erosive forces of water, have thinner soils. Indirectly, slope degree affects fire severity and frequency, as steeper slopes often experience higher wind speeds leading to shorter trees (Gardiner et al. 2016) and favouring deep-rooted species due to soil stability issues (Moore et al. 2018).

Aspect influences solar radiation, precipitation, and wind, impacting vegetation composition, structure, and growth (Burnett et al. 2008; Moeslund et al. 2013; Qin et al. 2021). In temperate regions of the Northern Hemisphere, equator-facing aspects such as south and southwest slopes experience significantly warmer microclimates than north- and east-facing slopes (Perring 1959). This temperature difference influences species composition and diversity across various mountain ranges (Yang et al. 2016; Heydari et al. 2021; Qin et al. 2021). The dryness of equator-facing slopes increases fire severity, affecting vegetation structure and species richness (Beaty and Taylor 2001; Rogeau and Armstrong 2017; Gowda et al. 2019).

Slope aspect can interact with other topographical variables, such as slope position, to influence vegetation structure (Gallardo-Cruz et al. 2009). The lower part of a slope, or hill base, is generally cooler, moister, and receives less intense solar radiation than the upper slope (Singh and Singh 1992). It also has lower evapotranspiration rates and deeper soils with higher organic matter capable of retaining runoff water better than other positions in a mountain landscape (Daws et al. 2002). Anthropogenic activity further increases differences in diversity patterns between opposite slopes and positions (Paudel and Vetaas 2009) as humans preferentially use more mesic sites with deeper soils for agricultural activity, thereby depleting the vegetation from such sites (Thadani 1999).

In this chapter, we demonstrate the impact of slope aspect and position on species distribution, vegetational attributes such as diversity and tree basal cover, tree water relations, and net assimilation rate of leaves. An elevational sequence of forests from the foothills to 2000 m elevation in the central Himalayas is used as an example in this chapter. Apart from vegetation, the chapter focuses on tree water relations, and photosynthesis rates are analysed in relation to contrasting aspects (North and South) and slope positions like hill base and ridge top in each of the forests. Different forest types served the purpose of replicates to explain differences in aspect and slope position (Tyagi 2006; Tyagi et al. 2023a, b).

Species Distribution in Relation to Topographical Features

Distribution of many tree species of the Himalayas is associated with particular slope aspects and habitats, such as shady ravines, damp slopes, ridges, water courses, and others. Species distribution in relation to slope aspect is in relative terms. For example, *Pinus roxburghii* occurs more frequently on warm and dry Southern (S) aspect and ridge tops as it most effectively outcompetes other co-occurring species on these drought-prone sites. *P. roxburghii* is often present on ridge top of a mountain slope which could have oaks at the hill base. On the other hand, *Quercus lamellosa* can be found on all slope aspects and ridges, but does better on more moist Northern aspect in the drier central Himalayas, though in the much wetter conditions of the eastern Himalayas this species is found on the dryer South-facing slopes (Stainton 1972). *Q. glauca* which is more sporadically distributed than other Himalayan oaks thrives best in shady ravines, along water courses. Several deciduous species like maple and horse chestnuts perform best along streams in areas where

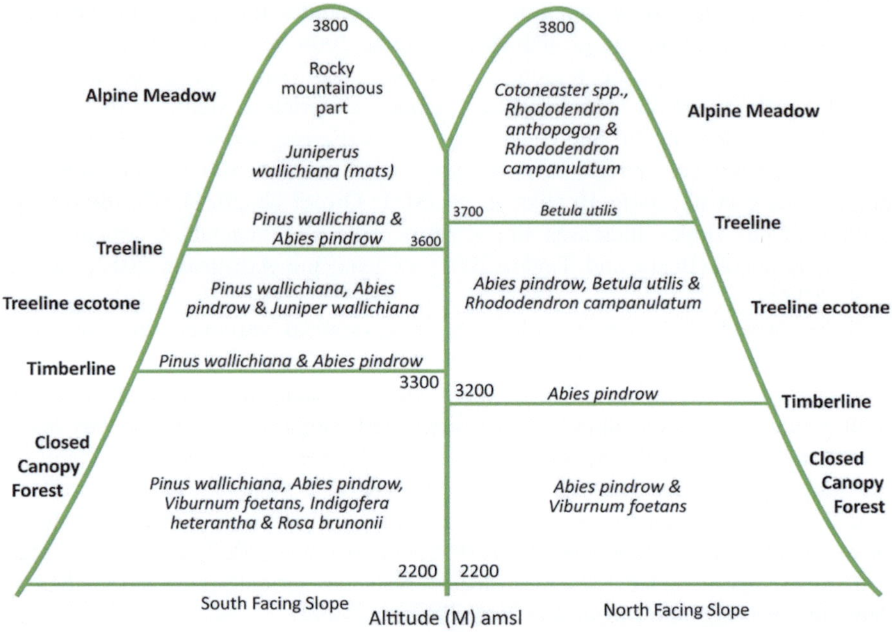

Fig. 11.1 Treeline vegetation patterns on the north-facing and south-facing slopes in Kashmir (derived from FTR-IHTRP, 2021, by Prof. Zafar Reshi and his associates). *Timberline is defined as the upper limit of forests with at least 30% crown cover (closed canopy); and Treeline is a line connecting the uppermost trees—often isolated and scattered single-stemmed trees (>2 m height).* (Singh et al.2021)

evergreen forests otherwise dominate. Near a treeline in Kashmir valley, N slopes are seen covered with *Betula utilis* with undergrowth of *Rhododendron campanulatum,* while the drier S slope aspect is under the domination of blue pine (*P. wallichiana*) along with juniper mat on ground. Both pines and junipers are known to have wide safety margins with respect to hydraulic loss (Choat et al. 2012). Silver fir, on the other hand, often occurs on slopes where birch dominates (Nanda et al. 2018). A comprehensive sampling of plant diversity in the Daksum-Sinthan Top site in Kashmir recorded 423 plant species (including non-seed plants) belonging to 291 genera and 120 families (Fig. 11.1). At all levels, the north-facing slopes (cool and moist) were richer than the south-facing slope (warm and dry), the species, genera, and family numbers being 280, 209, and 87 on the north aspect and 250, 180, and 85 on the south aspect, respectively. The two aspects differ in dominant forest species and elevational ranges. The treeline was higher on the south aspect. On the relatively drier south-facing slope, *P. wallichiana* was the characteristic treeline species (Fig. 11.1).

A study in the Daksum Sinthan Top area (between 2200 and 3800 m) and Drang Kongdoori (between 2300 and 3800 m) of Kashmir by Nanda (2024) also found that the total number of species was higher on the north (N) and south (S) slope aspects. However, species richness exhibited variation between the slopes. The north-facing slope demonstrated higher species richness across all groups, with the exception of bryophytes. The number of conifer species was equivalent on both slopes and sites. In terms of species richness, Asteraceae and Lamiaceae were the most significant families.

Steep south-facing slopes on limestone rocks are often dominated by *Cupressus torulosa* at around 2200 m elevations in the monsoonal outer ranges (Stainton 1972), and the areas with gentle slopes around the cypress forest remain under oak forests (*Q. floribunda* and *Q. leucotrichophora*).

In Chopta transect of Uttarakhand forest, forest composition changes rapidly from riverine sites to upland sites. Along the river, alder (*Alnus nepalensis*) dominates, and as we move up, *Betula alnoides* replaces alder, and further up *Q. leucotrichophora* and *Q. floribunda* monopolize the slopes. Occurrence of deciduous forests, such as *Aesculus-Juglans-Acer,* is common on river flats in the west midlands of Nepal (Stainton 1972) (Table 11.1). Alder is particularly associated with landslides. *Populus ciliata* is also a common colonizer of open barren sites.

Table 11.1 General distribution of tree species in relation to aspects

Forest species	Elevational range (m)	Aspect
Aesculus indica	1800–3100	Northern
Q. leucotrichophora	1000–2400	All aspects, more common on cool and moist northern aspects
Q. floribunda	1900–2900	All aspects but more on northern aspect
Q. semecarpifolia	2200–3500	All aspects, more on northern aspect
Q. lanata	1500–2400	Southern aspect
P. wallichiana	1600–3500	Southern aspect
P. roxburghii	500–2700	More on southern aspect Grows on bare-rock to an extent seldom seen in other species (Troup 1921)
Cedrus deodara	1800–3000	On both aspects. More common on southern in the moister outer ranges
Abies pindrow	2000–3500	Northern, but at higher elevations found on all aspects
Shorea robusta	200–1200	Southern
Schima wallichii	200–2000	Northern and southern both
Castanopsis indica	1000–2900	Northern and southern both
Anogeissus latifolia	Up to 1200	Southern
Q. lamellosa	1600–2800	Thrives best on north aspect but can be found on any aspect and also ridges
Q. glauca	800–2000	Shady ravines/water courses/damp sites
Abies spectabilis	2500–3000	Southern
Tsuga Dumosa	2100–3600	Northern and southern, but generally on southern
Cupressus torulosa	3000–3500	Southern
Picea smithiana	2100–3600	Southern

Source: Troup (1921), Champion and Seth (1968), Stainton (1972), and Miehe et al. (2015)

When *P. roxburghii* and *Q. leucotrichophora* occur together, the oak tends to occupy northern slope aspect and the pine on southern aspect. When *Q. leucotrichophora* and *Q. floribunda* are present, the latter tends to occupy cooler and moist sites.

In the Himalayas, topographically driven sites with deep shade, such as hollows, frequently exhibit distinct species compositions. In these low-light environments, canopy species may become less abundant, while understory species increase in density. *Q. floribunda* (canopy) and *Machilus duthei* (understory) represent a common example of such deeply shaded sites in the temperate zone.

Topographical features play a crucial role in shaping the composition of Himalayan forests, with various plant species exhibiting distinct preferences for specific aspects and landforms. The complex terrain of the Himalayas, characterized by steep slopes, varied elevations, and diverse microclimates, creates a mosaic of habitats that support a rich array of vegetation. For instance, north-facing slopes often harbour moisture-loving species due to reduced sun exposure and higher humidity levels, while south-facing slopes tend to support more drought-tolerant vegetation. Ridges and valleys also influence species distribution, with some plants thriving in the sheltered conditions of ravines and others adapted to the exposed environments of mountain crests. Additionally, elevation gradients contribute to vertical zonation of forest types, with species composition changing markedly as altitude increases. This intricate relationship between topography and forest composition underscores the importance of landscape heterogeneity in maintaining biodiversity and ecosystem functioning in the Himalayan region.

Effect of Slope Aspect on Leaf Characters

Leaf characters may change with change in aspect. In a central Himalayan transect, the leaf area on average was found larger on the warmer south (S) slope aspect than on the cooler north (N) aspect (Table 11.2; Singh and Negi 2018). However, because

Table 11.2 Leaf characteristics of dominant tree species of western Himalayan forests

	Leaf area (cm^2 $leaf^{-1}$) of mature leaf		Leaf mass (g $leaf^{-1}$) of mature leaf		Leaf mass per unit area (LMA) (g m^{-2})	
Species	N aspect	S aspect	N aspect	S aspect	N aspect	S aspect
S. robusta	169.6	175.1	2.05	2.1	120.87	119.93
Mallotus phillippensis	58.3	60.1	0.4	0.43	68.61	71.55
P. roxburghii	6.9	7.8	0.18	0.2	260.87	256.41
Myrica esculenta	26.6	29.9	0.34	0.37	127.82	123.75
Q. leucotrichophora	31.6	34.6	0.43	0.46	136.08	132.95
R. arboreum	39.8	42.7	1.66	1.75	417.09	409.84
Q. floribunda	16.7	18.3	0.34	0.36	203.59	196.72
Machilus duthei	32.4	35.3	0.45	0.49	138.89	138.81
Mean	**47.74**	**50.48**	**0.73**	**0.77**	**184.23**	**181.24**

Derived from Singh and Negi (2018)

of larger leaf area, leaf mass per leaf was also higher on S aspect than on N aspect; however, leaf mass per unit area (LMA) was greater on N aspect than S aspect.

Ackerly and Cornwell (2007) observed that plants on south-facing slopes were characterized by high leaf mass area (LMA) and smaller leaves. Similarly, Li et al. (2021) found a significant increase in LMA and a reduction in leaf size (LS) in herbaceous species on south-facing slopes in a subalpine meadow of the eastern Tibetan Plateau. Wright et al. (2018) reported that globally, leaf size at drier sites was primarily influenced by daytime energy balance, whereas at wetter sites, it was limited by nighttime energy balance. They noted that leaves were smaller at warmer, drier sites, in hotter dry regions, and in colder, wetter environments. This suggests that smaller leaves are more adaptive to environmental conditions by effectively regulating temperature, particularly under high light intensity and heat (Wang et al. 2019). Ackerly et al. (2002) also found that plants on south-facing slopes in coastal California forests had higher LMA and smaller LS, while Li et al. (2019) reported that north-facing slopes in alpine steppes supported larger leaves at the population level.

In a study examining the impact of slope aspect on plant traits and biomass allocation in *Gentiana hexaphylla* at the eastern edge of the Qinghai-Tibet Plateau, Xue et al. (2018) discovered that plants on north-facing slopes had significantly more branches compared to those on partial shady, east, and west-facing slopes. Contrastingly, Bonser and Aarssen (1994) reported that plants tended to increase branch numbers under strong solar radiation, opposite to the findings of Xue et al. (2018). Additionally, there were more diverse pollinator communities in low-elevation populations compared to those at higher elevations (Bingham and Orthner 1998; Arroyo et al. 1982).

Effect of Slope Aspect on Vegetation Attributes

In a study by Tyagi et al. (2023a, b), six forest types (ranging from sub-tropical forest types of the foothills (<500 m) to temperate forests (>2500 m elevation)) in Uttarakhand Himalayas were studied to understand the effect of difference in slope, aspect, and topographical position on soil and tree-water relation and vegetation attributes. Each forest type was investigated on two slope aspects (Northern and Southern) and three positions on each slope aspect, viz. hill base (HB), mid-slope (MS), and hill top (HT). The six study forest types (sal, sal-chir pine, chir pine, banj oak, tilonj oak, and kharsu oak) together showed the presence of 218 species: 44 trees, 63 shrubs, and 111 herbs. The species numbers on N slope aspect (total 186 species; 33 trees, 46 shrubs, and 107 herbs) were significantly ($p < 0.05$) higher than those on S slope aspect (158 species; 29 trees, 39 shrubs, and 90 herbs) (Table 11.3), because of higher average species richness for trees, shrubs, and herbs on the N slope aspect as compared to the S slope aspect ($p < 0.01$).

Of the 186 species that occurred in mesic aspect of the six forests, 94 species were unique to HB stand, which included 12 tree species, 22 shrub species, and 60 herb species not found in MS and HT (Table 11.3). Species unique to HB (not found in MS and HT) were 50.5% of the total species recorded (186) in mesic aspect type of

Table 11.3 Total number of plant species across six forests in relation to slope aspects (north (N) or south (S)) and slope positions (HB = hill base, MS = mid-slope, and HT = hill top) by growth form

Growth form	Total species (no.)	Total species number					Unique species number (Percent)		Unique species contribution number (Percent)				
		N	S	HB	MS	HT	Both aspects	All slope position	N	S	HB	MS	HT
Tree	44	33	29	28	27	24	26 (59.1%)	22 (50%)	15 (45.5%)[a] (57.7%)[b]	11 (34.5%)[a] (42.3%)[b]	8 (28.6%)[a] (36.4%)[b]	6 (22.2%)[a] (27.3%)[b]	8 (33.3%)[a] (36.4%)[b]
Shrub	63	46	39	47	41	33	41 (65.1%)	26 (41.3%)	24 (52.2%)[a] (58.5%)[b]	17 (43.6%)[a] (41.5%)[b]	15 (31.9%)[a] (57.7%)[b]	5 (12.2%)[a] (19.2%)[b]	6 (18.2%)[a] (23.1%)[b]
Herb	111	107	90	92	83	72	25 (22.5%)	32 (28.8%)	21 (19.6%)[a] (84%)[b]	4 (4.4%)[a] (16.0%)[b]	23 (25.0%)[a] (71.9%)[b]	7 (8.4%)[a] (21.9%)[b]	2 (2.8%)[a] (6.3%)[b]
Total	218	186	158	167	151	129	92 (42.2%)	80 (36.7%)	60 (32.3%)[a] (65.2%)[b]	32 (27.3%)[a] (34.8%)[b]	46 (27.5%)[a] (57.5%)[b]	18 (11.9%)[a] (22.5%)[b]	16 (12.4%)[a] (20.0%)[b]

Source: Tyagi et al. (2023a)
Notes: [a]Percent of total species present; [b]percent of unique species
Unique species are the species found only on one aspect type (either N or S) or one hill position (either HB or MS or HT)

six forests. On dry aspect, the number of such species was 77 (9 tree species, 16 shrubs, and 52 herbs) which comes to 48.4% of total species (Table 11.3).

Except for sal and tilonj oak forest, species occurring only in HB were more in mesic aspect than dry aspect. The aspect difference was larger in total species in kharsu oak, pine-sal, and pine forests. Species richness across various forest types and aspects shows distinct patterns. In Sal forests, shrub richness is almost similar between aspects, while trees and herbs are significantly richer on the moist northern aspect compared to the dry southern aspect (Fig. 11.2). Similarly, in pine-sal and

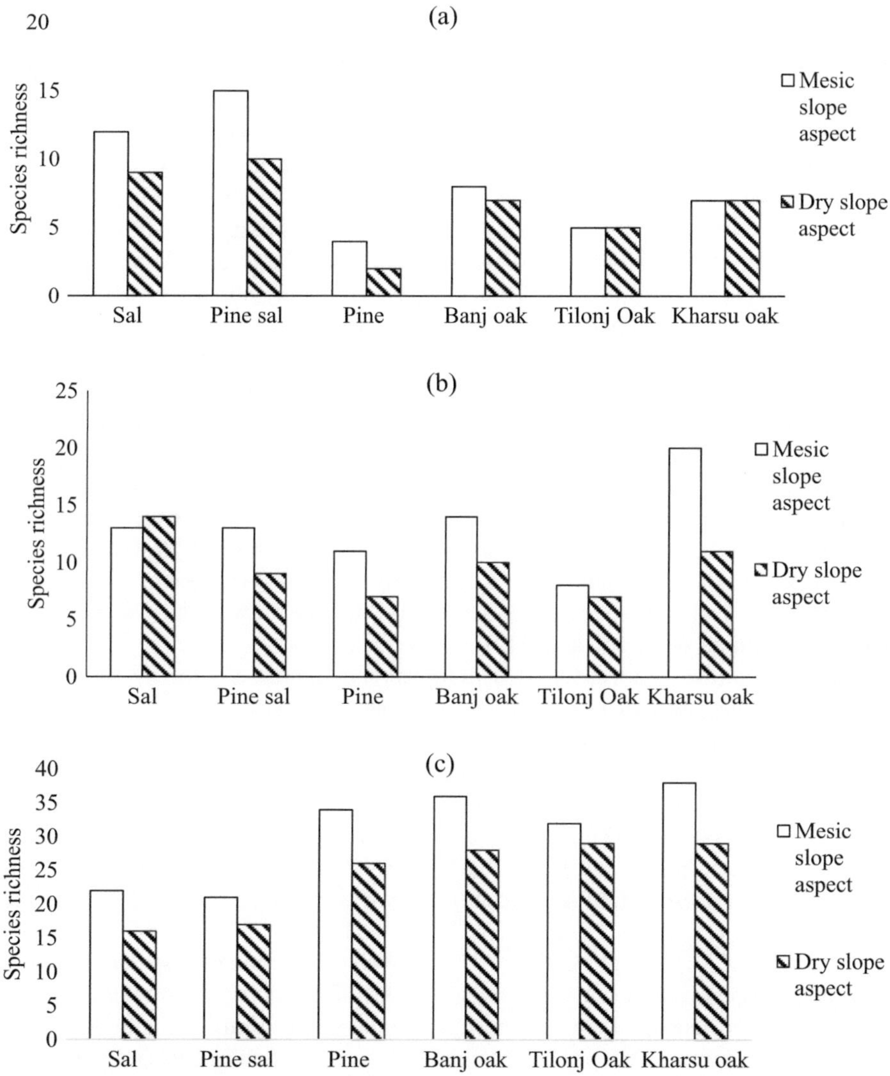

Fig. 11.2 Species richness of (**a**) Trees, (**b**) Shrubs, and (**c**) Herbs in mesic and dry aspects across all the study forest types (derived from Tyagi 2006). The effect of N aspect (mesic) on species richness of shrubs and herbs tended to increase with elevation

pine forests, tree, shrub, and herb richness is higher on the moist northern aspects with a notable difference in herb richness (Fig. 11.2). In banj oak forests, tree, shrub, and herb richness is higher on the moist northern aspect. The dry south aspect was particularly poor in shrubs, values being comparable to those of shrub species poor pine forest.

Of 54 species richness values (6 forest sites, 3 growth forms, and 3 slope positions), 47 (87%) values were higher on N slope aspect than S slope aspect (Table 11.4).

For species diversity (H'), the N slope aspect values were higher in 44 (81.5%) instances. In the sal forest, tree diversity ranged from 0.87 at the south-facing hilltop to 1.71 at the north-facing hill base, while shrub diversity varied between 1.79 and 2.14 and herb diversity from 1.75 and 2.74. In the pine-sal mixed forest, tree diversity ranged from 1.41 at the north hilltop to 1.98 at the north hill base, with shrub diversity between 1.33 and 2.04, and herb diversity from 1.59 to 2.49. The pine forest exhibited zero tree diversity at south and north hilltops, increasing to 0.71 at the north hill base, with shrub diversity from 0.67 to 1.83, and herb diversity ranging from 2.12 to 2.78. In the banj-oak forest, tree diversity ranged from 0.30 at the south slope to 0.62 at the north base, shrub diversity between 1.31 and 1.84, and herb diversity from 2.46 to 2.84. Thus, banj oak was only slightly richer than chir pine forest in shrubs and herbs diversity. The tilonj oak forest showed the minimum tree diversity at the south hilltop and maximum at the north hilltop, shrub diversity from 1.08 to 1.70, and herb diversity from 2.34 to 2.87. In the kharsu oak forest, tree diversity was the lowest at the south hilltop and highest at the north base, with shrub diversity ranging from 1.36 to 2.51, and herb diversity from 2.47 to 2.85. Overall, the north aspects generally supported higher diversity, while the south hilltops exhibited lower diversity across the studied forest types. Beta diversity was somewhat higher on S slope aspect (4.46 for trees, 4.25 shrubs, and 3.70 for herbs) than N slope aspect (3.88, 3.63, and 3.61, respectively). Of the 218 species, 92 (42.2%) were unique to just one aspect (i.e. either N slope or S slope) with the N slope having more such unique species (Table 11.3).

The total number of species across all forest types decreased from the more mesic HB (167 species) to MS (151 species) with least on the dryer HT (129 species) (Table 11.3). Among the slope positions, species diversity and species richness for trees, shrubs, and herbs were significantly higher for HB ($p < 0.05$) than for the others. This pattern was consistent with that of average species richness which declined from HB to HT in each growth form: from 7.33 to 4.83 for trees, 11.5 to 7.67 for shrubs, and from 20.83 to 16.67 for herbs. The tree, shrub, and herb species richness varied significantly among slope positions ($p < 0.05$). Contrarily, beta diversity increased from HB to HT: from 3.82 to 4.97 for trees, 4.09 to 4.30 for shrubs, and 3.56 to 4.32 for herbs. The number of species confined to a single slope position (unique) was 80 (36.7% of 218 species), of which 22 were trees, 26 shrubs, and 32 herbs. With regard to slope positions, the unique species richness was higher in HB in 21 instances compared to two in MS and one in HT (Table 11.4). Of the total unique species (80), 46 (57.5%) occurred at HB, 18 (22.5%) at MS, and only 12 (16%) at HT (Table 11.3).

Table 11.4 Plant species richness and the number of unique species in relation to slope aspect (N = north slope aspect and S = south slope aspect) and slope positions (HB = hill base, MS = mid-slope and HT = hill top) in the six forest sites

		Species richness							Unique species distribution by slope aspects and slope positions					
		HB		MS		HT			HB		MS		HT	
Forest site	Growth form	N	S	N	S	N	S	Nos of unique species	N	S	N	S	N	S
Sal forest	Trees	8	6	7	4	6	4	7	2	1	1	0	2	1
	Shrubs	10	9	9	7	7	7	12	2	4	0	2	1	3
	Herbs	18	13	13	9	9	6	3	2	0	1	0	0	0
	Total	36	28	29	20	22	17	22	6	5	2	2	3	4
Sal pine forest	Trees	10	8	7	7	7	6	12	4	3	2	1	2	0
	Shrubs	9	8	7	6	6	4	7	3	1	2	0	0	1
	Herbs	16	11	4	10	9	5	6	3	2	0	0	1	0
	Total	35	27	18	23	22	15	25	10	6	4	1	3	1
Pine forest	Trees	4	2	3	2	1	1	1	1	0	0	0	0	0
	Shrubs	7	3	5	3	4	2	4	3	0	1	0	0	0
	Herbs	22	18	16	15	12	9	6	5	1	0	0	0	0
	Total	33	23	24	20	17	12	11	9	1	1	0	0	0
Banj oak forest	Trees	6	4	4	3	3	3	6	2	0	1	1	0	2
	Shrubs	8	5	7	5	5	4	8	1	1	3	1	1	1
	Herbs	31	22	24	18	18	15	6	5	1	0	0	0	0
	Total	45	31	35	26	26	22	20	8	2	4	2	1	3
Tilonj oak forest	Trees	4	3	2	4	2	4	0	0	0	0	0	0	0
	Shrubs	6	5	4	3	4	3	4	2	1	1	0	0	0
	Herbs	20	19	19	12	14	13	11	3	1	2	3	0	2
	Total	30	27	25	19	20	20	15	5	2	3	3	0	2
Kharsu oak forest	Trees	5	4	6	5	3	2	1	0	0	0	0	1	0
	Shrubs	13	7	13	5	8	4	12	4	3	2	0	2	1
	Herbs	23	22	16	15	17	15	18	6	5	6	0	1	0
	Total	41	33	35	25	28	21	31	10	8	8	0	4	1

Source: Tyagi et al. (2023a)

Table 11.5 Plant species richness and unique species by the forest sites in relation to slope positions

Forest site	Growth form	Distribution of species richness by slope position			Nos of unique species	Distribution of unique species by slope position		
		HB	MS	HT		HB	MS	HT
Sal forest	Trees	10	9	8	7	3	1	3
	Shrubs	17	14	13	12	6	2	4
	Herbs	18	13	9	11	8	2	1
	Total	45	36	30	30	17	5	8
Sal pine forest	Trees	14	11	8	12	7	3	2
	Shrubs	11	9	8	7	4	2	1
	Herbs	19	14	9	10	8	1	1
	Total	44	34	25	29	19	6	4
Pine forest	Trees	4	3	1	1	1	0	0
	Shrubs	8	6	4	4	3	1	0
	Herbs	23	18	14	14	13	1	0
	Total	35	27	19	19	17	2	0
Banj oak forest	Trees	7	5	5	6	2	2	2
	Shrubs	9	11	6	9	3	4	2
	Herbs	34	26	24	7	7	0	0
	Total	50	42	35	22	12	6	4
Tilonj oak forest	Trees	4	4	4	1	0	1	0
	Shrubs	7	5	5	6	4	1	1
	Herbs	28	26	23	12	5	5	2
	Total	39	35	32	19	9	7	3
Kharsu oak forest	Trees	5	6	3	1	0	0	1
	Shrubs	17	15	10	12	7	2	3
	Herbs	33	26	21	22	15	6	1
	Total	55	47	34	35	22	8	5

Source: Tyagi et al. (2023a)
HB hill base, *MS* mid-slope, *HT* hill top

The proportional contribution of HB to unique herb species was about 72% (Table 11.3), while for unique tree species all the three positions contributed similarly. Within each of the six forests, species richness and unique species number generally decreased from HB to HT (Table 11.5).

The total tree density (sum of densities of all species present) in forest stands generally ranged between 500 and 700 trees ha^{-1}. The species density of trees, shrubs, and herbs were significantly higher on N slope aspect ($p < 0.01$) than S slope aspect (Fig. 11.3).

On average across the slopes of all forests, the total tree density was highest in HB, followed by HT and MS (Fig. 11.4). The tree, shrub, and herb species density varied significantly among slopes ($p < 0.05$) and between aspects ($p < 0.05$). The average tree density across all forest sites was higher at HB (614.2 ± 33.7 ind. h^{-1}),

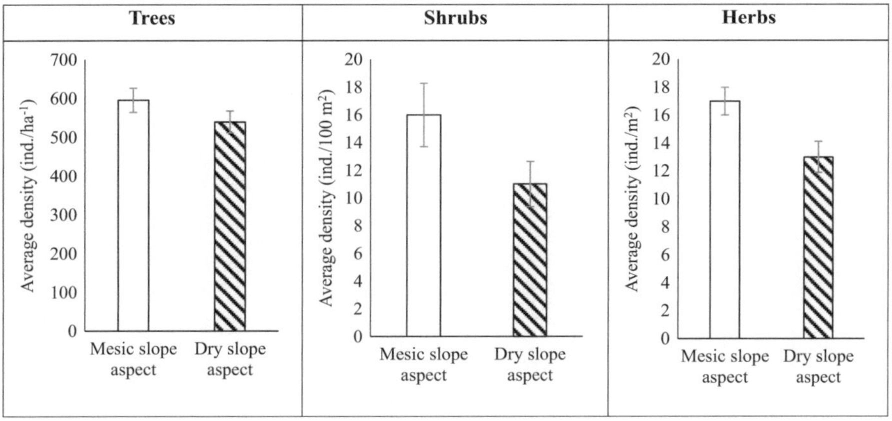

Fig. 11.3 Average density of trees, shrubs, and herbs in mesic and dry slope aspects across topographical positions of all the six study forests (derived from Tyagi 2006). Average of a slope aspects is of 18 values, i.e. 3 topographical positions × six forest types = 18

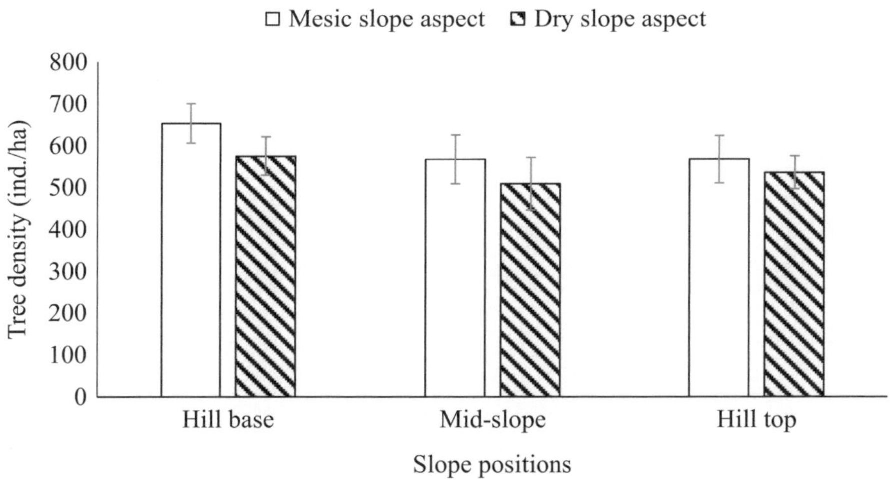

Fig. 11.4 Average tree density on different slope positions across all aspects in the study forest types (each histogram represents the average of six stands). (Derived from Tyagi 2006)

than HT (550.8 ± 33.2 ind. h^{-1}) and MS (537.5 ± 41.8 ind. h^{-1}). At HB, it was either more on mesic aspect (in four forests) or equal on two aspect types (in pine and tilonj oak forests).

The average density across all the six forest types decreased from HB to HT for shrubs (7.50–4.85 ind./100 m^{-2}), as well as for herbs (17.7–9.25 ind./m^{-2}). Pine-Sal forests exhibited higher densities of both shrubs and herbs on the moist northern aspects than on the dry southern aspects (Fig. 11.5). Pine forests had lower densities of shrubs and herbs on the dry southern slope compared to the mesic northern slope,

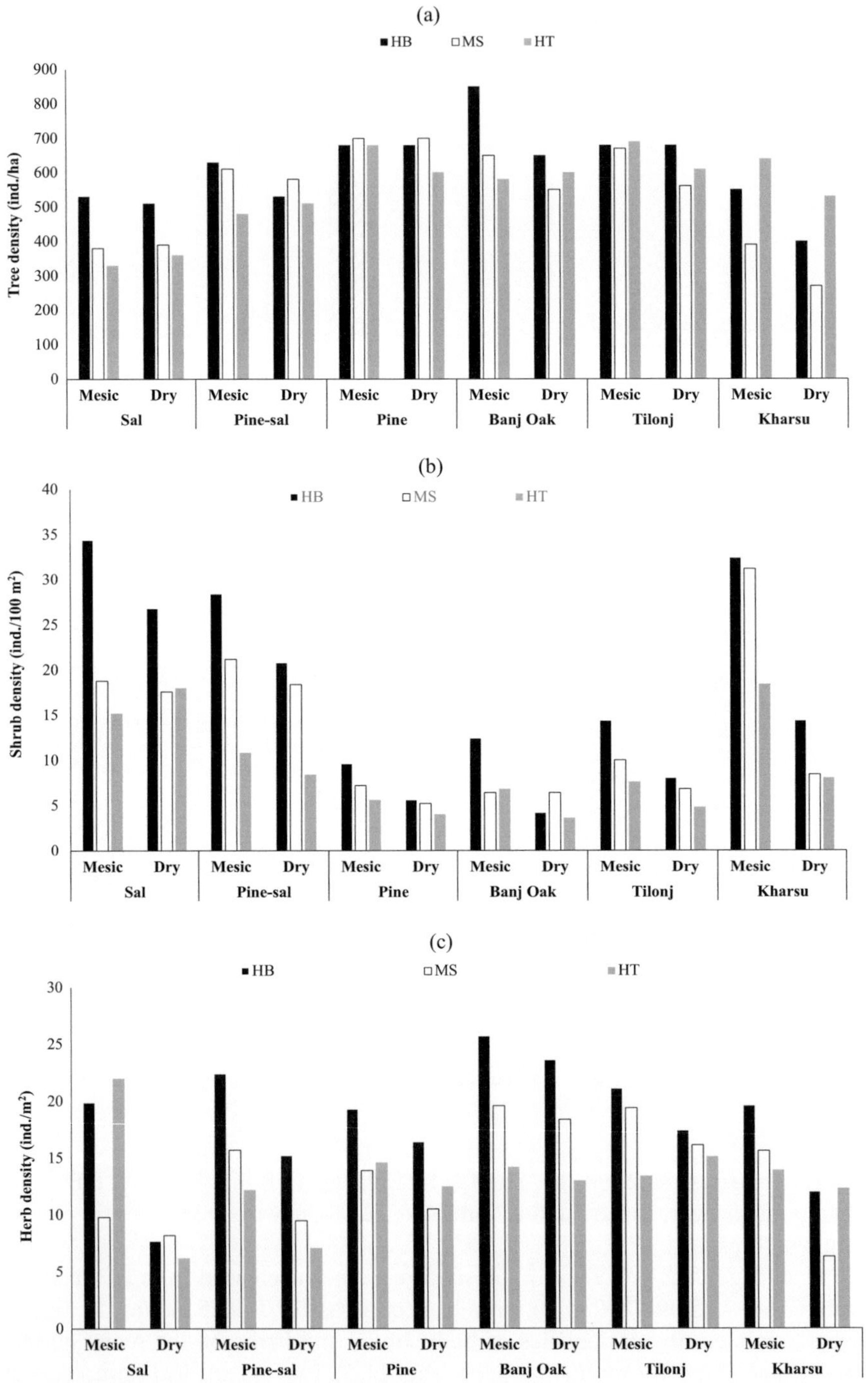

Fig. 11.5 Species density of (**a**) Tree, (**b**) Shrubs, and (**c**) Herbs in mesic and dry aspects across all the study forest types. (Derived from Tyagi 2006)

with high densities generally occurring at hill base (HB) except for trees, which show similar densities at hill base (HB) and mid-slope (MS). In Banj Oak forests, densities of trees, shrubs, and herbs are higher on the relatively moist northern aspect, higher at hill base (HB) and decreasing upslope (Fig. 11.5). The difference in densities between aspects is most pronounced at hill base (HB), declining towards the hill top (HT) for both shrubs and herbs. Tilonj Oak forests highlighted a distinct richness of shrubs on the moist aspect, while herb density peaked at hill base (HB) and mid-slope (MS) in the northern aspect and at hill top (HT) in the dry southern aspect (Fig. 11.5). Tree density remains consistent at hill base (HB) across both aspects, but declines from mid-slope (MS) to hill top (HT) in the dry southern aspect (Fig. 11.5). Lastly, in Kharsu Oak forests, shrub density in the moist aspect was approximately twice that in the dry aspect.

With few exceptions, in all slope positions of the study forests, total tree basal area (TBA) was higher on mesic aspects than dry aspects, the difference being particularly higher in pine-sal forest and banj oak forest. The TBA differed between aspects ($p < 0.001$). Out of 18 values (6 forest sites, and 3 slope positions) of TBA, 16 (88.88%) were higher on N than S, the difference being particularly greater in chir pine (*P. roxburghii*)—sal (*S. robusta*) forest and banj oak (*Q. leucotrichophora*) forest. The average of TBA across all the sites was higher for HB (30.9 ± 3.2 m^2h^{-1}; ranging between 35.2 at N aspect and 26.7 m^2h^{-1} at S aspect) than for the other two positions (MS, 26.7 ± 3.0 m^2h^{-1}; ranged between 31.2 at N aspect and 22.2 m^2h^{-1} at S aspect, and HT, 27.9 ± 2.2 m^2h^{-1}; ranged between 29.8 at N aspect and 26.0 m^2h^{-1} at S aspect), but differences were not statistically significant (Fig. 11.6).

Research across various mountainous regions highlights the influence of slope aspect on vegetation characteristics. Sharma et al. (2010) found that northern slopes in Garhwal Himalayas had higher basal cover (74.4 m^2 ha^{-1}), dispersion (0.85), and tree diversity (1.81). Similarly, Louhaichi et al. (2021) reported greater biomass production on north-facing slopes in Tajikistan, consistent with earlier findings (Louhaichi et al. 2013). Bhardwaj et al. (2021) observed aspect- and altitude-driven differences in tree composition in the mid-Himalayan ranges, with *Cedrus deodara*

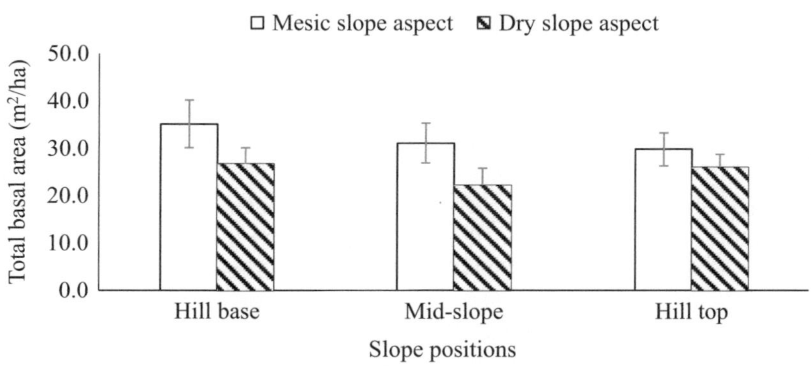

Fig. 11.6 Total basal area in different positions across the slope aspects of the six study forest types. (Derived from Tyagi 2006)

and *Q. leucotrichophora* dominant on northern slopes and *P. roxburghii* on southern slopes. In southeastern China, Yang et al. (2020) also noted higher biomass and vegetation coverage on north-facing slopes.

The HT sites are particularly vulnerable to climate change due to restricted upward movement space for species migration. In contrast, the interconnected habitats at HBs support more species and wider distribution ranges, enhancing species conservation. At HBs, 63.3% of species were unique to the northern aspect and 36.7% to the southern aspect (Tyagi et al. 2023a). HB positions of N aspect could serve as refugia for species affected by warming-induced droughts, providing buffering against climate change impacts (Morelli et al. 2016; Krawchuk et al. 2020). These sites, with locally elevated moisture levels, act as drought refugia, supporting species persistence amid climatic shifts (Krawchuk et al. 2020).

Effect of Slope Aspect on Soil Water Potential

In the monsoonal climate of Uttarakhand, soil Ψ continuously declines from the monsoon peak (July–August) to pre-monsoon period (April–May), with some minor variations. In the study by Tyagi et al. (2023a, b), pre-monsoon Ψ_{soil} values indicated that the soil in chir pine forests was the driest, with values falling below −3 MPa repeatedly across most aspects and slope positions (except at the hill base (HB) of the north-facing slope). In contrast, in the moist sal forest only the hilltop (HT) of the south-facing slope had Ψ_{soil} below −3 MPa. Soils in banj oak and tilonj oak forests were also quite dry, with pre-monsoon Ψ_{soil} values below −3 MPa, except at the HB of both slope aspects. Of the 30 Ψ_{soil} values of the pre-monsoon season (5 forest sites × 2 slope aspects × 3 positions on each slope aspect), 17 were below −3 MPa. At the HB position (5 forests × 2 slopes), only one value (in the chir pine forest) was less than −3 MPa, compared to 6 values at mid-slope (MS) and 10 at HT.

In 17 of the 18 comparisons (6 forests × 3 topographical positions), Ψ_{soil} was higher on the N slope aspect, with one instance showing equal values. Across all topographical positions (HB, MS, HT), Ψ_{soil} consistently declined from HB to HT (Fig. 11.7), the percentage decline being from 34% to 78% across the different forest types.

The warmer S slope aspect is expected to lose more soil water than N slope aspect. Furthermore, the drier S slope is relatively more erodible.

During the monsoon, Ψ_{soil} occurred within limited range from −0.11 MPa at HB of North slope aspect in sal forest to −0.69 MPa at HT of south slope aspect of pine forest. This variation is likely to be due to the differences in soil drainage characteristics. The difference in Ψ_{soil} magnified several folds from monsoon to pre-monsoon period: from −1.20 MPa at HB of sal forest to −4.84 MPa at HT of pine forest, with seasonal drop between monsoon and pre-monsoon values in Ψ_{soil} being −1.15 MPa and −4.84 MPa, respectively. Even in the case of HB which was invariably more moist, the aspect-driven difference in Ψ_{soil} was consistently evident. A gradient of decreasing Ψ_{soil} was discernable as following: $HB_{mesic} > HB_{dry} > MS_{mesic} > MS_{dry} > HT_{mesic} > HT_{dry}$.

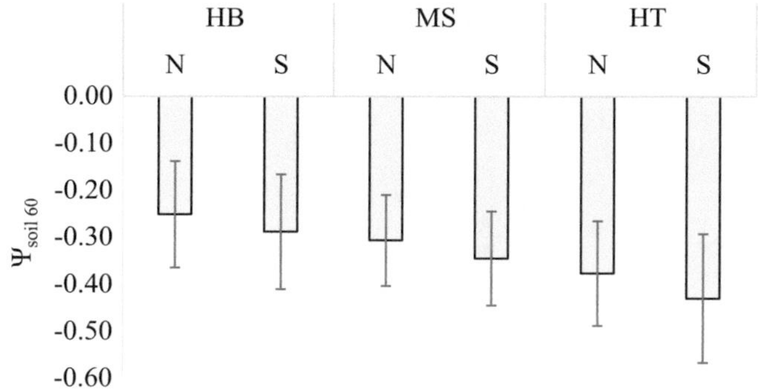

Fig. 11.7 A comparison in Ψ_{soil} in N (north) and S (south) aspect of three slope positions (i.e. HB = Hill base, MS = Mid-slope, and HT = Hill top) across the six study forest types (derived from Tyagi 2006). Generally, monsoon Ψ_{soil} was higher and decreased less on N slope aspect than S aspect

Higher biomass, as indicated by larger total basal area (TBA) at hill bases (HB), would typically be expected to lead to drier soils due to increased evapotranspirational loss. However, in the Himalayan context, soils at HB are actually moister (higher Ψ_{soil}), underscoring the significant role of topographical position in hydrology. This finding contrasts with the results of an experimental study in Spain by Oorthuis et al. (2021), where vegetation effects were predominant in influencing soil moisture. In the Himalayas, it appears that topographical effects override vegetation impacts on hydrology. The region's broken and shattered rocks significantly influence forest hydrology by facilitating the easy passage of water (Singh and Singh 1992), of which HB is the principal beneficiary.

A significant positive correlation ($p < 0.001$) between monsoon and pre-monsoon Ψ_{soil60} across all forests (Fig. 11.8) indicates that the effect of monsoon moisture persists during much of the year and is also indicative of the intrinsic hydrological differences in sites occupied by different species.

Seasonal decreases in Ψ_{soil} are more pronounced on south-facing slopes (2.89 ± 0.17 MPa) compared to north-facing slopes (2.46 ± 0.17 MPa), and more at hilltops (HT, 3.21 ± 0.20 MPa) than at HB (2.20 ± 0.19 MPa). This suggests that the extent of soil moisture retention and the subsequent rate of moisture loss are heavily dependent on both slope aspect and topographical position.

Effect of Slope Aspect on Tree Water Potential

In all forest types, ψ_{pd} was the highest during monsoon, whereafter it continuously declined (Fig. 11.9) until pre-monsoon (March to May) when it was at its lowest. During the pre-monsoon, when temperatures are high and precipitation sporadic and scanty, trees with elongating new shoots are most vulnerable to water-stress. We examined here the effect of slope aspect and position on daily change (difference

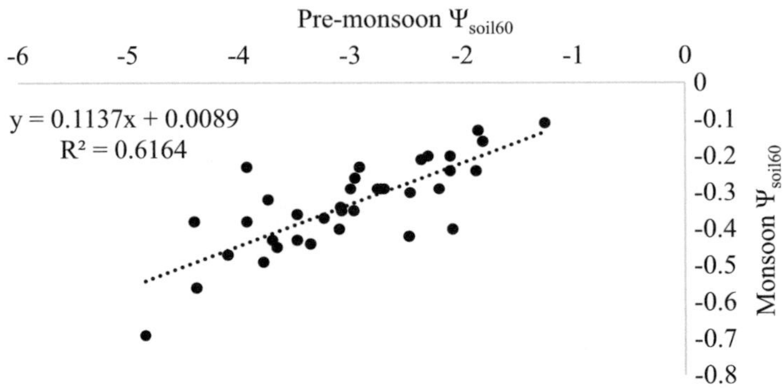

Fig. 11.8 Relationship between pre-monsoon Ψ_{soil60} and monsoon Ψ_{soil60} across all the study forests. (Derived from Tyagi 2006)

between predawn- and mid-day tree ψ), when ψ_{pd} was the lowest. This is also influenced by the species trait with regard to keeping stomatal responses to water stress.

The pre-monsoon (dry season) daily Ψ change was the highest in oak trees (from 1.25 to 1.50 MPa), the lowest in sal trees (0.63–0.73 MPa), and it was intermediate in *P. roxburghii* (0.86–1.07 MPa). Sal maintains high Ψ_{PD} despite high stomatal conductance, indicating that its deep roots have access to water. Chir pine is relatively shallow rooted (Pathak et al. 2021), hence it loses more water than sal, despite lower stomatal conductance and more water conservation (isohydric) strategies.

The highest seasonal daily change varied with predawn water potential (Ψ_{PD}) across different forest types. In sal, chir pine-sal, and chir pine forests, it increased with rising Ψ_{PD} from HT to HB, whereas in oak forests, it followed a reverse trend, increasing with decreasing Ψ_{PD} from HB to HT (Fig. 11.10). Thus, oaks seem to keep stomata open and fix carbon even in a dry condition (anisohydric). As shown in Fig. 11.11, for each species Ψ_{PD} values and daily changes were generally on the higher side of the range on N aspect than S aspect. HB (indicated as diamond) showed the highest value for both Ψ_{PD} and daily ψ changes. In brief, *sal is able to sustain high stomatal conductance as well as* Ψ_{PD} *presumably because of the access of its roots to soil water; the pine depends on a wider safety margin, thus closing stomatal while* Ψ_{PD} *is high, and oaks continue to keep stomatal conductance despite decreasing* Ψ_{PD}. *Q. leucotrichophora* sustained an increasing daily change even beyond Ψ_{PD} −1 MPa (from −1.34 MPa to −1.5 MPa). Oaks are known for narrow safety margin globally (Choat et al. 2012), partly because of Ψ_{50} (when the loss hydraulic conductivity is 50% of when Ψ_w is zero) and partly because of the trait of keeping stomata even under stress.

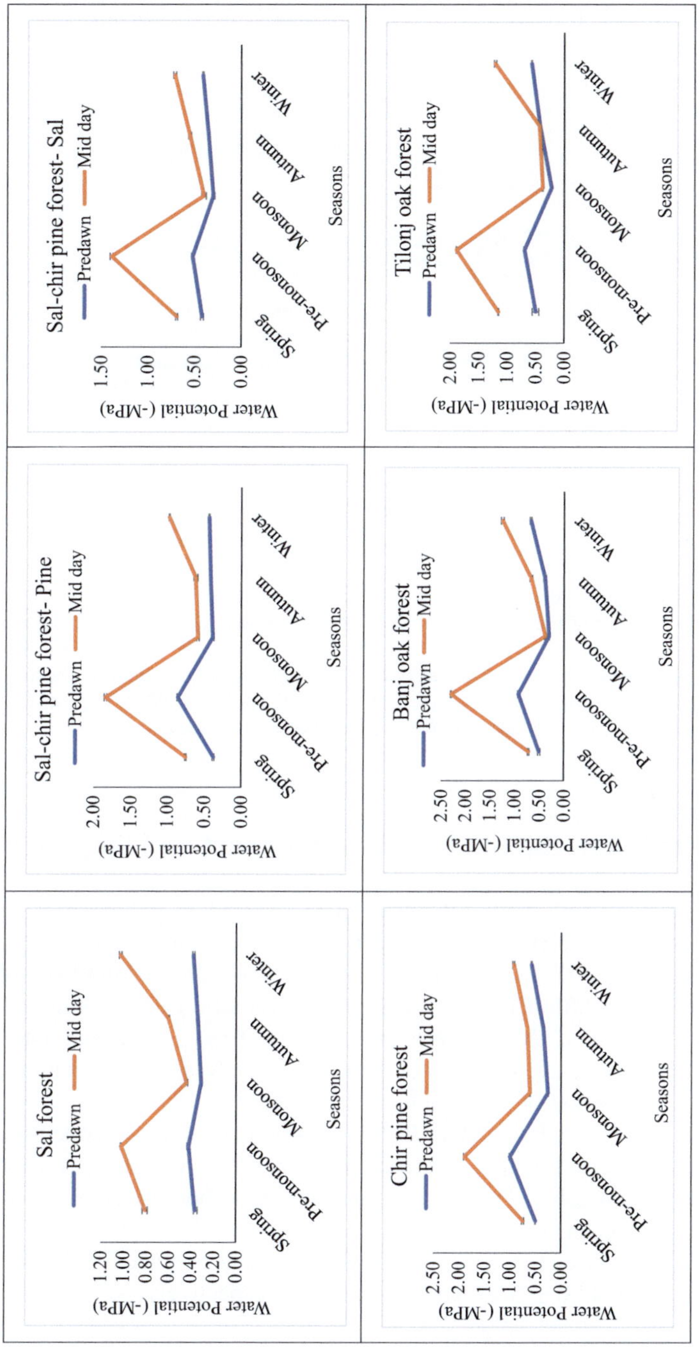

Fig. 11.9 A seasonal pattern in tree water potential of study forest types by comparison between predawn (Ψ_{PD}) and midday (Ψ_{MD}) for hill base (HB) of south slope aspect. (Derived from Tyagi 2006)

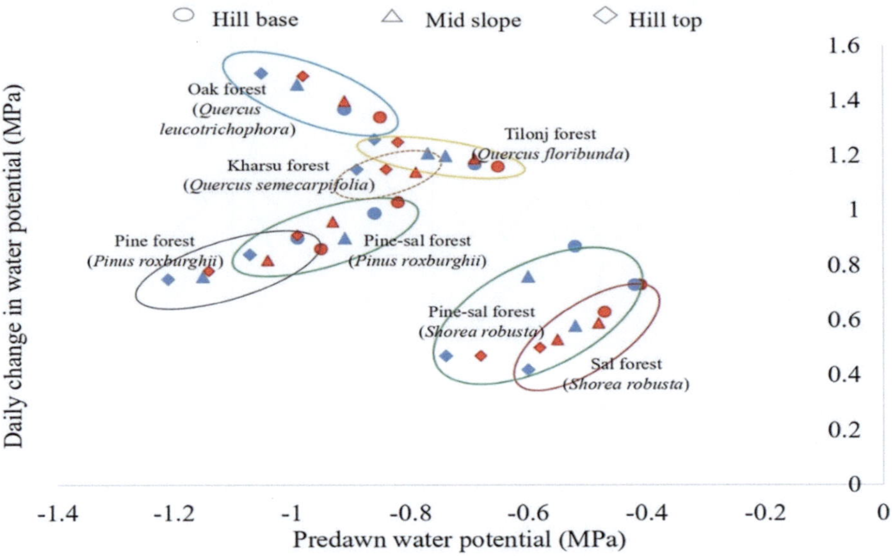

Fig. 11.10 The highest daily change in tree water potential ($\psi_{predawn}$- $\psi_{afternoon}$, MPa) of tree species, recorded during pre-monsoon plotted against the $\psi_{predawn}$ (Source: Tyagi et al. 2023a). Values are shown in relation to slope aspect and topographical position: red color is N slope aspect and blue is S slope aspect in each forest type, on a slope aspect the circle is for hill base (HB), triangle for mid-slope (MS), and rectangle for hill top (HT) position. While daily change decreases with decreasing Ψ_{PD}, in banj oak it shows a reverse pattern (left top)

Net Assimilation Rate in Relation to Aspect and Slope Position

In plants, there is a daily trade-off between CO_2 absorption and loss of water through stomata. Roughly 400 H_2O units are lost for every CO_2 unit gain (Taiz and Zeiger 2010). The choices between losing too much of water and risking hydraulic failure and capturing too little CO_2 and getting non-structural carbohydrate are depleted (Méndez-Alonzo et al. 2022). Here, the analysis of net assimilation rate (NAR) is based on four broadleaved species of the transect analysed for vegetation and water relations in the context of topographical variations.

Overall, Net Assimilation Rate (NAR) was higher on the N slope aspect than S slope aspect, and on a slope, it decreased from HB to HT across all forests. The consistency was notable. The NAR in morning was about twice as much as in the afternoon. The mean NAR (μ mol m^{-2} s^{-1}) across all seasons on N aspect was 5.03 ± 0.22, compared to 4.61 ± 0.22 on S aspect. The average for HB, MS, and HT was 5.13 ± 0.29, 4.80 ± 0.27, and 4.61 ± 0.26, respectively.

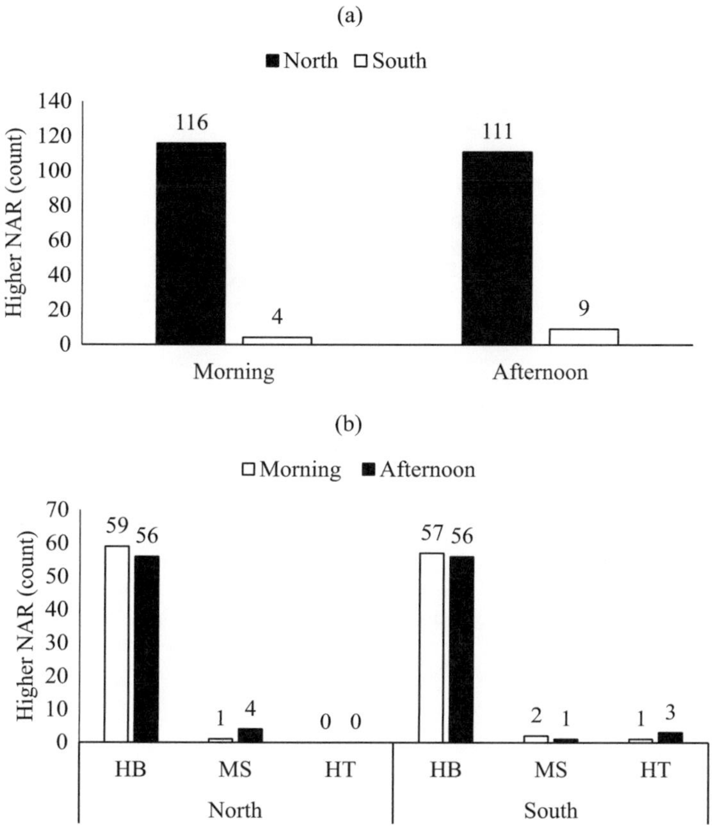

Fig. 11.11 Number of instances of Net Assimilation Rate (NAR) values (4 species/sites × 5 seasons × 2 slope aspects × 3 slope positions) (**a**) on North and South slope aspect and (**b**) at three hill slope positions, viz. HB = Hill base, MS = Mid slope, and HT = Hill top on north and south aspect in morning and afternoon. (Derived from Tyagi et al. 2024)

Temperature, soil moisture, and sunlight are major drivers of NAR in natural conditions. While N slope aspect is relatively cooler and moist, S aspect is warmer, drier, and brighter. Consequently, NAR was almost always higher on the North slope and for the Hill Base as compared to south aspect or other slope positions (The 120 values included 4 species/sites × 5 seasons × 2 slope aspects × 3 slope positions) (Fig. 11.11).

How common is the favourable effect of N aspect in terms of tree photosynthesis? In mountains of semi-arid climate of southwestern USA, while in *Pinus ponderosa* NAR rate on N aspect was double that on S aspect, *Pseudotsuga menziesii* had almost the same rate on N and S aspects (Murphy et al. 2020). In *Robinia pseudoacacia,* the photosynthesis rate was found to be significantly higher on N

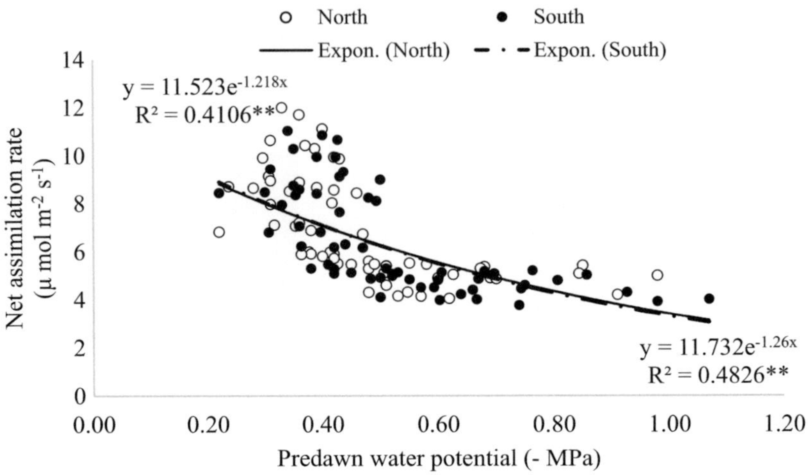

Fig. 11.12 Relationship between predawn tree water potential and morning net assimilation rate (NAR) of all the study tree species namely, Sal (*S. robusta*) in foot hill sal forest, mountain sal with chir pine forest, banj oak (*Q. leucotrichophora*), and tilonj oak (*Q. floribunda*) in north and south aspect during five study seasons, i.e. pre-monsoon, monsoon, autumn, winter, and spring (Source: Tyagi et al. 2024). Among the seasons, the highest values both of NAR and tree water potential are in autumn and monsoon. The significance level is: $p < 0.001$ (**). NAR (μ mol m^{-2} s^{-1}) ranged from 3.77 μ mol m^{-2} s^{-1} to 11.99 μ mol m^{-2} s^{-1}

aspect than S aspect because of high soil moisture and nitrogen. North slope aspect in Himalayas is known to favour the occurrence of several species preferring high moisture. At N slope aspect, HB is known to have also more nutrients and soil depth (Zheng et al. 2021; Pathak et al. 2021). When all seasonal values of NAR are plotted against Ψ_{PD} of all tree species, we find a significantly positive correlation ($p < 0.001$; Fig. 11.12).

Across all the species, seasons, slope aspects, and slope positions, NAR ranged from 3.92 ± 0.06 μ mol m^{-2} s^{-1} (banj oak, S aspect, MS, pre-monsoon) to 11.99 ± 0.05 μ mol m^{-2} s^{-1} (sal, N, HB, autumn) during morning and from 1.74 ± 0.01 μ mol m^{-2} s^{-1} (banj oak, HT, S aspect, spring) to 5.44 ± 0.11 μ mol m^{-2} s^{-1} (sal, HB, N aspect, autumn) in afternoon. Among the forests, the maximum seasonal NAR occurred in Autumn, when soil moisture deposited in monsoon was still high; days were bright and temperature still fairly warm, followed by monsoon, characterized by warm, wet, and cloudy conditions. Sunlight could be occasionally a limiting factor for photosynthesis (for details see Table 11.6).

Table 11.6 Average (mean ± SE) and range of Net assimilation rate (NAR) (μ mol m^{-2} s^{-1}) of study species across five seasons in relation to aspect (N = North; S = South) and slope positions (HB = Hill base; MS = Mid slope; and HT = Hill top)

	Sal (*S. robusta*) in foothill sal forest	Sal (*S. robusta*) in mountain sal with chir pine forest	Banj oak (*Q. leucotrichophora*) in banj oak forest	Tilonj (*Q. floribunda*) in tilonj oak forest
Across all slope aspects and positions-morning	**7.47 ± 0.44** 5.07 (S, MS, Pre-monsoon) to 11.99 (N, HB, Autumn)	**6.74 ± 0.37** 4.45 (S, HT, Pre-monsoon) to 10.63 (N, HB, Autumn)	**7.47 ± 0.44** 3.92 (S, MS, Pre-monsoon) to 8.94 (N, HB, Autumn)	**7.47 ± 0.44** 3.77 (S, HT, Winter) to 10.29 (N, HB, Autumn)
Across all slope aspects and positions-afternoon	**3.80 ± 0.16** 2.58 (S, HT, Winter) to 5.44 (N, HT, Autumn)	**2.73 ± 0.12** 2.03 (N, HT, Winter) to 4.11 (N, HB, Autumn)	**2.84 ± 0.15** 1.74 (S, HT, Spring) to 4.79 (N, HB, Autumn)	**2.80 ± 0.15** 1.87 (S, HT, Winter) to 4.46 (N, HB, Autumn)

Derived from Tyagi et al. (2024)

Conclusions

Given the range of forest types in central Himalayan transect and parameters analysed, the consistency in patterns is remarkably high, i.e. higher values on N aspect than S aspect and at HB than MS and HT for species richness, species uniqueness, tree water potential, and net assimilation rate at leaf level. Higher values at HB than other slope positions emphasize that moisture is a limiting factor even when slope is mesic. Being largely a subsurface flow system, hill bases are richer in moisture and nutrients. That the NAR was the highest in autumn, not in monsoon, suggests that excessively moist condition of monsoon is not always most favourable for plant performances and light limitation could act as a negative factor. The subsurface flow of hydrology not only contributes to growth of plants at HB, but also to much of lower elevation areas, where sal grows. The effect of favourable conditions of HB and N aspect is seen in a number of parameters: species richness particularly of herbs, presence of unique species, water potential and retention of water across the seasons, and NAR at leaf level. HBs seem to contribute considerably to the resilience of Himalayan forest ecosystems by providing shelter to several species represented by small number of individuals and retaining soil and water which is the key driver of photosynthesis and growth. Though the effect of slope aspect and position on ecosystems is often referred to, this chapter for the first time has provided requisite evidences drawn from a range of ecological characters of several forest types and trees.

References

Ackerly DD, Cornwell WK (2007) A trait-based approach to community assembly: partitioning of species trait values into within- and among-community components. Ecol Lett 10(2):135–145. https://doi.org/10.1111/j.1461-0248.2006.01006.x

Ackerly D, Knight C, Weiss S, Barton K, Starmer K (2002) Leaf size, specific leaf area and microhabitat distribution of chaparral woody plants: contrasting patterns in species level and community level analyses. Oecologia 130(3):449–457. https://doi.org/10.1007/s004420100805

Arroyo MTK, Primack RB, Armesto JJ (1982) Community studies in pollination ecology in the high temperate Andes of Central Chile. I. Pollination mechanisms and altitudinal variation. Am J Bot 69(1):82–97

Beaty RM, Taylor AH (2001) Spatial and temporal variation of fire regimes in a mixed conifer forest landscape, Southern Cascades, California, USA. J Biogeogr 28(8):955–966

Bennie J, Hill MO, Baxter R, Huntley B (2008) Influence of slope and aspect on long-term vegetation change in British chalk grasslands. J Ecol 96(3):532–541

Bhardwaj DR, Tahiry H, Sharma P, Pala NA, Kumar D, Kumar A (2021) Influence of aspect and elevational gradient on vegetation pattern, tree characteristics and ecosystem carbon density in Northwestern Himalayas. Land 10(11):1109

Bingham RA, Orthner AR (1998) Efficient pollination of alpine plants. Nature 391(6664):238–239

Bonser SP, Aarssen LW (1994) Plastic allometry in young sugar maple (Acer saccharum): adaptive responses to light availability. Am J Bot 81(4):400–406

Broxton PD, Troch PA, Lyon SW, Troch LP (2009) On the role of aspect to quantify water transit times in small mountainous catchments. Water Resour Res 45(8)

Burnett BN, Meyer GA, McFadden LD (2008) Aspect, slope, and vegetation controls on sediment transport, storage, and yield in two small, mountainous catchments, northern New Mexico. Geomorphology 83(1–2):74–89

Cantlon JE (1953) Vegetation and microclimates on north and south slopes of Cushetunk Mountain, New Jersey. Ecol Monogr 23(3):241–270

Casanova MT, Brock MA, Ganf GG (2000) Does water regime affect the sensitivity of aquatic and semi-aquatic plants to water level fluctuation? Aquat Bot 68(3):235–241

Champion HG, Seth SK (1968) A revised survey of Forest types of India. Govt. of India Press, New Delhi, p 404

Choat B, Jansen S, Brodribb T et al (2012) Global convergence in the vulnerability of forests to drought. Underw Nat 491:752–755. https://doi.org/10.1038/nature11688

Daws MI, Mullins CE, Burslem DFRP, Paton SR, Dalling JW (2002) Topographic position affects the water regime in a semideciduous tropical forest in Panama. Plant Soil 238(1):79–90

Gallardo-Cruz JA, Meave JA, González EJ, Lebrija-Trejos E, Romero-Romero MA, Pérez-García EA (2009) Predicting tropical dry forest successional attributes from topography: the case of the Yucatan Peninsula. Ecosyst 12(5):829–847

Gardiner B, Byrne K, Hale S, Kamimura K, Mitchell SJ, Peltola H, Ruel JC (2016) A review of mechanistic modelling of wind damage risk to forests. Forestry 89(2):158–171

Geroy IJ, Gribb MM, Marshall HP, Chandler DG, Benner SG, McNamara JP (2011) Aspect influences on soil water retention and storage. Hydrol Process 25(25):3836–3842

Gowda JH, Vázquez DP, Segura AM, Lombardi G (2019) Fire severity and vegetation responses in montane ecosystems of northwestern Patagonia, Argentina. For Ecol Manag 439:121–128

Gutiérrez-Jurado HA, Vivoni ER, Méndez-Barroso LA, Watts CJ (2007) Seasonal and interannual variations in water and energy fluxes in a semiarid basin in Mexico. J Arid Environ 70(4): 533–553

Heydari M, Cheraghi J, Omidipour R, Mirab-balou M, Pothier D (2021) Beta diversity of plant community and soil mesofauna along an elevational gradient in a mountainous semi-arid oak forest. Community Ecol 22:165–176. https://doi.org/10.1007/s42974-021-00046-7

Holland PG, Steyn DG (1975) Vegetational responses to latitudinal variations in slope angle and aspect. J Biogeogr 2(3):179–183

Jucker T, Bongers F, Brodribb TJ, Field TS, Fyllas NM (2018) Topography shapes the structure, composition and function of tropical forest landscapes. J Ecol 106(3):1337–1347

Krawchuk MA, Moritz MA, Parisien MA, Dorn JV, Hayhoe K (2020) Global pyrogeography: the current and future distribution of wildfire. PLoS One 5(10):e1001102

Li L, Zhang T, Zhao C, Li Y, Li Y, Mu C (2019) Leaf and stem traits variation of *Stellera chamaejasme* Linn. with slope aspect in alpine steppe. Ecol Res 34(1):119–126

Li XE, Song X, Zhao J, Lu H, Qian C, Zhao X (2021) Shifts and plasticity of plant leaf mass per area and leaf size among slope aspects in a subalpine meadow. Ecol Evol 11(20):14042–14055

Louhaichi M, Nubekov A, Madaminov A, Barbara R (2013) The influence of geo-morphological landscape patterns on vegetation characteristics in Tajikistan grasslands. In: Revitalising Grasslands to Sustain our Communities: Proceedings, 22nd International Grassland Congress, 15–19 September 2013, Sydney, Australia, 2013. New South Wales Department of Primary Industry pp 860–862.

Louhaichi M, Rajabov T, Moyo H, Belgacem AO (2021) Effect of slope aspect on vegetation characteristics in mountain rangelands of Tajikistan: considerations for future ecological management and restoration. Afr J Range Forage Sci 39(4):320–328. https://hdl.handle.net/20.500.11766/13565

Lozano-García B, Parras-Alcántara L, Brevik EC (2016) Impact of topographic aspect and vegetation (native and reforested areas) on soil organic carbon and nitrogen budgets in Mediterranean natural areas. Sci Total Environ 544:963–970

Méndez-Alonzo R, Olson ME, Paz H, Reyes-García C et al (2022) Studies of plant hydraulics and water relations in Mexican environments: adaptation, physiology, and applications. Bot Sci 100: 325–345

Miehe S, Miehe G, Miehe S, Böhner J et al (2015) Vegetation ecology. In: Miehe G, Pendry CA (eds) Nepal: an introduction to the natural history, ecology and human environment in the Himalayas—a companion to the Flora of Nepal, chapter 16, 1st edn. The Royal Botanical Garden Edinburgh, Edinburgh, pp 385–472

Moeslund JE, Arge L, Bøcher PK, Dalgaard T, Ejrnæs R, Skov F (2013) Topographically controlled soil moisture is the primary driver of local vegetation patterns across a lowland region. Ecosphere 4(7):1–25

Moore PL, Fairbairn MJ, O'Connell MJ, Hudson MD (2018) Deep rooting tree species enhance ecosystem stability in afforested catchments in SW England. For Ecol Manag 410:164–175

Morelli TL, Daly C, Dobrowski SZ et al (2016) Managing climate change refugia for climate adaptation. PLoS One 11(8):e0159909

Murphy PC, Knowles JF, Moore DJ et al (2020) Topography influences species-specific patterns of seasonal primary productivity in a semiarid montane forest. Tree Physiol 40(10):1343–1354

Nanda SA (2024) Community structure, taxonomic, functional and phylogenetic plant diversity along the elevation gradients in Kashmir Himalaya. School of Biological Sciences, Department of Botany. University of Kashmir, Srinagar, J&K, India

Nanda SA, Reshi ZA, Ul-Haq M, Lone A, Mir SA (2018) Taxonomic and functional plant diversity patterns along an elevational gradient through treeline ecotone in Kashmir. Trop Ecol 59(2): 211–224

Oorthuis R, Vaunat J, Hürlimann M et al (2021) Slope orientation and vegetation effects on soil thermo-hydraulic behavior. An experimental study. Sustainability 13(1):14

Pathak GC, Joshi H, Singh RD, Tewari A et al (2021) Vertical root distribution in Himalayan trees: about half of roots occur below 30 cm, the generally sampled depth. Trop Ecol 62:479–491

Paudel PK, Vetaas OR (2009) Leaf traits in contrasting slope aspects and land use types of the trans-Himalaya. Flora 204(8):536–546

Perring FH (1959) Topographical gradients of chalk grassland. J Ecol 47(1):155–158

Pook EW, Moore CWE (1966) The influence of aspect on the composition and structure of dry sclerophyll forest on Black Mountain, Canberra, ACT. Aust J Bot 14(2):223–242

Qin Y, Feng Q, Adamowski JF, Zhu M, Zhang X (2021) Community level response of leaf stoichiometry to slope aspect in a montane environment: a case study from the Central Qilian Mountains, China. Glob Ecol Conserv 28:e01703

Rasmussen C, Tabor NJ (2007) Applying a quantitative pedogenic energy model across a range of environmental gradients. Soil Sci Soc Am J 71(1):171–182

Rogeau MP, Armstrong GW (2017) A critical look at historical fire return interval data for Alberta's mixedwood boreal forest. For Ecol Manag 401:176–185

Sharma S, Rawat YS, Singh JS (2010) Forest resource use patterns in relation to socioeconomic status: a case study in a village in Garhwal Himalaya, India. MRD 30(2):131–140

Singh P, Negi GCS (2018) Seasonal changes in leaf area and leaf mass of dominant trees of Western Himalayan forests. Indian Forester 144(2):182–190

Singh JS, Singh SP (1992) Forests of Himalaya: structure, functioning, and impact of man. Gyanodaya Prakashan, Nainital

Singh SP, Singh RD, Gumber S (2021) Interpreting mountain treelines in a changing world. Central Himalayan Environment Association and International Centre for Integrated Mountain Development

Stainton JDA (1972) Forests of Nepal, vol 22. John Murray, London

Sundqvist MK, Giesler R, Wardle DA (2013) Within- and across-species responses of plant traits and litter decomposition to elevation across contrasting vegetation types in subarctic tundra. PLoS One 8(8):e72404

Taiz L, Zeiger E (2010) Plant physiology, 5th edn. Sinauer Associates, Sunderland, p 782

Thadani R (1999) Disturbance, microclimate and the competitive dynamics of tree seedlings in banj oak (*Quercus leucotrichophora*) forests of the central Himalaya, India. Yale University

Troup RS (1921) The Silviculture of Indian trees. Vol. I to III. Oxford University Press

Tyagi V (2006) Forest species composition and diversity in relation to water status in an altitudinal transect in Kumaun Himalaya. PhD Thesis. Kumaun University, Nainital

Tyagi V, Singh SP, Singh RD, Gumber S et al (2023a) Influence of slope position and aspect on the vegetation attributes and tree-water relations in forests of the Central Himalayas. J Mt Sci 20: 2592–2602. https://doi.org/10.1007/s11629-023-7967-z

Tyagi V, Singh SP, Singh RD, Gumber S (2023b) Chir pine and banj oak responses to pre-monsoon drought across slope aspects and positions in Central Himalaya. Environ Monit Assess 195(2): 258

Tyagi V, Singh SP, Gumber S, Singh RD (2024) Consistent response to topographical variation in net assimilation rate across the central Himalayan broadleaved forests. Ecol Front 44(3): 580–588

Wang L, Benton J, Xu X (2011) The influence of topography on soil water dynamics in a simulated ecosystem in Northern Mongolia. J Arid Environ 75(6):472–480

Wang C, He J, Zhao TH, Cao Y, Wang G, Sun B, Yan X, Guo W, Li MH (2019) The smaller the leaf is, the faster the leaf water loses in a temperate forest. Front Plant Sci 10:58

Wright IJ, Dong N, Maire V, Prentice IC et al (2018) Global climatic drivers of leaf size. Science 357(6354):917–921. https://doi.org/10.1126/science.aal4760

Xue J, He J, Wang L, Gao J, Wu Y (2018) Plant traits and biomass allocation of *Gentiana hexaphylla* on different slope aspects at the eastern margin of Qinghai-Tibet plateau. Appl Ecol Environ Res 16(2):1835–1853

Yang Y, Shen Z, Han J, Zhongyong C (2016) Plant diversity along the eastern and western slopes of Baima Snow Mountain, China. Forests 7(4):89

Yang J, El-Kassaby YA, Guan W (2020) The effect of slope aspect on vegetation attributes in a mountainous dry valley, Southwest China. Sci Rep 10(1):16465

Zheng JY, Zhao JS, Shi ZH, Wang L (2021) Soil aggregates are key factors that regulate erosion-related carbon loss in citrus orchards of southern China: bare land vs. grass-covered land. Agric Ecosyst Environ 309:107254

Forest Hydrological Processes in the Lesser Himalaya: A Critical Review

12

L. Adrian Bruijnzeel, Mark Mulligan, and Jun Zhang

Abstract

This chapter critically reviews our knowledge of hydrological processes in the chief zonal forest types within the Lesser Himalaya, notably the *Shorea*-dominated forests of the tropical lowland and foothills, the subtropical pine and *Castanopsis*-dominated forests at intermediate elevations, and the temperate oak- or conifer-dominated forests above ~2000 m. Processes at the hillslope and headwater catchment scale are compared between forest types. Next, the soil and water impacts of various forest uses and interventions are addressed, along with those of forest regrowth after agricultural abandonment and (pine) reforestation. Overall, quantitative knowledge of forest hydrological processes in the region is limited, particularly on forest water uptake and runoff generation. The chapter offers various recommendations for further research, notably on changes in vegetation water use during forest succession, the hydrology of cloud-affected forests with potentially high water yields, and the water dynamics of agroforestry systems. To document impacts of climate change on forest hydrological processes, it is recommended to establish 'critical zone observatories' (CZOs) in which the stocks and flows of water are monitored. Maximum synergy may be achieved by locating such CZOs at sites with past or ongoing research that together capture the chief geological and climatic variations along the mountain chain.

L. A. Bruijnzeel (✉) · M. Mulligan
Department of Geography, King's College London, London, UK
e-mail: sampurno70@outlook.com

J. Zhang
Environmental Modelling, Sensing & Analysis, TNO, Petten, The Netherlands

S. P. Singh, R. Thadani (eds.), *Himalayan Forests in a Changing World*,
https://doi.org/10.1007/978-981-95-0978-2_12

Keywords

Evapotranspiration · Forest management · Hillslope hydrology · Infiltration · Runoff generation · Spring discharge · Water budget

Introduction

The Himalayan mountain chain and adjacent Tibetan Plateau together form one of the world's major 'water towers' that feeds many of Asia's largest rivers and provides water to over a billion people in the lowlands of India, Pakistan, Bangladesh, and China. While the vulnerability of the higher, ice-covered parts of the region to global warming and the resulting changes in downstream water provisioning are subject to intensive research (Shakti et al. 2014; Wijngaard et al. 2017; Immerzeel et al. 2020; Budhatoki et al. 2021), the hydrological implications of the major changes in land use and land cover over the last five decades in the 'Lesser Himalaya' have received far less scientific attention (Ghimire et al. 2014a; Qazi et al. 2017). Conventionally, the Lesser Himalaya (LH) represents the area between two major geological fault lines, the Main Boundary Thrust (MBT) and the Main Central Thrust (MCT; Searle and Treloan 2019). The MBT separates the Siwalik foothills (called *Chure* or *Churia* in Nepal) in the south from the central Middle Mountains (also referred to as the *Mahabharath Lekh* or the 'Mid-Hills'), whereas the MCT marks the boundary with the High Mountains or Greater Himalaya to the north. The term LH as used here includes the Siwaliks as well as the lower slopes of the High Mountains up to ~3000 m asl (Fig. 12.1).

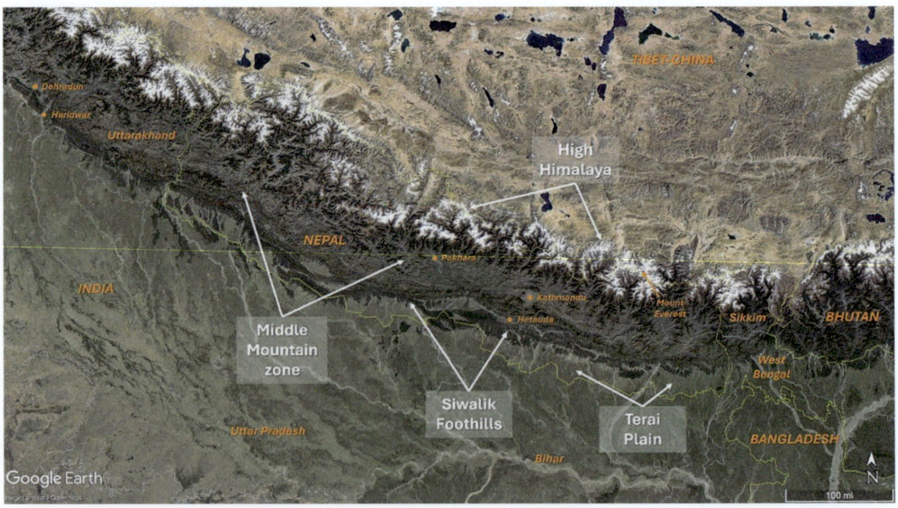

Fig. 12.1 Broad physiographic zones referred to in this chapter. (After Bruijnzeel et al. 2024)

The potential hydrological importance of changes in land use /cover in the LH may well be large: >80% of streamflow emanating annually between the Sutlej valley in the west and Arunachal Pradesh in the east derives from rain falling within the LH itself; that is, <20% of all streamflow consists of glacial and snowmelt originating in the High Himalaya (Bookhagen and Burbank 2010).

Until the mid-1980s, it was generally assumed that the widespread degradation and loss of hill forests in support of agricultural cropping and grazing in the LH were largely responsible for the flooding and sedimentation problems experienced on the plains of northern India and Bangladesh (Eckholm 1975; Bowonder 1982; cf. Ives 1987). This view was subsequently shown to be untenable in light of the overriding influence of geoclimatic factors on regional flooding and sediment production—notably the steep topography, the often shallow soils and fractured nature of the underlying rocks, coupled with regular tectonic activity and, especially, a highly seasonally concentrated rainfall regime (Hamilton 1987; Bruijnzeel and Bremmer 1989; Hofer and Messerli 2006; Bookhagen and Burbank 2010).

However, the adverse *local* hydrological effects of advanced forest degradation and loss, such as accelerated erosion and reduced groundwater recharge and dry-season streamflow (Bartarya 1989; Gardner and Gerrard 2002; Schreier et al. 2006), did require remedial measures at the time. Hence, both to regreen denuded hillslopes and to meet subsistence energy needs of hill farmers (fuelwood), major reforestation programmes were initiated in the early 1980s that targeted degraded lands across the Middle Hills of Nepal and NW India. For this, pine species were used mostly in view of their ability to survive nutrient-poor and seasonally dry soil conditions (Gilmour et al. 1990), although in Uttarakhand emphasis subsequently shifted towards the planting of oaks (R. Thadani, pers. comm.). Furthermore, the unsustainable extraction of large amounts of animal fodder and litter material from native broad-leaf forests in support of traditional agriculture (Tiwari 1988; Schreier et al. 2006) as well as harvesting of timber and fuelwood became better regulated following the expansion of community-based forest management, with gradual positive effects on forest structure, density, and diversity, particularly in Nepal (Chakraborty 2001; Webb and Gautam 2001; Niraula et al. 2013; Pokharel et al. 2018).

A related development concerns increased abandonment of rain-fed agriculture across the western and central parts of the LH (Paudel et al. 2012; Upadhyay et al. 2021), both as a result of 'internal' migration from the highlands (>1400 m) to lowland towns and roadheads (Bhawana et al. 2017) and the emergence in recent years of a 'remittance economy' following large-scale labour outmigration from rural areas to urban centres and abroad (Paudel et al. 2022; Upadhyay et al. 2021). Combined, these developments have caused a gradual re-greening of the Nepalese LH in terms of forest area and density (Ghimire 2017; Gao et al. 2023). Conversely, population increase, urbanization, and peri-urban agricultural intensification have caused forest loss in the Himalayan foothills and valleys of NW India (Mishra and Chaudhari 2015).

Due to the strongly seasonal character of precipitation, water has always been at a premium during the lean season (March–May) across much of the central LH (Merz et al. 2003; Agarwal et al. 2014; Sharma et al. 2016). Traditionally, the majority of people in the Middle Mountains relied on springs associated with forested areas for drinking water and other domestic needs, whereas headwater streams are widely used to irrigate rice fields of the lower slopes and valley bottoms (Tiwari and Joshi 2012; Sharma et al. 2016; Gurung et al. 2019). However, spring discharges have been declining throughout the LH for decades, with climate change (increased temperatures, intensifying monsoon rains, decreasing winter precipitation), indiscriminate road construction and tectonic changes in geological substrates (both disrupting subsurface hillslope drainage patterns), forest degradation (livestock grazing, fire, and litter harvesting causing surface degradation and reduced rainfall infiltration), pine encroachment into degraded broad-leaved forest, and pine plantation establishment being among the most frequently cited potential reasons for the hydrological decline (Valdiya and Bhattarya 1991; Tambe et al. 2012; Ghimire et al. 2014a; Chapagain et al. 2017; Poudel and Duex 2017; Adhikhari et al. 2020; Badu et al. 2022; Tarafdar and Dutta 2023; Pandit et al. 2024). At the same time, demands for water are rising in areas with rapid urbanization and peri-urban agricultural intensification, causing acute scarcity during the latter part of the dry season (Tiwari and Joshi 2012; Poudel and Duex 2017; Adhikhari et al. 2020; Upadhyay et al. 2021).

The fact that both forest degradation *and* forest expansion or (pine) tree planting have been blamed for diminished spring and stream discharges in the LH underscores the need for improved understanding of the hydrological functioning of different forest types and of the impacts of forest management activities and uses. Likewise, there is an urgent need to disentangle effects of climatic warming and rainfall intensification on spring and stream discharges from those incurred by changes in land cover or use. This requires adequately parameterized hydrological models (preferably physically based) for which quantitative knowledge is required, in turn, of key forest hydrological processes and soil characteristics (Dass et al. 2023; Bruijnzeel 2023; Norouzi-Moghanjoghi et al. 2025).

This chapter, therefore, aims to critically review the knowledge base on the hydrological functioning of the chief zonal forest types within the LH, notably the *Shorea*-dominated forests of the subtropical foothills below ~1000 m and adjacent Terai plain (vegetation category HSBF in Fig. 12.2), the subtropical *Chir* pine- and *Schima–Castanopsis* forests of intermediate elevations (~1000 to ~2000 m; categories HSPF and EHBF, respectively), and the more temperate and partly cloud-affected oak- or conifer-dominated forests above ~2000 m (category WHBL).

Key forest hydrological processes (rainfall interception by canopy and forest floor, transpiration, infiltration, hillslope runoff generation) are discussed and compared between forest types, paying ample attention to methodological issues. Next, the soil and water impacts of such traditional forms of forest use and management as lopping branches, harvesting of litter, grass and understorey vegetation, and cattle grazing are addressed, along with the hydrological consequences of deliberate thinning, forest regrowth following agricultural abandonment (including post-slash & burn), and (pine) reforestation of degraded land.

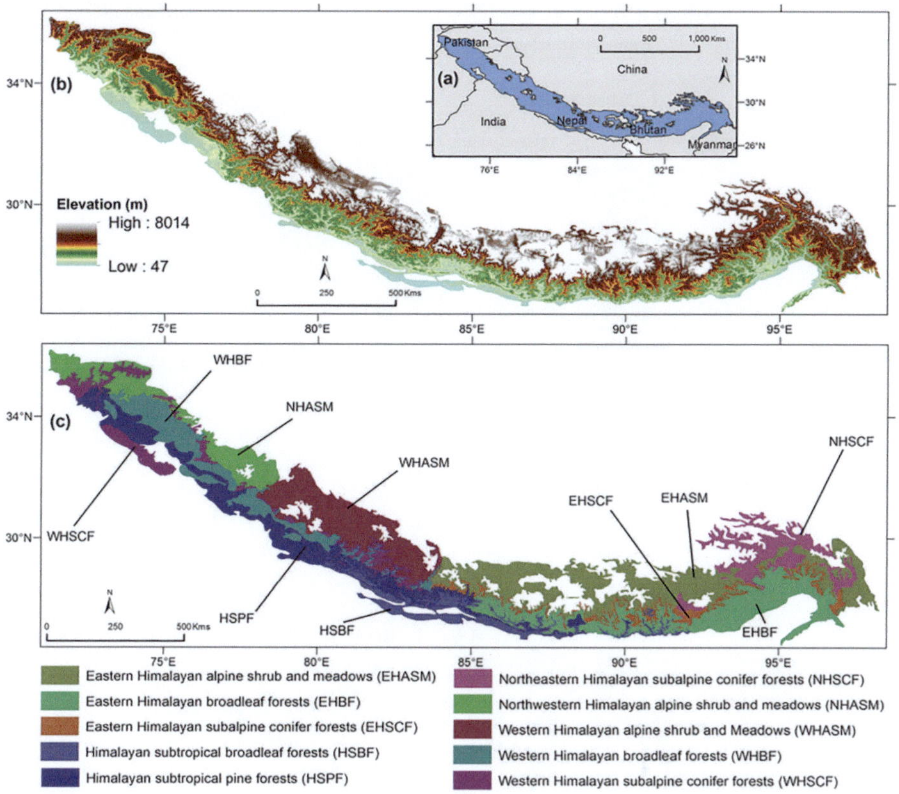

Fig. 12.2 Broad vegetation types across the Himalaya. (After Mishra and Mainali 2017)

The chapter concludes with several recommendations for further hydrological research, notably the quantification of: (i) changes in vegetation water use and soil physical attributes during succession from shrub to closed-canopy forest and beyond (mostly in the Mid-Hills); (ii) 'occult' inputs from fog and low cloud to 'mossy' forests at higher elevations with potentially high water yields; (iii) possibilities for boosting spring discharges through enhanced infiltration; and (iv) impacts of climate change on forest hydrological functioning across elevations.

The Hillslope Forest Hydrological Cycle

The principal components of the hillslope forest hydrological cycle (after Bruijnzeel 2000) are illustrated in Fig. 12.3. Rainfall is the chief precipitation input to most forests in the LH, supplemented by snow in winter at higher elevations (mostly >2500 m; Lang and Barros 2004; Eeckman et al. 2019) and 'occult' precipitation (passing fog captured by the canopy) in the montane fog belt between ~2000 m and ~3500 m (Wangda and Ohsawa 2006; Miehe et al. 2015). Part of the precipitation

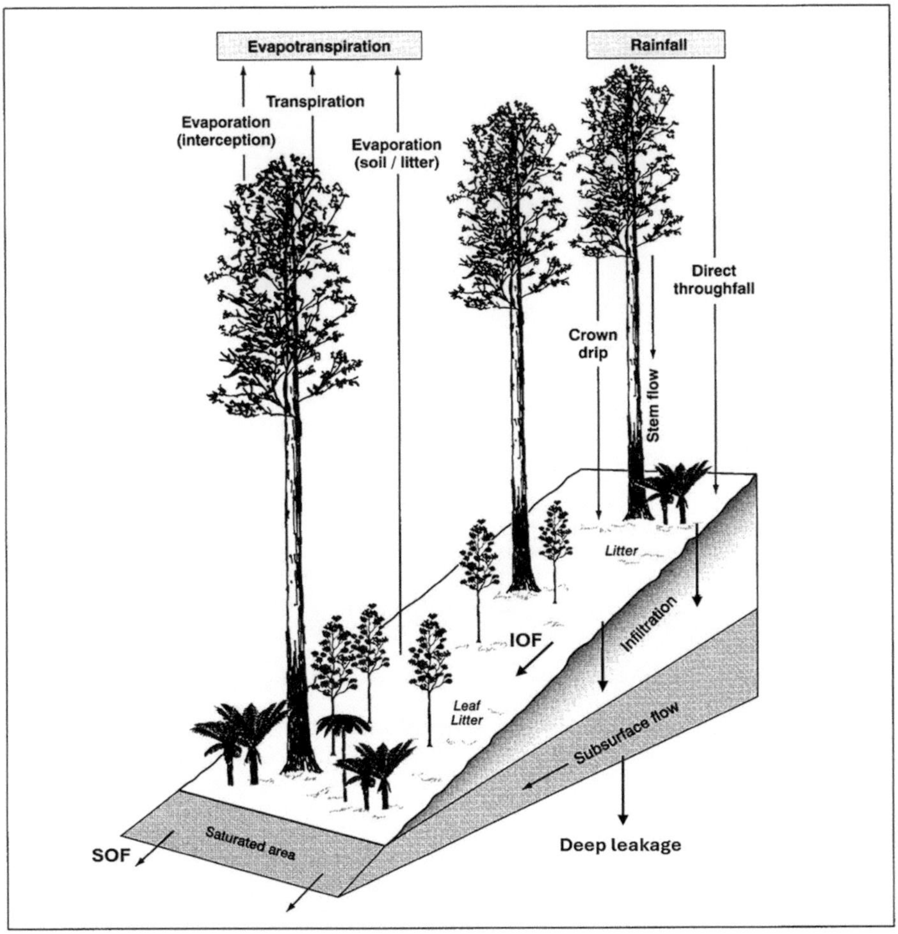

Fig. 12.3 The hillslope hydrological cycle. (Adapted from Vertessy et al. 1998)

reaches the forest floor directly through gaps in the canopy ('direct' throughfall). Another (often small) part reaches the soil via tree trunks as 'stemflow', whereas a substantial portion of incident precipitation is intercepted by the canopy and evaporated back to the atmosphere ('interception loss'). The remainder reaches the forest floor as crown drip once the storage capacity of the canopy is exceeded. Because direct throughfall and crown drip cannot be determined separately in the field, the two are conventionally taken together and referred to as stand 'throughfall'. The sum of throughfall and stemflow is termed 'net precipitation'. It is usually substantially smaller than incident precipitation, unless occult inputs are large enough to compensate the losses incurred by wet-canopy evaporation (Bruijnzeel et al. 2005).

If the intensity of net precipitation exceeds the infiltration capacity of the soil, the unabsorbed excess runs off as 'infiltration-excess' overland flow (IOF). Due to the generally high absorption capacity of forest soils, IOF is rarely observed in undisturbed forests, unless a dense clayey substrate is present (Elsenbeer et al. 1992) or stemflow amounts are excessive (Herwitz 1986). Not all water infiltrating into the soil will emerge again as spring- or streamflow. A substantial part is taken up by plant roots and returned to the atmosphere via 'transpiration'. The term 'evapotranspiration' (ET) is often used to denote the sum of transpiration (evaporation from a dry canopy, E_t), interception loss (evaporation from a wet canopy, E_i), and evaporation from the litter and soil surface (E_s). The latter term is often small, especially in dense forests where little radiation reaches the forest floor, atmospheric humidity is high, and wind speeds low (Roberts et al. 2005; Oishi et al. 2020).

Water percolating through the soil is deflected laterally upon meeting an impermeable layer or underlying rock. The resulting slow lateral drainage ('throughflow') feeds foot-slope springs and riparian groundwater bodies maintaining the 'baseflow' of streams. In seasonal climates like that of the LH, baseflows from small headwater catchments typically reach a minimum during the latter part of the dry season (March–May), especially in the western half of the mountain chain and below the snow-affected zone (Qazi et al. 2017). Further east and higher up (e.g. easternmost Nepal and Sikkim), the seasonal low in flows occurs around March before precipitation picks up in April–May, as do contributions from snowmelt (Ries 1994; Kumar et al. 2024b).

However, during times of (high) rainfall, infiltrating water may take one of several faster pathways to the nearest stream channel, depending on changes in soil hydraulic conductivity with depth, slope morphology (concave, straight, or convex), and soil wetness. So-called 'saturation-excess' overland flow (SOF) occurs when rain falls on an already saturated soil. Near-saturated conditions are often found in valley bottoms, hillslope hollows, and on concave foot-slopes near streams where throughflow tends to converge (Fig. 12.3; Ward 1984; Kumar et al. 2024b). SOF may also occur on slopes away from hollows and foot-slope depressions if an impeding layer or rock of low permeability is present close to the surface and rainfall is sufficiently large to saturate the soil (Elsenbeer et al. 1992; Qazi and Bruijnzeel 2023). Once the soil is thoroughly wetted, 'macropores' (root channels, animal burrows, etc.) become activated by new rainfall, and both amounts and the traveling speed of throughflow increase rapidly ('subsurface stormflow', SSSF; Ward 1984; Graham et al. 2010). As a result of contributions by SSSF and SOF, forest stream discharges usually increase rapidly during times of intense rainfall (*cf.* Fig. 12.11). This increase above baseflow levels is usually called 'stormflow' (or 'quickflow'), whereas the highest instantaneous discharge is referred to as 'peak flow'. Peak discharges for forested catchments may be reached during the rain event itself or afterwards, depending on catchment characteristics (notably soil depth and catchment size), antecedent soil wetness, and the duration, intensity, and quantity of the rainfall (Ward 1984; Graham et al. 2010). The total volume of water produced as streamflow from a catchment over a given period of time (month, season or year) is conventionally called catchment 'water yield'.

Hydrological Processes in Major Himalayan Forest Types

Rainfall Interception by the Canopy

With quantitative data available for more than 50 Himalayan forest sites (both natural and planted), rainfall interception loss (IL) is one of the better studied forest hydrological processes in the region. Appendix 12.1 lists the fractional amounts of throughfall (TF), stemflow (SF), and interception loss as a percentage of incident precipitation (*P*) for 46 selected studies, along with several basic forest structural attributes.

Methodological Considerations It is pertinent to note that nearly all studies focused on the monsoon season (June–September), during which the bulk of annual precipitation is generally received (Bookhagen and Burbank 2010; Böhner et al. 2015). Because the value of IL/*P* per event typically decreases as the amount of rainfall increases, interception fractions during the post-monsoon and winter seasons (when most events are small) tend to be (much) higher (Singh 1987; Badu et al. 2022). Also, leaf shedding in most subtropical and warm-temperate forests in the LH tends to peak around April–May, causing temporarily low values in leaf area index (LAI) (Behera et al. 2015; Joshi and Sarkoti 2020; Sheikh et al. 2020) when the first intensive showers mark the transition to the upcoming rainy season (Böhner et al. 2015; Kumar et al. 2024b). The net effect of these seasonal variations is that annual values of IL/*P* may be 2–3% higher than the monsoon-based values listed in Appendix 12.1, at least for evergreen broad-leaved and coniferous forests (Singh et al. 1983; Singh 1987; Ghimire et al. 2012; Badu et al. 2022).

Further, very few Himalayan interception studies used a truly random sampling approach with a sufficiently large number of TF collectors (e.g. Dabral and Subba Rao 1969; Ghimire et al. 2012). Both are generally considered important for adequate characterization of stand-scale TF in species-rich natural forests, particularly broad-leaved ones where spatial variability of TF is highest (Czarnowski and Olszewki 1970; Lloyd and Marques-Filho 1988; Holwerda et al. 2006). Although TF collectors were moved regularly to new positions in many studies to take spatial variations into account (notably 'drip points' where TF is concentrated and even may exceed *P*; Lloyd and Marques-Filho 1988), a majority positioned collectors deliberately beneath trees *while avoiding gaps in the canopy* (see notes in Appendix 12.1). In doing so, these studies effectively measured crown drip, not stand-scale TF (Fig. 12.3). As a result, stand-scale TF becomes underestimated and IL correspondingly overestimated, depending on the magnitude of the gap fraction. Pathak et al. (1985) and Loshali and Singh (1992) acknowledged this shortcoming and adjusted measured amounts of crown drip by adding presumed amounts of direct precipitation (Fig. 12.3) to derive stand-scale TF. Needless to say, doing so introduces additional uncertainty.

Lastly, the majority of interception studies (~80%) are located in the (somewhat drier) north-western Indian Himalaya, with concentrations around Dehradun (Dabral and Subba Rao 1969; Ghosh et al. 1980; Tomar et al. 2022) and Nainital (Pathak et al. 1985; Loshali and Singh 1992). Eastern Himalayan forests are represented by two sites in east central Nepal (Ghimire et al. 2012; Badu et al. 2022), two locations in southern Sikkim (Rai and Sharma 1998; Jain et al. 2000), and two fog-affected, high-elevation forests in Bhutan (Om 2016). Similar studies in forests of western Nepal and most of NE India (Arunachal Pradesh, Meghalaya) appear to be lacking.

Results Keeping the above caveats in mind, the following tentative conclusions may be drawn from Appendix 12.1 regarding the partitioning of monsoon rainfall by the main forest types:

- The lowest TF (53–67.5%) and relatively high SF (3–7.5%) were generally reported for mature stands of *subtropical hardwoods* (*Shorea robusta, Tectona grandis, Acacia katchecu*) in the Indian foothills (vegetation category HSBF in Fig. 12.2). This resulted in fairly high interception fractions (21–35%) with a tentative average for mature *Shorea* forest stands of 29.5 ± 5% ($n = 3$ only); NB: coppiced *Shorea* showed distinctly lower interception losses (16.5–18%; $n = 2$), plus high SF (9.5%) in a dense stand with numerous thin stems.
- *Younger subtropical hardwood plantations* (12–17 years old) had remarkably high SF (8–19%) which gradually diminished to values below ~3.5% in older (~50 years) stands.
- TF- and SF for *subtropical pine forests* (HSPF) varied between 74.5–83% and 0.3–1.0%, respectively, yielding relatively stable *IL/P* fractions (average: 21.3 ± 2.7%; $n = 8$).
- TF- and SF in *subtropical mixed broad-leaved forests* (oak-dominated in the western parts (WHBF below ~1750 m asl) and *Schima–Castanopsis* forest in the east, EHBF) had range of 73–80.5% and 0.3–2%, respectively, giving an average *IL/P* of 24 ± 1%; $n = 6$).
- *Temperate mixed broad-leaved forests* (mostly oaks, WHBF above ~1900 m) showed considerable variation in TF (64–87%) and SF (0.1–12%), with an overall average *IL/P* of 23.5 ± 8% ($n = 7$); very high SF reported for mixed temperate broad-leaved forests at higher elevations in Sikkim and Bhutan (10–12%)—but not for similar forests in the western Himalaya (0.1–0.6%)—merits further investigation.
- Results for *temperate coniferous forests* (Blue pine, *Cryptomeria,* Deodar, *Abies*; WHSCF or planted) were rather variable too, possibly due to methodological issues in some cases (excessively low TF and very high SF for sites in Kashmir and Sikkim); excluding the latter two yielded a tentative average *IL/P* of 25.5 ± 6.5% ($n = 4$ only); more work is needed.
- High TF (up to 91%) and low *IL/P* (~6.5%) were derived for two *temperate cloud-affected forests* in Bhutan (one oak-dominated, the other mixed coniferous and broad-leaved), likely due to (unmeasured) inputs of passing fog stripped by the canopy and suppressed wet-canopy evaporation under such conditions; high SF (12%) in the broad-leaved forest requires further scrutiny but agrees with observations for fog-ridden forests in central Myanmar (Zaw and Oue 2024).

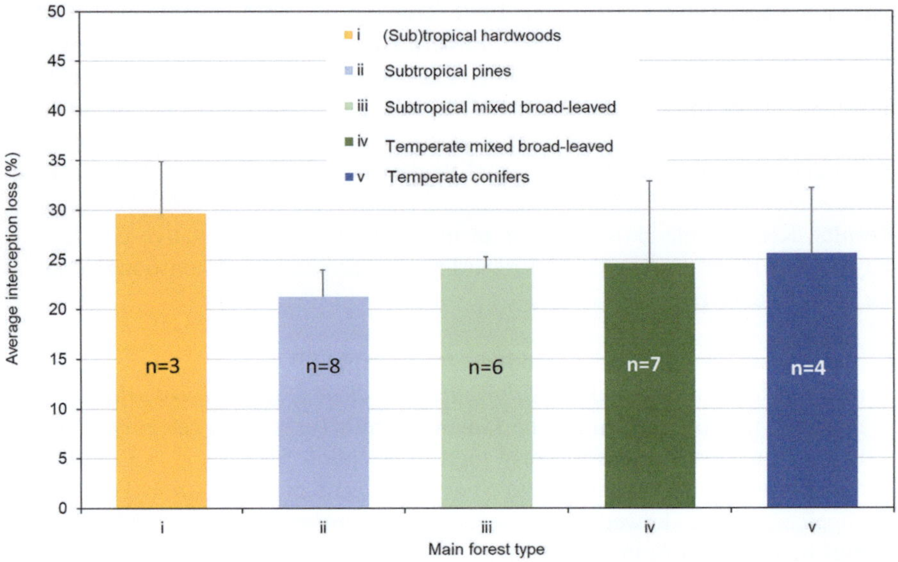

Fig. 12.4 Tentative average rainfall interception losses (%) and their standard deviations during the monsoon for the main Himalayan forest types distinguished in this chapter: (i) Subtropical *Shorea*-dominated forest; (ii) subtropical pine forest; (iii) subtropical mixed broad-leaved forest; (iv) temperate mixed broad-leaved forest; and (v) temperate coniferous forest. (Original data are listed in Appendix 12.1)

Figure 12.4 summarizes the average monsoonal interception fractions derived for the respective broad forest types. Despite the possible limitations of the original data, average IL/*P*-values per forest type broadly match the respective average values of leaf biomass and stand basal area for these forest types in the western Himalaya (Singh et al. 1994).

The relatively high average IL/*P* inferred for the tall *Shorea*-dominated forests of the Terai plain and Siwalik foothills (category (i) in Fig. 12.4) likely reflects their often high LAI (up to 7.7; Behera et al. 2015). Also, since rainfall intensities at lower elevations can be high (Sharma et al. 2021), an additional factor may be enhanced evaporation of fine splash droplets generated when high-energy raindrops hit the canopy (Murakami 2006; Dunkerley 2009). Similarly, the relatively low average IL/*P* for the subtropical *Pinus roxburghii* forests (category ii) matches their generally low LAI (usually <2; Ghimire et al. 2012; Singh et al. 2014a; Badu et al. 2022). The relatively high LAI (4–5.25; Ghimire et al. 2012; Badu et al. 2022) of the subtropical mixed evergreen broad-leaved forests (category iii) and the frequent occurrence of relatively small rainfall events (<5 mm) at intermediate elevations (Gardner and Gerrard 2003; Ghimire et al. 2012; Mukherjee et al. 2018; Badu et al. 2022) may be responsible for the somewhat higher overall IL/*P* derived for this forest type. The same factors likely play a role in the temperate mixed oak forests above ~1900 m (category iv), particularly the lower rainfall intensities prevailing at higher elevations (Pathak et al. 1984; Singh et al. 1994; Om 2016). Lastly, although subject to broadly similar climatic conditions as the temperate oak forests, temperate Himalayan coniferous forests (category v) may intercept a larger fraction of precipitation. The data-set is very small, however (four studies for four different conifers) with another two studies reporting deviating results in the form of excessively high stemflow and extremely low crown drip (Rai and Sharma 1998; Wani and Manhas 2012; see footnotes to Appendix 12.1). Nevertheless, the higher overall IL/*P* fraction derived for this forest category may reflect the higher water retention capacity of densely set needles (Calder 1990; Singh 1987) compared to broad-leaved species, as well as the generally moderate stemflow production of conifers. The latter has been attributed to their drooping branch habit, particularly in older stands, leading intercepted water away from the trunks (Levia and Germer 2015). Interestingly, comparatively modest interception values were reported for plantations of *P. wallichiana* and *Cedrus deodara* in the NW Himalaya (Singh 1987; Singh et al. 1983) at an elevation (2440 m) where fog may be pronounced (Böhner et al. 2015; Miehe et al. 2015). Further interception studies in temperate Himalayan coniferous forests are not only desirable, but should also include the interception dynamics of snow and fog (see Sect. 12.3.2).

Predicting Interception Loss Under Himalayan Conditions Relationships between interception losses for the sites listed in Appendix 12.1 and stand basal area, LAI, and canopy cover fraction are shown in Fig. 12.5. Given the lack of a clear relation between interception fraction and stand basal area (Fig. 12.5a), the latter is not a useful proxy for estimating IL/*P* in Himalayan forests. LAI for subtropical mixed broad-leaved forest (Fig. 12.5b) and canopy cover fraction across forest types (Fig. 12.5c) are both positively related to IL/*P*, however. Such preliminary relationships may offer possibilities for deriving broad spatiotemporal patterns of interception loss using remotely sensed information on LAI (Nieschulze et al. 2009; Oliveira et al. 2024). Such information is available for forests in the LH and the adjacent Terai (Tripathi et al. 2014; Srinet et al. 2019; Gao et al. 2023). Nevertheless, the database linking site LAI to IL/*P* is still very small and largely limited to subtropical mixed evergreen broad-leaved forests and a few plantations (Fig. 12.5b). As such, there is a clear need for additional work relating adequately measured interception losses to seasonal changes in LAI (cf. Tanaka et al. 2015, 2017) for most forest categories, notably for *Shorea*-dominated semi-deciduous forest and temperate oaks and conifers.

Because the magnitude of IL/*P* per event depends on event size, the overall value for a given forest will effectively reflect the size distribution of precipitation events during the study period (Helvey and Patric 1965; Badu et al. 2022). Also, where seasonal changes in LAI are large, an overall average IL/*P* value will not be adequate year-round. Hence, the use of seasonally applicable regression equations linking amounts of TF, SF, or IL per storm to *P* provides a more reliable way to predict

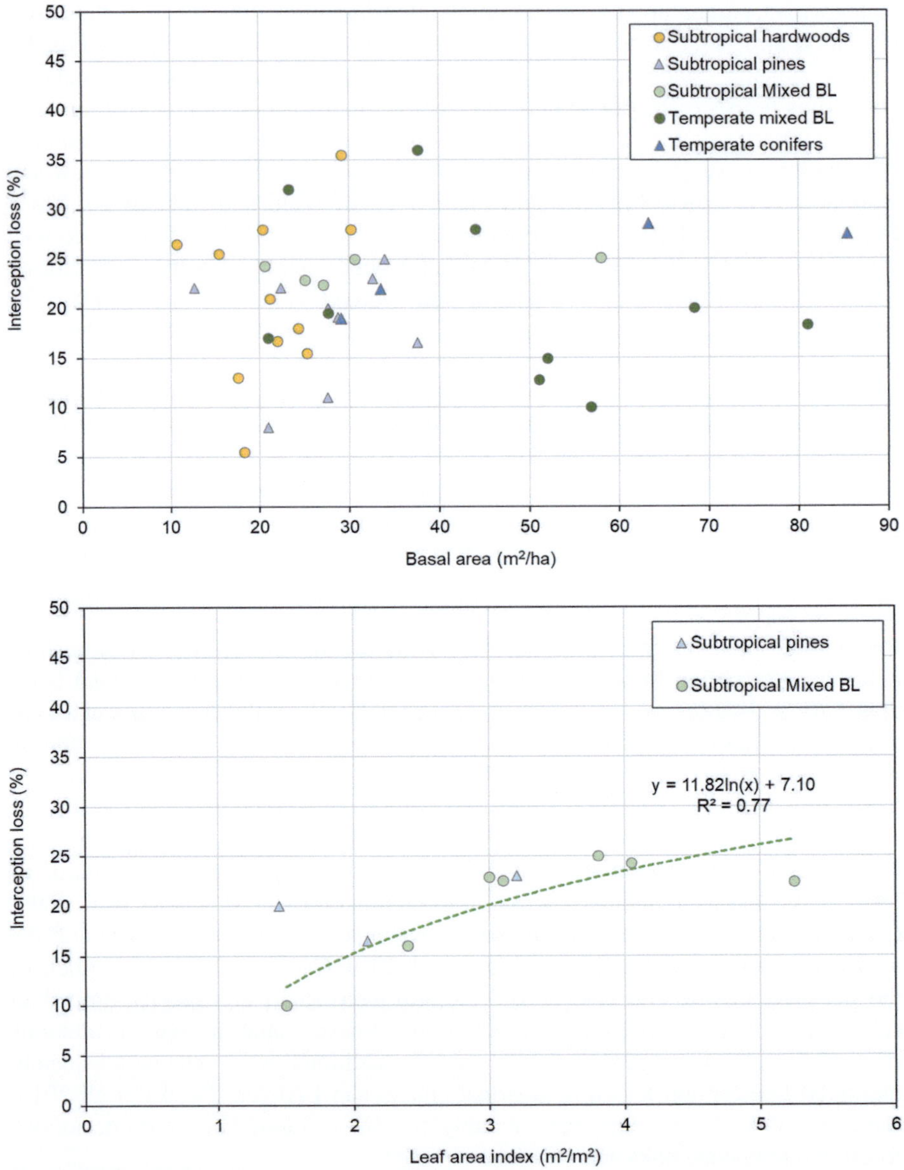

Fig. 12.5 Relationships between interception loss (% of incident rainfall) and forest structural attributes for Lesser Himalayan forests: (**a**) Basal area ($m^2\ ha^{-1}$); (**b**) Leaf area index ($m^2\ m^{-2}$); and (**c**) Canopy cover fraction (dimensionless). Original basal area and canopy cover fraction data listed in Appendix 12.1 and Table 12.6

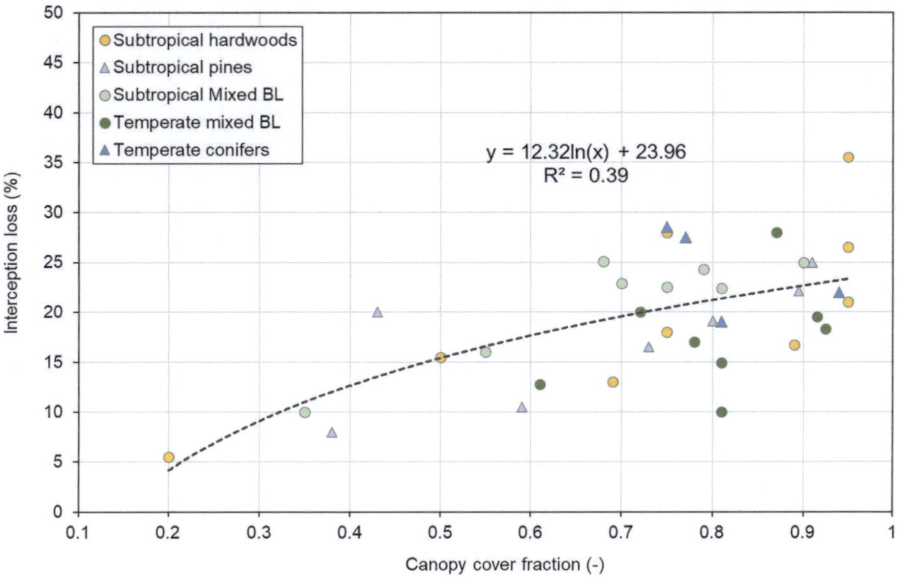

Fig. 12.5 (continued)

seasonal and annual interception losses. However, such regressions remain site-specific and their results are best extrapolated to similar forests and climatic conditions.

A more rigorous approach concerns the analytical interception model advanced by Gash (1979) and subsequently adapted by Gash et al. (1995) to take variations in canopy density (or LAI: Van Dijk and Bruijnzeel 2001) into account. Without going into detail, Gash-type models use four parameters to characterize forest structure: canopy gap fraction or direct throughfall coefficient (p), canopy capacity (S), stemflow coefficient (p_t), and trunk storage capacity (S_t). The most important of these, S, is defined as the amount of water present on the canopy once TF (canopy drip) ceases. It is traditionally derived by determining the negative Y-intercept of a line with slope ($1-p_t$) drawn through the highest data points (to avoid evaporation effects) in a plot of TF versus P (Leyton et al. 1967). Jackson (1975) later introduced a more objective way of deriving p and S by separately plotting TF against P for small storms that are not able to fully saturate the canopy and larger ones that do. The two stemflow-related parameters can be derived from a similar plot of SF versus P (Gash and Morton 1978). Lastly, the model requires the ratio between the mean evaporation rate from the wetted canopy per m^2 of ground surface ($\overline{E}$) or canopy ($\overline{E}_c$) and the mean rainfall rate onto the saturated canopy ($\overline{R}$). Gash-type models have been used successfully in numerous settings (Muzylo et al. 2009; Oliveira et al. 2024), including the Nepalese Himalaya (Ghimire et al. 2012). Importantly, progress has been made with applications over larger areas using remotely sensed vegetation information (Wallace et al. 2013; Oliveira et al. 2024).

Several Himalayan studies presented regression equations linking TF, SF, or IL to *P* (e.g. Pathak et al. 1985; Singh 1987; Ghimire et al. 2012; Tomar et al. 2022). Among these, only Ghimire et al. (2012) applied the Jackson method to derive *p* and *S* (for a *Schima–Castanopsis* forest and a planted pine forest; Table 12.1). Because the recommendation of Leyton et al. (1967) to only use TF data representing events with minimum evaporation was not followed in any of the other studies, their values of *S* may have been overestimated (Table 12.1). In conclusion, representative values for the respective interception model parameters (notably *S* and $\overline{E}/\overline{R}$) for the main Himalayan forest types are urgently needed to allow spatiotemporal modelling of interception loss patterns along the lines of Oliveira et al. (2024).

Fog Interception

Background Rainfall interception represents a loss of water to the ecosystem via evaporation from a wet canopy, thereby reducing amounts of precipitation reaching the forest floor (Appendix 12.1). Conversely, interception of passing low cloud (fog) by a forest canopy or isolated trees affects the water budget positively (Bruijnzeel et al. 2005; Oliveira et al. 2014). Fog blown against or deposited upon vegetation may reach the ground as 'fog drip' (Dawson 1998; Bruijnzeel et al. 2011), marking an extra input of water. Fog is also absorbed via foliar uptake (Limm et al. 2009; Eller et al. 2013). Although amounts involved are generally small, its physiological significance can be substantial, especially when soils are dry (Simonin et al. 2009; Eller et al. 2015). Also, fog presence typically causes strong reductions in transpiration (Eller et al. 2013; Alvarado-Barrientos et al. 2014) and can even reverse the direction of sap flow in tree stems, thereby replenishing soil moisture (Burgess and Dawson 2004; Eller et al. 2013).

Fog absorption proved important for survival of epiphytic lichens, ferns, and bryophytes in cloud-affected forest under monsoonal conditions at ~2300 m in SW China where fog occurred on average on 212 days per year (Liu et al. 2021). Similarly, drip from local radiation fog in seasonal rain forests at lower elevations (700–750 m) in southern Yunnan was essential for the survival of seedlings during the dry season, although adult trees did not rely on this extra source (Liu et al. 2010; Liu et al. 2014).

Fog interception and associated drip can represent a significant extra input depending on the frequency, density, and duration of the fog, prevailing wind speeds as well as the structure and topographic exposure of the vegetation (Bruijnzeel et al. 2011; Jarvis and Mulligan 2011). Along with rainfall, most of these factors also govern occurrence, diversity, and biomass of bryophytes, and a distinct increase in the latter is generally reported upon entering the cloud belt on tropical mountains (Frahm and Gradstein 1991; Wolf 1993). In the Nepalese Himalaya, epiphytic fern

Table 12.1 Preliminary values of forest structural parameters for selected Himalayan forests as used in interception modelling. S = canopy saturation value (mm), p = direct throughfall fraction (–), S_t = trunk storage capacity (mm), p_t = stemflow coefficient (–), $\overline{E}/\overline{R}$ = ratio (–) between the average wet-canopy evaporation rate, $\overline{E}$ and average rainfall intensity, $\overline{R}$. BA = stand basal area ($m^2\ ha^{-1}$) and c = canopy cover fraction (–); *UK* Uttarakhand, *HP* Himachal Pradesh

Forest type	Location	Elev.	S	p	S_t	p_t	$\overline{E}/\overline{R}$	BA	c	Remarks
Hill *Shorea robusta*[1]	Chorgalia, UK	330	0.66	0.11	0.08	0.012	–	22	0.89	Regressions include all rainfall events*
Secondary *Shorea* (~50 years)[2]	Dehradun, UK	760	0.61	0.50	0.10	0.019	0.36	25.3	0.50	Ibidem*
P. roxburghii, some oak[1]	Kalona, UK	1250	0.19	0.62	0.012	0.004	–	21	0.38	Ibidem*
P. roxburghii (25 years)[3]	Dhulikhel, Nepal	1580	0.67	0.27	0.033	0.007	0.27	37.6	0.73	Separate regressions for P < or > 1.1 mm
P. roxburghii[1]	Baldiyakhan, UK	1700	1.36	0.09	0.010	0.004	–	54	0.91	Regressions include all rainfall events*
Mixed oak & pine[1]	Ranikhet, UK	1900	0.40	0.22	0.003	0.003	–	21	0.78	Ibidem*
Schima–Castanopsis[3]	Dhulikhel, Nepal	1500	0.89	0.19	0.020	0.017	0.36	27.1	0.81	Separate regressions for P < and > 0.9 mm
Mixed oaks (mostly *tilonj*)[1]	Nainital, UK	2000	1.55	0.19	0.030	0.0024 s	–	52	0.81	Regressions include all rainfall events*
Mixed oaks (*rianj, tilonj*)[1]	Kilbury, UK	2120	0.27	0.07	0.072	0.039	–	81	0.93	Ibidem*
P. wallichiana (20 years)[4]	Shimla, HP	2400	5.37	0.19	0.27	0.030	0.13	29.1	0.81	Ibidem*

[1]Pathak et al. (1985); [2]Tomar et al. (2022); [3]Ghimire et al. (2012); [4]Singh (1987); *Tilonj* = *Q. floribunda*; *Rianj* = *Q. lanuginose*; *Ideally, values of S are derived for non-composite rainfall events of limited duration rather than all storms (Leyton et al. (1967))—further, Pathak et al. (1985) adjusted TF to include direct TF in gaps before deriving regressions of stand TF against P. Resulting values of S need to be treated with caution, therefore

diversity peaks around 2000 m (Bhattarai et al. 2004), whereas that of mosses and liverworts occurs around 2500 and 2800 m, respectively (Grau et al. 2007). Foliose lichens are most diverse between 2200 and 2600 m in Uttarakhand versus 2600–3100 m in Sikkim (Nanda et al. 2021). These observations agree with reports of a Himalayan cloud belt (and cloud-affected forests) occurring mostly between ~2000 and ~3400 m within which Miehe et al. (2015) distinguished three zones for the area east of the Kali Gandaki (Central Nepal): a lower belt dominated by oaks (notably *Quercus lamellosa* and *Q. lanata*) with epiphytic orchids, ferns, and foliose lichens between ~2000 and 2500 m; a middle belt with hemlock (*Tsuga dumosa*) and *Q. semecarpifolia* with pending mosses between ~2500 and 3000 m; and an upper belt with fir, birch, and *Rhododendron* between ~3000 and 3500 m where beard mosses (*Usnea* spp.) and epiphytic liverworts prevail. Although monsoon rainfall totals decrease with elevation within the zone, atmospheric humidity is generally high year-round and radiation impeded by cloud shading, with the most persistently humid conditions prevailing in the upper cloud belt (Miehe et al. 2015; Om 2016).

Actual measurements of fog climatology and the hydrological impacts of fog across the LH are effectively lacking,[1] but the low values of IL/*P* for mixed oak forest at 2460 m and *Tsuga–Quercus* forest at 3260 m in Bhutan (Om 2016; Appendix 12.1) may reflect the effect of unmeasured inputs via fog interception. Similarly, modest interception fractions have been reported for cloud-affected montane forests in SW China and central Myanmar (9–11%; Liu et al. 2003; Song et al. 2017; Zaw and Oue 2024) and N Thailand (9.4%; Tanaka et al. 2005) that are structurally and floristically (at *genus* level) comparable with the *Castanopsis* forests of the eastern Himalaya. Crown drip and stemflow were frequently observed during dense fog in montane forest in otherwise subhumid central Myanmar, to the extent that fog-induced drip made up 27% of all throughfall while fog-induced stemflow contributed nearly 40% of all stemflow (Zaw and Oue 2024).

Estimated Fog Interception Across the Himalaya In the absence of actual measurements of fog interception from the region, Fig. 12.6 shows the average frequency of cloud occurrence during the four main seasons (monsoon, post-monsoon, winter, and 'summer'), while Fig. 12.7 presents a first approximation of the spatial distribution of annual fog inputs along the range as modelled by WaterWorld 2 (v2.92) (Mulligan 2013; Box 12.1).

[1] A Grünow-type passive fog gauge operated during the monsoon of 1963 at Thodung (3100 m) in eastern Nepal captured 22% more precipitation (i.e., fog and wind-driven rain) than an adjacent standard rain gauge (Ries 1994)

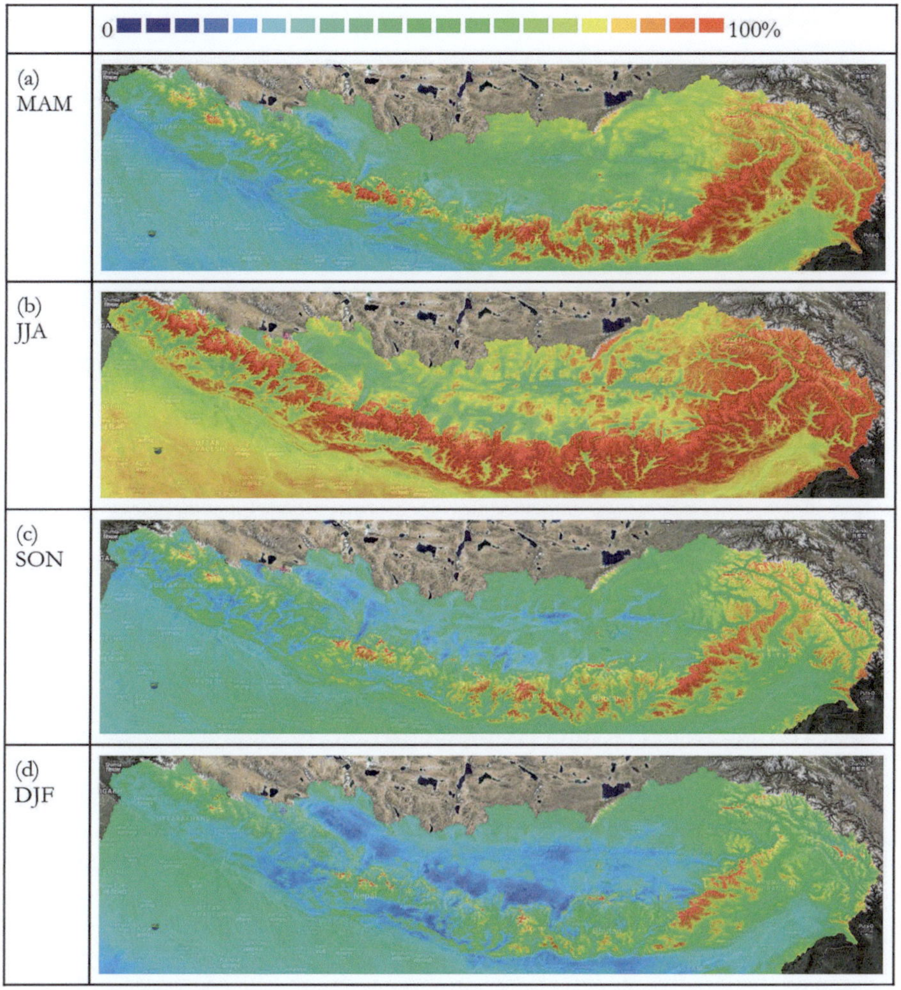

Fig. 12.6 Cloud frequency (% of observations) patterns for the Ganges—Brahmaputra River Basin for periods related to (**a**) summer (March–May), (**b**) monsoon (June–August), (**c**) post-monsoon (September–November), and (**d**) winter (December–February). All maps are masked to exclude elevations >3500 m asl. (Data: Wilson and Jetz (2016). Map data: © Google)

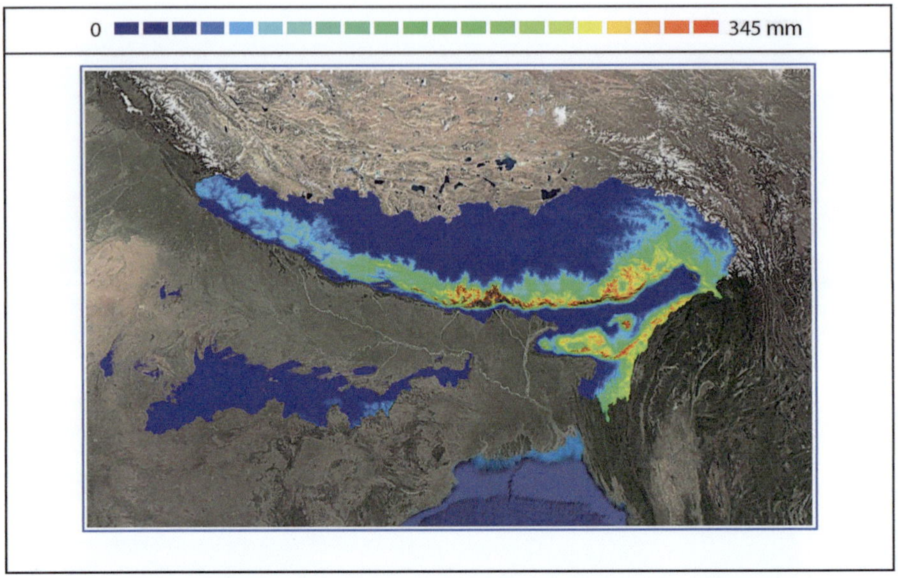

Fig. 12.7 Modelled annual fog inputs (mm) using WaterWorld for elevations below 3500 m across the Ganges—Brahmaputra River Basin and surrounding area. (Map data: © Google)

Box 12.1 Modelling Occult Precipitation Inputs to Montane Forest Ecosystems

WaterWorld is a spatially explicit model developed originally for the computation of the water budget of tropical montane areas, emphasizing a physically based representation of water inputs via wind-driven rain and fog interception and losses via ET. These are modelled on a raster pixel basis at a spatial resolution of 1 ha or 1 km^2. To model regional fog interception across the LH, monthly and diurnal patterns of cloud frequency were derived using the MODIS MOD35 satellite cloud mask product (Mulligan 2006; Wilson and Jetz 2016). The model employs four time steps each month that are designed to characterize the diurnal cycle within that month, thus requiring 48 time steps in total to represent a full year. Long-term mean values of temperature and atmospheric humidity at each time step are used to compute the associated level of the cloud base. Combined with the satellite-observed cloud frequencies and spatial information on vegetation structure and its fog-intercepting capacity (using LAI as a proxy), plus wind speeds, relative humidity (as a proxy for cloud liquid water content), and topographical exposure, the model computes fog interception. Details are given by Mulligan (2013); for the equations describing and linking the respective processes, see Mulligan and Burke (2005) and www.policysupport.org/waterworld. Although absolute values are indicative only, model applications have produced reliable geographical patterns (Bruijnzeel et al. 2011).

The four panels in Fig. 12.6 illustrate the seasonal course of cloud frequencies across the Himalayan region. Cloud frequency is greatest in magnitude and extent during June–August (JJA; panel 12.6b) and is seen to gradually retreat in extent towards higher ground during the pre-monsoon (March–May, MAM) and post-monsoon (September–November, SON) seasons, especially in the western Himalaya (panels 12.6a and c). By December–February (winter, DJF), only the highest peaks plus wider areas in Sikkim, Bhutan, and Arunachal Pradesh still experience high cloud frequencies (panel 12.6d). As such, the cloud frequency environment of the western Himalaya is distinctly different from that of the eastern part.

A similar contrast emerges for the magnitude of annual fog inputs as calculated at 1 km spatial resolution using WaterWorld for elevations below 3500 m (Fig. 12.7). Fog interception as a result of deposition (under low wind speeds) and impaction on protruding vegetation (under higher wind speeds) is greatest in the eastern Himalaya and NE India, particularly on slopes and ridges in the *Mahabharat* range that are exposed to moist air masses from the Bay of Bengal (Fig. 12.7).

Modelled annual fog inputs are greatest in eastern Nepal, Sikkim, southern Arunachal Pradesh, and Bhutan, with maximum values reaching ~345 mm year^{-1} (Fig. 12.7). Seasonal changes follow cloud frequency, with the highest values derived for March, July, and August in different areas of the western Himalaya versus May, July, August, October, and November in different parts of the eastern Himalaya (data not shown). The significance of these fog inputs varies with the seasonality of other forms of precipitation (mostly rainfall, some snow). Fog inputs as a percentage of total precipitation are greatest in November across most of the Himalaya and in December for the eastern part when rainfall is low (data not shown).

Litter-Layer Interception and Evaporation

Background Amounts of precipitation reaching the forest floor are modified upon hitting the litter layer. Like the canopy, the litter layer also possesses a certain water retention capacity (C_{min}; Putuhena and Cordery 1996; Sato et al. 2004). Generally speaking, this capacity needs to be exceeded before significant infiltration into the underlying soil will occur (Walsh and Voigt 1977; Tobón et al. 2000). Water retained by litter is available for evaporation, typically at low rates (0.15–0.4 mm day^{-1}) due to the low radiation, high atmospheric humidity, and low wind speeds beneath closed-canopied vegetation (Deguchi et al. 2008; Oishi et al. 2020; Table 12.2).

Values of C_{min} are conventionally derived by exposing litter material (usually leaves, occasionally other components as well; Putuhena and Cordery 1996; Van Stan et al. 2017) to simulated rain and determining the moisture content once drainage has ceased (Walsh and Voigt 1977). The magnitude of C_{min} generally increases with litter mass and, to a lesser extent, rainfall intensity (Putuhena and Cordery 1996; Sato et al. 2004; Zhu et al. 2021; Deng et al. 2023), but tends to decrease with slope gradient (Zhao et al. 2022). Broad-leaved litter typically retains more water per unit mass than does needle litter (Dabral et al. 1963; Putuhena and

Table 12.2 Litter layer interception (LI) fractions of incident precipitation (*P*) and average daily evaporative loss (E_L) during the monsoon for selected Himalayan forests. Results for similar forest types elsewhere added for comparison. *LM* litter mass (kg m^{-2}), C_{min} litter retention capacity (mm)

Forest types	Elev. (m)	LI/P (%)	E_L (mm d^{-1})	LM (kg m^{-2})	C_{min} (mm)	C_{min}/LM (mm kg^{-1} m^{-2})	Remarks
Himalayan forests (all around Dehradun or Nainital)							
Shorea (coppice)[1,2]	330	9.8[a]	0.93	0.71	2.3 ± 0.4	3.24	Three trays in the open; mean of two monsoons; C_{min} based on 7 events
Shorea (36 years)[3]	640	8.0[a]	–	1.18[x]	1.6 ± 0.6	1.36	One tray in the open; litter refreshed each time; C_{min} based on 13 events
Tectona (35 years)[3]	700	8.4[a]	–	1.22[x]	1.7 ± 0.8	1.39	Ibidem
Pinus roxburghii (26 years)[3]	640	6.7[a]	–	1.89[x]	1.4 ± 0.5	0.74	Ibidem
P. roxburghii[4]	1600	8.3[a]	0.66	–	–	–	Three trays in the open; mean of two monsoons
P. roxburghii[1,2]	1700	7.2[a]	0.73	0.66	1.4 ± 0.3	2.12	Three trays in the open; mean of two monsoons; C_{min} based on 5 events
P. roxburghii, some oak[1,2]	1250	7.2[a]	0.70	0.63	2.1 ± 0.7	3.33	Ibidem
P. roxburghii, some oak[4]	1750	9.5[a]	0.76	–	–	–	Three trays in the open; mean of two monsoons
Mixed oak, some pine[1,2]	1900	8.8[a]	0.66	0.70	1.9 ± 0.7	2.71	Ibidem; mean of two monsoons; C_{min} based on 6 events
Mixed oaks (*banj, tilonj*)[4]	1980	11.4[a]	0.93	–	–	–	Ibidem; mean of two monsoons
Mixed oaks (mostly *tilonj*)[1,2]	2000	9.4a	1.05	0.67	1.9 ± 0.2	2.84	Ibidem; mean of two monsoons; C_{min} based on 5 events
Mixed oaks (*rianj, tilonj*) [1]	2120	9.2[a]	0.93	0.62	1.9 ± 1.0	3.06	Ibidem
Other locations							
Pinus caribaea (15 years), Fiji[5]	82	9.3[b]	0.47[b]	1.35[y]	2.4	1.78	Four trays in situ; field-based drying curves; $n = 317$ days
P. patula (16 years), S Africa[6,7]	1065	12.1[b]	0.21[b]	3.34[y]	4.2	1.26	Two trays dug into the ground; 3 years; numerous very small storms

Mixed BL, Coweeta, SE USA[8]	945	3.6[d]	0.13[d]	0.68	1.45[a]	2.13	Field-based wetting-drying curves; $n = 153$ days; regrowth forest
Ibidem (35 years)[9]	770	3.1[d]	0.24[d]	0.93[y]	2.0	2.15	Ibidem; *Liriodendron–Quercus* regrowth after selective harvest
Ibidem (85 years)[9,10]	730	3.2[d]	0.22[d]	0.91[y]	2.0	2.20	Ibidem; mature mixed oak–maple forest; E_L during summer = 0.14 mm d^{-1}
Ibidem (>200 years)[9,10]	1205	3.9[d]	0.31[d]	0.96[y]	2.1	2.19	Ibidem; annual sub-canopy eddy covariance-based $E = 0.16$ mm d^{-1}
Cryptomeria (25 years), S Japan[11,12]	70	5.6[b]	0.21[b]	1.73[y]	0.82	0.48y	Weighing lysimeter; wetting-drying curves; 3 years
Lithocarpus (20 years), S. Japan[11,12]	70	5.8[b]	0.22[b]	2.07[y]	1.74	0.84	Ibidem
Lowland rain forest SP, Colombia[13]	250	11[b]	(1.0)[c]	7.65[z]	16.3	2.13	Daily (n = 45 trays, 1 year), continuous soil moisture, field saturation
Lowland rain forest MT, Colombia[13]	250	7[b]	(0.74)[c]	4.95[z]	7.6	1.54	Daily (n = 30 trays, 1 year), continuous soil moisture, field saturation
P. radiata (10–15 years), SE Australia[14,15]	975	11[b]	0.21[b]	2.05[y]	1.54[y]	0.97	Catchment-wide LI (6 years: 1988–1993), LM, and C_{min}
Native eucalypts, SE Australia[14,15]	975	8[b]	0.15[b]	0.77[y]	0.91[y]	1.13	Ibidem
Pine, South Central China[16]	–	–	–	0.5	0.55	1.10	Lab experiment; *P. massoniana* litter at rainfall intensity $R = 50$ mm h^{-1}
Broad-leaf, South Central China[16]	-	–	–	0.5	0.75	1.50	Ibidem; *Magnolia grandiflora* litter at $R = 50$ mm h^{-1}

[1]Pathak et al. (1985); [2]Pathak et al. (1989); [3]Dabral et al. (1963);[4]Loshali and Singh (1992); [5]Waterloo (1994); [6,7]Bulcock and Jewitt (2012a, 2012b); [8]Helvey (1964); [9]Brantley et al. (2019) and S.T. Brantley, pers. comm.; [10]Oishi et al. (2018); [11]Sato et al. (2003); [12]Sato et al. (2004); [13]Tobón et al. (2000); [14]Putuhena and Cordery (1996); [15]Putuhena and Cordery (2000); [16]Hao et al. (2022); [a]June–September (monsoon); [b]Annual value; [c]Uptake via roots in litter layer; [d]mid-May–mid-September; [x]Average amount of litter in sampling tray; [y]Including twigs/bark; [z]Including F + H layers but excluding fine roots; *C_{min}/LM values pertain to leaf/needle material only unless indicated otherwise; *Banj* = *Quercus leucotrichophora*; *Tilonj* = *Q. floribunda*; *Rianj* = *Q. lanuginosa*

Cordery 1996; Zhao et al. 2022; Ilek et al. 2024). Similarly, broad-leaved litter has been shown to be four times more effective than pine needle litter at suppressing evaporation from wetted mineral soil under Himalayan field conditions (Kumar and Verma 1989). Of particular relevance to steep, forested hillslopes is the finding that slopes ≥15° generate significantly more lateral (downslope) flow within the litter at the expense of vertical infiltration, particularly in large-leaved litter (Li et al. 2021; Zhao et al. 2022). Pertinently, although litter retention capacity increases with the degree of decomposition (Ilek et al. 2024), virtually all interception studies have used undecomposed litter.

Himalayan Results Interception by leaf litter (LI) has been studied in various Indian forests by exposing litter samples in trays placed in the open to natural rainfall (Dabral et al. 1963; Pathak et al. 1985, 1989; Loshali and Singh 1992). Hence, the magnitude of C_{min} and the intercepted fraction of rain (LI/*P*) derived in this way depend on the number and size of sampled events, with smaller events exhibiting a larger intercepted fraction and vice versa (Walsh and Voigt 1977; Pathak et al. 1989). Table 12.2 summarizes results from the region along with comparative data for selected forests with similar climatic conditions elsewhere.

Values of C_{min} for Himalayan leaf litters (1.35–2.3 mm) are within the range for comparable forest types elsewhere (0.55–4.2 mm), as are corresponding values of LI/*P* (7.2–11.4% versus 4.4–12%, respectively; Table 12.2). Again, C_{min} for broad-leaved litters (*Shorea*, teak, oaks) exceeded those for pine litter, in agreement with general findings (Walsh and Voigt 1977; Ilek et al. 2024), although standard deviations were considerable. However, normalizing C_{min} for litter mass for better comparability between sites showed the Himalayan values to be higher than those obtained for comparable litter types and climatic conditions elsewhere (Table 12.2). The same applied to the average evaporation rates from the respective litter types (Table 12.2), suggesting that the Himalayan results may represent overestimates. Possible reasons include placement of litter samples in the open rather than beneath the canopy which will increase insolation and ventilation levels to which the samples are exposed, and hence potential losses via evaporation (Helvey 1964; Deguchi et al. 2008). In addition, leaving a gap between the litter sample (placed on a wire mesh) and the water-collecting tray underneath (Dabral et al. 1963; Pathak et al. 1989) disturbs the natural drainage from undecomposed litter into the semi-decomposed layer below (Tobón et al. 2000). This 'interface effect' causes litter interception to be overestimated (Helvey and Patric 1965; Sato et al. 2004; Bulcock and Jewitt 2012a). Lastly, (large amounts of) fresh oven-dried (rather than air-dry) leaves were used for every new measurement in the experiments of Dabral et al. (1963).

Alternatively, the use of forest litter wetting and drying curves based on in situ moisture measurements is considered to better represent field conditions (Waterloo 1994; Tobón et al. 2000; Sato et al. 2003; Brantley et al. 2019). Application of the average daily summer-time sub-canopy evaporation rates listed in Table 12.2 for cool-temperate (0.13–0.24 mm d^{-1} in SE US, ~21.5 °C; Helvey 1964; Brantley et al. 2019; Oishi et al. 2020), warm-temperate (~0.3 mm d^{-1} in South Central Japan,

~24 °C; Deguchi et al. 2008), and tropical conditions (~0.45 mm d^{-1} in Fiji, ~26 °C; Waterloo 1994) suggests that, depending on site elevation, evaporative losses from Himalayan litter during the monsoon (June–September) may total 20–55 mm instead of the 80–130 mm inferred by Pathak et al. (1985). Support for this comes from below-canopy pan evaporation totals of less than 30 mm during the wet season (May–October) at Ailao Shan (Yunnan, 2475 m; Song et al. 2017). Without significant precipitation inputs during most of the remaining year, and given the exponential decline of litter moisture content with time since precipitation (Helvey 1964; Waterloo 1994), the corresponding seasonal evaporation totals from dried-up litter are likely to be low as well, despite the higher evaporative demand of the atmosphere outside the monsoon season (Böhner et al. 2015). Field observations of the water dynamics of Himalayan forest floors using improved methodology (Tobón et al. 2000; Sato et al. 2003) are needed for conclusive results.

Transpiration and Evapotranspiration

Methodological Considerations Globally, amounts of water taken up and evaporated by plants (transpiration, E_t) constitute the largest fraction of the total evaporative loss (ET) from forest ecosystems (Jasechko et al. 2013; Coenders-Gerrits et al. 2014). This holds true for the forests of the LH and adjacent plain as well, although there is very little *direct* evidence. The vast majority of Himalayan forest evaporation studies derived estimates of *total* ET only (Table 12.3 below), mostly via catchment water budgets in which ET is taken to equal the difference between precipitation inputs (*P*) and gauged streamflow outputs (*Q*). Such estimates should be interpreted with caution given that ungauged subterranean water losses (catchment leakage) will cause ET determined in this way to be overestimated. Much of the LH is broken up tectonically and many rocks are fractured (Rai 1993; Andermann et al. 2012; Tarafdar et al. 2019). Also, high spatial variability of precipitation in mountainous terrain presents a challenge to accurately estimate catchment-wide water inputs (Hamilton 1954; Kathyar and Striffler 1984; Stewart et al. 2017; Kumar et al. 2021; Jiang et al. 2023), whereas occult contributions need to be taken into account in the cloud-affected zone. Furthermore, differences in rainfall between successive years can cause major contrasts in amounts of water stored in the soil and in groundwater bodies which need to be incorporated in the budget equation as well (Shimizu et al. 2015; Shiraki et al. 2017; Eeckman et al. 2019). Apart from these uncertainties, the inherently 'lumped' water budget approach does not allow distinctions between evaporation components (cf. Fig. 12.3), although a first approximation of the sum of transpiration and evaporation from the forest floor (i.e. $E_t + E_s$) may be obtained by subtracting measured interception losses, IL, from total apparent ET (derived, in turn, from *P* minus *Q* in most cases).

Table 12.3 Estimates of annual evapotranspiration (ET), interception loss (IL), and the sum of transpiration and soil evaporation ($E_t + E_s$) for selected Himalayan forest catchments and sites. P = annual precipitation input; Q = annual streamflow output (water yield) or inferred site drainage (values bracketed since not measured directly); Q/P = runoff coefficient (%). Annual totals rounded off to the nearest 5 mm. Mean daily values of ET and ($E_t + E_s$) added to facilitate comparisons between sites and forest types with transpiration values in bold being used in Fig. 12.8. Methodological abbreviations: *BREB* Bowen ratio—energy budget method, *CWB* catchment water budget, *ECV* eddy covariance method, *OP & CP* open- & closed-path sensors, *SWB* site water budget, *UK* Uttarakhand, *HP* Himachal Pradesh. Eastern Himalayan vegetation: BL = mixed evergreen broad-leaved forest

Location	Size (ha)	Elev. (m asl)	P (mm)	Q (mm)	Q/P (%)	ET (mm)	ET (mm d^{-1})	IL (mm)	($E_t \pm E_s$) (mm)	($E_t \pm E_s$) (mm d^{-1})	Remarks
(Sub-)tropical deciduous hardwoods (mostly Shorea)											
Kaziranga NP, Assam[1]	Plot	80	1835	(605)	(36.5)	1165	3.2	–	–	–	ECV-CP, 80% closure; 3 years; inundated
Barkot, Dehradun UK[2]	Plot	415	1115	(420)	(38)	[950]*	[2.6]*	[255]#	695#	1.9	ECV-OP, 73% closure; excl. Wet canopy
Deurali Kholsi, W C Nepal[3]	50	675	(855)	(355)	–	(500)	1.65	–	–	–	CWB, 10 months excl. 2 monsoon months
Rajpur, Dehradun UK[4, 5]	6.5	895	2425	1015	41.5	*1415*	3.9	~435	~980	2.7	CWB, W1, 4 years; *high leakage*
Subtropical to warm-temperate pines (mostly P. roxburghii)											
FRI Campus, Dehradun UK[6]	Plot	640	2245	(1245)	(55)	1000	2.7	–	–	–	BREB; year 2011; advected heat
Kosi-Almora, Kumaun UK[1,7]	Plot	1215	~995	(35)	(3.5)	[960]	[2.65]	~230[g]	730	2.0	ECV-OP, 2016; IL for nearby Papoli forest
Bhetagad, Kumaun UK[8] (48%)	260	1460	1075	290	27	[785]	[2.7]$	–	–	–	CWB, 3 years; 52% agriculture
Bemunda WS HP[9,10]	1754	1500	2235	1210	54	1025	2.8	~425[j]	~600	1.65	CWB, Blue pine; some leakage?
Dhulikhel, East C Nepal[11]	Plot	1580	1425	(845)	(59.5)	575	1.6	235	340	**0.93**	SWB; sap-flow, IL; Es estimated; 1 year
Siahi Devi, Kumaun UK[12]	102	1780	1460	305	21	1155@	3.15@	160	995@	2.7@	CWB, wet year; *Q* underestimated?

Khulgad, Kumaun UK[13, 14]	29.6	1805	970	165	17	805	2.2	–	–	–	CWB, 1991–93;
Binsar, Kumaun UK[15]	38.5	1960	1175	165$^{@}$	14	1010$^{@}$	2.75$^{@}$	270	740$^{@}$	2.0$^{@}$	CWB, 1 year; *Q* underestimated?
Temperate oak forest (Q. leucotrichophora)											
Arnigad, Garhwal UK[16,17]	285	1940	2920	1615	55	1265	3.5	~730	~535^{q}	**~1.46**	CWB, 3 years; high rainfall zone
Khulgad, Kumaun UK[13,14]	53	1945	970	70	7.4	900	2.5	–	–	–	CWB, 1991–93;
Siahi Devi, Kumaun UK[12]	87.6	1980	1430	450	31.5	980$^{@}$	2.7$^{@}$	455	525$^{@}$	**1.44**$^{@}$	CWB, wet year; *Q* underestimated?
Binsar, Kumaun UK[15]	42.3	2040	1175	510$^{@}$	43.5	665$^{@}$	1.8$^{@}$	420	245$^{@}$	**0.67**$^{@}$	CWB, 1 year; Q underestimated?
Sitlakhet, Kumaun UK[7]	91	2070	970	330$^{\$}$	34$^{\$}$	640$^{\$}$	1.75$^{\$}$	~310	~330$^{\$}$	**~0.9**$^{\$}$	CWB, 1 year; 58% oak forest
Eastern Himalayan forests and surrounding region											
Dhulikhel, E C Nepal[11] (BL)	Plot	1500	1330	(805)	(60.5)	525	1.44	295	230	**0.63**	SWB; sapflow, IL; Es estimated; 1 year
PG site, W Bhutan[18] (mixed oaks)	Plot	2460	1125	(770)	(68.5)	355	1.0	80	275	**0.75**	SWB modelled; slightly fog-affected
Chhukarpo Khola, E Nepal[19] (BL)	23.5	2600	2325	1645^{+}	71	680	1.85	–	–	–	CWB, 1 year; disturbed; 1/third grazed
Kharikhola, E Nepal[20] (BL)	1820	2900	2290	1745	76	545	1.49	–	–	–	CWB, 2015/16; 75% BL-forest; fog?
TG site, W Bhutan[18] (hemlock-oak)	Plot	3260	1185	(875)	(74)	310	0.85	70	240	0.65	SWB modelled; fog-affected
Tauche, E Nepal[20] (not forest)	465	>3700	730	500	68	230	0.63	–	–	–	CWB, 2015/16; rock/ shrub/grass 33% each

(continued)

Table 12.3 (continued)

Location	Size (ha)	Elev. (m asl)	P (mm)	Q (mm)	Q/P (%)	ET (mm)	ET (mm d^{-1})	IL (mm)	($E_t \pm E_s$) (mm)	($E_t \pm E_s$) (mm d^{-1})	Remarks
Mae Mo, N Thailand[21,22] (teak)	Plot	380	1335	(360)	(27)	975	2.68	120	855	2.34	ECV-OP, 7 years; IL, mature deciduous
Xishuangbanna, Yunnan[23,24] (BL)	Plot	750	1325	(675)	(51)	650	1.78	280	370	1.0	ECV-OP, 2 years; IL, dry-season fog
Kog-Ma, NW Thailand[25] (BL)	8.63	1340	2130	1350	63.5	780	2.14	195	585	**1.6**	CWB, IL, 12 years; 2.3% fog
Ailao Shan, Yunnan[26] (BL)	Plot	2475	1635	(775)	(47.5)	865	2.36	145	720	**2.0**	ECV-OP, IL; 5 years; some fog

[1]Deb Burman et al. (2021); [2]Srinet et al. (2022); [3]Balla (1988); [4]Subba Rao et al. (1973); [5]Ghosh et al. (1980); [6]Singh et al. (2013); [7]Daniel et al. (2022); [8]Sharma et al. (2007); [9]Puri et al. (1982); [10]Singh and Gupta (1987); [11]Ghimire et al. (2014b); [12]Kumar (1990); [13]Rai (1993); [14]Rawat et al. (1997); [15]Kumar and Verma (1991); [16]Qazi et al. (2017); [17]Qazi (2021); [18]Om (2016) and re-interpreting original data for IL as 7% (oaks) and 6% (hemlock); [19]Ries (1994) for the period mid-June 1990—mid-June 1991; [20]Eeckman et al. (2017); [21]Igarashi et al. (2015); [22]Tanaka et al. (2015); [23]Dou et al. (2007); [24]Xin et al. (2006); [25]Tanaka et al. (2005); [26]Song et al. (2017)

*Corrected for energy budget imbalance (Giambelluca et al. 2009; Shimizu et al. 2015); #Assuming measured evaporation (1.9 mm d^{-1}) represented ($E_t + E_s$); adding unmeasured IL (18–28% of P, Fig. 12.4) gave ET-values of 895–1005 mm (950 mm on average); @Streamflow measured once a day during rainy season, hence Q possibly underestimated and ET (and thus $E_t + E_s$) overestimated accordingly; $Value for forest-only, assuming ET cropland = 600 mm (Rawat et al. 2000); +Approximate value—Q read from annual hydrograph (mid-June 1990—mid-June 1991)

Direct estimates of (E_t + E_s) may be derived from measurements above the evaporating canopy using the eddy covariance technique (ECV) in conjunction with quantification of the energy budget (Deb Burman et al. 2019). Applications in the often steep and complex topography of the LH require various additional corrections (Mukherjee et al. 2020), whereas the widely used open-path gas analyzers of EVC systems function poorly during intense rainfall (Oishi et al. 2020; Srinet et al. 2022). Hence, ECV is best used under (and representative of) dry canopy conditions.

Alternatively, E_t can be determined from the rate of sap-flow ascent in tree stems combined with knowledge of sapwood cross-sectional areas. The chief limitation of the technique in species-rich forests concerns upscaling of the measurements (made by necessity on individual trees and species) to stand level (McJannet et al. 2007; Oishi et al. 2008; Ghimire et al. 2014b). In addition, separate measurements need to be made of evaporation from the understorey and forest floor (Oishi et al. 2020).

Estimates of E_t and ET for a given forest type vary between sites as a function of climatic (precipitation, temperature, and radiant energy) and edaphic (soil depth, water retention capacity) conditions as well as stand characteristics (age, density, presence or absence of understorey). To eliminate climatic differences between sites, it is helpful to normalize E_t and ET by dividing them by the site's potential ET (PET). Different definitions and formulations exist for PET that yield different absolute results (Allen et al. 1998; Dorji et al. 2016; Dai et al. 2023), but applying a given PET to all examined sites allows better comparability, while the role of forest structural factors can be highlighted by examining relationships between (normalized) E_t or ET and forest basal area or LAI (McJannet et al. 2007; Ghimire et al. 2022). Here, the FAO-version of the PET for an ideal, well-watered grassland (ET_0; Allen et al. 1998) was used.

Himalayan Results ECV-based estimates of seasonal and annual ET were derived for a semi-deciduous lowland (80 m asl) tropical forest in Assam (Deb Burman et al. 2019 and 2021), a subtropical *Shorea robusta* forest at Barkot near Dehradun (415 m; Srinet et al. 2022), and a 20–25-year-old *P. roxburghii* stand near Almora in the Kumaun Himalaya (1217 m; Deb Burman et al. 2021). Singh et al. (2013) and (2014a) documented the seasonal course of the radiation budget and ET for a vigorous young stand of *Chir* pine with a *Lantana* understorey at Dehradun (640 m). No explicit distinction was made between wet- and dry-canopy evaporation in any of these studies, although periods of high rainfalls were generally excluded. Sap-flow sensors were used in several higher-elevation forests in Sikkim (Kumar et al. 2023, 2024a, b) and Bhutan (Om 2016; Wangdi 2016; Gadermaier 2018) to study the response of individual trees to changes in environmental conditions (weather, soil moisture), but annual stand-scale values were not derived. Hence, the only documented year-round Himalayan cases to date concern the comparative water budget work of Ghimire et al. (2014b) in native *Schima–Castanopsis* forest and mature planted pines near Dhulikhel (east central Nepal) along with new comparative micrometeorological observations in oak and pine forests near Nainital by Mohanty et al. (2025).[2] Apart from soil water budget studies in two cloud-affected, high-elevation oak and hemlock forests in Bhutan (Om 2016), all further estimates of Himalayan forest ET represent catchment water budgets. Table 12.3 lists the results ($n = 23$) along with data for several forests from the wider region experiencing broadly comparable climatic conditions.

The highest annual ET (1165 mm or 3.2 mm day^{-1} on average) was reported for the *(semi-deciduous) tropical lowland forest* at Kaziranga NP, Assam (Deb Burman et al. 2021). This estimate should include wet- and dry-canopy evaporation (since a closed-path sensor was used) and thus may represent 'true' total ET. It is worth noting, however, that the forest was inundated during the monsoon (Deb Burman et al. 2019). Hence, overall ET may have been boosted by evaporation from the free water surface. In comparison, ET reported for the *deciduous subtropical Shorea forest* at Barkot (695 mm or 1.9 mm day^{-1}; Srinet et al. 2022) may be an underestimate. Not only was 27% of the energy budget unaccounted for, but evaporation during 'bad weather' (heavy rain) was excluded from the analysis as well (Srinet et al. 2022). Further, subtracting the (high) interception loss derived for a nearby similar *Shorea* forest (38%; Tomar et al. 2022) from this 'total' ET would give an unrealistically low value of ~0.75 mm day^{-1} for ($E_t + E_s$). Approximate alternative estimates of annual ET for the Barkot forest converge around 950 mm (2.45–2.8 mm day^{-1}; see footnotes in Table 12.3). Some support for the latter estimate comes from the 975 mm (2.7 mm day^{-1}) of evaporation derived for a mature teak forest in northern Thailand (Mae Mo site in Table 12.3) experiencing near-identical radiation and rainfall inputs as the Barkot forest. The physiology and ecology of teak and *Shorea* are very similar (Behera et al. 2012, 2015), hence their water use is likely comparable (Dabral et al. 1965). Further work (including adequate quantification of wet-canopy evaporation rates) is needed for *Shorea*,[3] also because of the practical implications for regional groundwater recharge.

Reported annual evaporative losses from Himalayan *subtropical to warm-temperate pine forests* range between 525 mm for a mature plantation at 1580 m in east central Nepal and 1155 mm for the Siahi Devi catchment at 1780 m in Kumaun (Table 12.3 and footnotes). A tentative overall average ET of ~925 ± 185 mm year^{-1} was derived (~2.54 ± 0.51 mm day^{-1}; $n = 7$). Subtracting an average interception loss for pine forest of 21.3% at the group-average

[2] Annual ET-values for the Nainital forests have not been published yet but Kumar et al. (2025) presented two years' worth of sapflow observations for the pine forest. Kumar et al. (2024a) derived a mean daily E_t (1.29 mm d^{-1}) for low-statured regrowth over a six-month period covering winter and warm pre-monsoon conditions at 2100 m in Sikkim; comparative sap-flow measurements have also been made by the GB Pant National Institute of Himalayan Environment in Kumaun pine and oak forests; analysis was still ongoing at the chapter's proof stage (September 2025; Dr. Sandipan Mukherjee, pers. comm.)

[3] Measurements of interception loss (using numerous throughfall collectors) were initiated by the Nepalese Institute of Forestry in a mature *Shorea* forest at 400 m elevation near Hetauda at the start of the 2024 monsoon (Dr. Chandra Ghimire, pers. comm.)

precipitation of 1500 mm (Fig. 12.4) yielded an average annual value for ($E_t + E_s$) of ~605 mm or ~ 1.65 mm day^{-1}. Repeating the exercise for the five pine forest studies listed in Table 12.3 that also *measured* IL gave comparable values of ~675 ± 210 mm year^{-1} and 1.85 ± 0.6 mm day^{-1}, respectively. Thus, ($E_t + E_s$) make up at least two-thirds (65–73%) of total ET from subtropical Himalayan pine forests. However, the fact that the two studies employing sap-flow sensors to measure pine water use also reported the lowest E_t (Ghimire et al. 2014b; Kumar et al. 2025) suggests that the cited annual and daily average values may have been inflated by catchment leakage. Additional direct measurements of E_t in pine forests remain essential[2].

Direct measurements of ET for *subtropical oak forests* are currently lacking,[4] whereas reported catchment-based annual estimates for five *warm-temperate oak-forests* in the NW Himalaya vary roughly two-fold (640–1265 mm (Table 12.3). With ET_0 varying relatively little between the respective sites, part of this variability likely reflects differences in catchment leakage (see footnotes to Table). Hence, the inferred average ~885 ± 260 mm (~2.4 ± 0.8 mm day^{-1}) is likely an overestimate and the more moderate apparent ET totals derived for two of the catchments (Binsar and Sitlakhet: 640–665 mm) may be more representative. However, precipitation amount (and thus interception loss) plays a role as well: the highest ET by far was derived for the wettest catchment (Arnigad; Table 12.3). Due to the high variation in ET between catchments, the corresponding estimates for ($E_t + E_s$) are equally variable (0.7–1.6 mm day^{-1}; Table 12.3). Again, given potential leakage issues for some of the higher estimates, the lower value may be more realistic, also in view of similar average daily values (0.62–0.75 mm day^{-1}) measured for two eastern Himalayan mixed broad-leaf forests (Table 12.3).

Despite this variation, values of mean daily ($E_t + E_s$) for oak- and similar broad-leaf forests in the wider region correlated positively with stand basal area ($R^2 = 0.64$; $n = 10$; Fig. 12.8), suggesting some influence of forest structure and stand density on the magnitude of E_t (cf. McJannet et al. 2007; Ghimire et al. 2022). Again, additional direct estimates of E_t for Himalayan forests across the subtropical and warm-temperate belts are needed, also in light of the continuing debate on the hydrological role of pines versus oaks (see **Box 12.2** below)[2,4].

Tentative estimates of ET for *higher-elevation Himalayan forests* (>2400 m; mostly wet *Rhododendron* forest grading into drier *kharsu* oak forest and conifers with increasing elevation) may be derived from hydrological observations in the Chhukarpo Khola (Ries 1994) and Kharikhola (Eeckman et al. 2017) catchments in NE Nepal, along with plot-scale water budget work in oak- and hemlock-dominated

[4]The carbon and water exchanges of a regenerating oak forest at 1650 m elevation in the Pithoragarh District of Kumaun have been studied using eddy covariance equipment by the GB Pant National Institute of Himalayan Environment. To date (October 2025), only results for carbon have been published (Lohani et al. 2023). New comparative observations on oak and pine evaporation in the Nainital area using the Bowen Ratio Energy Balance method (Mohanty et al. 2025) qualitatively confirm the assessment offered in Box 12.2.

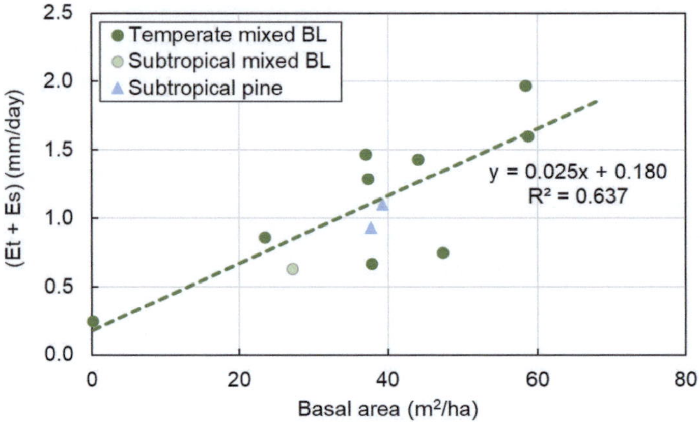

Fig. 12.8 Tentative relationship between stand basal area and mean daily rates of transpiration (E_t) plus evaporation from the forest floor (E_s) in warm-temperate Himalayan oak-dominated forests and *Castanopsis—Lithocarpus* forests from the wider region (see Table 12.3 for data). Average Et-values for 30-year-old broad-leaf forest in Sikkim (December–May only; Kumar et al. 2024a) and 25-year-old *Chir* pine in Kumaun Himalaya (Kumar et al. 2025) added for comparison

forests at 2450 and 3260 m in Bhutan (Om 2016).[5] All these forests were affected to some extent by fog (Ries 1994; Wangdi 2016; Eeckman 2017). Despite potential issues with catchment leakage and/or inadequate quantification of precipitation inputs, ET estimates for the two Nepalese forests were plausible at 1.5–-1.85 mm day^{-1} (Table 12.3). Furthermore, both forests gave very high water yields (71–76% of conventionally measured precipitation; Table 12.3). The latter may reflect both unmeasured occult inputs (Fig. 12.7) and suppressed evaporation (Ries 1994) within the cloud belt (cf. Bruijnzeel et al. 2011). Similarly high water yields (68.5–74% of *P*) were implied by the low apparent evaporation totals derived for the two Bhutanese forests: 1.0 mm day^{-1} for the mixed oak forest and 0.85 mm day^{-1} for the mixed coniferous–broad-leaf forest. Corresponding amounts of ($E_t + E_s$) were 0.75 and 0.65 mm day^{-1}, respectively (Om 2016; Table 12.3). Pertinently, and in contrast to the *Shorea*, pine and oak forests of lower elevations—for which ET/ET$_0$ was close to 0.85 on average—values for the two Bhutanese cloud forests ranged from 0.40 to 0.51 which likely reflects the effect of unmeasured occult inputs (Bruijnzeel et al. 2011). Lastly, Eeckman et al. (2017) derived the water budget for the non-forested (equal fractions under shrub, steppe, and rock/ice) Tauche catchment above the cloud belt (>3700 m) in the Everest region. Apparent mean

[5] Illien et al. (2021) presented graphic rainfall and stream water-level data for three consecutive years for the largely forested Kahule Khola (35 km^2) in east central Nepal (1200–3400 m), while Kumar et al. (2024b) presented annual stream discharge hydrographs for several high-elevation catchments with temperate to subalpine vegetation in Sikkim. Neither study converted stream discharge data to annual streamflow totals, thereby preventing estimates of seasonal and annual ET for these little studied Himalayan forest types to be derived

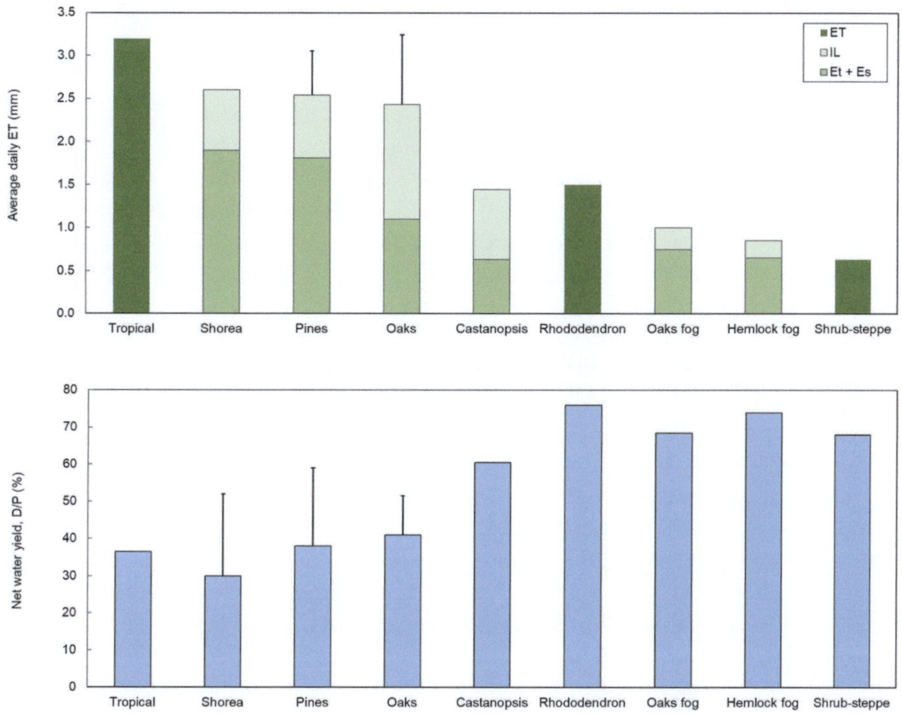

Fig. 12.9 (**a**) Average daily evapotranspiration rates (ET) and (**b**) approximate relative water yields (streamflow or drainage output expressed as a percentage of precipitation) for the examined main forest types: (i) Lowland tropical semi-deciduous forest; (ii) *Shorea*-dominated subtropical forest; (iii) subtropical to warm-temperate pine forests; (iv) warm-temperate oak forest; (v) subtropical *Castanopsis*-dominated mixed evergreen broad-leaf forest; (vi) wet Rhododendron forest; (vii) fog-affected temperate mixed oak forest; (viii) fog-affected cool-temperate mixed hemlock–oak forest; and (ix) shrub—steppe vegetation

daily ET was 230 mm year^{-1} (0.63 mm day^{-1}) and relative water yield (*Q/P*) slightly lower at 68% (Table 12.3).

Figure 12.9 summarizes the changes in average daily ET for the main forest types with increasing elevation, together with approximate estimates of the corresponding fractions of 'effective' precipitation feeding groundwater, springs, and streams. While some numbers are evidently more tentative than others (since necessarily based on a single study for five of the nine distinguished vegetation types!), a clear decline in ET with elevation emerges, along with a gradual increase in relative water yield, especially at higher, cloud-affected elevations in the eastern Himalaya.

Box 12.2 Oaks, Pines, and Springs: A Case of Not Seeing the Soil for the Trees

Throughout the Middle Mountains, the presence of mixed broad-leaved forest (oak-dominated in the west, *Schima–Castanopsis* forest in the east) is considered to maintain springs and streams, whereas *Chir* (*P. roxburghii*) is seen as causing these to decline or even dry up. Indeed, the average daily rate of soil water uptake (plus evaporation from the forest floor) derived for (mostly subtropical) pine forest (1.8 ± 0.6 mm) is markedly higher than that for the (warm-temperate) oak forest (1.1 ± 0.4 mm). Partly, this is due to the lower radiation loads and temperatures of the moister north-facing slopes favoured by oaks compared with the drier and sunnier south-facing slopes where pines prevail, but there are inherent physiological contrasts between the two forest types as well (Zobel et al. 2001; Ghimire et al. 2014b). However, interception losses are slightly higher on average for oak forests (~24%) than for the more open pine forests (21.3 ± 2.7%). The net effect is that overall average daily water losses (total ET) essentially do not differ between pine and oak forests at 2.54 ± 0.51 and 2.43 ± 0.81 mm day^{-1}, respectively, although the relative water yield for oaks (41 ± 10% of rainfall) may slightly exceed that for pines (37 ± 21%) due to differences in rainfall between individual sites and the small sample size per group. *Without a major difference in total ET from the two forest types (~40 mm year*$^{-1}$*), what other factors underlie their opposite hydrological reputations?*

Apart from lower insolation levels, north-facing slopes in the LH also tend to have deeper soils than their south-facing counterparts (Kumar and Verma 1990; Rai 1993; Shrestha 1997) due to the predominant tilting of rock strata towards the north (Searle and Treloan 2019). Also, soils beneath oak forest are often more clayey than beneath pines (Kumar and Verma 1990; Raina and Gupta 2009; Daniel et al. 2022). Thus, soil water retention and supply are better under oak than under pine. Further, the denser oak canopy suppresses evaporation from the forest floor (Kumar and Verma 1989) and provides a more favourable microclimate for litter decomposition which, together with the better quality of oak litter, stimulates decomposition (Chaturvedi and Singh 1987; Kumar and Tewari 2014; Joshi and Garkoti 2020). This produces deeper humus layers and further increases moisture retention (Kumar and Verma 1990; Tyagi et al. 2013). Together, the greater moisture storage capacity of the deeper soils of oak forests plus the lower transpiration rate of oak cause the outflow of water from oak forests during the dry season to be both higher and to last longer (Kumar and Verma 1990).

Conversely, the harsher microclimate of the drier, south-facing slopes beneath more open pine forest (Kumar and Verma 1989) combines with lower-quality needle litter to slow down rates of decomposition (Chaturvedi and Singh 1987) and humus formation. Also, the shallower soils under pine

(continued)

Box 12.2 (continued)
likely generate more frequent saturation-excess overland flow during the monsoon (see Sect. 12.3.5) which may further impoverish the soil as nutrients and finer soil particles are washed off by surface runoff (Naudiyal and Schmerbeck 2017). The lower water storage and retention capacity of the shallower and sandier soils and the higher transpiration by pine together lead to a faster and shorter outflow of water during the dry season. If left undisturbed for a sufficiently long period, *Chir* forests can develop deep humus layers as well (Raina and Gupta 2009) and gradually give way to mixed pine-oak forest and, ultimately, oak-dominated forest (Naudiyal and Schmerbeck 2018). In practice, however, succession is often interrupted by fire and other disturbances, with adverse hydrological effects (see Sect. 12.4.1).

Seasonal Evaporation Patterns Seasonal variations in precipitation, solar radiation, temperature, and atmospheric vapour pressure deficit (VPD) across the LH are pronounced, with forest types such as *Shorea*, *Chir,* and various oaks exhibiting major seasonal fluctuation in LAI in response (Singh et al. 2013; Behera et al. 2015; Deb Burman et al. 2019; Joshi and Sarkoti 2020). Hence, E_t is expected to vary seasonally as well. Incoming solar (or net) radiation generally constitutes the main driver of E_t with LAI and VPD as major regulating variables (Ghimire et al. 2014b; Singh et al. 2014b; Deb Burman et al. 2019; Srinet et al. 2022). Transpiration rates typically increase with increasing VPD but soon level off once a threshold value is reached upon which stomata begin to close. Depending on prevailing radiation levels, VPD-threshold values have been reported to lie between 1.0 and 1.3 kPa for tropical lowland semi-deciduous forest (Deb Burman et al. 2019) and for well-watered pines in the foothills (Singh et al. 2014b). Lower threshold values (~0.5 kPa) were derived for *Schima–Castanopsis* and pine trees at intermediate elevations (1500–1600 m; Ghimire et al. 2014b) and for various oaks (but not understorey *Rhododendron* spp.) within the lower cloud belt of Bhutan, with an even lower threshold of ~0.25 kPa reported for fog-ridden hemlock forest at 3260 m (Gadermaier 2018).

Interestingly, with the exception of some shallow-rooted pioneer species in the high rainfall zone of Sikkim (Kumar et al. 2023), none of the cited studies found E_t to be significantly affected by soil moisture levels at any time of the year. Apparently, this is due to a combination of strong stomatal control by the trees in response to rising VPD and continued access by roots to moisture stored in deeper soil layers and fractured rock during times of low rainfall (Poudyal et al. 2004; Ghimire et al. 2014b; Pathak et al. 2021; Srinet et al. 2022; Kumar et al. 2023). Medium and coarse roots have been described as being abundant in deeper subsoil (120–155 cm) beneath mature *banj* oak forests (*Q. leucotrichophora*) in Garhwal Himalaya (Sharma et al. 2006), while Pathak et al. (2021) showed ~50% of the total root mass for ten common tree species in Kumaun to occur below 30 cm depth. The importance of moisture uptake from deeper layers (down to 100 cm) during the dry season was demonstrated further by Song et al. (2017) for a temperate evergreen mixed broad-leaf forest in SW China. Likewise, despite the generally lower fine root biomass (Chand et al. 2022) and shallower rooting habit of *Chir* pine (Pathak et al. 2021), no significant effect of soil moisture status down to 75 cm depth on E_t was detected by Ghimire et al. (2014b) at the end of the dry season in a mature pine plantation in east central Nepal. Further work could usefully employ stable isotope tracers to determine the relative importance of soil moisture stored at different depths for continued transpiration at different times of the year (cf. Stahl et al. 2013; Yang et al. 2015; Ghmire et al. 2022).

With soil water status apparently exerting limited control on forest transpiration rates in the LH, seasonal patterns of E_t largely reflect variations in radiation, temperature, and VPD, as well as LAI. Radiation levels are generally lowest in winter and peak during the pre-monsoon between mid-March and mid-June at most elevations (Fig. 12.10; Singh et al. 2014a; Ghimire et al. 2014b; Srinet et al. 2022), but remain high outside of the winter months in lowland Assam (Deb Burman et al. 2019). Consequently, VPD levels generally also peak in the pre-monsoon, with comparatively lower values during the remainder of the year—particularly during the monsoon (Fig. 12.10). Somewhat elevated VPD levels occur in September–October as well as at some sites in the western Himalaya (Singh et al. 2014a; Deb

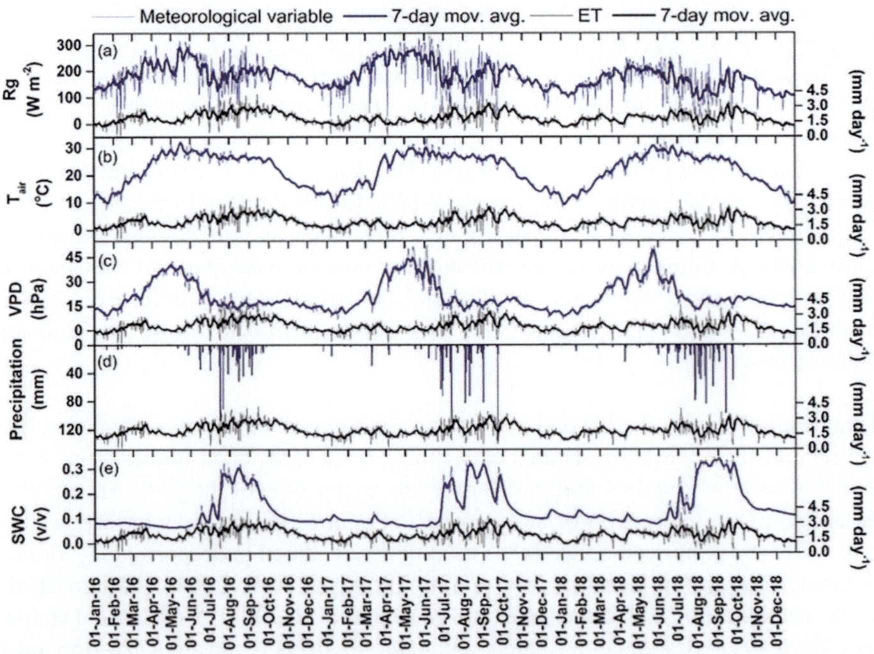

Fig. 12.10 Seasonal course of chief climatic variables and evaporation from *Shorea* forest at Barkot (Dehradun). R_g short-wave radiation, T_{air} air temperature, *VPD* vapour pressure deficit, and *SWC* volumetric soil water content down to 100 cm. (After Srinet et al. 2022)

Burman et al. 2021), whereas high values have also been reported for the latter part of the monsoon in Assam (Deb Burman et al. 2021) and during winter at higher elevations in Bhutan (Wangdi 2016). Again, LAI generally reaches a minimum around March–May when many tree species partly or wholly shed their leaves or needles (Ghimire et al. 2014b; Singh et al. 2013; Behera et al. 2015; Deb Burman et al. 2019; Joshi and Sarkoti 2020; Sheikh et al. 2020).

Although seasonal variation in the main climatic variables is marked, corresponding fluctuations in forest evaporation rates ($E_t + E_s$) are less pronounced (Fig. 12.10), mostly because of the strong stomatal control by the plants in response to increases in VPD. Furthermore, the season with the highest evaporative demand overlaps with the period of minimum LAI (March–May), thereby further reducing evaporative losses (Ghimire et al. 2014b; Srinet et al. 2022). Nevertheless, relatively high evaporation rates have been observed during sunny spells shortly after pre-monsoon rainfall events (Ghimire et al. 2014b; Chatterjee et al. 2018; Srinet et al. 2022; Kumar et al. 2024a,b).

Runoff Generation

Rainfall in the LH being highly seasonal, it is of interest to examine the runoff response of undisturbed forested hillslopes and headwater catchments before discussing hydrological effects of forest disturbance and conversion. Illien et al. (2021) advanced a conceptual model of the seasonal filling and emptying of the groundwater reservoir (subsoils, regolith, and fractured rock) that provides the bulk of annual streamflow in the LH. Pertinently, rainfall in the pre-monsoon season (March to mid-June) does not produce visible increases in baseflow (which continues to recede), but rather replenishes soil water reserves that were depleted during the preceding dry period. Only after a certain moisture storage threshold is exceeded (usually by late June–early July; Mathur et al. 1976; Subba Rao et al. 1985) do baseflows rise rapidly and does streamflow respond more directly to precipitation as the monsoon progresses, while soil moisture levels stabilize at a high level (Fig. 12.11). By this stage, precipitation inputs and streamflow outputs are strongly coupled until the rains subside in October and groundwater outflow and baseflows begin their long recession (Ries 1994; Hessel et al. 2007a, 2007b; Qazi et al. 2017; Illien et al. 2021).

Infiltration-excess overland flow (IOF) occurrence is governed by the interaction between the intensity of rain and the capacity of the soil to absorb it (i_c). Table 12.4 lists i_c-values for selected forest sites across the LH ($n = 23$). With reported i_c ranging from 13 mm h^{-1} to >2000 mm h^{-1} but median rainfall intensities in the Middle Mountains and higher up often being ~3 mm h^{-1} or less (Pathak et al. 1984; Ries 1994; Gardner and Gerrard 2003; Ghimire et al. 2012; Om 2016; Qazi et al. 2017; Badu et al. 2022), it is not surprising that the majority of Himalayan plot-scale runoff studies observed only minor 'surface runoff' under (more or less) undisturbed forested conditions (often <0.5% of rainfall; Table 12.5).

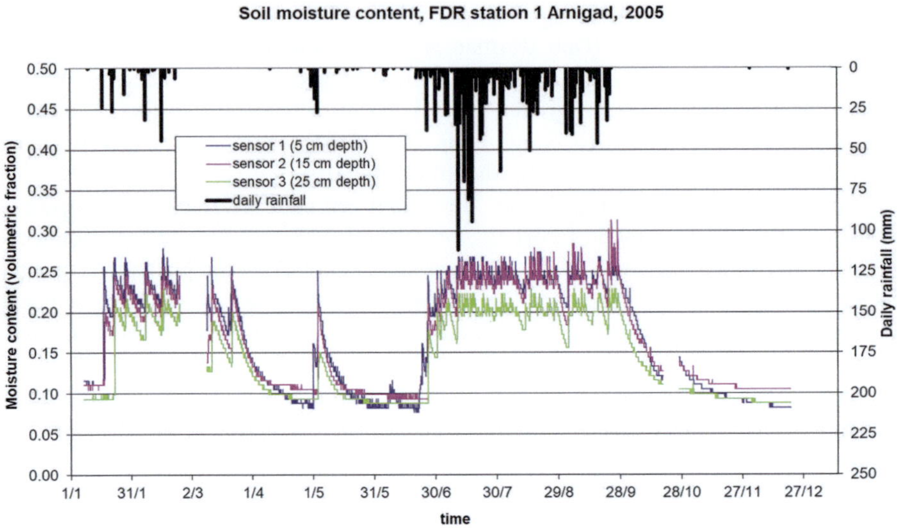

Fig. 12.11 Seasonal variation in soil moisture contents at different depths in the Arnigad catchment, Uttarakhand. (After Hessel et al. 2007a, b). Values at full saturation vary between 40 and 60 volume per cent depending upon depth and site location (Tyagi et al. 2013)

Methodological Considerations An important caveat regarding published values for surface runoff on Himalayan hillslopes is that these are invariably based on measurements made on relatively small bounded plots covering only part of a hillslope (Tables 12.5 and 12.7). Measured runoff amounts per unit area typically decrease as plot length increases (Gomi et al. 2008; Moreno de las Heras et al. 2010), mainly because of spatial variations in soil attributes that allow runoff to re-infiltrate in more permeable patches on the way down (Woolhiser et al. 1996; Nanda et al. 2019; Koppe et al. 2022). Thus, caution is needed when comparing runoff data for differently sized plots. Also, averaged point measurements of i_c are often lower than 'block permeabilities' of whole hillslopes because of this re-infiltration (Wen and Gonzalez-Hernandez 1996; Chappell et al. 1998; Cheng et al. 2023).

Results The very low surface runoff reported for many Himalayan forest sites (Table 12.5) may represent a plot edge effect (e.g. rain moving within the litter layer over short distances above the runoff collecting gutter; Li et al. 2021; Zhao et al. 2022) or simply that only a very small part of the plot contributed (e.g. a central natural drainage line; Pathak et al. 1984). Either way, these very low runoff coefficients imply that virtually all rainfall reaching the forest floor in *undisturbed* Himalayan forests infiltrates and continues towards the nearest stream (at least initially) via subsurface pathways (Pathak et al. 1984; Gilmour et al. 1987; Hessel et al. 2007a; Ghimire et al. 2013a). The latter can be fast (subsurface stormflow SSSF via temporarily activated macropores and other 'preferential flow paths') or slow (through the soil matrix; Sarkar et al. 2008; Chouksey et al. 2017). An example of

Table 12.4 Surface infiltration capacities (i_c) and related soil attributes for selected more or less undisturbed Himalayan forests. BD = topsoil bulk density; SOC = topsoil organic carbon content. Soil textural indicators: *Cl* clay, *L* loam, *Sa* sand, *Si* silt, *DRI* double-ring infiltrometer, *SRI* single-ring infiltrometer, *CHLP* constant-head lab permeameter, *FHLP* falling-head lab permeameter, *UK* Uttarakhand, *HP* Himachal Pradesh

Forest type	Location	Elevation (m asl)	i_c (mm h^{-1})	BD (g cm^{-3})	SOC (%)	Clay (%)	Silt (%)	Sand (%)	Texture	Remarks
Young regrowth (5 years)[1]	Assam valley, NE India	55	78 (w)	1.14	–	12.5	18	69.5	Sa L	DRI, $n = 5$; wet soil; dense shrub
Idem[1]	Idem	55	156 (d)	1.14	–	12.5	18	69.5	Sa L	Idem; dry soil; sparse vegetative cover
Idem[1]	Idem	55	135 (w)	1.14	–	12.5	18	69.5	Sa L	18 m X 6 m plot; wet soil; dense shrub
Idem[1]	Idem	55	62 (d)	1.14	–	12.5	18	69.5	Sa L	Idem; dry soil; sparse vegetative cover
Shorea (virgin forest)[2]	Dehradun, UK	515	122	1.03	1.8	19.5	39	41.5	L	Hood infiltrometer; $n = 6$
Shorea forest[3]	Dehradun, UK	535	60	(high)	–	–	–	–	Si L	SRI, $n = 6$; good litter cover
Shorea forest[3]	Dehradun, UK	840	58	(high)	–	–	–	–	Si L	SRI, $n = 6$; good litter cover
Shorea regrowth (66 years)[4,5]	Dehradun, UK	550	28	1.20	2.0	23	37	40	L	DRI; $n =$? but plural
Shorea regrowth (23 years)[6]	Likhu Khola, C Nepal	800	66	1.17	1.6	8	20	72	Sa L	DRI; $n =$? but plural; good litter cover
Lantana shrub[7–9]	Arnigad, Garhwal UK	1055	393	1.16	1.7	17	15	68	37 Sa L	DRI, $n = 6$–8; gravelly soil
Schima–Castanopsis forest[10]	Dhulikhel, E C Nepal	1500	333	0.93	4.1	29.5	46.5	24	Cl L	Disk infiltrometer, $n = 11$
Schima–Castanopsis forest[11,12]	Chautara, E C Nepal	1660	370	0.90	6.2	20	55	25	Si L	SRI, $n = 25$
Pinus roxburghii (35 years)[13]	Roshi Khola, E C Nepal	1550	84.5	1.11	2.3	13.5	72.5	14	Si L	CHLP, $n = 6$; mixed understorey

(continued)

Table 12.4 (continued)

Forest type	Location	Elevation (m asl)	i_c (mm h^{-1})	BD (g cm^{-3})	SOC (%)	Clay (%)	Silt (%)	Sand (%)	Texture	Remarks
P. roxburghii forest (native)[14]	Siahi Devi, Kumaun UK	1780	21.5	1.37	1.4	10	21	69	Sa L	DRI, $n = 8$ (tests of 2 hours)
P. roxburghii forest (native)[15,16]	Binsar, Kumaun UK	1960	42	1.18	1.1	13	17	70	Sa L	DRI, $n = 8$; monsoon season
Banj oak forest[7–9]	Arnigad, Garhwal UK	1810	437	1.11	3.6	36	27	41	Cl L	DRI, $n = 6$–8; mature forest
Banj oak forest[17]	Jijli, Garhwal UK	1895	786	1.14	–	3	41	55	Sa L	DRI, $n = 4$; site no. 2; gravelly
Banj oak forest[14]	Siahi Devi, Kumaun UK	1980	42	0.96	2.6	21	20	59	Sa Cl L	DRI, $n = 8$ (tests of 2 hours)
Banj oak forest[15,16]	Binsar, Kumaun UK	2040	112	0.82	2.8	28	19	53	Sa Cl L	DRI, $n = 9$; monsoon season
Mixed oak forest[18]	W Bhutan, PG site	2460	>1680	0.42	12	45	48	7	Si Cl	FHLP, $n = 4$
Mixed conifers (*Abies, Picea*)[19]	Kotgarh, Simla HP	2650	28.5	1.15	3.9	33	39	26	Si Cl L	DRI, $n = 50$
Rhododendron–oak–*Abies*[20]	Kharikhola, E Nepal	2675	13	0.72	–	1	26.5	72.5	L Sa	SRI, $n = 2$; wet forest*
Kharsu oak forest[20]	Kharikhola, E Nepal	2755	285	0.60	–	1	26	73	L Sa	SRI, $n = 2$; dry forest**
Mixed conifers (*Cedrus, Picea*)[21]	Kufri, Simla HP	2700	20	1.0	3.9	28	38	27	Cl L	DRI, $n = 36$; post-snow
Hemlock–oak forest[18]	W Bhutan, TG site	3260	>2085	0.54	8.4	38.5	33	28.5	Cl L	FHLP, $n = 4$
Rhododendron–Abies forest[20]	Kharikhola, E Nepal	3550	61	0.84	–	1	9.5	90.5	Sa	SRI, $n = 1$; mossy forest***

[1]Sarkar et al. (2008); [2]Kar et al. (2023); [3]Patnaik and Virdi (1962); [4]Mandal and Jayaprakash (2009); [5]Kaushal et al. (2020); [6]Gardner and Gerrard (2002); [7,8]Hessel et al. (2007a, b); [9]Sharma et al. (2006); [10]Ghimire et al. (2013a, b); [11]Gilmour et al. (1987); [12]Ghimire et al. (2014c); [13]Badu et al. (2022); [14]Kumar (1990); [15,16]Kumar and Verma (1990) and (1991)); [17]Qazi (2019); [18]Om (2016); [19]Singh and Gupta (1989); [20]Eeckman (2017); [21]Mathur et al. (1982); *Banj* oak = *Quercus leucotrichophora*; *Kharsu* oak = *Q. semecarpifolia*; *Mean of Judith and Razen sites; **Mean of Nuage and Remi sites; ***Anemone site

Table 12.5 Surface runoff (OF) production during the monsoon season in more or less undisturbed Himalayan forests. *P* = monsoon rainfall total; BD = topsoil bulk density; SOC = topsoil organic carbon content; K_{sat} = topsoil saturated hydraulic conductivity. Textural indicators: *Cl* clay, *L* loam, *Sa* sand, *Si* silt, *UK* Uttarakhand

Forest type	Location	Elev. (m asl)	*P* (mm)	OF/P (%)	BD (g cm^{-3})	SOC (%)	K_{sat} (mm h^{-1})	Soil texture	Remarks
Western Himalaya									
Hill *Shorea robusta*[1]	Chorgalia UK, India	330	1153	0.80	–	~1.2[2]	–	Sa L	Three, 25 m^2 plots; 1981/82
S. robusta, 23-yr-old regrowth[3]	Likhu Khola, W C Nepal	~800	2140	4	1.17	1.6	66	Sa L	One, 34 m^2 plot (Sal 7); 1992
S. robusta, 24-yr-old regrowth[3]	Likhu Khola, W C Nepal	~800	2210	**9.5**	1.14	1.2	36	L Sa	One, 25 m^2 plot (Sal 6); 1993; **incl. SOF?**
S. robusta, 50-yr-old regrowth[4]	Dehradun UK, India	715	2460	**8.5**	1.16	0.8	**6.5**	L	Three, 400 m^2 plots; 2014
P. roxburghii, some oak[1]	Kalona UK, India	1250	1180	1.24	–	–	–	Sa L	Three, 25 m^2 plots; 1981/82
P. roxburghii[1]	Baldiyakhan UK, India	1700	1245	0.67	–	~2.4[2]	–	L Sa	Three, 25 m^2 plots; 1981/82
P. roxburghii[5]	Siahi Devi UK, India	1780	1460	0.52	1.37	1.4	21.5	Sa L	One, 25 m^2 plot; 1986/87
Mixed oaks[5]	Siahi Devi UK, India	1980	1430	0.49	0.96	2.6	42	Sa Cl L	One, 25 m^2 plot; 1986/87
Oak (mostly *banj*)[6]	Jijli, Garhwal UK, India	1895	1085	**9**	1.14	–	322	Sa L	One, 9 m^2 plot (Site 2); 2018; **incl. SOF**
Oak (mostly *banj*)[1]	Ranikhet UK, India	1900	915	0.42	–	–	–	Sa L	Three, 25 m^2 plots; 1981/82
Mixed oaks (mostly *tilonj*)[1]	Nainital UK, India	2000	1345	0.39	–	–	–	L Sa	Three, 25 m^2 plots; 1981/82
Mixed oaks (*rianj*, *tilonj*)[1]	Kilbury UK, India	2120	1235	0.48	–	~3.4[2]	–	Sa	Three, 25 m^2 plots; 1981/82

(continued)

Table 12.5 (continued)

Forest type	Location	Elev. (m asl)	*P* (mm)	OF/P (%)	BD (g cm^{-3})	SOC (%)	K_{sat} (mm h^{-1})	Soil texture	Remarks
Open oak (*tilonj*) forest[6]	Nainital UK, India	2050	1500	0.44	–	–	–	–	Natural drainage, ~50 m^2; two monsoons
Mixed, semi-dense oak forest[7]	Nainital UK, India	2050	1500	0.49	–	–	–	–	Natural drainage, ~240 m^2; two monsoons
Dense oak (*tilonj*) forest[7]	Nainital UK, India	2050	1500	0.44	–	–	–	–	Natural drainage, ~30 m^2; two monsoons
Aesculus indica[8]	Nanda Devi UK, India	2400	(227)	0.25	–	–	–	L	Three, 25 m^2 plots; 14 events (August)
Q. semecarpifolia[8,9]	Nanda Devi UK, India	2600	(227)	0.31	0.72	4.3	–	Sa Cl L	Three, 25 m^2 plots; 14 events (August)
Abies pindrow (Silver fir)[8,9]	Nanda Devi UK, India	2500	(227)	tr.	0.59	3.8	–	Cl L	Three, 25 m^2 plots; 14 events (August)
Eastern Himalaya and surrounding region									
'Derelict' *Shorea* forest[10]	Jhargram, W Bengal	125	950	3.2	–	–	–	–	Three, 300 m^2 plots; 18 events 2.5–147 mm
Shorea coppice[10]	Jhargram, W Bengal	125	950	3.3	–	–	–	–	Three, 300 m^2 plots; 18 events 2.5–147 mm
Schima–Castanopsis[11]	Dhulikhel, E C Nepal	1500	710	2.5	0.93	4.1	333	Cl L	One, 75 m^2 plot; monsoon of 2011; SOF?
P. roxburghii (35 years)[12]	Roshi Khola, E C Nepal	1550	950	3.9	1.11	2.3	84.5	Si L	Three, 18 m^2 plots; monsoons of 2015/16
Alnus, Juglans, Quercus[13]	Mamlay, Sikkim, India	2275	?	(2.6)	–	–	–	Sa L	Three, 30 m^2 plots; 6 events / 2 monsoons
Rhododendron forest[14]	Surma Khola, E Nepal	2435	1020	5.7	–	7.3	–	Sa L	One, 14 m^2 plot; monsoon of 1991

Seasonal rain forest[15,16]	Xishuangbanna, Yunnan	600	1425	0.53	0.99	~2.2	~145	Sa L	Three, 10 m^2 plots; monsoon of 2017
Evergreen montane forest[17]	Ailao Shan, Yunnan	2475	1965	0.43	–	12.9	–	L	Two, 10 m^2 plots; year with av. P (1998)
Post-*jhum* regrowth (10 years)[18,19]	Burnihat, Meghalaya	125	1420	**13***	1.39	2.65	–	Sa L	Four, 400 m^2 plots; 2 monsoons; ***mostly SOF?**
Post-*jhum* regrowth (10 years)[20,21]	Shillong, Meghalaya	1500	1460	**16.5***	1.06	1.8	–	Sa	Four, 400 m^2 plots; 2 monsoons; ***mostly SOF?**

[1]Pathak et al. (1984); [3]Gardner and Gerrard (2002); [4]Tomar et al. (2022); [5]Kumar (1990); [6]Qazi (2019); [7]Pandey et al. (1983); [8]Negi et al. (1998a); [9]Adhikari (1992); [10]Ray (1971); [11]Ghimire et al. (2013a, b); [12]Badu et al. (2022); [13]Sharma and Rai (2004); [14]Ries (1994); [15]Zhu et al. (2018); [16]Chen et al. (2021); [17]Liu et al. (2003); [18]Toky and Ramakrishnan (1981); [19]Ramakrishnan and Toky (1981); [20,21]Mishra and Ramakrishnan (1983a, b); *Banj* = *Q. leucotrichophora*; *Tilonj* = *Q. floribunda*; *Rianj* = *Q. lanuginosa*;

rapid SSSF leading to equally rapid rises in streamflow during monsoon rains has been reported for the oak-forested Arnigad catchment in Uttarakhand (Hessel et al. 2007a). The area's sandy soils become gravelly with depth (Sharma et al. 2006), have a thick organic layer, and can accommodate large amounts of water (Tyagi et al. 2013). As a result, drainage is rapid and the soil never reaches complete saturation (Fig. 12.11). Average storm runoff responses during three successive monsoon seasons were only 3.5–7% of rainfall, with an absolute maximum of 21.7% for an extreme event delivering 167 mm of rain during an exceptionally wet period (Qazi et al. 2017).

However, this situation does not apply to all Lesser Himalayan forests: significant amounts of plot-scale runoff have been reported for (regenerating) *Shorea* forests in west central Nepal and oak forests in Garhwal (4–9.5%; Gardner and Gerrard 2002; Qazi and Bruijnzeel 2023) as well as for mixed broad-leaved or pine forests in the eastern Himalaya (2.5–5.7%; Ries 1994; Ghimire et al. 2013a; Badu et al. 2022; Table 12.5). Higher values (13–16.5%) were also observed under 10-year old regrowth following slash-and-burn cultivation in wet Meghalaya (Toky and Ramakrishnan 1981; Mishra and Ramakrishnan 1983a; Table 12.5). A remnant effect of previous disturbance and loss of permeable topsoil due to erosion cannot be excluded in the case of the cited regenerating *Shorea* forests, pine forests, and the two Meghalayan regrowth sites (Table 12.5). However, average values of i_c for the Garhwal oak forest (320 mm h^{-1}; Qazi and Bruijnzeel 2023), *Schima–Castanopsis* forest in east central Nepal (335–370 mm h^{-1}; Gilmour et al. 1987; Ghimire et al. 2013a), and 10-year-old regrowth at various locations in Meghalaya (100–150 mm h^{-1}; Zwartendijk et al. 2024) all exceeded the vast majority of prevailing rainfall intensities. Generation of IOF at these sites is extremely unlikely, therefore, suggesting the observed runoff consists of saturation-excess overland flow (SOF) instead.

Perched groundwater levels were not measured at any of these sites, but soil moisture recordings at several sites showed saturated conditions to coincide with OF occurrence (Ghimire et al. 2016; Daniel et al. 2022; Qazi and Bruijnzeel 2023). Figure 12.12 presents two examples from Garhwal where SOF occurred frequently on steep slopes, both under well-developed oak forest and beneath *Lantana* shrub. Both these sites had highly permeable, but shallow soils with dense rock blocking percolation at shallow depth. SOF is typically generated once a site-specific antecedent soil moisture storage cum rainfall threshold value is exceeded. Interestingly, although soil depths (and thus water storage capacities) were comparable between the two sites, the threshold value for the shrubland was substantially lower than that for the oak forest (Fig. 12.12b). The lower interception loss (and thus higher net precipitation) of shrubland compared to denser forest (Table 12.6) likely plays a role here (Qazi and Bruijnzeel 2023).

Further work is required to ascertain how widespread SOF occurrence in the LH may be in time and space, but it most likely occurs where soils are shallow or an impeding layer is found at shallow depth, in combination with high or prolonged rainfall (i.e. around the peak or latter half of the monsoon) (Gardner and Gerrard 2002; Singh 2009; Ghimire et al. 2016; Eeckman 2017; Nanda et al. 2018; Daniel

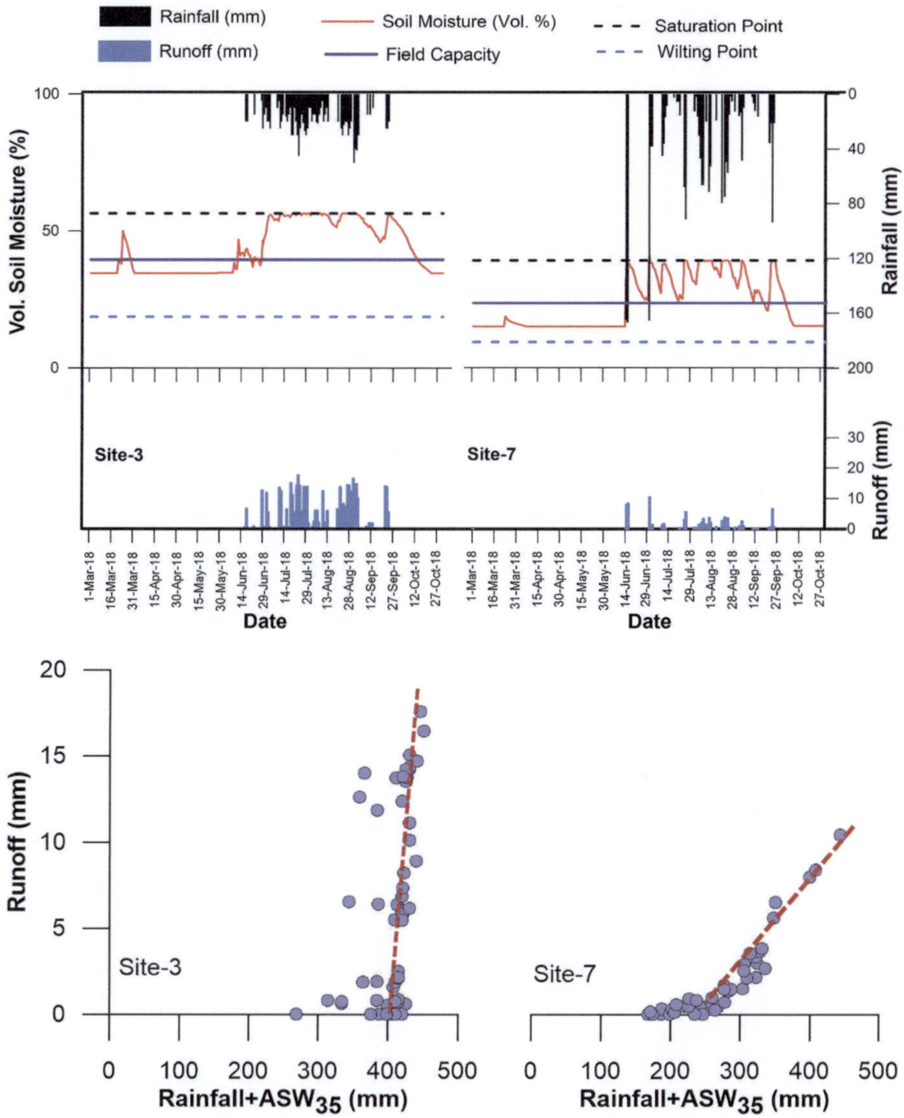

Fig. 12.12 Response of soil moisture content and overland flow to precipitation in (**a**) dense oak forest (Site 3) and open forest with shrub (Site 7) in the Jijli catchment, Tehri Garhwal, India; (**b**) contrasting threshold behaviour for SOF occurrence at the two sites (based on unpublished data of NQ Qazi). ASW_{35} = antecedent soil water storage in the top 35 cm of soil

et al. 2022; Qazi and Bruijnzeel 2023). Topographic position plays a role as well with SOF occurring most frequently on concave foot-slopes where subsurface flows converge to create saturated conditions (Fig. 12.13; Kumar and Sterk 2005; Singh 2009; cf. Dunne 1978). Lastly, land-cover types with reduced interception loss

Table 12.6 Changes in throughfall (TF), stemflow (SF), and interception loss (IL) fractions of incident precipitation (*P*) due to lopping or thinning in selected Himalayan forests. *N* = stand density (ha^{-1}), BA = basal area (m^2 ha^{-1}); *c* = canopy cover fraction. *UK* Uttarakhand

Forest type	Location	Age	Elev.	TF/P	SF/P	IL/P	*N*	BA	*c*	Methodological remarks
Shorea (coppiced)[1]	Dehradun UK, India[5]	36	640	72.5	9.5	18	2250	~20.6	~0.75	11 random gutters
Idem, thinned 22%[1]	Dehradun UK, India[5]	36	640	81	6	13	1750	~17.6	~0.69	11 random gutters
Shorea (regenerating)[2]	Dehradun UK, India[7]	~50	715	58/68*	3.7	38.5/28.5*	442	20.4	0.75	9 fixed gauges beneath trees
Ibidem, mildly disturbed[2]	Dehradun UK, India[7]	~50	760	63/82*	2.5	34/15.5*	432	25.3	**0.50?**	9 fixed gauges beneath trees
Ibidem, heavily disturbed[2]	Dehradun UK, India[7]	~50	830	65/93*	1.7	33/5.5*	317	18.3	**0.20?**	9 fixed gauges beneath trees
Banj oak (lightly disturbed)[3]	Jijli, Garhwal UK, India	Natural	1370	74.5	(0.5)	~25	980	30.6	~0.88	4 fixed gauges; LAI = 3.8; site 3
Banj oak (idem)[3]	Jijli, Garhwal UK, India	Natural	1090	77	(0.5)	~22.5	700	–	~0.75	4 fixed gauges; LAI = 3.1; site 1
Banj oak (<40% trees; shrub)[3]	Jijli, Garhwal UK, India	Natural	1275	83.5	(0.5)	~16	520	–	~0.55	4 fixed gauges; LAI = 2.4; site 4
Banj oak (<10%; 90% shrub)[3]	Jijli, Garhwal UK, India	Natural	1220	89.5	(0.5)	~10	340	–	~0.33	4 fixed gauges; LAI = 1.5; site 7

[1]Ghosh et al. (1980); [2]Tomar et al. (2022); [3]Qazi (2019); *UK* Uttarakhand; *Approximate value after inclusion of direct throughfall in gaps following method of Pathak et al. (1985). Original values represent crown drip rather than stand-scale TF; *Banj* oak = *Q. leucotrichophora*;

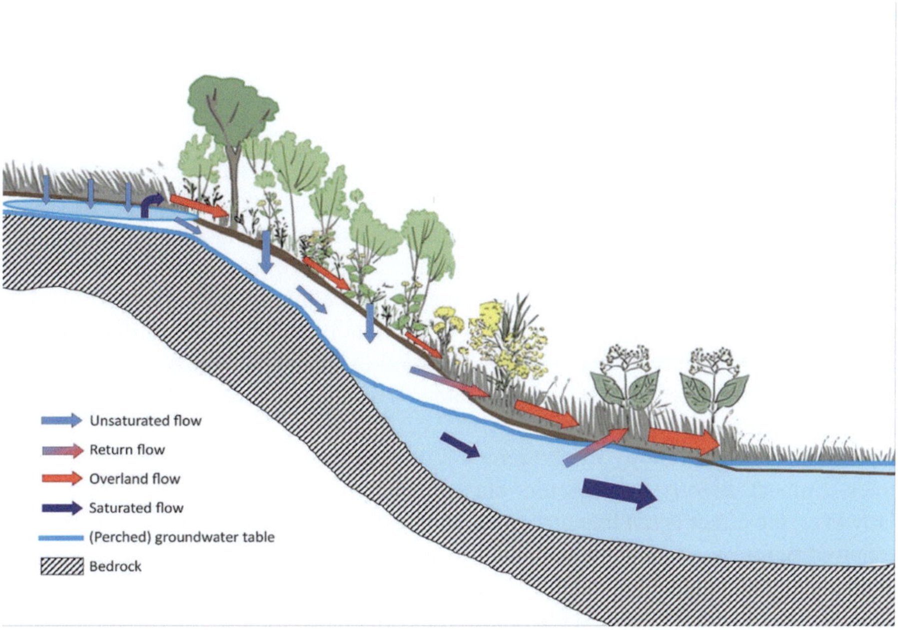

Fig. 12.13 Hillslope profile illustrating different runoff pathways (indicated by arrows) during the monsoon as influenced by soil depth and slope morphology. (Based on field descriptions by Kumar and Sterk (2019) for the *Shorea*-forested Pasta catchment, Siwaliks, Dehradun)

(e.g. shrubland, grassland, cloud forests; Ries 1994; Qazi and Bruijnzeel 2023) may also be more amenable to SOF generation (Fig. 12.12b).

Overland flow occurrence in the LH has been viewed traditionally as representing IOF (Pathak et al. 1984; Negi et al. 1998a; Sharma and Rai 2004) which can be remedied by vegetation restoration and/or soil conservation measures (Gardner and Gerrard 2002, 2003; Tiwari et al. 2009). However, SOF occurrence cannot be reduced simply by improving surface infiltration (Zwartendijk et al. 2024). Hence, to avoid erosion by exfiltrating return flow and hillslope SOF (Fig. 12.13), it is vital to maintain a protective vegetation cover in such locations (Kumar and Sterk 2005; cf. Toky and Ramakrishnan 1981; Mishra and Ramakrishnan 1983a).

Contributions by SOF to hillslope or headwater catchment storm runoff across the LH remain to be quantified, however (but see Eeckman 2017; Nanda et al. 2018), as are (to a lesser extent) changes in soil hydraulic conductivity with depth (Gilmour et al. 1987; Sarkar et al. 2008; Ghimire et al. 2013a; Om 2016; Daniel et al. 2022) and, especially, *along slopes* (Kumar and Sterk 2005; Nanda et al. 2019). The same holds for the identification of typical (average) storm runoff fractions associated with the main forest types and geological substrates in the region (cf. Bruijnzeel and Bremmer 1989; Hessel et al. 2007a; Qazi et al. 2017; Eeckman et al. 2019; Kumar et al. 2024b).

Hydrological Effects of Forest Disturbance, Conversion, and Restoration

Disturbance Impacts

Although the intensity of forest use in much of the central LH has decreased in recent decades following the decline in upland agricultural activities, forests remain a primary source for a range of products that support the livelihoods of millions of rural people (Chakraborty et al. 2017; Badu et al. 2019). At the same time, forest springs and streams are important sources of water that require protection. This section, therefore, examines the impacts of regionally common forest disturbances (e.g. lopping, litter harvesting, grazing, burning) on forest hydrological functioning before exploring to what extent deteriorated streamflow regimes may be improved again through forest restoration.

Opening up the canopy by the removal of branches (firewood, fodder) or entire trees (timber) directly impacts amounts of precipitation reaching the forest floor. However, the effect is not linear and major reductions in crown cover fraction (c) are required before a substantial reduction in interception is reached (Fig. 12.5c). For example, removing 22% of the trees in a coppiced *Shorea* forest increased throughfall by 8.5% while reducing interception loss by 5% only because of a concurrent reduction in stemflow after thinning (Ghosh et al. 1980). Based on the Himalayan data collated in Table 12.6 and Fig. 12.5c, successively reducing c from an initial value of 70–90% to ~50%, 30% and 20% will reduce an initial average interception loss of 20–25% to values around 15%, 10%, and 5%, respectively. Importantly, such effects are temporary only as new branches and leaves are produced with time. In addition, understorey biomass generally increases rapidly as well after thinning (Raturi and Subba Rao 1985; Stogsdill et al. 1992).

Reducing canopy leaf area will also affect stand transpiration (Srinet et al. 2022; Ghimire et al. 2022). Although no statistically significant relationship was found between LAI and ($E_t + E_s$) for the evergreen broad-leaf forests examined in Fig. 12.8, removal of whole trees will affect a stand's basal area. This can be expected to reduce tree-stratum E_t in an approximately linear fashion, at a tentative rate of ~0.25 mm day^{-1} for every 10 m^2 ha^{-1} of basal area reduction (Fig. 12.8). The effect will be even more temporary than that for interception loss, however, because of rapid post-thinning increases in understorey biomass. Also, residual trees typically show increased growth and water uptake in the first few years after thinning as their expanding root systems exploit the increased moisture availability afforded by the larger amounts of net precipitation (Stogsdill et al. 1992; Momiyama et al. 2021). Indeed, Subba Rao et al. (1985) did not find any significant effect on fortnightly streamflow totals after thinning a *Shorea* forest by 20%, suggesting gains in throughfall were roughly compensated by increased ($E_t + E_s$). The only measurable hydrological effect concerned a temporary 8.6% increase in peak discharges during rain shortly after thinning. This heightened runoff response likely reflects temporarily wetter soil conditions (Subba Rao et al. 1985), particularly in the riparian zone (Subba Rao et al. 1973). Similarly, removing trees by various degrees from a warm-temperate coniferous forest in Japan hardly affected stream baseflows due to the high moisture retaining capacity of the soil (Momiyama et al. 2021).

While the effects of lopping branches and removing individual trees are comparatively minor, other common practices in Himalayan forests like litter harvesting, deliberate burning (to stimulate grass growth), and cattle grazing can have more serious consequences (Naudiyal and Schmerbeck 2017). Removal of understorey, and especially litter, exposes bare soil to the highly erosive impact of crown drip (Hall and Calder 1993; Nanko et al. 2008), thereby promoting slaking and crusting of surface soil, reducing soil infiltration capacity, and enhancing IOF and erosion (Yadav and Vasisth 1989; Gardner and Gerrard 2002; Ghimire et al. 2013a, 2014c; Lacombe et al. 2018; Song et al. 2020). Prescribed (or unintended) burning of understorey vegetation and forest litter essentially produces the same deleterious effects on surface soils (Toky and Ramakrishnan 1981; Ribolzi et al. 2017), as do grazing animals that compact surface soils and diminish protective grass cover (Patnaik and Virdi 1962; Gardner and Gerrard 2002; Ghimire et al. 2014c; Tomar et al. 2022). The runoff and soil physical data compiled in Table 12.7 for various disturbed Himalayan sites further illustrate these points. Surface runoff production following forest soil disturbance by grazing, litter removal, or clearing and burning of forest is greatly increased, thereby rendering former subsurface flow-dominated systems (Table 12.5) increasingly OF-dominated (Fig. 12.14).

Depending on the fraction of surface runoff that re-infiltrates on the way downslope (Kumar and Sterk 2005; Nanda et al. 2019), the typically increased amounts following forest disturbance and conversion (Table 12.7) increase stormflow volumes and peak discharges in headwater streams (Gupta et al. 1974, 1975; Rana and Subehia 1996; Qazi et al. 2017; Table 12.8). However, rapid (near-)surface runoff also represents a loss to the site water budget as it no longer contributes to replenishment of local soil moisture and groundwater reserves, with potential reductions in dry-season discharges of downslope springs and streams as a result (Valdiya and Bartarya 1989; Bruijnzeel 1989; Rawat et al. 1997; Qazi et al. 2017).

Badu et al. (2022) estimated that losses via increased surface runoff from two fairly intensively used forests in east central Nepal reduced local groundwater recharge by 12–14% (100–115 mm year^{-1}). For a pine plantation in the same region that was subject to heavy grazing and litter harvesting (Ghimire et al. 2014a), the foregone infiltration and recharge were estimated to have increased to 27% (230 mm year^{-1}). Several studies listed in Table 12.7 reported even larger amounts of surface runoff (e.g. up to 515–730 mm year^{-1} for variously disturbed *Shorea* forests in the rainy Nepalese Likhu Khola and Indian Dehradun areas). Although the impacts of such large amounts on local groundwater recharge are likely substantial, a lack of information on the amounts of water lost via evaporation from these forests and, in some cases, the precise nature of the runoff generating mechanism (e.g. possibly partial SOF at Likhu Khola; Gardner and Gerrard 2002) precludes a full quantitative evaluation of the effect.

There can be little doubt, however, that advanced surface degradation in the LH may not only produce greatly increased stormflow volumes at the local headwater scale, but also lead to seriously reduced baseflows in the dry season (Table 12.8).

Table 12.7 Surface runoff (OF) production in disturbed Himalayan forests. *P* = monsoon rainfall total; BD = topsoil bulk density; SOM = topsoil organic matter content; K_{sat} = topsoil saturated hydraulic conductivity. Textural indicators: *Cl* clay, *L* loam, *Sa* sand, *Si* silt, *UK* Uttarakhand

Forest type	Location	Elev. (m asl)	*P* (mm)	OF/*P* (%)	BD (g cm^{-3})	SOC (%)	K_{sat} (mm h^{-1})	Soil Texture	Plot size (m^2)	Disturbance level/type
Western Himalaya and surrounding region										
S. robusta, 15-yr-old[1]	Dhading, W C Nepal	500	1200	9	1.20	2.5	90	Sa L	24 (*n* = 4)	Medium (fodder, litter, some grazing)
S. robusta shrub[2]	Likhu Khola, W C Nepal	~700	2210	**33**	1.33	0.55	54	Sa L	54 (*n* = 1)	High (fodder, grazing)
S. robusta regrowth[2]	Likhu Khola, W C Nepal	~800	2210	**31**	1.29	0.42	12	L Sa	25 (*n* = 1)	High (fodder, grazing)
S. robusta[3]	Salyan District, W Nepal	700	~830	**10**	–	–	41.5	Sa L	20 (n = 1)	High (burned); 7 events only
S. robusta, 50-yr-old[4]	Dehradun, UK	~760	2295	9.5	1.57	~0.6	4.6	L	400 (*n* = 3)	Medium (lopping, grazing)
S. robusta, 50-yr-old[4]	Dehradun, UK	~830	2350	**22**	1.80	0.42	4.3	L	400 (*n* = 3)	High (lopping, grazing)
P. roxburghii[3]	Salyan District, W Nepal	870	~830	**19**	–	–	37.5	Sa L	20 (*n* = 1)	High (litter); 7 events only
P. roxburghii[5]	Durga Gad, UK	1200	(195)	**16**	–	1.6	–	L	25 (*n* = 3)	High (litter, fuelwood)
P. roxburghii[6]	Bhetagad, UK	1460	1435	**17**	1.43	1.3	58	Sa L	100 (*n* = 1)	High (grazing, litter)
P. roxburghii[7]	Jijli, Garhwal, UK	1470	1680	(9)	1.66	–	280	Sa	**1** (Site 6)	High (repeated fire); **SOF**
Open *banj* oak forest[7]	Jijli, Garhwal, UK	1100	1880	(24)	1.17	–	120	Sa L	**1** (Site 5)	High (lopping); *soil undisturbed*; **SOF**

Ibidem (90% shrub)[7]	ijli, Garhwal, UK	1220	1860	(10)	1.41	–	835	Sa L	**1** (Site 7)	High (tree loss); *soil undisturbed*; **SOF**
P. wallichiana[8]	Hilkot, N Pakistan	1705	1340	4	1.45	–	–	L	100 ($n = 1$)	Medium (grazing, fuelwood)
Eastern Himalaya and surrounding region										
Shorea coppice[9]	Jhargram, W Bengal	80	950	5	–	–	–	–	300 ($n = 1$)	High (burned twice a year)
Regenerating mixed BL[10]	Roshi Khola, C Nepal	1615	950	8.5	1.0	4.7	50.5	Si L	18 ($n = 3$)	Medium (litter, fodder); 20 years old
Ibidem/planted pine[10]	Roshi Khola, C Nepal	1585	950	7.5	1.18	3.3	26.5	Si L	18 ($n = 3$)	Medium (litter, fodder); pines 35 years
Planted *P. roxburghii*[11]	Dhulikhel, C Nepal	1580	880	**15.5**	1.24	1.7	26	L	75 ($n = 1$)	Medium (litter, grazing); pines 25 years
Mixed evergreen BL[12]	Mamlay, Sikkim	950	?	(4.5)	–	–	–	Si L	30 ($n = 3$)	'Disturbed'; only 6 'typical' events
Post-*jhum* regrowth[13,14]	Burnihat, Meghalaya	125	1420	**19***	1.48	1.8	–	Sa L	40 ($n = 4$)	Five-years-old regrowth; ***mostly SOF?**
Post-*jhum* regrowth[15,6]	Shillong, Meghalaya	1500	1460	**27***	1.09	1.6	–	Sa	40 ($n = 4$)	Five-years-old regrowth; ***mostly SOF?**

[1]Tiwari et al. (2009); [2]Gardner and Gerrard (2002); [3]Wester (2013); [4]Tomar et al. (2022); [5] GCS Negi, pers. comm.; [6]Kothyari et al. (2004); [7]Qazi (2019); [8]Zokaib and Naser (2011); [9]Ray (1971); [10]Badu et al. (2022); [11]Ghimire et al. (2013a, b); [12]Sharma and Rai (2004); [13]Toky and Ramakrishnan (1981); [14]Ramakrishnan and Toky (1981); [15,16]Mishra and Ramakrishnan (1983a, b)

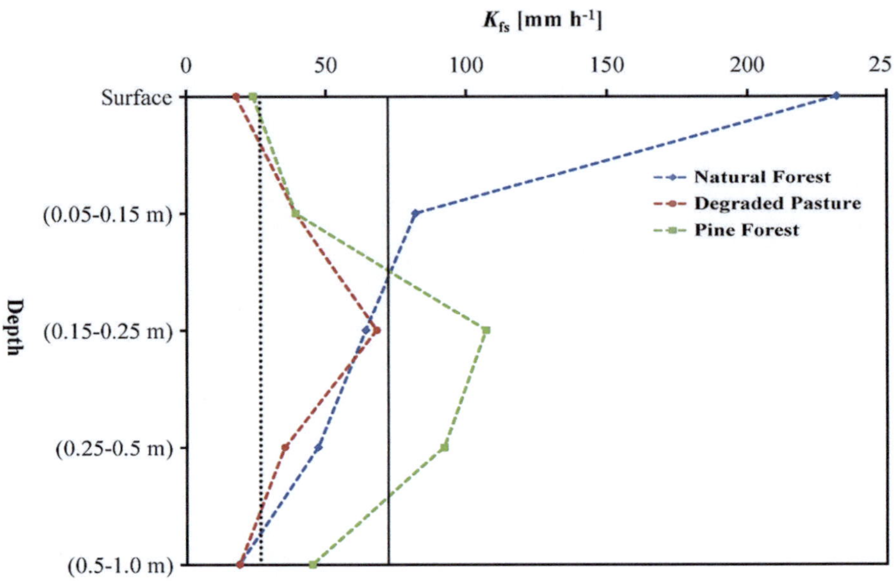

Fig. 12.14 Changes in soil hydraulic conductivity (K_{fs}) with depth in a natural mixed evergreen broad-leaved forest, a mature but intensively used *Chir* pine plantation, and a degraded pasture near Dhulikhel, east central Nepal (based on data presented in Ghimire et al. 2013a). The dashed and solid vertical lines denote the median and 95th percentile of the maximum five-minute rainfall intensity, respectively. Note strongly reduced conductivities down to 15 cm depth in the grazed pasture and pine plantation

Storm runoff from the near-undisturbed Arnigad oak forest catchment discussed earlier (Fig. 12.11) was ~6% of precipitation for a year with near-average rainfall. Conversely, the seriously degraded adjacent Bansigad catchment had 32% storm runoff that year. Further, while nearly 90% of streamflow at Arnigad consisted of baseflow, this dropped to 51% for Bansigad, implying nearly half of all flows from the degraded area consisted of (mostly sediment-laden) stormflow. In addition, despite receiving nearly 3000 mm of rainfall annually, the Bansigad stream generally ran dry in the pre-monsoon when water needs are typically highest (Qazi et al. 2017).

Restoration Impacts

The adverse effects of advanced land degradation often lead to calls for forest restoration, either through natural or assisted regeneration of degraded natural forest settings (Poffenberger 2015; Lewis et al. 2019) or tree planting in deforested terrains (Gilmour et al. 1990; Bastin et al. 2019). Broadly speaking, adding trees to a degraded landscape has two main hydrological effects: a comparatively rapid increase in rainfall interception and soil water uptake as the new vegetation matures, and a more gradual improvement of soil infiltration and water retention capacities.

Table 12.8 Streamflow responses to Himalayan forest degradation or conversion to other land-cover types. Q_t = total water yield; Q_b = baseflow; Q_s = stormflow. Annual or monsoonal streamflow totals rounded off to the nearest 5 mm. *PCE* paired catchment experiment, *UP* Uttar Pradesh, *HP* Himachal Pradesh, *UK* Uttarakhand

Location	Land use	Size (ha)	P (mm)	Q_t (mm)	Q_t/P (%)	Q_b (mm)	Q_b/Q_t (%)	Q_s (mm)	Q_s/P (%)	Soil depth (cm)
Siwalik zone										
M-II, Chandigarh UP (PCE)[1]	Sparse shrub/grass (baseline)	2.0	460*	–	–	–	–	125	**27**	
M-II, Chandigarh UP (PCE)[1]	Tree planting/terraced/bunded (years 3–7)	2.0	460*	–	–	–	–	45	10	
LT2, Chandigarh UP (PCE)[2]	Sparse shrub/grass, overgrazed (baseline)	1.9	715*	–	–	–	–	50	7	
LT5, Chandigarh UP (PCE)[2]	Tree planting/terraced/bunded (year 2)	1.9	715*	–	–	–	–	12	1.7	
Bater WS, Solan HP[3]	Shrubby *S. robusta* (degraded); normal P	13.4	1110	–	–	–	–	275	**25**	Shallow
Bater WS, Solan HP[3]	Shrubby *S. robusta* (degraded); wet year	13.4	1650	–	–	–	–	625	**38**	Shallow
Selakui, Dehradun UK (PCE)[4]	Shrubland (5–11 years old), non-grazed	1.45	725*	–	–	–	–	65	9	Very deep
Selakui, Dehradun UK (PCE)[4]	Planted Eucalypts, dense shrub (3–5 years)	1.45	605*	–	–	–	–	45	7.4	Very deep
Mid-Hills										
Dugar Gad, Garhwal UK[5]	9% pine, 53% barren, 38% crops	306	1700	700	41	225	32	475	**28**	60[$]
Srikot Gad, Garhwal UK[5]	50% oak, 12% barren, 38% crops	286	2365	430	18	135	32	290	12.5	
Naula, Khulgad, Kumaun UK[6]	85% oak, 15% pine (1991–1993)	53	1115	80	7	40	50	40	3.6	
S Rautela, Khulgad, Kumaun UK[6]	Natural *P. roxburghii*	29.6	1120	110	10	50	44	65	5.5	110
Khunt, Khulgad, Kumaun UK[6]	Rainfed cropping	75	715	145	20	35	25	105	15	
Kaneli, Khulgad, Kumaun UK[6]	Barren land (shrub, grass)	60	690	240	35	70	28.5	175	**25**	

(continued)

Table 12.8 (continued)

Location	Land use	Size (ha)	P (mm)	Q_t (mm)	Q_t/P (%)	Q_b (mm)	Q_b/Q_t (%)	Q_s (mm)	Q_s/P (%)	Soil depth (cm)
Arnigad, Mussoorie UK[7+]	83% oak, 12% housing, 5% crops	285	2880	1605	56	1435	89.5	170	6	80–155
Bansigad, Mussoorie UK[7+]	65% degraded oak, 33% barren/ housing	190	2930	1930	66	990	48.5	940	**32**	<0.8
Surrounding region										
Chittagong Hill Tract, Bangladesh[8]	Natural regrowth post-*jhum* (6–7 years)	0.89	3085	–	–	–	–	945	**31**	50–100
Kog-Ma, N Thailand[9]	Mixed evergreen montane forest	8.63	1870	1230	65.5	1155	94	75	4	530

[1]Kaushal et al. (1975); [2]Gupta et al. (1975); [3]Rana and Subehia (1996); [4]Mathur et al. (1976); [5]Joshi et al. (1998); [6]Rawat et al. (2000); [7]Qazi et al. (2017); [8]Gafur et al. (2003); [9]Shiraki et al. (2017); [*]Runoff-generating rains only; [$]Depth to saprolite (Tarafdar et al. 2019); [+]Arnigad/Bansigad: data for year with near-average rainfall

Fractional interception losses (%) and leaf biomass for tropical regrowth tend to stabilize within ~10 years (Zimmermann et al. 2013; Brown and Lugo 1990), by which time the LAI will often have reached the threshold value of 4–5 where transpiration rates begin to plateau (Ghimire et al. 2022). However, full soil physical recovery and re-development of soil humus may well take multiple decades, depending on the initial soil degradation level (Lozano-Baez et al. 2019; Zhang et al. 2019; Zwartendijk et al. 2024; Bruijnzeel et al. 2025).

Regarding forestation impact on streamflow, the increases in ET and infiltration following forestation will both cause a reduction in the *total* amount of streamflow (i.e. water yield; Hou et al. 2023). However, in areas with strongly seasonal rainfall like the LH, the distribution of streamflow throughout the year (flow regime) is more important than annual water yield (Bruijnzeel and Bremmer 1989). Effects of forestation on *baseflows and dry-season flows* will vary, depending on the trade-off between gains from improved infiltration (manifesting as reduced stormflows at the catchment scale) on the one hand, and the extra losses via increased evaporation on the other. This balance is referred to as the 'evaporation - infiltration trade-off' (Bruijnzeel 1989). Where the increase in evaporative losses exceeds the gain in infiltration—for instance, when foresting non-eroding soils with an already sufficiently high infiltration capacity to accommodate most rainfalls—streamflow will be lower than before throughout the year (Jackson et al. 2005; Scott et al. 2005). Conversely, where gains via improved infiltration after forestation more than compensate the extra ET—for instance, after restoring highly degraded soils under conditions of ample rainfall—dry-season flows can be boosted (Chandler and Walter 1998; Scott et al. 2005; Krishnaswamy et al. 2013). Nevertheless, given the temporal discrepancy between rapidly increasing ET and the often less rapid recovery of soil infiltration capacity after forest establishment, dry-season flows may be reduced more often initially than that they are increased (Bruijnzeel 2004; Scott et al. 2005).

What, then, are the prospects for boosting dry-season spring and stream discharges through forest restoration and forestation under the climatic and soil conditions prevailing in the LH? Planting fast-growing pines on heavily compacted grassland in the Chautara area in east central Nepal increased i_c from 39 mm h^{-1} in the pasture to 51 mm h^{-1} under adjacent pines after 5 years, with a further increase to 183 mm h^{-1} measured in a nearby 12-year-old pine stand (Gilmour et al. 1987; Fig. 12.15). Taking these numbers at face value implies an initial improvement in i_c of ~2.4 mm h^{-1} per year of pine plantation development on average, possibly accelerating to ~19 mm h^{-1} $year^{-1}$ between age five and twelve. Zwartendijk et al. (2024) derived an average increase in i_c of 3.0 ± 0.6 mm h^{-1} $year^{-1}$ in rainy Meghalaya during the first 50 years of mixed coniferous and broad-leaved natural regrowth following abandonment of slash-and-burn cultivation. No acceleration in soil physical recovery after the initial stage was observed, possibly because grazing was absent and starting values for i_c were already fairly high (70 mm h^{-1}; Zwartendijk et al. 2024). Such findings suggest, however, that the inferred rate of i_c improvement for the 12-year-old pine forest at Chautara may represent an overestimate, likely because of a less degraded starting point compared to that for the nearby 5-year-old stand (Gilmour et al. 1987; cf. Lozano-Baez et al. 2019). Ghimire

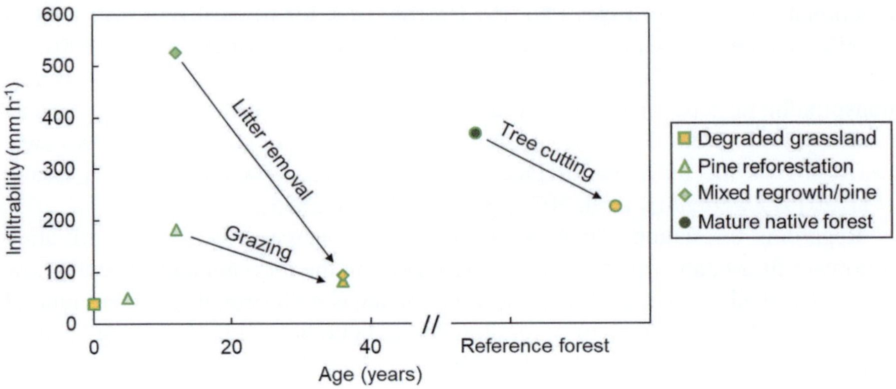

Fig. 12.15 Changes in surface infiltration capacity (i_c) with time following reforestation of heavily grazed grassland with pines in the Chautara area, Nepal. Original data from Gilmour et al. (1987). Also shown are reductions in i_c after ~20 years of grazing or litter harvesting in pine reforestation, and after manual removal of timber from undisturbed native forest, respectively, based on repeat measurements after 25 years by Ghimire et al. (2014c)

et al. (2014c) conducted repeat measurements of i_c at the Chautara sites after 25 years. Instead of the expected increases in i_c, major *reductions* were observed that were attributed to long-term grazing or litter harvesting (in the respective pine stands) or manual logging and timber harvesting (native forest) since the measurements of Gilmour et al. (1987) (Fig. 12.15). These findings underscore the importance of forest protection during vegetation development if soil hydrological functions are to be adequately restored.

The actual degree to which increased infiltration capacity will lead to reduced hillside overland flow depends strongly on prevailing rainfall intensities. Where these are predominantly low (as in much of the higher LH), raising i_c will not make much difference except during more intense storms (Pathak et al. 1984; Gilmour et al. 1987; Ghimire et al. 2013a). Further, hillslope response to rainfall is also governed by slope morphology and soil depth (Fig. 12.13). Where soils are shallow for geological reasons (e.g. on steep slopes subject to mass wasting) or due to past erosion following deforestation, opportunities to store any extra water as afforded by improved post-restoration infiltration will be inherently limited (Bruijnzeel 2004; Zhang et al. 2019).

Field-based information regarding the *effect of forestation on headwater storm runoff* in the region is limited to early work in the Indian Siwalik foothills (Gupta et al. 1974, 1975; Mathur et al. 1976); comparative catchment data for the Middle Mountain zone appear to be lacking. Stormflow from a small (1.45 ha) shrubland catchment near Dehradun drained by an ephemeral stream (running between late June and September) was 9.1 ± 5.9% of runoff-generating rainfall (range: 2.7–17.5%). Converting the shrubland to a eucalypt plantation reduced runoff slightly to 7.4 ± 4.5% (range: 2–11%) once the trees were 3 to 5 years old and a dense understorey had returned. The lower runoff response was ascribed mostly to the higher ET from the fast-growing vegetation rather than to diminished overland flow (Mathur et al. 1976). The catchment was not grazed (i.e. infiltration capacity was maintained), while the consistently low runoff coefficients suggested the bulk of the runoff originated in a fixed part of the catchment, presumably as valley-bottom SOF[6] (cf. Dickinson and Whiteley 1970; Qazi et al. 2017). Importantly, the associated reduction in seasonal stormflow (~21 mm) will be much smaller than the anticipated (but not measured) increase in ET under eucalypts (cf. Sharda et al. 1988). Hence, baseflows would be expected to decline. However, the absence of perennial flow (implying substantial subterranean water losses) precluded derivation of forest ET from the catchment's water budget, and thus of the effect of plantation establishment and growth on baseflows (cf. Zhang et al. 2018).

Larger reductions in stormflows can be expected after foresting more degraded catchments in the Siwaliks where forest grazing is common, infiltration capacities often low (Patnaik and Virdi 1962), runoff flashy, and erosion rampant (Ghimire et al. 2013b). Two paired-catchment experiments were conducted in the 1960s and 1970s near Chandigarh to quantify the effect of contour trenching and tree planting (*Eucalyptus, Acacia*) on storm runoff and peak discharges from small, degraded headwater catchments covered with sparse shrub and grass. Storm runoff from a 2.0 hectare catchment was reduced from 27% of runoff-generating precipitation prior to the intervention to 10% in years two to six after planting and trenching, with the cumulative difference in runoff over these 5 years (392.5 mm or 78 mm year^{-1}), implying an overall reduction of 63% (Patnaik et al. 1974; Kaushal et al. 1975). Peak discharges were similarly reduced (61%). Stormflow reduction during the second monsoon after imposing a similar treatment on another small (1.93 ha) but hydrologically much less responsive catchment (pre-intervention runoff response to rainfall, 4%) was 29% (Gupta et al. 1974, 1975). Hence, *the more advanced the initial soil degradation, the more may be gained from successful vegetation restoration in terms of stormflow reduction and infiltration gain*. Likewise, the higher the rainfall, the greater this potential gain (Kaushal et al. 1975; Scott et al. 2005; Bruijnzeel et al. 2025). Again, in the absence of measurements of the associated (changes in) vegetation water use or valley-bottom groundwater levels, it is not possible to quantify the net effect of such interventions on groundwater recharge and baseflow. However, no mention was made of increases in post-rainfall streamflow duration following trenching and tree planting in the Chandigarh experiments.

Comparative studies of the effects of reforesting degraded land in the Middle Mountains on flows from headwater catchments or 'spring-sheds' appear to be lacking. Ghimire et al. (2014a) presented annual plot-scale water budgets for a heavily grazed pasture (degraded baseline), an intensively used pine reforestation (25 years), and a near-undisturbed mixed evergreen broad-leaved forest (ecological optimum) in east central Nepal. Losses via evaporation and surface runoff were quantified, with the added notion that the slope exposition of the pine forest was

[6] A very similar storm runoff coefficient (10%) was reported for a substantially larger (70.4 ha) well-forested Siwalik headwater catchment in the same area, confirming the dominant runoff-generating role of the riparian zone (Bansal et al. 1975)

Table 12.9 Annual amounts of overland flow (OF) and evapotranspiration (ET) for degraded pasture, an (intensively used) mature pine plantation, and near-undisturbed natural broad-leaved forest near Dhulikhel (Nepal), together with the resultant gains in infiltration and losses via extra evaporation following conversion of pasture to planted pine or natural broad-leaved forest. All values rounded off to the nearest 5 mm. (Original data from Ghimire et al. (2014a))

Site	OF (mm)	ET (mm)	Infiltration gain (mm)	Loss via extra ET (mm)	Trade-off (mm)
Degraded pasture	305	225	–	–	–
Pine forest	220	590	85	−365	−280
Natural forest	35	520	270	−295	−25

towards the south-west and that of the broad-leaved forest towards the north-west (Table 12.9).

The results summarized in Table 12.9 may be taken as a first approximation of what to expect hydrologically after foresting severely degraded land in the Middle Mountain zone under moderate rainfall (1425 mm). Evaporation from the pine reforestation and the native forest exceeded that from the grassland by 295–-365 mm year^{-1}, whereas infiltration gains remained lower at 85–270 mm year^{-1}, particularly in the grazed pine plantation. As a result, the hydrological trade-off of planting pines was heavily negative (−280 mm year^{-1} after 25 years), although the extent of the effect could be reduced to −60 mm year^{-1} if losses via surface runoff would be eliminated by improved forest management (e.g. termination of grazing and litter harvesting). A near-neutral trade-off (−25 mm year^{-1}) was inferred for a full return to mature broad-leaved forest, even more so in the absence of surface runoff (+10 mm year^{-1}; Ghimire et al. 2014a; see footnotes to Table 12.9).

It is unclear, however, how long full hydrological recovery might take (Fig. 12.15) and to what extent this would translate into *seasonally* improved flows. Further work into evaporation - infiltration trade-offs is needed, both at the 'springshed' and headwater scale. Ideally, such work should target the vegetation at different stages of growth and throughout the year (cf. Ghimire et al. 2022; Kumar et al. 2024a,b). A pan-tropical modelling study of the potential impacts of foresting degraded land on dry-season flows indicated a modest positive effect for the moister parts of the eastern Himalaya, while large increases were inferred for Meghalaya, provided soils were deep enough to store the extra infiltrated water (Bruijnzeel et al. 2025). In view of the possibly large amounts of SOF generated in these parts (Toky and Ramakrishnan 1981; Mishra and Ramakrishnan 1983a; Zwartendijk et al. 2024), these model-based inferences need to be backed up by field observations.

Given the marginal increases in flows following forest thinning or vegetation restoration according to the above examples, and the long period of time that may be required to recover soil physical attributes sufficiently, people have often resorted to additional mechanical measures to boost infiltration. Examples include infiltration pits, contour trenches, tree planting holes, and other ways to collect hillslope surface runoff (Negi et al. 1998b; Tambe et al. 2012). Alternatively, one may consider land-cover types that combine adequate soil protection with a potentially lower vegetation water use, such as various forms of agroforestry (Wallace et al. 2005; Saha et al. 2007; Kumar et al. 2018; Kaushal et al. 2021).

Conclusions and Recommendations for Further Research

Having examined the forest hydrological literature for the Lesser Himalaya as published over the last sixty years, can firm conclusions be drawn regarding the water use (ET) and runoff production associated with the main zonal forest types? Further, what are the soil and water impacts of forest disturbance and what are the chances of success when trying to boost declining springs by vegetation management or manipulation? And lastly, what are the chief hydrological knowledge gaps hampering optimum forest and water resources management in the region?

Summary of Findings

Reliable quantitative forest hydrological data from the region proved to be scant, with several forest types having been addressed by a single or a few studies at best (e.g. temperate coniferous forests, cloud-affected forests, even lowland tropical deciduous forests). Furthermore, where multiple studies are available for specific forest types (e.g. subtropical pine forest, warm-temperate oak forest), results tend to be highly variable (Appendix 12.1 and Table 12.3). To some extent, this relates to differences in local climate, stand density, or growth stage between sites, but methodological issues often played a role as well. Examples include: (i) overestimation of rainfall interception losses because of a non-random throughfall sampling design that avoided gaps in the canopy (Appendix 12.1); (ii) overestimation of litter interception losses by exposing samples in the open (rather than beneath a canopy) and by breaking the naturally occurring moisture tension gradient (Table 12.2); (iii) excluding rainy periods when measuring evaporation using micrometeorological methods, thereby underestimating total evaporative losses (Table 12.3); and (iv) overestimation of ET in catchment water budgets because of unknown leakage losses (Table 12.3). As a result, a certain degree of subjectivity could not be avoided when assessing the plausibility of individual studies (see footnotes to tables).

Despite these uncertainties, a clear declining trend with elevation was found for mean daily ET and for the sum of mean daily transpiration and soil evaporation ($E_t + E_s$) (Fig. 12.9a). Notably, overall ET from *Chir* pines and warm-temperate oaks did not differ as much as perceived by many in the region, partly because the higher transpiration by pine is compensated somewhat by generally lower interception losses (Figs. 12.9a and 12.4). Other important (but often overlooked) factors governing the different hydrological behaviour of pine and oak forests include soil depth (generally greater on north-facing slopes carrying oak) and micro-climate (higher insolation levels and temperatures on south-facing slopes with pine).

Significant relationships were derived linking mean interception loss (% of precipitation) to stand leaf area index ($R^2 = 0.77$; Fig. 12.5b) or crown cover fraction ($R^2 = 0.39$; Fig. 12.5c) across forest types. No such relationships were evident for specific forest types, however, because of the large variability between sites. Although interception fractions did not exhibit a clear relationship with stand basal area across forest types (Fig. 12.5a), basal area proved to be a strong determinant of mean daily (E_t + E_s) for mixed evergreen montane broad-leaved forests ($R^2 = 0.64$; Fig. 12.8).

Fog interception has not been measured explicitly in the Lesser Himalaya, but modelled annual totals reached up to ~345 mm year^{-1} for exposed slopes and ridges in south-eastern Nepal, Sikkim, Bhutan, and southern Arunachal Pradesh; values for the drier western Himalaya were much smaller (Fig. 12.7). Not only do these occult inputs represent an extra source of moisture for plants during rainless periods, but foggy conditions also suppress transpiration. Hence, water yields associated with fog-affected forests (> 2400 m asl) are typically higher than those from similar forests below the cloud belt (Fig. 12.9b; Table 12.3). Reliable measurements of the total water inputs to—and evaporation from—fog-affected forests in the region are urgently needed for a better understanding of their hydrological role.

To our knowledge, no single study has quantified the hydrological response to rainfall along entire Himalayan hillslopes as the monsoon season progresses, although Kumar and Sterk (2005) presented a momentary snapshot of runoff flow paths along a *Sal*-forested slope in the Indian Siwalik zone (Fig. 12.13). Illien et al. (2021) described the seasonal filling and emptying of the groundwater reservoir (subsoil, regolith, and fractured rock) that provide the bulk of annual streamflow in the LH. Soil moisture stores are filled rapidly after the onset of the monsoon and stabilize at a high level despite continued high precipitation inputs (Fig. 12.11). Upon exceeding a moisture threshold (around the end of June—early July), precipitation inputs and streamflow outputs—and thus the intermediary groundwater body feeding the stream—become tightly coupled until the rains subside in October and the soil and groundwater stores begin their long recession until the next pre-monsoon season.

This predominantly subsurface hillslope runoff behaviour is confirmed by very low amounts of overland flow in many more or less undisturbed forests across the drier western and central Himalaya (Table 12.5). Larger plot-scale surface runoff amounts have been observed in the wetter eastern Himalaya and Meghalaya (Table 12.5) which may represent saturation-excess overland flow (SOF) once soils have saturated at the height of the monsoon, even on steep slopes (Fig. 12.12). A key question is, therefore, how widespread and important SOF may be across the LH. This has important implications for surface erosion, headwater peak discharges, and the delineation of areas to be protected. SOF occurrence also influences the achievable degree of stormflow reduction by forest restoration (Badu et al. 2022; Zwartendijk et al. 2024).

Forest thinning has only minor and short-lived hydrological effects, unless a major fraction of the canopy is removed (Table 12.6; Fig. 12.5c). More intensive interventions like cattle grazing and repeated removal of the litter cause major adverse changes in hydrological behaviour, including serious reductions in soil infiltration capacity (Fig. 12.14) and corresponding increases in overland flow (Table 12.7) and stormflow (Table 12.8). In extreme cases, the extra losses of water via increased surface runoff caused serious reductions in groundwater recharge and stream baseflows (Table 12.8).

With time, reforesting degraded land improves soil infiltration capacities (Fig. 12.15) and reduces surface runoff on hillslopes (Table 12.7) as well as headwater stormflows (Table 12.8). However, no convincing case of restored baseflow in the LH has been documented to date as measured reductions in surface runoff or stormflow proved insufficient to compensate the higher losses via evaporation after forestation, possibly because amounts of rainfall were low (various experiments in the Indian Siwaliks; cf. Bruijnzeel et al. 2025). One thing seems certain: full recovery of hillslope hydrological functioning may require at least several decades of *uninterrupted* forest regeneration (Table 12.9; Fig. 12.15).

Recommendations for Further Research

Given the general scarcity of solid quantitative hydrological information for Lesser Himalayan forests, the list of topics meriting further research is both long and varied. However, seeing the gradual decline of regional spring discharges, the increased pressure on peri-urban water resources due to continued migration from the uplands to regional towns and cities, and the recovery of forests across the Siwaliks and Middle Mountains of the central Himalaya under gradually warmer conditions and intensifying rainfall (Shrestha et al. 2012; Upgupta et al. 2015; Mishra and Chaudhari 2015; Upadhyay et al. 2021), several pressing research questions emerge, including:

- How does vegetation water use (all evaporation components) evolve following agricultural abandonment as shrubs give way to trees and, eventually, closed-canopy forest?
- What is the contribution to regional water yields by cloud-affected forests (> 2400 m asl)? How important are occult inputs (fog and low cloud)?
- Where might increases in infiltration following forestation boost stream baseflows and where will increased evaporation dominate the trade-off?
- How important is saturation-excess overland flow as a form of surface runoff and where is it the most prevalent?
- *And above all: How will continued climate change affect forest water use, runoff response to rainfall, and water yields at different elevations?*

These questions are discussed briefly below along with suggestions for suitable research approaches and field locations where pertinent.

Changes in Vegetation Water Use During Succession Important changes in the magnitude of specific evaporation components (interception loss, transpiration) have been observed in long-term studies of forest regrowth under warm-temperate conditions following wildfire (Vertessy et al. 2001; Benyon et al. 2017) or clear-cutting (Swank et al. 2001; Oishi et al. 2020). Similar changes would be expected for a typical western Himalayan succession from *Chir* pine to mixed forest to oak-dominated forest (Naudiyal and Schmerbeck 2017; Verma and Garkoti 2019). Mixed broad-leaved understoreys are developing in mature pine plantations following reduced forest usage pressure in central Nepal (Badu et al. 2022) along with a general expansion and recovery of foot-hill and mid-montane broad-leaved forests across the country (Ghimire 2017; Gao et al. 2023).

Associated changes in evaporation and soil hydrological behaviour could be studied at the plot scale using a chronosequence approach involving differently aged vegetations with comparable initial soil conditions and land-use histories (Vertessy et al. 2001; Oishi et al. 2020; Lozano-Baez et al. 2019; Zwartendijk et al. 2024). Measured evaporation rates may be compared with remotely sensed evaporation products (Zhang et al. 2016; Salazar-Martínez et al. 2022) to allow extrapolations in time and space (Purohit et al. 2022; Sett et al. 2023). Replicated studies are needed to cover contrasting conditions in climate, both along the mountain range (Bookhagen and Burbank 2010) and between elevations (Miehe 2015).

Hydrological Role of Cloud-Affected Forests Hydrologically speaking, cloud-affected forests (~2400–3600 m asl) are among the least studied Himalayan forests. Based on scant local experience (Eeckman et al. 2017), regional water balance modelling (Pandeya 2013), and reports from (sub)tropical mountain ranges elsewhere (Bruijnzeel et al. 2011; Zhan et al. 2020; Zaw and Oue 2024), water yields from cloud forests likely exceed the water production potential of all other Himalayan forest types (Fig. 12.9), but modelled occult inputs and 'hot spots' of fog interception (Fig. 12.7) need to be validated along with the degree of transpiration suppression during times of fog (Alvarado-Barrientos et al. 2014; Gadermaier 2018). Ideally, measurements of occult inputs and forest water use should take place within gauged catchments to allow validation of model-based predictions of streamflow (Mulligan 2013). Finally, cloud forests and associated water budgets are particularly vulnerable to regional warming via lifting of the cloud base (Los et al. 2021). The frequency of low-cloud has declined across the montane tropics in the last 25 years (Guzman et al. 2024). This phenomenon has not been studied explicitly in the Himalaya, although increased vegetation browning has been observed above 1600 m in the western Himalaya, particularly in coniferous and sub-alpine forests (Mishra and Chaudhari 2015). Further work is needed.

Infiltration Trade-Offs In contrast to the eastern Himalaya (Fig. 12.15; Zwartendijk et al. 2024), no comparative observations of changes in soil infiltration capacity (i_c) with time since plantation establishment or vegetation recovery seem available for the western Himalaya. To obtain *changes in real time,* one could re-visit sites where such measurements were made in the past (Tables 12.5 and 12.7). Similarly, re-measuring the Chautara pine plantations in central Nepal studied earlier in 1986 (Gilmour et al. 1987) and 2011 (Ghimire et al. 2014c) could yield information on the speed of soil hydrological recovery as forest usage in the area has declined markedly in recent years, allowing a mixed broad-leaved understorey to develop (Badu et al. 2022).

'Translating' changes in i_c into quantitative predictions of reductions in surface runoff remains challenging, however. Variable rainfall intensities, changes in soil hydraulic conductivity with depth, and soil water storing capacity—all of which determine how quickly the soil may saturate and generate saturation-excess flow alongside 'regular' surface runoff—all play a role (Fig. 12.12). With most of the latter-type information effectively missing for most, if not all, Himalayan locations, the use of physically based models that take soil macropores into explicit account (Cheng et al. 2017; Chouksey et al. 2017; Schwemle et al., 2024) is still a bridge too far. In practice, changes in infiltration and storm runoff associated with land-cover or climate change are predicted using models with semi-empirical routines for infiltration, such as SWAT (Kumar et al. 2016; Sooryamol et al. 2022) or W3RA-LUM (Peña-Arancibia et al. 2019), despite being criticized (e.g. Ogden et al. 2017). Bruijnzeel et al. (2025) applied W3RA-LUM to identify areas across the tropics where foresting degraded land might boost stream baseflows due to a positive trade-off between changes in infiltration and evaporation. Modest increases in dry-season flows were inferred for parts of the eastern Himalaya and Meghalaya, but not further west—in line with experimental results from the Indian Siwaliks (Table 12.8). It will be of great practical and scientific interest to establish field sites (including springsheds) where such predictions of improved flows after forest restoration under wetter conditions in the Himalaya can be validated (cf. Bruijnzeel et al. 2024). However, in view of the high water use of a fully developed forest cover, alternative land uses such as agroforestry systems that combine adequate soil protection and infiltration with a potentially more moderate vegetation water use (and a host of other products; Kumar et al. 2018) should be considered as well. Thus far, most related research has focussed on the plant production and carbon sequestration potential of agroforestry systems along with the capacity to reduce surface runoff and erosion (Kaushal et al. 2021; Panwar et al. 2022; Jinger et al. 2023). More work is needed to quantify the water dynamics of different agroforestry systems, notably in relation to plant water use and deep percolation.

Importance and Distribution of Saturation-Excess Overland Flow Although traditionally not considered in Himalayan forest hydrology, there is increasing evidence that saturation-excess overland flow (SOF) may be important during the monsoon, even in undisturbed forests on steep hillslopes where infiltration-excess overland flow (IOF) is unlikely. Understanding spatiotemporal patterns of SOF is important for the delineation of hillsides in need of protection to avoid erosion (Fig. 12.13), but also for a realistic assessment of the degree to which surface runoff can be reduced by vegetation restoration.

Predicting extent and magnitude of hillside SOF in Himalayan forests requires additional field studies to better characterize soil macropore networks (Chouksey et al. 2017; Zwartendijk et al. 2024) and parameterize suitable hydrological models (Tague and Band 2004; Sarkar and Dutta 2012; Schwemle et al. 2024). Stable isotope applications may help distinguish different hillslope runoff pathways (Zhao et al. 2013, 2018; van Meerveld et al. 2019).

Climate Change Impacts on Forest Hydrological Functioning Given the emphasis in other chapters on climate change in the Himalaya and its various impacts on regional vegetation, this chapter has focussed on forest hydrological processes as documented to date. Without going into detail, a continued rise in temperatures and intensification of rainfall will have major consequences for the hydrological functioning of these forests. Higher temperatures imply greater atmospheric evaporative demands. Coupled with potentially longer rainless periods, this will put additional stress on the vegetation and may lead to the replacement of more moisture-loving species by more drought-tolerant ones (Das et al. 2021). Further, if the regional cloud base rises markedly, this must affect the hydrology of cloud forests in a major way. Increases in the frequency of intense rainfall render adequate infiltration and water retention by forest soils even more important, also from the perspective of erosion and flood prevention. As stated previously, potential hydrological impacts of climate and land-cover change in the LH are conventionally addressed using semi-empirical models. Needless to say, such exercises can benefit from improved characterization of soil and vegetation attributes based on field measurements rather than prescribed look-up tables. It is hoped that the material collated here will contribute, although much remains to be done.

Outlook: Towards Stepped-Up Regional Research Cooperation? Given the difficult working conditions in much of the LH, field-based measurements as referred to above will be limited in practice to a few judiciously chosen locations. Bruijnzeel (2015) suggested establishing a regional Himalayan network of 'Critical Zone Observatories' (CZOs) with long-term observations of the flows of water (and nutrients and sediment carried) as governed by atmospheric, biotic, soil, and geological factors (Fig. 12.16; Lin 2010; Brantley et al. 2017). Advantages of this approach include shared measurement protocols for better comparability of results between sites and increased integration across disciplines (White et al. 2015; Richardson 2017; Pujari et al. 2020)—in stark contrast to the monodisciplinary nature of most studies reviewed here and the methodological issues identified in a number of cases (see text and footnotes to Tables).

CZOs are also particularly suitable for testing new field technologies such as the application of geophysical techniques for characterizing soil water dynamics (Illien et al. 2021) and rooting patterns (Balwant et al. 2022). By locating (some of these) CZOs in places with previous relevant work, funds may be used optimally, although

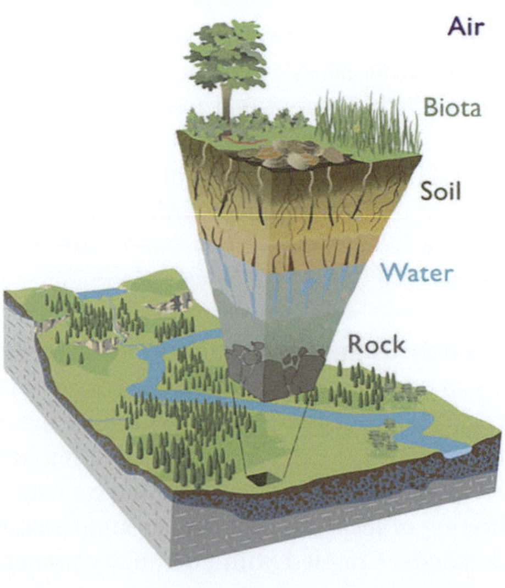

Fig. 12.16 Conceptual model of the Critical Zone linking atmosphere, biosphere, geosphere, and hydrosphere and extending from the top of the vegetation down to the bottom of the groundwater aquifer. (After Richardson 2017)

this would introduce an initial bias towards the Indian Himalaya where most work has been conducted to date. Complementary follow-up studies could include interception measurements at the Kaziranga, Barkot, and Kosi-Almora flux tower sites to help separate evaporation under wet and dry canopy conditions (Deb Burman et al. 2021; Srinet et al. 2022); quantifying transpiration (sapflow sensors) and groundwater dynamics (valley-bottom piezometry) in *Shorea* forest near Dehradun (Siwaliks) to complete the catchment water budget (Mathur et al. 1976; Subba Rao et al. 1985); observing perched water table dynamics in the pine and broad-leaved forests at Dhulikhel studied by Ghimire et al. (2014b) to better understand Mid-Hills hillslope runoff responses (cf. Zwartendijk et al. 2020, 2023); and initiate throughfall and stemflow studies in Nepalese (e.g. Kharikhola; Eeckman et al. 2019), Sikkimese (upper Tistha; Kumar et al. 2024b), and Bhutanese (Om 2016) fog-affected forests to quantify fog interception (cf. Schmid et al. 2011; Zhan et al. 2020; Zaw and Oue 2024). At any rate, exciting times lie ahead for regional forest hydrologists (cf. Bruijnzeel et al. 2024; Pant et al. 2024).

Acknowledgements We dedicate this chapter to all those researchers who collected the field data upon which this review is based. The senior author is most grateful to the following persons (in alphabetical order) for providing valuable background information on research sites, methods or results, for help with procuring semi-obscure literature, unpublished theses and dissertations, reports or data, diagrams, and/or photographic material: Bhupendra Adhikari (Wildlife Institute of India, Dehradun), Steven Brantley (USDA Forest Service), Denzil Daniel (Indian Institute of Technology, Roorkee), Pramit Deb Burman (Indian Institute of Tropical Meteorology, Pune), Shalini Dhyani (National Environmental Engineering Research Institute, Nagpur, India), Josef Gadermaier and Peter Hietz (University of Natural Resources and Life Sciences—BOKU, Vienna), Chandra Prasad Ghimire (SCION, New Zealand), John Gash and Alan Jenkins (Centre for Ecology & Hydrology, UK), Rudi Hessel (Wageningen University & Research Centre, Netherlands), Luc Illien (Helmholtz Center Potsdam, Germany), Rajesh Kaushal (ICAR, Indian Institute of Soil and Water Conservation, Dehradun), Otto Klemm (Universität Münster, Germany), Jagdish Krishnaswamy (Indian Institute for Human Settlements, Bengaluru), Tomo'omi Kumagai (Nagoya University, Japan), Kireet Kumar (GB Pant Institute for Himalayan Environment and Sustainable Development, Almora, India), Suresh Kumar (Indian Institute for Remote Sensing, Dehradun), Wenjie Liu (Xishuangbanna Tropical Botanical Garden, Yunnan, China), Tim McVicar (CSIRO, Canberra, Australia), Girish Negi (GB Pant Institute, Almora), Colin Pendry (Edinburgh Botanical Garden, UK), Shive Prakash Rai (University of Varanasi, India), Anneke de Rouw (Sorbonne Université, Paris), Eklabya Sharma (formerly with ICIMOD, Lalitpur, Nepal), Run Tang (Nanjing University of Information Science and Technology, China), Krishna Raj Tiwari (Tribhuvan University, Kirtipur, Nepal), Conrado Tobón (Universidad Nacional de Colombia, Medellin), John Van Stan (Cleveland State University, US), Norbu Wangdi (Bhutan Ecological Society, Thimphu), and Svenja Wittpoth (Universität Freiburg, Germany). Special thanks to Rajiv Pandey (Indian Council of Forestry Research and Education, Dehradun), Sandipan Mukherjee (GB Pant Institute, Leh, India), Nuzhat Qazi (HER Himalayan Ecosystem Services Trust, Indore, India), and Bob Zwartendijk (InHolland, Alkmaar, Netherlands) for great help in various ways, and to the Editors for their continued trust, pertinent feedback and, above all, patience.

Appendix 12.1

Appendix 12.1 Throughfall (TF), stemflow (SF), and interception loss (IL) as fractions of incident precipitation (P) during the monsoon (June–September) for selected Himalayan forests. N = stand density (ha^{-1}), BA = basal area ($m^2\ ha^{-1}$); c = canopy cover fraction. Bold IL/P-values were used to derive the tentative averages per main forest type shown in Fig. 12.4. *UK* Uttarakhand, *HP* Himachal Pradesh

Forest type	Location	Age	Elev.	TF/P	SF/P	IL/P	N	BA	c	Methodological remarks
(Sub)tropical moist forest (N = 12)										
Acacia katchecu (planted)[1*]	Arabari, W Bengal, India	17	~50	~52.5	~19	~28	1675	30.2	–	36 random troughs
Shorea (planted) [1*]	Arabari, w Bengal, India	17	~50	~66.5	~8	**~26**	1678	15.5	–	12 random troughs
Tectona (planted)[2]	Pantnagar UK, India	12	245	75.5	8	16	625	29.5	0.95?	8 fixed gauges below trees
Terminalia (planted)[2]	Pantnagar UK, India	12	245	81	9.3	9.5	625	29.7	0.95?	8 fixed gauges below trees
Shorea robusta (coppice)[3]	Chorgalia UK, India	Natural	330	82.5	0.85	**16.5**	600	22	0.89	36 roving gauges incl. Gaps
Acacia katchecu (planted)[4]	Dehradun UK, India	27	640	67.5	4.2	28.5	574	10.7	0.95	8 roving troughs below trees
Shorea (planted)[4]	Dehradun UK, India	37	640	55/ 58[@]	7	38/ **35**[@]	668	29.2	0.95	8 roving troughs below trees
Shorea (coppice)[5]	Dehradun UK, India	36	640	72.5	9.5	**18**	2250	20.6	0.75	11 random gutters
Tectona (planted)[6]	Dehradun UK, India	35	700	73	6	21	472	21.2	0.95	8 random roving troughs
Shorea (regenerating)[7]	Dehradun UK, India	~50	715	58/ 68[@]	3.7	38.5/ **22**[@]	442	20.4	0.62	9 fixed gauges below trees
Sacred grove (semi-deciduous)[8]	Nambol, Manipur, India	Natural	~800	53	3	44	1960	7.6	-	10 fixed gauges below trees
Sacred grove (evergreen)[8]	Wangoi, Manipur, India	Natural	~800	60.5	4.7	35	2500	9.9	–	10 fixed gauges below trees
Subtropical coniferous forest (Chir pine) (N = 10)										

Pinus roxburghii (planted)[6]	Dehradun UK, India	26	640	74.5	3.6	**22**	1156	22.4	0.95	8 random roving troughs
P. roxburghii, some oak[3]	Kalona, UK, India	Natural	1250	91.5	0.4	8	325	21	**0.38**	36 roving gauges incl. Gaps
P. roxburghii[9]	Papoli UK, India	Natural	1525	76.5	0.3	**23**	–	48.6[a]	–	5 gutters, total 2 m^2; 1 year
P. roxburghii (planted)[10]	Roshi Khola, C Nepal	~35	1550	79.5	0.8	**20**	1385	27.7	**0.43**	12 roving gauges
P. roxburghii (planted)[11]	Dhulikhel, C Nepal	25	1580	83	0.5	**16.5**	855	37.6	0.73	10 roving gauges; 2 troughs
P. roxburghii[12]	Nainital UK, India	Natural	1600	77.5	0.3	**22**	310	12.7	0.90	10 roving gauges incl. gaps
P. roxburghii[3]	Baldiyakhan UK, India	Natural	1700	74.5	0.3	**25**	525	34[b]	0.91	36 roving gauges incl. gaps
P. roxburghii, some oak[12]	Nainital UK, India	Natural	1750	80.5	0.3	**19**	550	28.8	0.80	10 roving gauges incl. gaps
P. roxburghii[13]	Siahi Devi UK, India	Natural	1780	81.5	1.0	17.5/ 11[@]	415	24.2	**0.59**	8 roving gauges excl. gaps
P. roxburghii[14]	Binsar UK, India	Natural	1960	–	–	**23**	624	32.6	–	8 roving gauges excl. gaps?
Subtropical mixed broad-leaf forest (oaks in W Himalaya, Schima–Castanopsis in E Himalaya) (N = 6)										
Q. leucotrichophora (disturbed)[15]	Jijli, Garhwal UK, India	Natural	1370	74.5	(0.5)	**25**	980	30.6	~0.90	4 fixed gauges; LAI = 3.8; site 3
Q. leucotrichophora[16]	Pauri District UK, India	Natural	1580	74.5	0.5	**25**	–	–	–	9 fixed gutters below trees
Schima–Castanopsis[11]	Dhulikhel, C. Nepal	Natural	1500	76	1.4	**22.5**	1870	27.1	0.81	15 roving gauges; 2 troughs
Schima–Castanopsis & pines[10]	Roshi Khola, C. Nepal	~35	1585	76	1.3	**23**	2090	25.1	0.70	12 roving gauges

(continued)

Appendix 12.1 (continued)

Forest type	Location	Age	Elev.	TF/*P*	SF/*P*	IL/*P*	*N*	BA	*c*	Methodological remarks
Schima–Castanopsis (regrowth) [10]	Roshi Khola, C. Nepal	~20	1615	74	1.6	**24.5**	2415	20.6	0.79	12 roving gauges
Castanopsis–Symplocos forest[17]	Sikkim, India	Natural	1750	73	2	**25**	–	58	0.68	9 roving gauges
Temperate broad-leaf forest (mostly oaks) (N = 11)										
Q. leucotrichophora, some pine[3]	Ranikhet UK, India	Natural	1900	83	0.3	**17**	425	21	0.78	36 roving gauges incl. gaps
Q. leucotrichophora[13]	Siahi Devi UK, India	Natural	1980	67	0.5	32/ **28**[@]	600	44	0.87	8 roving gauges excl. gaps
Mixed oaks (*banj, tilonj*)[12]	Nainital UK, India	Natural	1980	80.5	0.1	**19.5**	710	27.7	0.92	10 roving gauges inc. gaps
Mixed oaks (*tilonj*) [3]	Nainital UK, India	Natural	2000	84.5	0.4	**15**	400	52	0.81	36 roving gauges below trees
Q. leucotrichophora[14]	Binsar UK, India	Natural	2040	–	–	**36**	632	37.7	–	8 roving gauges excl. gaps?
Q. leucotrichophora[9]	Sitlakhet UK, India	Natural	2070	67.5	0.6	**32**	1254[c]	23.3[c]	–	5 gutters, total 2 m^2; 1 year
Mixed oaks (*rianj, tilonj*) [3]	Kilbury UK, India	Natural	2120	81	0.9	**18**	430	81	0.93	36 roving gauges below trees
Alnus, Juglans, Quercus[18]	Mamlay, Sikkim, India	Natural	2275	80	**10?**	(10)	–	56.8	0.81	5 roving gauges below trees
Aesculus indica[19]	Nanda Devi UK, India	Natural	2400	87	0.2	(13)[$]	135	51	0.61	9 roving gauges below trees[$]
Q. semecarpifolia (*kharsu*)[19]	Nanda Devi UK, India	Natural	2600	80	0.1	(20)[$]	355	68.4	0.72	9 roving gauges below trees[$]
Q. griffithii/lanata[20,21]	W Bhutan	Natural	2460	81	**12**?	(7)[d]	667	47.3	–	10 fixed gauges; fog influence

Temperate coniferous forest (N = 7)										
Cedrus deodara[16]	Tarkeswhar UK, India	Natural	1805	65.5	0.3	**34**	–	–	–	9 fixed gutters below trees
Pinus wallichiana (Blue pine) [22]	Srinagar, Kashmir, India	29	~2000	26.5/ 45[@]	36.5?	37/ 28[@]	700	63.3	0.75	12 fixed troughs; faulty TF,SF?
Cryptomeria japonica (planted) [18]	Mamlay, Sikkim, India	~40	2000	33/ 79[@]	18/7[#]	49/ 14[@]	?	42.5	**0.31**	5 roving gauges beneath trees
P. wallichiana (planted) [23]	Shilaroo, Shimla HP, India	20	2400	78.5	2.8	**19**	1200	29.1	0.81	8? fixed collectors in rows
Cedrus deodara (planted) [24]	Shilaroo, Shimla, India	35	2400	74.5	3.4	**22**	515	33.5	0.94	8 fixed random collectors
Abies pindrow (Silver fir) [19]	Nanda Devi UK, India	Natural	2500	72.5	tr.	**27.5**	245	85.5	0.77	9 roving gauges below trees[$]
Tsuga dumosa, kharsu[20,21]	W Bhutan	Natural	3260	92	1.9	6	400	77.8	–	10 fixed gauges; fog influence

[1]Ray (1970); [2]Anand et al. (2022); [3]Pathak et al. (1985); [4]Dabral and Subba Rao (1969); [5]Ghosh et al. (1980); [6]Dabral and Subba Rao (1968); [7]Tomar et al. (2022); [8]Gupta and Usharani (2009); [9]Daniel et al. (2022); [10]Badu et al. (2022); [11]Ghimire et al. (2012); [12]Loshali and Singh (1992); [13]Kumar (1990); [14]Kumar and Verma (1991); [15]Qazi (2019); [16]Jana et al. (2021); [17]Jain et al. (2000); [18]Rai and Sharma (1998); [19]Negi et al. (1998a); [20]Om (2016) using data for 2015; [21]Wangdi (2016); [22]Wani and Manhas (2012); [23]Singh and Gupta (1987); [24]Singh et al. (1983); *UK* Uttarakhand, *HP* Himachal Pradesh; *representative of Gangetic plain conditions; [@]Approximate value after including direct throughfall in gaps following Pathak et al. (1985)[3]; [#]Stemflow for mature *Cryptomeria* plantation in equally wet monsoonal SW Japan after Shimizu et al. (2015); [$]Values for one wet-season month only. *Banj* = *Q. leucotrichophora*; *Tilonj* = *Q. floribunda*; *Rianj* = *Q. lanuginosa*; *Kharsu* = *Q. semecarpifolia*; [a]Lohani et al. (2023); [b]Alternative value following Joshi et al. (2021); [c]Gosain et al. (2015) and taking reported average Diameter at Breast Height (48.3 cm) to be Girth at Breast Height; [d] Taking SF of 12% as measured in 2014 at face value and combining with TF measured in 2015

References

Adhikari BS (1992) Biomass, productivity and nutrient cycling in *kharsu* oak and silver fir forests in central Himalaya. Dissertation, University of Kumaun, Naini Tal, India

Adhikhari S, Gurung A, Chauhan R et al (2020) Status of springs in mountain watershed of Western Nepal. Water Policy

Agarwal A, Agrawal NK, Nema RK (2014) Life line: the springs of Uttarakhand, India. Int J Innov Res Sci Eng Tech 3(4):11553–11560

Allen RG, Pereira LS, Raes D, Smith M (1998) Crop evapotranspiration—guidelines for computing crop water requirements. FAO Irrig Drain Paper 56:1–300. http://www.fao.org/docrep/X0490E/X0490E00.htm

Alvarado-Barrientos MS, Holwerda F, Asbjornsen H et al (2014) Suppression of transpiration due to cloud immersion in a seasonally dry Mexican weeping pine plantation. Agric For Meteo 186: 12–25. https://doi.org/10.1016/j.agrformet.2013.11.002

Anand R, Chaturvedi S, Kumar A et al (2022) Rainfall partitioning in important multipurpose tree species of Himalayan foothills. Trop Ecol 63:480–484. https://doi.org/10.1007/s42965-022-0022105

Andermann C, Longuevergne L, Bonnet S et al (2012) Impact of transient groundwater storage on the discharge of Himalayan rivers. Nat Geosci 5:127–132. https://doi.org/10.1038/NGEO1356

Badu M, Nuberg I, Ghimire CP et al (2019) Negative trade-offs between community forest use and hydrological benefits in the forested catchments of Nepal's mid-hills. Mnt Res Dev 39:R22–R32. https://doi.org/10.1659/MRD-JOURNAL-D-18-00066.1

Badu M, Ghimire CP, Bruijnzeel LA et al (2022) Net precipitation, infiltration and overland flow production in three types of community-managed forest in East Central Nepal. Trees For People 8:100218. https://doi.org/10.1016/j.tfp.2022.100218

Balla MK (1988) Establishment of a paired watershed study to estimate the impact of cultivating marginal lands on the production of sediments and runoff in Nepal. Unpubl Report. Institute of Forestry, Pokhara, Nepal, p 13

Balwant P, Jyothi V, Pujari PR et al (2022) Tree root imaging by electrical resistivity tomography: geophysical tools tp improve understanding of deep root structure and rhizospheric processes. Trop Ecol 63:319–324. https://doi.org/10.1007/s42965-021-00213-x

Bastin JF, Finegold Y, Garcia C et al (2019) The global tree restoration potential. Science 365:76–79. https://doi.org/10.1126/science.aax0848

Bansal RC, Babu R, Srivastava MM (1975) Hydrological behaviour of small watersheds under different land uses—Dehra Dun. In: Annual Report 1975. Central Soil and Water Conservation Research and Training Institute, Dehra Dun, India, pp 13–15

Bartarya SK (1989) Hydrogeology, geo-environmental problems and watershed management strategies in a central Himalayan river basin, Kumaun, India. In: Kreek J, Haigh MJ (eds) Headwater control. IUFRO/WASWC/CSVIS, Plzen, Czechoslovakia, pp 308–318

Behera SK, Mishra AK, Sahu N et al (2012) The study of microclimate in response toBiodivdifferent plant community associations in a tropical moist deciduous forest from northern India. Biodivers Conserv 21:1159–1176. https://doi.org/10.1007/s10531-012-0230-5

Behera SK, Behera MD, Tuli R (2015) An indirect method of estimating leaf area index in a tropical deciduous forest of India. Ecol Indic 58:356–364. https://doi.org/10.1016/j.ecolind.2015.05.038

Benyon RG, Nolan RH, Hawthorn SND, Lane PNJ (2017) Stand-level variation in evapotranspiration in non-water-limited eucalypt forests. J Hydrol 551:233–244. 10.1016.j.jhydrol.2017.06.002

Bhattarai KR, Vetaas OR, Grytnes JA (2004) Fern species richness along a central Himalayan elevational gradient. Nepal J Biogeogr 31:389–400. https://doi.org/10.1046/j.0305-0270.2003.01013.x

Bhawana KC, Wang T, Gentle P (2017) Internal migration and land use and land cover changes in the Middle Mountains of Nepal. Mnt Res Dev 37:446–455. https://doi.org/10.1659/MRD-JOURNAL-D-17-00027.1

Böhner J, Miehe G, Miehe S, Nagy L (2015) Climate and weather. In: Miehe G, Pendry C (eds) Nepal: an introduction to the natural history, ecology and human environment of the Himalayas. Royal Botanic Garden of Edinburgh, UK, pp 23–90
Bookhagen B, Burbank DW (2010) Toward a complete Himalayan hydrological budget: spatio-temporal distribution of snowmelt and rainfall and their impact on river discharge. J Geophys Res 115:F03019. https://doi.org/10.1029/2009JF001426
Bowonder B (1982) Deforestation in India. Int J Env Stud 18:223–236
Brantley SL, Eissenstat DM, Marshall JA et al (2017) Reviews and syntheses: on the roles trees play in building and plumbing the critical zone. Biogeosciences 14:5115–5142. https://doi.org/10.5194/bg-14-5115-2017
Brantley ST, Miniat CF, Bolstad PV (2019) Rainfall partitioning varies across a forest age chronosequence in the southern Appalachian Mountains. Ecohydrology 12:e2081. https://doi.org/10.1002/eco.2081
Brown S, Lugo AE (1990) Tropical secondary forests. J Trop Ecol 6:1–32. https://www.jstor.org/stable/2559366
Bruijnzeel LA (1989) (De)forestation and streamflow in the tropics: a closer look. J Trop For Sci 1: 229–243. https://www.jstor.org/stable/43594578
Bruijnzeel LA (2000) Forest hydrology. In: Evans JC (ed) The forests handbook: an overview of forest science. Blackwell Scientific, Oxford, pp 301–343. https://doi.org/10.1002/9780470757062.ch12
Bruijnzeel LA (2004) Hydrological functions of tropical forests: not seeing the soil for the trees? Agric Ecosyst Environ 104:185–228. https://doi.org/10.1016/j.agee.2004.01.015
Bruijnzeel LA (2015) Towards a Himalayan network of critical zone observatories. Seminar presented at the International Centre for Integrated Mountain Development, Kathmandu, Nepal, 20 February 2015. https://www.researchgate.net/publication/358668589 (abstract) and https://www.researchgate.net/publication/358668573 (presentation)
Bruijnzeel LA (2023) Forest hydrological processes in the Lesser Himalaya and management impacts: what do we know? In: Sun G et al (eds) International symposium on the forests–water–livelihood nexus in the Lesser Himalaya, Pokhara, Nepal, 29–30 September 2023, pp 13–17 https://www.researchgate.net/publication/374661424
Bruijnzeel LA, Bremmer CN (1989) Highland-lowland interactions in the Ganges Brahmaputra River Basin: a review of published literature. ICIMOD Occas Paper 11:136. https://lib.icimod.org/record/24847
Bruijnzeel LA, Eugster W, Burkhardt R (2005) Fog as a hydrological input. In: Anderson MG, McDonnell JJ (eds) Encyclopedia of hydrological sciences. J Wiley, Chichester, pp 559–582. https://doi.org/10.1002/0470848955.hsa041
Bruijnzeel LA, Mulligan M, Scatena FN (2011) Hydrometeorology of tropical montane cloud forests: emerging patterns. Hydrol Proc 25:465–498. https://doi.org/10.1002/7974
Bruijnzeel LA, Peña-Arancíbia JL, Sheil D et al (2025) Potential for improved groundwater recharge and dry-season flows through Forest landscape restoration on degraded lands in the tropics. For Ecosyst 2025. https://doi.org/10.1016/j.fecs.2025.1003760
Bruijnzeel LA, Sun G, Zhang J et al (2024) Forests, water, and livelihoods in the Lesser Himalaya. Eos 105. https://doi.org/10.1029/2024EO240217
Budhatoki A, Babel MS, Shrestha S et al (2021) Climate change impact on water balance and hydrological extremes in different physiographic regions of the West Seti River Basin, Nepal. Ecohydrol Hydrobiol 21:79–95. https://doi.org/10.1016/j.ecohyd.2020.07.001
Bulcock HH, Jewitt GPW (2012a) Field data collection and analysis of canopy and litter interception in commercial forest plantations in the KwaZulu-Natal Midlands, South Africa. Hydrol Earth Syst Sci 16:3717–3728. https://doi.org/10.5194/hess-16-3717-2012
Bulcock HH, Jewitt GPW (2012b) Modelling canopy and litter interception in commercial forest plantations in South Africa using the variable storage gash model and idealised drying curves. Hydrol Earth Syst Sci 16:4693–4705. https://doi.org/10.5194/hess-16-4693-2012

Burgess SSO, Dawson TE (2004) The contribution of fog to the water relations of *Sequoia sempervirens* (D. Don): foliar uptake and prevention of dehydration. Plant Cell Env 27:1023–1034. https://doi.org/10.1111/j.1365-3040.2004.01207.x

Calder IR (1990) Evaporation in the uplands. J Wiley, Chichester, p 148

Chakraborty RN (2001) Stability and outcomes of common property institutions in forestry: evidence from the Tarai in Nepal. Ecol Econ 36(2):341–353. https://doi.org/10.1016/S0921-8009(00)00237-8

Chakraborty A, Joshi PK, Sachdeva K (2017) Capturing forest dependency in the central Himalayan region: variations between oak (*Quercus* spp.) and pine (*Pinus* spp.) dominated forest landscapes. Ambio 47:504. https://doi.org/10.1007/s13280-017-0947-1

Chandler DG, Walter MF (1998) Runoff responses among common land uses in the uplands of Matalom, Leyte, Philippines. Trans Am Soc Agric Eng 41:1635–1641. https://researchgate.net/publication/259622808

Chand HB, Joshi G, Bhatta RP et al (2022) Fine root biomass differs significantly across different forest types and soil depth in central Himalaya, India. Curr Sci 123:194–201. https://www.researchgate.net/publication/362248890

Chapagain PS, Ghimire M, Shrestha S (2017) Status of natural springs in the Melamchi region of the Nepal Himalayas in the context of climate change. Environ Dev Sustain. https://doi.org/10.1007/s10668-017-0036-4

Chappell NA, Franks SW, Larenus J (1998) Multi-scale permeability estimation for a tropical catchment. Hydrol Proc 12:1507–1523. https://doi.org/10.1002/(SICI)1099-1085(199807)12:9<1507::AID-HYP653>3.0.CO;2-J

Chatterjee A, Roy A, Chakraborty S et al (2018) Biosphere atmosphere exchange of CO_2, H_2O vapour and energy during spring over a high-altitude Himalayan forest in Eastern India. Aerosol Air Qual Res 18:2704–2719. https://doi.org/10.4209/aaqr.2017.12.0605

Chaturvedi OPC, Singh JS (1987) A quantitative study of the forest floor biomass, litter fall and nutrient return in a *Pinus roxburghii* forest in Kumaun Himalaya. Vegetatio 71:97–106. https://www.jstor.org/stable/20038168

Chen C, Zou X, Singh AK et al (2021) Effects of hillslope position on soil water infiltration and preferential flow in tropical forest in Southwest China. J Environ Manag 299:113672. https://doi.org/10.1016/j.jenvman.2021.113672

Cheng YY, Ogden FL, Zhu J (2017) Earthworms and tree roots: a model study of the effect of preferential flow paths on runoff generation and groundwater recharge in steep, saprolitic, tropical lowland catchments. Water Resour Res 53. https://doi.org/10.1002/2016WR020258

Cheng Z, Zhang J, Yu B et al (2023) Stormflow response and "effective" hydraulic conductivity of a degraded tropical *Imperata* grassland catchment as evaluated with two infiltration models. Water Resour Res 59:e2022WR033625. https://doi.org/10.1029/2022WR033625

Chouksey A, Lambey V, Nikam BR et al (2017) Hydrological modelling using a rainfall simulator over an experimental hillslope plot. Hydrology 4:17. https://doi.org/10.3390/hydrology4010017

Coenders-Gerrits AMJ, van der Ent RJ, Bogaard TA et al (2014) Uncertainties in transpiration estimates. Nature 506:E1–E2. https://doi.org/10.1038/nature12925

Czarnowski MS, Olszewki JL (1970) Number and spacing of rainfall-gauges in a deciduous forest stand. Oikos 21:48–51. https://www.jstor.org/stable/3543837

Dabral BG, Subba Rao BK (1968) Interception studies in *chir* and teak plantations – New Forest. Indian For 94:541–551. https://doi.org/10.36808/if/1968/v94i7/26032

Dabral BG, Subba Rao BK (1969) Interception studies in Sal (*Shorea robusta*) and Khair (*Acacia catechu*) plantations – New Forest. Indian For 95:314–323. https://doi.org/10.36808/if/1969/v95i10/26087

Dabral BG, Nath P, Swarup R (1963) Some preliminary investigations on the rainfall interception by leaf litter. Indian For 89:112–124. https://doi.org/10.36808/if/1963/v89i2/24856

Dabral BG, Yadav JSP, Sharma DR (1965) Soil moisture studies in chir pine, teak and sal plantations at New Forest, Dehra Dun. Indian For 91:701–713. https://doi.org/10.36808/if/1965/v91i10/25154

Dai W, Subedi R, Jin K, Hao L (2023) Spatiotemporal variation of potential evapotranspiration and meteorological drought based on multi-source data in Nepal. Nat Haz Res 3:271–279. https://doi.org/10.1016/j.nhres.2023.04.007

Das A, Menon T, Ratnam J et al (2021) Expansion of pine into mid-elevation Himalayan oak forests: patterns and drivers in a multiple-use landscape. For Ecol Manage 497:119491. https://doi.org/10.1016/j.foreco.2021.119491

Daniel D, Gupta S, Sen S (2022) Role of soil in regulating runoff processes in pine- and oak-dominated headwater catchments of the Western Himalayas. Paper presented at the 2022 AGU fall meeting, Chicago, 12–16 December 2022 (abstract no. H44H-02)

Dass B, Daniel D, Saxena N et al (2023) Informing watershed management in data-scarce Indian Himalayas. Water Security 19:100138. https://doi.org/10.1016/j.wasec.2023.100138

Dawson TE (1998) Fog in the California redwood forest: ecosystem inputs and use by plants. Oecologia 117:476–485. https://www.jstor.org/stable/4222190

Deb Burman PK, Sarma D, Morrison R et al (2019) Seasonal variation of evapotranspiration and its effect on the surface energy budget closure at a tropical forest over north-East India. J Earth Syst Sci 128:127. https://doi.org/10.1007/s1204-019-1158-x

Deb Burman PK, Launiainen S, Mukherjee S et al (2021) Ecosystem-atmosphere carbon and water exchanges of subtropical evergreen and deciduous forests in India. For Ecol Manag 495: 119371. https://doi.org/10.1016/j.foreco.2021.119371

Deguchi A, Hattori S, Daikoku K, Park HT (2008) Measurement of evaporation from the forest floor in a deciduous forest throughout the year using microlysimeter and closed-chamber systems. Hydrol Proc 22:3712–3723. https://doi.org/10.1002/hyp.6974

Deng W, Zheng X, Xiao S et al (2023) Effects of leaf type, litter mass and rainfall characteristics on the interception storage capacity of leaf litter based on process simulation. J Hydrol 624:129943. https://doi.org/10.1016/j.jhydrol.2023.129943

Dickinson WT, Whiteley H (1970) Watershed areas contributing to runoff. Int Assoc Hydrol Sci Publ 1:12–26. https://www.researchgate.net/publication/238797114

Dorji U, Olesen JE, Seidenkrantz MS (2016) Water balance in the complex mountainous terrain of Bhutan and linkages to land use. J Hydrol Region Stud 7:55–68. https://doi.org/10.1016/j.ejrh.2016.05.001

Dou J, Zhang Y, Yu G et al (2007) Interannual and seasonal variations of energy and water vapor fluxes above a tropical seasonal rain forest in Xishuangbanna, SW China. Acta Ecol Sinica 27: 3099–3019. https://doi.org/10.1016/S1872-2032(07)60064-X

Dunkerley DL (2009) Evaporation of impact water droplets in interception processes: historical precedence of the hypothesis and a brief literature overview. J Hydrol 376:599–604. https://doi.org/10.1016/j.jhydrol.2009.08.004

Dunne T (1978) Field studies of hillslope flow processes. In Hillslope Hydrology, MJ Kirkby (Ed), J Wiley, chichester, UK, 227–293

Eckholm EP (1975) The deterioration of mountain environments. Science 189:764–770. https://doi.org/10.1126/science.189.4205.764

Eeckman J (2017) Caractérisation des systèmes hydro-climatiques á l'échelle locale dans le Himalaya népalais. Dissertation. Université de Montpellier, France, p 231. https://www/researchgate.net/publication/322339362

Eeckman J, Chevallier P, Boone A et al (2017) Providing a non-deterministic representation of spatial variability of precipitation in the Everest region. Hydrol Earth Syst Sci 21:4879–4893. https://doi.org/10.5194/hess-21-4879-2017

Eeckman J, Nepal S, Chevallier P et al (2019) Comparing the ISBA and J2000 approaches for surface flows modelling at the local scale in the Everest region. J Hydrol 569:705–719. https://doi.org/10.1016/j.jhydrol.2018.12.022

Eller CB, Lima AL, Oliveira RS (2013) Foliar uptake of fog water and transport belowground alleviates drought effects in the cloud forest tree species, *Drimys brasiliensis* (Winteraceae). New Phytol. https://doi.org/10.1111/nph.12248

Eller CB, Burgess SSO, Oliveira RS (2015) Environmental controls in the water use patterns of a tropical cloud forest tree species, *Drimys brasiliensis* (Winteraceae). Tree Phys 35. https://doi.org/10.1093/treephys/tpv001

Elsenbeer H, Cassel K, Castro J (1992) Spatial analysis of soil hydraulic conductivity in a tropical rain forest catchment. Water Resour Res 28:3201–3214. https://doi.org/10.1029/92WR01762

Frahm JP, Gradstein SR (1991) An altitudinal zonation of tropical rain forests using bryophytes. J Biogeogr 18:669–676. https://www.researchgate.net/publication/271689885

Gadermaier J (2018) Effect of artificial drought stress on sap flow of different tree species in the eastern Himalaya. Master's thesis, University of Natural Resources and Life Sciences (BOKU), Vienna, Austria, p 74

Gafur A, Jensen JR, Borggaard OK, Petersen L (2003) Runoff and losses of soil and nutrients from small watersheds under shifting cultivation (*Jhum*) in the Chittagong Hill Tracts of Bangladesh. J Hydrol 274:30–46. https://doi.org/10.1016/S0022-1694(02)00351-7

Gao S, Wang L, Hao L, Sun G (2023) Community forestry dominates the recent land greening amid climate change in Nepal. Environ Res Lett 18:104014. https://doi.org/10.1088/1748-9326/acf8de

Gardner RAM, Gerrard AJ (2002) Relationships between runoff and land degradation on non-cultivated land in the Middle Hills of Nepal. Int J Sustain Dev World Ecol 9:59–73. https://doi.org/10.1080/13504500209470103

Gardner RAM, Gerrard AJ (2003) Runoff and soil erosion on cultivated rainfed terraces in the Middle Hills of Nepal. Appl Geogr 23:23–45. https://doi.org/10.1016/S0143-6228(02)00069-3

Gash JHC (1979) An analytical model of rainfall interception by forests. Quart J Royal Meteo Soc 105:43–55. https://doi.org/10.1002/qj.49710544304

Gash JHC, Morton AJ (1978) An application of the Rutter model to the estimation of the interception loss from Thetford forest. J Hydrol 38:49–58. https://doi.org/10.1016/0022-1694(78)90131-2

Gash JHC, Lloyd CR, Lachaud G (1995) Estimating sparse forest rainfall interception with an analytical model. J Hydrol 170:79–86. https://doi.org/10.1016/0022-1694(95)02697-N

Ghimire M (2017) Historical land covers change in the Chure–Tarai landscape in the last six decades: drivers and environmental consequences. In: Li A et al (eds) Land cover change and its eco-environmental responses in Nepal. Springer, Singapore, pp 109–147. https://doi.org/10.1007/978-981-10-2890-8_5

Ghimire CP, Bruijnzeel LA, Bonell M (2016) Realistic prediction of infiltration-excess overland flow occurrence under contrasting land covers conditions in the Nepalese Middle Hills. AGU Chapman Conference on Emerging Issues in Tropical Ecohydrology, Cuenca, Ecuador, 5–9

Ghimire CP, Bruijnzeel LA, Lubczynski MW, Bonell M (2012) Rainfall interception by natural and planted forests in the Middle Mountains of Central Nepal. J Hydrol 475:270–280. https://doi.org/10.1016/j.jhydrol.2012.09.051

Ghimire CP, Bonell M, Bruijnzeel LA et al (2013a) Reforesting severely degraded grassland in the Lesser Himalaya of Nepal: effects on soil hydraulic conductivity and overland flow production. J Geophys Res Earth Surf 118:2528–2545. https://doi.org/10.1002/2013JF002888

Ghimire SK, Higaki D, Bhattarai TP (2013b) Estimation of soil erosion rates and eroded sediment in a degraded catchment of the Siwalik hills, Nepal. Land 2:370–391. https://doi.org/10.3390/land2030370

Ghimire CP, Bruijnzeel LA, Lubczynski MW, Bonell M (2014a) Negative trade-off between changes in vegetation water use and infiltration recovery after reforesting degraded pasture land in the Nepalese Lesser Himalaya. Hydrol Earth Syst Sci 18:4933–4949. https://doi.org/10.5194/hess-18-4933-2014

Ghimire CP, Lubczynski MW, Bruijnzeel LA, Chavarro-Rincon D (2014b) Transpiration, canopy conductance and decoupling coefficient of two contrasting forest types in the lesser Himalaya of Central Nepal. Agric For Meteo 197:76–90. https://doi.org/10.1016/j.agrformet.2014.05.012

Ghimire CP, Bruijnzeel LA, Bonell M et al (2014c) The effects of sustained forest use on hillslope soil hydraulic conductivity in the Middle Mountains of Central Nepal. Ecohydrology 7:478–495. https://doi.org/10.1002/eco.1367

Ghimire CP, van Meerveld HJ, Zwartendijk BW et al (2022) Vapour pressure deficit and solar radiation are the major drivers of transpiration in montane tropical secondary forests in eastern Madagascar. Agric For Meteo 326:109159. https://doi.org/10.1016/j.agrformet.2022.109159

Ghosh RC, Subba Rao BK, Ramola BC (1980) Interception studies in sal (*Shorea robusta*) coppice forest. Indian For 106(8):513–525. https://doi.org/10.36808/if/1980/v106i8/11214

Gilmour DA, Bonell M, Cassells DS (1987) The effects of forestation on soil hydraulic properties in the Middle Hills of Nepal. Mnt Res Dev 7:239–249. https//www.jstor.org/stable/3673199

Giambelluca TW, Martin RE, Asner GP et al (2009) Evapotranspiration and energy balance of native wetmontane cloud forest in Hawai'i. Agric For Meteo 149:230–243. https://doi.org/10.1016/j.agrformet.2008.08.004

Gilmour DA, King GC, Applegate GB, Mohns B (1990) Silviculture of plantation forest in Central Nepal to maximise community benefits. For Ecol Manag 32:173–186. https://doi.org/10.1016/0378-1127(90)90169-C

Gomi T, Sidle RC, Ueno M et al (2008) Characteristics of overland flow generation on steep forested hillslopes of Central Japan. J Hydrol 361:275–290. https://doi.org/10.1016/j.jhydrol.2008.07.045

Gosain BG, Negi GCS, Dhyani PP et al (2015) Ecosystem services of forests: carbon stock in vegetation and soil components in a watershed of Kumaun Himalaya, India. Int J Ecol Environ Sci 41:177–188. https://www.researchgate.net/publication/294506921

Grau O, Grytnes JA, Birks HJB (2007) A comparison of altitudinal species richness patterns of bryophytes with other plant groups in Nepal, Central Himalaya. J Biogeogr 34:1907–1915. https://doi.org/10.1111/j.1365-2699.2007.01745.x

Graham CB, Woods RA, McDonnell JJ (2010) Hillslope threshold response to rainfall: a field based forensic approach. J Hydrol 393:65–76. https://doi.org/10.1016/j.jhydrol.2009.12.015

Gupta A, Usharani L (2009) Rainfall partitioning in a tropical forest of Manipur, north East India. Trop Ecol 50:355–358. https://www.researchgate.net/publication/257921317

Gupta RK, Mishra PR, Shankar P et al (1974) Studies on the effect of different land treatments on water yield in the Siwalik–Chandigarh. In: Annual Report 1974. Central Soil and Water Conservation Research and Training Institute, Dehra Dun, India, pp 18–21

Gupta RK, Kaushal RC, Dayal SKN, Mishra PR (1975) Study on the effect of closure with afforestation and gully control measures on runoff and peak discharge – Chandigarh. In: Annual Report 1975. Central Soil and Water Conservation Research and Training Institute, Dehra Dun, India, pp 15–17

Gurung A, Adhikhari S, Chauhan R et al (2019) Water crises in a water-rich country: case studies from rural watersheds of Nepal's mid-hills. Water Policy 21:826–847. https://doi.org/10.2166/wp.2019.245

Guzman JA, Hamann HF, Sanchez-Azofeifa GAS (2024) Multi-decadal trends of low-clouds at the tropical montane cloud forests. Ecol Indic 158:111599. j.ecolind.2024.111599

Hall RL, Calder IR (1993) Drop size modification by forest canopies: measurements using a disdrometer. J Geophys Res Atm 98(D10):18465–18470. https://doi.org/10.1029/93JD01498

Hamilton EN (1954) Rainfall sampling on rugged terrain. Tech Bull 1096, USDA, Washington DC, p 41

Hamilton LS (1987) What are the impacts of Himalayan deforestation on the Ganges-Brahmaputra lowlands and delta? Assumptions and facts. Mnt Res Dev 7:256–263. https://doi.org/10.2307/3673202

Helvey JD (1964) Rainfall interception by hardwood forest litter in the southern Appalachians. US For Serv Res Paper SE-8, p. https://research.fs.usda.gov/treesearch/3151

Helvey JD, Patric H (1965) Canopy and litter interception of rainfall by hardwoods of Eastern United States. Water Resour Res 1:193–206. https://doi.org/10.1029/WR001i002p00193

Herwitz SR (1986) Infiltration-excess caused by stemflow in a cyclone-prone tropical rainforest. Earth Surf Proc Landforms 11:401–412. https://doi.org/10.1002/esp.3290110406

Hessel R, Gupta MK, Singh Datta P et al (2007a) Rainfall, soil moisture content and runoff in a small catchment in the Indian Himalayas. Int J Ecol Environ Sci 33:115–128. https://www.researchgate.net/publication/40101323

Hessel R, Gupta MK, Singh Datta P, Gelderman E (2007b) Application of the LISEM soil erosion model to a forested catchment in the Indian Himalayas. Int J Ecol Environ Sci 133:129–142. https://www.researchgate.net/publication/40101326

Hofer T, Messerli B (2006) Floods in Bangladesh. History, dynamics and rethinking the role of the Himalayas. United Nations University, Tokyo, p 450. https://digitallibrary.un.org/record/587111?v=pdf

Holwerda F, Scatena FN, Bruijnzeel LA (2006) Throughfall in a Puerto Rican lower montane rain forest: a comparison of sampling strategies. J Hydrol 327:592–602. https://doi.org/10.1016/j.jhydrol.2005.12.014

Hou Y, Wei X, Zhang M et al (2023) A global synthesis of hydrological sensitivities to deforestation and forestation. For Ecol Manag 529:120718. https://doi.org/10.1016/j.foreco.2022.120718

Igarashi Y, Katul GG, Kumagai T et al (2015) Separating physical and biological controls on long-term evapotranspiration fluctuations in a tropical deciduous forest subjected to monsoonal rainfall. J Geophys Res Biogeosci 120:1262–1278. https://doi.org/10.1002/2014JG002767

Ilek A, Blonska E, Miszewski K et al (2024) Investigating water storage dynamics in the litter layer: the impact of mixing and decay of pine needles and oak leaves. Forests 15:350. https://doi.org/10.3390/f15020350

Illien L, Andermann C, Sens-Schönfelder C et al (2021) Subsurface moisture regulates Himalayan groundwater storage and discharge. AGU Adv 2:e2021AV000398. https://doi.org/10.1029/2021AV000398

Immerzeel WW, Lutz AF, Andrade M et al (2020) Importance and vulnerability of the world's water towers. Nature 577:364–369. https://doi.org/10.1038/s41586-019-1822-y

Ives JD (1987) The theory of Himalayan environmental degradation: its validity and application challenged by recent research. Mnt Res Dev 7:189–199. https://doi.org/10.2307/3673192

Jackson IJ (1975) Relationships between rainfall parameters and interception by tropical forest. J Hydrol 24:215–238. https://doi.org/10.1016/0022-1694(75)90082-7

Jackson RB, Jobbaggy EB, Avissar R et al (2005) Trading water for carbon with biological carbon sequestration. Science 310:1944–1947. https://doi.org/10.1126/science.11192820

Jain A, Rai SC, Sharma E (2000) Hydro-ecological analysis of a sacred lake watershed system in relation to land-use / cover change from Sikkim Himalaya. Catena 40:263–278. https://doi.org/10.1016/S0341-8162(00)00086-2

Jana P, Dasgupta S, Todaria NP (2021) Throughfall and stemflow nutrient flux in deodar and oak forests, Garhwal Himalaya, India. Water Supply. https://doi.org/10.2166/ws.2021.009

Jarvis A, Mulligan M (2011) The climate of cloud forests. Hydrol Proc 25:327–343. https://doi.org/10.1002/hyp.7847

Jasechko S, Sharp Z, Gibson JJ et al (2013) Terrestrial water fluxes dominated by transpiration. Nature 496:347–350. https://doi.org/10.1038/nature11983

Jiang Y, Yang K, Qi Y et al (2023) TPHiPr: A long-term (1979–2020) high-accuracy precipitation dataset (1/30° , daily) for the Third Pole region based on high-resolution atmospheric modeling and dense observations. Earth Syst Sci Data 15:621–638. https://doi.org/10.5194/essd-15-621-2023

Jinger D, Kaushal R, Kumar R et al (2023) Degraded land rehabilitation through agroforestry in India: achievements, current understanding and future prospectives. Front Ecol Evol 11: 1088796. https://doi.org/10.3390/fevo.2023.1088796

Joshi RK, Sarkoti CG (2020) Litter dynamics, leaf area index and forest floor respiration as indicators for understanding the role of Nepalese alder in white oak forests in central Himalaya, India. Ecol Indics 111:106065. https://doi.org/10.1016/j.ecolind.2020.106065

Joshi V, Negi GCS, Kumar K (1998) Surface runoff, soil loss and land use studies in two micro-catchments of the western Himalaya, India. In: Chalise SR et al (eds) Ecohydrology of high mountain areas. ICIMOD, Kathmandu, pp 467–472

Joshi VC, Negi VS, Bisht D et al (2021) Tree biomass and carbon stock assessment of subtropical and temperate forests in the Central Himalaya, India. Trees For People 6:100147. https://doi.org/10.1016/j.tfp.2021.100147

Kar SK, Patra S, Singh RM et al (2023) Impact of land-use reformation on soil hydraulic properties and recovery potential of conservation tillage in Indias's North-Western Himalayan region. Ecohydrol Hydrobiol 23:290–303. https://doi.org/10.1016/j.ecohyd.2022.12.008

Kathyar VS, Striffler WD (1984) Rainfall variation in a small Himalayan watershed. Indian J Soil Conserv 12:57–64. http://krishi.icar.gov.in/jspui/handle/123456789/42149

Kaushal RC, Dayal SKN, Mishra PR (1975) Study on the effect of closure with afforestation and gully control measures on runoff and peak discharge – Chandigarh. In: Annual Report 1975. Central Soil and Water Conservation Research and Training Institute, Dehra Dun, India, pp 57–59

Kaushal R, Singh I, Thapliyal et al (2020) Rooting behaviour and soil properties in differen bamboo species of Western Himalayan foothills, India. Sci Rep 10:4966. https://doi.org/10.1038/s41598-020-61418-z

Kaushal R, Mandal D, Panwar P et al (2021) Chapter 20 – Soil and water conservation benefits of agroforestry. In: Forest Resources Resilience & Conflicts 2021, pp 259–275. https://doi.org/10.1016/B978-0-12-822931-6.00020-4

Koppe E, Schneider FJA, Londero AL et al (2022) Soil water infiltration evaluation from punctual to hillslope scales. Environ Monitor Assess 194:300. https://doi.org/10.1007/s10661-022-09893-x

Kothyari BP, Verma PK, Joshi BK, Kothyari UC (2004) Rainfall–runoff-soil and nutrient loss relationships for plot size areas of bhetagad watershed in Central Himalaya, India. Agric Ecosyst Environ 293:137–150. https://doi.org/10.1016/j.jhydrol.2004.01.011

Krishnaswamy J, Bonell M, Venkatesh B et al (2013) The groundwater recharge response and hydrological services of tropical humid forest ecosystems to use and reforestation: support for the "infiltration – evapotranspiration trade-off hypothesis". J Hydrol 498:191–209. https://doi.org/10.1016/j.jhydrol.2013.06.034

Kumar S (1990) Studies on some aspects of vegetation and hydrology in pine-oak forests of Kumaun Himalaya. Dissertation, Kumaun University, Almora, India

Kumar S, Sterk G (2005) Processes and soil erosion assessment at hills scale in the lesser Himalayas, India. In: Singhal DC et al (eds) Hydrological perspectives for sustainable development. Allied Publishers, New Delhi, pp 420–427

Kumar S, Tewari LM (2014) Pattern of litter fall and litter decomposition in a *Quercus leucotrichophora* A. Camus forest in Kumaun Himalaya. Int J Biodiv Conserv 61:108–114. https://doi.org/10.5897/IJBC2013.0617

Kumar S, Verma KR (1989) Effect of pine oak forest's litter on evaporation and water conservation in Kumaun hills. In: Khulbe RD (ed) Perspectives in aquatic biology. Papyrus Publishing House, New Delhi, pp 260–264

Kumar S, Verma KR (1990) Effects of soil characteristics and grasses on infiltration at two watersheds in Kumaun Himalaya. Paper presented at the International Symposium of Environmental Impact on Biosystems, Madras, December 1989

Kumar S, Verma KR (1991) The impact of vegetal cover on streamflow, with special reference to Kumaun Himalaya. In: Bhatt JD, Pande RK (eds) Ecology of the mountain waters. Ashish Publishing House, New Delhi, pp 348–358

Kumar S, Singh A, Shrestha DP (2016) Modelling spatially distributed surface runoff generation using SWAT-VSA: a case study in a watershed of the north-west Himalayan landscape. Model Earth Syst Environ 2:202. https://doi.org/10.1007/s40808-016-0249-9

Kumar BM, Handa AK, Dhyani SK et al (2018) Agroforestry in the Indian Himalayan Region: an overview. In: Gordon AM, Newman SM, Coleman BRW (eds) Temperate agroforestry systems. CABI International, Wallingford, UK, pp 153–172

Kumar M, Hodnebrog O, Daloz AS et al (2021) Measuring precipitation in Eastern Himalaya: ground validation of eleven satellite, model and gauge interpolated gridded products. J Hydrol 599:126252. https://doi.org/10.1016/j.jhydrol.2021.126252

Kumar M, Joseph G, Bhutia Y, Krishnaswamy J (2023) Contrasting sap flow characteristics between pioneer and late-successional tree species in secondary tropical montane forests of eastern Himalaya, India. J Exp Bot. https://doi.org/10.1093/jxb/erad207

Kumar M, Bhutia Y, Varma GR et al (2024a) Role of transpiration in modulating ecosystem services in secondary tropical montane forests of eastern Himalaya in India. Ecohydrol Hydrobiol 25. https://doi.org/10.1016/j.ecohyd.2024.04.001

Kumar M, Sen S, Badiger S, Krishnaswamy J (2024b) Forest hydrology in high tropical mountains: Contrasting ecohydrological processes in temperate broadleaf and subalpine conifer mixed tropical montane forests of Eastern Himalaya, India. https://www.researchgate.net/publication/381995456

Kumar S, Khanna J, Mohanty JR, Varma G, Daniel D, Krishnaswamy J, Sen S (2025) Sap flux and stand scale transpiration measurements of a Chir pine forest in the west-central Himalayas of India – radional and seasonal variations and comparison with co-located hydrometeorological and evaporation measurements. https://doi.org/10.2139/ssrn.5377276

Lacombe G, Valentin C, Sounyafong P et al (2018) Linking crop structure, throughfall, soil surface conditions, runoff and soil detachment: 10 land uses analyzed in Northern Laos. Sci Total Environ 616–617:1330–1338. https://doi.org/10.1016/j.scitotenv.201.10.185

Lang TJ, Barros AP (2004) Winter storms in the Central Himalayas. J Meteo Soc Jap 82:829–844. https://doi.org/10.2151/jmsj.2004.829

Levia DF, Germer S (2015) A review of stemflow generation dynamics and stemflow-environment interactions in forests and shrublands. Rev Geophys 53:673–714. https://doi.org/10.1002/2015RG000479

Lewis S, Wheeler C, Mitchard E, Koch A (2019) Restoring natural forests is the best way to remove atmospheric carbon. Nature 568(7750):25–28. https://doi.org/10.1038/d41586-019-01026-8

Leyton L, Reynolds ERC, Thompson FB (1967) Rainfall interception in forests and moorland. In: Sopper WE, Lull HW (eds) International symposium on forest hydrology. Pergamon Press, Oxford, pp 163–177

Li K, Zhao L, Hou R et al (2021) Effect of leaf distribution pattern on the interception storage capacity of leaf litter under simulated rainfall conditions. Hydrol Proc 35:e14022. https://doi.org/10.1002/hyp.14022

Limm EB, Simonin KA, Bothman AG, Dawson TE (2009) Foliar water uptake: a common water acquisition strategy for plants of the redwood forest. Oecologia 161:449–459. https://doi.org/10.1007/s00442-009-1400-3

Lin HS (2010) Earth's critical zone and hydropedology: concepts, characteristics, and advances. Hydrol Earth Syst Sci 14:25–45. https://doi.org/10.5194/hess-14-25-2010

Liu WY, Fox JED, Xu Z (2003) Nutrient budget of a montane evergreen broad-leaved forest at Ailao Mountain National Nature Reserve, Yunnan, Southwest China. Hydrol Proc 17:1119–1134. https://doi.org/10.1002/hyp.1184

Liu WJ, Liu WY, Li P et al (2010) Dry season water uptake by two dominant canopy tree species in a tropical seasonal rainforest of Xishuangbanna, SW China. Agric For Meteo 150:380–388. https://doi.org/10.1016/j.agrformet.2009.12.006

Liu WJ, Li P, Duan W, Liu WY (2014) Dry-season water utilization by trees growing on thin karst soils in a seasonal tropical rainforest of Xishuangbanna, Southwest China. Ecohydrology 7: 927–935. https://doi.org/10.1002/eco.1419

Liu LL, Yang B, Lu HZ et al (2021) Dry-season fog water utilization by epiphytes in a subtropical montane cloud forest of Southwest China. Water 13:3237. https://doi.org/10.3390/w13223237
Lloyd CR, Marques Filho ADO (1988) Spatial variability of throughfall and stemflow measurements in Amazonian rainforest. Agric For Meteo 42:63–73. https://doi.org/10.1016/0168-1923(88)90067-6
Lohani P, Mukherjee S, Sekar KC et al (2023) Impact of monsoon season rainfall spells on the ecosystem carbon exchanges of Himalayan Chir-Pine and Banj-Oak-dominated forests: a comparative assessment. Environ Monitor Assess 195:287. https://doi.org/10.1007/s10661-023-11297-4
Los SO, Street-Perrott FA, Loader NJ, Froyd CA (2021) Detection of signals linked to climate change, land-cover change and climate oscillators in Tropical Montane Cloud Forests. Remote Sens Environ 260:112431. https://doi.org/10.1016/j.rse.2021.112431
Loshali DC, Singh RP (1992) Partitioning of rainfall by three Central Himalayan forests. For Ecol Manag 53:99–105. https://doi.org/10.1016/0378-1127(92)90037-A
Lozano-Baez SE, Cooper M, Meli P et al (2019) Land restoration by tree planting in the tropics and subtropics improves soil infiltration, but some critical gaps still hinder conclusive results. For Ecol Manag 444:89–95. https://doi.org/10.1016/j.foreco.2019.04.046
Mandal D, Jayaprakash J (2009) Water repellency of soils in the lower Himalayan regions of India. Curr Sci 96:148–152. https://www.jstor.org/stable/24104743
Mathur HN, Babu R, Joshie P, Singh B (1976) Effect of clearfelling and reforestation on runoff and peak rates in small watersheds. Indian For 102:219–226
Mathur RN, Singh RP, Gupta VK (1982) Comparative study of infiltration in soils under forest cover and agriculture in temperate climate. Indian Forester 108:648–653
McJannet D, Fitch P, Disher M, Wallace JS (2007) Measurements of transpiration in four tropical rainforest types of North Queensland, Australia. Hydrol Proc 21:3549–3564. https://doi.org/10.1002/hyp.6576
Merz J, Nakarmi G, Shrestha SK et al (2003) Water: a scarce resource in rural watersheds of Nepal's Middle Mountains. Mnt Res Dev 23:41–49. https://www.jstor.org/stable/3674534
Miehe G (2015) Ecological transects. In: Miehe G, Pendry C (eds) Nepal: an introduction to the natural history, ecology and human environment of the Himalayas. Royal Botanic Garden of Edinburgh, UK, pp 347–384
Miehe G, Miehe S, Böhner J et al (2015) Vegetation ecology. In: Miehe G, Pendry C (eds) Nepal: an introduction to the natural history, ecology and human environment of the Himalayas. Royal Botanic Garden of Edinburgh, UK, pp 385–472
Mishra NB, Chaudhari G (2015) Spatio-temporal analysis of trends in seasonal vegetation productivity across Uttarakhand, Indian Himalayas, 2000–2014. Appl Geogr 56:29–41. https://doi.org/10.1016/j.apgeog.2014.10.007
Mishra NB, Mainali KP (2017) Greening and browning of the Himalaya: spatial patterns and the role of climatic change and human drivers. Sci Total Environ 587–588:326–339. https://doi.org/10.1016/j.scitotenv.2017.02.156
Mishra BK, Ramakrishnan PS (1983a) Slash and burn agriculture at higher elevations in North-Eastern India. I. Sediment, water and nutrient losses. Agric Ecosyst Environ 9:69–82. https://doi.org/10.1016/0167-8809(83)90007-5
Mishra BK, Ramakrishnan PS (1983b) Slash and burn agriculture at higher elevations in North-Eastern India. II. Soil fertility changes. Agric Ecosyst Environ 9:83–96. https://doi.org/10.1016/0167-8809(83)90008-7
Momiyama H, Kumagai T, Egusa T (2021) Model analysis of forest thinning impacts on the water resources during hydrological drought periods. For Ecol Manag 499:119593. https://doi.org/10.1016/j.foreco.2021.119593
Mohanty JR, Sen S, Krishnaswamy J (2025) Forest transition and its hydro-climatic impacts in the Indian Himalayas: inferences from field observations. EGU General Assembly 2025, abstract no. EGU25–12489. https://doi.org/10.5194/egusphere-egu25-12489

Moreno de las Heras M, Nicolau JM, Merino-Martin L, Wilcox BP (2010) Plot-scale effects on runoff and erosion along a slope degradation gradient. Water Resour Res 46:W04503. https://doi.org/10.1029/WR2009WR007875

Mukherjee S, Lohani P, Kumar K et al (2020) Assessment of new alternative scaling properties of the convective boundary layer: application to velocity and temperature spectra. Boundary-Layer Meteo 176:271–289. https://doi.org/10.1007/s10546-020-00525-w

Mukherjee S, Joshi R, Kumar K (2018) Compendium of meteorological data 2012–2016; field station: Kosi-Katarmal, Almora, Uttarakhand, Indian Himalayan Region. GB Pant National Institute for Himalayan Environment and Sustainable Development, Almora, India

Mulligan M (2006) MODIS cloud climatology. https://geodata.policysupport.org/modis-cloud-climatology

Mulligan M (2013) WaterWorld: a self-parameterising, physically-based model for application in data-poor but problem-rich environments globally. Hydrol Res 44(5):748–769. https://doi.org/10.2166/nh.2012.217

Mulligan M, Burke SM (2005) FIESTA: Fog Interception for the Enhancement of Streamflow in Tropical Areas. https://www.ambiotek.com/fiesta

Murakami S (2006) A proposal for a new forest canopy interception mechanism: splash droplet evaporation. J Hydrol 319:72–82. https://doi.org/10.1016/j.jhydrol.2005.07.002

Muzylo A, Llorens P, Valente F et al (2009) A review of rainfall interception modelling. J Hydrol 370:191–206. https://doi.org/10.1016/j.jhydrol.2009.02.058

Nanda A, Sen S, Jirwan V et al (2018) Understanding plot-scale hydrology of a Lesser Himalayan watershed – a field study and HYDRUS-2D modelling approach. Hydrol Proc 32:1254–1266. https://doi.org/10.1002/hyp.11499

Nanda A, Sen S, McNamara JP (2019) How spatiotemporal variation of soil moisture can explain hydrological connectivity of infiltration-excess dominated hillslope: observations from lesser Himalayan landscape. J Hydrol 579:124146. https://doi.org/10.1016/j.jhydrol.2019.124146

Nanda SA, Haq MU, Singh SP et al (2021) Species richness and β-diversity patterns of macrolichens along elevation gradients across the Himalayan Arc. Sci Rep 11:20155. https://doi.org/10.1038/s41598-021-99675-1

Nanko K, Mizugaki S, Onda Y (2008) Estimation of soil splash detachment rates on the forest floor of an unmanaged Japanese cypress plantation based on field measurements of throughfall drop sizes and velocities. Catena 72:348–361. https://doi.org/10.1016/j.catena.2007.07.002

Naudiyal N, Schmerbeck J (2017) The changing Himalayan landscape: pine-oak forest dynamics and the supply of ecosystem services. J For Res 28:431–443. https://doi.org/10.1007/s11676-016-0338-7

Naudiyal N, Schmerbeck J (2018) Linking forest successional dynamics to community dependence on provisioning ecosystem services from the Central Himalayan forests of Uttarakhand. Environ Manag 62:915–928. https://doi.org/10.1007/s00267-018-1087-5

Negi GCS, Rikhari HC, Garkoti SC (1998a) The hydrology of three high-altitude forests in Central Himalaya, India: a reconnaissance study. Hydrol Proc 14:343–350. https://doi.org/10.1002/(SICI)1099-1085(199802)12:2<343::AID-HYP578>3.0.CO;2-L

Negi GCS, Joshi V, Kumar K (1998b) Spring sanctuary development to meet household water demand in the mountains: a call for action. In: GB Pant Instititute for Himalayan Environment and Development (ed) Research for mountain development: some initiatives and accomplishments. Gyanodaya Prakashan, Nainital, pp 25–48

Nieschulze J, Erasmi S, Dietz J, Hölscher D (2009) Satellite-based prediction of rainfall interception by tropical forest stands of a human-dominated landscape in Central Sulawesi, Indonesia. J Hydrol 364:227–235. https://doi.org/10.1016/j.jhydrol.2008.10.024

Niraula RR, Gilani H, Pokharel BK, Qamer FM (2013) Measuring impacts of community forestry program through repeat photography and satellite remote sensing in the Dolakha district of Nepal. J Environ Manag 126:20. https://doi.org/10.1016/j.envman.2013.04.006

Norouzi-Moghanjoghi K, Fakhraei H, Valipour M et al (2025) Soil texture-based parameterisation and hydrological insights of a fully coupled surface and subsurface model at the Hubbard Brook Experimental Forest, USA. Hydrol Proc 39:e70045. https://doi.org/10.1002/hyp.70045
Ogden FL, Hawkins RP, Walter MT, Goodrich DC (2017) Comment on "Beyond the SCS-CN method: a theoretical framework for spatially lumped rainfall-runoff response" by M. S. Bartlett et al. Water Resour Res 53:WR020176. https://doi.org/10.1002/2016WR020176
Oishi AC, Oren R, Stoy PC (2008) Estimating components of forest evapotranspiration: a footprint approach for scaling sap flux measurements. Agric For Meteo 148:1719–1732. https://doi.org/10.1016/j.agrformet.2008.06.013
Oishi AC, Miniat FC, Novick KA et al (2018) Warmer temperatures reduce net carbon uptake, but do not affect water use, in a mature southern Appalachian forest. Agric For Meteo 252:269–282. https://doi.org/10.1016/j.agrformet.2018.01.011
Oishi AC, Denham SO, Brantley ST et al (2020) Quantifying the effects of stand age on components of forest evapotranspiration. Acta Hortic 1300:89–96. https://doi.org/10.17660/ActaHortic.2020.1300.12
Oliveira RS, Eller CB, Bittencourt PRL, Mulligan M (2014) The hydroclimatic and ecophysiological basis of cloud forest distributions under current and projected climates. Ann Bot 113:909–920. https://doi.org/10.1093/aob/mcu060
Oliveira S, Cunha J, Nobrega RLB et al (2024) Enhancing global rainfall interception loss estimation through vegetation structure modelling. J Hydrol 631:130672. https://doi.org/10.1016/j.jhydrol.2024.130672
Om K (2016) Hydrological characterization of two mountain ecosystems in Bhutan. Master's thesis, University of Natural Resources and Life Sciences (BOKU), Vienna, Austria, p 93
Pandey AN, Pathak PC, Singh JS (1983) Water, sediment and nutrient movement in forested and non-forested catchments in Kumaun Himalaya. For Ecol Manag 7:19–29. https://doi.org/10.1016/0378-1127(83)90054-3
Pandeya B (2013) Understanding hydrological ecosystem services produced by the indo-Gangetic basin and selected mountain catchments in the Himalaya. Dissertation, King's College London, p 252
Pandit A, Batelaan O, Pandey VP, Adhikari S (2024) Depleting spring sources in the Himalayas: environmental drivers or just perception? J Hydrol Region Stud 53:101752. https://doi.org/10.1016/j.ejrh.2024.101752
Pant N, Hagare D, Maheshwari B et al (2024) Rejuvenation of the springs of the Hindu Kush Himalayas through transdisciplinary approaches – a review. Water 16:3675. https://doi.org/10.3390/w16243675
Panwar P, Mahalingappa DG, Kaushal R et al (2022) Biomass production and carbon sequestration potential of different agroforestry systems in India: a critical review. Forests 13:1274. https://doi.org/10.3390/f13081274
Pathak PC, Pandey AN, Singh JS (1984) Overland flow, sediment output and nutrient loss from certain forested sites in the Central Himalaya, India. J Hydrol 71:239–251. https://doi.org/10.1016/0022-1694(84)90099-4
Pathak PC, Pandey AN, Singh JS (1985) Apportionment of rainfall in Central Himalayan forests (India). J Hydrol 76:319–332. https://doi.org/10.1016/0022-1694(85)90140-4
Pathak PC, Rawat YS, Singh RP (1989) Rainfall interception by leaf litter in Central Himalayan forests. Indian For 115:822–831. https://doi.org/10.36808/if/1989/v115i11/9126
Patnaik N, Virdi SS (1962) Field infiltration studies in Doon Valley. Irrig Power 19:1003–1012
Patnaik US, Gupta RK, Kaushal RC et al (1974) Study on the effect of closure with afforestation and gully control measures on runoff and peak discharge – Chandigarh. In: Annual report 1974. Central Soil and Water Conservation Research and Training Institute, Dehra Dun, India, pp 62–66
Pathak GC, Joshi H, Singh RD et al (2021) Vertical root distribution in Himalayan trees: about half of roots occur below 30 cm, the generally sampled depth. Trop Ecol 62:479–491. https://doi.org/10.1007/s42965-021-00159-0

Paudel KP, Dahal D, Shah R (2012) Abandoned agricultural land in Mid-Hills of Nepal: status, causes and consequences. IUCN Nepal, Kathmandu, p 42

Paudel NS, Ojha H, Bangade MR et al (eds) (2022) Revitalising community forestry in the changing socioeconomic context of Nepal. ENLIFT2 Program and Forest Action Nepal, Kathmandu, p 86. https://www.researchgate.net/publication/359500990

Peña-Arancíbia JL, Bruijnzeel LA, Mulligan M, van Dijk AIJM (2019) Forests as 'pumps' and 'sponges': assessing the impact of deforestation on dry-season flows across the tropics. J Hydrol 574:946–963. https://doi.org/10.1016/j.jhydrol.2019.04.064

Poffenberger M (2015) Restoring and conserving Khasi forests: a community-based REDD strategy from Northeast India. Forests 6:4477–4494. https://doi.org/10.3390/f6124382

Pokharel BK, Uprety DR, Niraula RR, Pokharel PR (2018) An assessment of the impact of silviculture and forest management regimes to forest cover change in the Churia region during 1992 to 2014. Banko Janakari Spec Issue 4:38–44. https://doi.org/10.3126/banko.v27i3.20540

Poudel DD, Duex TW (2017) Vanishing springs in Nepalese Mountains: assessment of water sources, farmers' perceptions, and climate change adaptation. Mnt Res Dev 37:35–46. https://www.jstor.org/stable/90001379

Poudyal K, Jha PK, Zobel DB, Thapa CB (2004) Patterns of leaf conductance and water potential of five Himalayan tree species. Tree Phys 24:689–699. https://doi.org/10.1093/treephys/24.6.689

Pujari PR, Jain V, Singh V et al (2020) Critical zone: an emerging research area for sustainability. Curr Sci 118:1487–1488. https://www.researchgate.net/publication/342421737

Puri DN, Juyal GP, Katiyar VS (1982) Hydrology of forest watershed – a case study. In: Proceedings of the international symposium on hydrological aspects of mountainous watersheds, Roorkee, India, 4–6 November 1982. Manglik Prakashan, Sharanpur, p VIII-28–35

Purohit S, Patel NR, Nikam BR (2022) Estimation of evapotranspiration fluxes from Sal (Shorea robusta Gaertn.f) forest using METRIC model: case study of Doon Valley, India. Geocarto Int 2021. https://doi.org/10.1080/10106049.2021.1920630

Putuhena WM, Cordery I (1996) Estimation of interception capacity of the forest floor. J Hydrol 180:283–299. https://doi.org/10.1016/0022-1694(95)02883-8

Putuhena W, Cordery I (2000) Some hydrological effects of changing forest cover from eucalypts to *Pinus radiata*. Agric For Meteo 100:59–72. https://doi.org/10.1016/S0168-1923(99)00086-6

Qazi N (2019) Hillslope hydrological functioning along a forest degradation gradient in the lesser Himalaya of India. Final Report on Project SR/WOS-A/EA-1016/2015. Dept Sci Technol, New Delhi, India

Qazi NQ, Bruijnzeel LA (2023) Changes in runoff response to rainfall along forest degradation gradients in the Lesser Himalaya of NW India. In: Sun G et al (eds) International symposium on the forests–water–livelihood nexus in the Lesser Himalaya, Pokhara, Nepal, 29–30 September 2023, pp 25–27 https://www.researchgate.net/publication/374661689

Qazi NQ, Bruijnzeel LA, Rai SP, Ghimire CP (2017) Impact of forest degradation on streamflow regime and runoff response to rainfall in the Garhwal Himalaya, Northwest India. Hydrol Sci J. https://doi.org/10.1080/02626667.2017.1308637

Rai SP (1993) Hydrological and geomorphological studies of the Syahidevi-Binsar area, District Almora. Dissertation, Kumaun University, Almora, India, p 160

Rai SC, Sharma E (1998) Hydrology and nutrient flux in an agrarian watershed of the Sikkim Himalaya. J Soil Water Conserv 53:125–132. https://www.researchgate.net/publication/282757775

Raina AK, Gupta MK (2009) Soil and vegetation studies in relation to parent material I n Kempty forest range, Uttarakhand. Indian For 135:331–341. https://www.i-scholar.in/index.php/indianforester/article/view/354

Ramakrishnan PS, Toky OP (1981) Soil nutrient status of hill agro-ecosystems and recovery patterns after slash and burn agriculture in North-Eastern India. Plant Soil 60:41–60. https://doi.org/10.1007/BF02377111

Rana RS, Subehia SK (1996) Rainfall-runoff-soil loss relationships for Sirsa catchment in submontane zone of Himachal Pradesh. Indian J Soil Conserv 14:169–171

Raturi AS, Subba Rao BK (1985) Effect of thinning on the vegetation cover characteristics in a Sal (*Shorea robusta*) coppice forest. Indian For 111:272–289. https://www.i-scholar.in/index.php/indianforester/article/view/10205

Rawat JS, Rai SP, Valdiya KS (1997) Response of hydrologic cycle to environmental degradation in the Khulgad watershed, Kumaun Himalaya. In: Shah SL (ed) Sustainable replicable eco-development in Central Himalaya. Shree Almora Book Department, Almora, pp 127–162

Rawat JS, Rawat G, Rai SP (2000) Impact of human activities on geomorphic processes in the Almora region, Central Himalaya, India. In: Slaymaker O (ed) Geomorphology: human activity and global environmental change. J Wiley, Chichester, pp 285–299

Ray MP (1970) Preliminary observations on stemflow etc. in *Alstonia scholaris* and *Shorea robusta* plantations at Arabari. West Bengal Indian For 96:482–493. https://www.i-scholar.in/index.php/indianforester/article/view/26285

Ray MP (1971) A preliminary study of run-off and soil-wash in run-off plots at Jhargram, West Bengal. Indian For 97:2–12. https://www.i-scholar.in/index.php/indianforester/article/view/26374

Ribolzi O, Evrard O, Huon S et al (2017) From shifting cultivation to teak plantation: effect on overland flow and sediment yield in a montane tropical catchment. Sci Rep 7:3987. https://doi.org/10.1038/s41598-017-04385-2

Richardson JB (2017) Critical Zone. In: White W (ed) Encyclopedia of geochemistry. Encyclopedia of earth sciences series. Springer, Cham. https://doi.org/10.1007/978-3-319-39193-9_355-1

Ries JB (1994) Soil erosion in the high mountain region, eastern central Himalaya. A case study in the Bamti/Bhandar/Surma area, Nepal. Freiburger Geogr Hefte 42:1–397

Roberts JM, Gash JHC, Tani M, Bruijnzeel LA (2005) Controls on evaporation in lowland tropical rainforest. In: Bonell M, Bruijnzeel LA (eds) Forests, water and people in the humid tropics. Cambridge University Press, Cambridge, pp 273–313. https://doi.org/10.1017/CBO9780511535666.019

Saha R, Tomar JMS, Ghosh PK (2007) Evaluation and selection of multi-purpose tree for improving soil hydro-physical behaviour under hilly eco-system of north east India. Agrofor Syst 69: 239–247. https://doi.org/10.1007/s10457-007-9044-y

Salazar-Martinez D, Holwerda F, Holmes T et al (2022) Evaluation of remote sensing-based evapotranspiration products at low-latitude eddy covariance sites. J Hydrol 610:127786. https://doi.org/10.1016/j.jhydrol.2022.127778

Sarkar R, Dutta S (2012) Field investigation and modeling of rapid subsurface stormflow through preferential pathways in a vegetated hillslope of Northeast India. J Hydrol Eng ASCE 17:333–341. https://doi.org/10.1061/(ASCE)HE.1943-5584.0000431

Sarkar R, Dutta S, Panigrahy S (2008) Effect of scale on infiltration in a macropore-dominated hillslope. Curr Sci 94(4):490–494. https://www.researchgate.net/publication/254446841

Sato Y, Otsuki K, Ogawa S (2003) Estimation of the litter interception loss in the evergreen forest. J Jap Soc Hydrol Water Resour 16:640–651. https://doi.org/10.3178/jjshwr.16.640

Sato Y, Kumagai T, Kume A et al (2004) Experimental analysis of moisture dynamics of litter layers – the effects of rainfall conditions and leaf shapes. Hydrol Proc 18:3007–3018. https://doi.org/10.1002/hyp.5746

Schmid S, Burkard R, Frumau KFA et al (2011) Using eddy covariance and stable isotope mass balance techniques to estimate fog water contributions to a Costa Rican cloud forest during the dry season. Hydrol Proc 25:429–437. https://doi.org/10.1002/hyp.7739

Schreier HP, Brown S, MacDonald JR (eds) (2006) Too little and too much: water and development in a Himalayan watershed. Institute for Resources and Environment, University of British Columbia, Vancouver, Canada, p 258. https://idl-bnc-idrc.dspacedirect.org/server/api/core/bitstreams/b14d407e-93f4-4b33-a682-1e3cea817ad5/content

Schwemle R, Leistert H, Steinbrich A, Weiler M (2024) RoGeR v3.0.5 – a process-based hydrological toolbox model in Python. Geosci Model Develop 17:5249–5262. https://doi.org/10.5194/gmd-17-5249-2024

Scott DF, Bruijnzeel LA, Mackensen J (2005) The hydrological and soil impacts of forestation in the tropics. In: Bonell M, Bruijnzeel LA (eds) Forests, water and people in the humid tropics. Cambridge University Press, Cambridge, pp 622–650. https://doi.org/10.1017/CBO9780511535666.032

Searle MP, Treloan PJ (2019) Introduction to Himalayan tectonics: a modern synthesis. Geol Soc London Spec Publ 483:1–17. https://doi.org/10.1144/SP483-209-20

Sett T, Nandy S, Patel NR et al (2023) Spatio-temporal dynamics of water use efficiency over forest ecosystems using time series satellite data and carbon flux measurements. For Ecol Manag 548: 121385. https://doi.org/10.1016/j.foreco.2023.121385

Shakti PC, Pradhananga D, Ma W, Wang P (2014) An overview of glaciers distribution in the Nepal Himalaya. Hydro Nepal J Water Energy Environ 13:20–27. https://doi.org/10.3126/hn.v13i.10034

Sharda VN, Samraj P, Chinnamani S et al (1988) Hydrological behaviour of the Nilgiri sub-watersheds as affected by bluegum plantations, part II. Monthly water balances at different rainfall and runoff probabilities. J Hydrol 103:347–355. https://doi.org/10.1016/0022-1694(88)90143-6

Sharma P, Rai SC (2004) Streamflow, sediment and carbon transport from a Himalayan watershed. J Hydrol 289:190–203

Sharma SD, Singh BR, Sitaula B, Gupta MK (2006) Influence of land cover on soil properties in a watershed in the Lesser Himalayas. Ann For 14:16–34

Sharma E, Bhuichar S, Xing M, Kothyari BP (2007) Land use change and its impact on hydro-ecological linkages in Himalayan watersheds. Trop Ecol 48:151–161. https://www.researchgate.net/publication/241612349

Sharma B, Nepal S, Gyawali D et al (2016) Springs, storage towers, and water conservation in the Midhills of Nepal. ICIMOD Working Paper 2016/3. ICIMOD, Kathmandu, p 45. https://doi.org/10.13140/RG.2.1.4142.4886

Sharma IP, Kumar S, Sharma JC (2021) Rainfall-runoff-soil loss relations under different land uses in the Mid-Hill region of Himachal Pradesh. Indian J Soil Conserv 28:91–97

Sheikh MA, Kumar M, Todaria NP, Pandey R (2020) Biomass and soil carbon along altitudinal gradients in temperate *Cedrus deodara* forests in Central Himalaya, India: implications for climate change mitigation. Ecol Indic 111:106025. https://doi.org/10.1016/j.ecolind.2019.106025

Shimizu T, Kumagai T, Kobayashi M et al (2015) Estimation of annual forest evapotranspiration from a coniferous plantation watershed in Japan (2): comparison of eddy covariance, water budget and sap-flow plus interception loss. J Hydrol 522:250–264. https://doi.org/10.1016/j.jhydrol.2014.12.021

Shiraki K, Tanaka N, Chatchai T, Suzuki M (2017) Water budget and rainfall to runoff processes in a seasonal tropical watershed in northern Thailand. Hydrol Res Lett 11:149–154. https://doi.org/10.3178/hrl.11.149

Shrestha DP (1997) Assessment of soil erosion in the Nepalese Himalaya, a case study in Likhu Khola valley, Middel mountain region. Land Husbandry 2:59–80. https://www.researchgate.net/publication/255578442

Shrestha UB, Gautam S, Bawa KS (2012) Widespread climate change in the Himalayas and associated changes in local ecosystems. PLoS One 7(5):e36741. https://doi.org/10.1371/journal.pone.0036741

Simonin KA, Santiago LS, Dawson TE (2009) Fog interception by Sequoia sempervirens (D. Don) crowns decouples physiology from soil water deficit. Plant Cell Environ 32:882–892. https://doi.org/10.1111/j.1365-3040.2009.01967.x

Singh RP (1987) Rainfall interception by *Pinus wallichiana* plantation in temperate region of Himachal Pradesh, India. Indian For 113:559–566. https://www.i-scholar.in/index.php/indianforester/article/view/9434

Singh A (2009) Characterizing runoff generation mechanism for modelling runoff and soil erosion in small watershed of Himalayan region. Masters thesis, International Institute for

Geo-Information Science and Earth Observation, Enschede, The Netherlands, and Indian Institute of Remote Sensing, Dehradun, India, p 86. https://www.iirs.gov.in/node/279

Singh RP, Sharma KC, Mathur HN et al (1983) Interception studies in *Cedrus deodara* Loudon plantation in Himachal Pradesh. Indian For 109:261–266. https://www.i-scholar.in/index.php/indianforester/article/view/10627

Singh SP, Adhikari BS, Zobel DB (1994) Biomass, productivity, leaf longevity, and forest structure in the Central Himalaya. Ecol Monogr 64:401–421. https://doi.org/10.2307/2937143

Singh N, Bhattacharya BK, Soni P et al (2013) Energy and water dynamics over young pine forest as influenced by climatic variability and land management practices. J Agrometeo 15:100–107. https://www.researchgate.net/publication/259625381

Singh N, Bhattacharya BK, Nanda MK et al (2014a) Radiation and energy balance dynamics over young chir pine (*Pinus roxburghii*) system in Doon of western Himalayas. J Earth Syst Sci 123: 1451–1465. https://doi.org/10.1007/s12040-014-0480-6

Singh N, Patel NR, Bhattacharya BK et al (2014b) Analyzing the dynamics and inter-linkages of carbon and water fluxes in subtropical pine (*Pinus roxburghii*) ecosystem. Agric For Meteo 197: 206–218. https://doi.org/10.1016/j.agrformet.2014.07.004

Song QH, Braekevelt E, Zhang YP et al (2017) Evapotranspiration from a primary subtropical evergreen forest in Southwest China. Ecohydrology 10:e1826. 1002/eco.1826

Sooryamol KR, Kumar S, Regina M, Raj AD (2022) Modelling climate change impact on soil erosion in a watershed of north-western lesser Himalayan region. J Sediment Environ 2022. https://doi.org/10.1007/s43217-022-00089-4

Song L, Boithias L, Sengtaheuanghoung O et al (2020) Understory limits surface runoff and soil loss in teak tree plantations of Northern Lao PDR. Water 12:2327. https://doi.org/10.3390/w120922327

Srinet T, Nandy S, Watham T et al (2022) Measuring evapotranspiration by eddy covariance method and understanding its biophysical controls in moist deciduous forest of northwest Himalayan foothills of India. Trop Ecol. https://doi.org/10.1007/s42965-021-00216-8

Srinet R, Nandy S, Patel NR (2019) Estimating leaf area index and light extinction coefficient using random forest regression algorithm in a tropical moist deciduous forest, India. Ecol Inform 52: 94–102. https://doi.org/10.1016/j.ecoinf.2019.05.008

Stahl C, Hérault B, Rossi V et al (2013) Depth of soil water uptake by tropical rainforest trees during dry periods: does tree dimension matter? Oecologia 173:1191–1201. https://doi.org/10.1007/s00442-013-2724-6

Stewart SB, Choden K, Fedrigo M et al (2017) The role of topography and the north Indian monsoon on mean monthly climate interpolation within the Himalayan Kingdom of Bhutan. Int J Climatol 37:897–909. https://doi.org/10.1002/joc.5045

Stogsdill WR, Wittwer RF, Hennessey TC, Dougherty PM (1992) Water use in thinned loblolly pine plantations. For Ecol Manag 50:233–245. https://doi.org/10.1016/0378-1127(92)90338-A

Subba Rao BK, Dabral BG, Ramola BC (1973) Quality of water from forested watersheds. Indian For 99:682–690. https://www.i-scholar.in/index.php/indianforester/article/view/10232

Subba Rao BK, Ramola BC, Sharda VN (1985) Hydrologic response of a forested mountain watershed to thinning: a case study. Indian For 111:418–431. https://www.i-scholar.in/index.php/indianforester/article/view/10232

Swank WT, Vose JM, Elliott KJ (2001) Long-term hydrologic and water quality responses following commercial clearcutting of mixed hardwoods on a southern Appalachian catchment. For Ecol Manag 143:163–178. https://doi.org/10.1016/S0378-1127(00)00515-6

Tambe S, Kharel G, Arrawata ML et al (2012) Reviving dying springs: climate change adaptation experiments from the Sikkim Himalaya. Mnt Res Dev 32:62–72. https://doi.org/10.1659/MRD-JOURNAL-D-11-00079.1

Tanaka N, Tantasirin C, Kuraji K et al (2005) Inter-annual variation in rainfall interception at a hill evergreen forest in northern Thailand. Bull Tokyo Univ For 113:11–44. https://www.researchgate.net/publication/29778106

Tanaka N, Levia DF, Igarashi Y et al (2015) Throughfall under a teak plantation in Thailand: a multifactorial analysis on the effects of canopy phenology and meteorological conditions. Int J Biometeorol 59:1145–1156. https://doi.org/10.1007/s00484-014-0926-1

Tanaka N, Levia DF, Igarashi Y et al (2017) What factors are most influential in governing stemflow production from plantation-grown teak trees? J Hydrol 544:10–20. https://doi.org/10.1016/j.jhydrol.2016.11.010

Tarafdar S, Dutta S (2023) Long-term decline in rainfall causing depletion in groundwater aquifer storage sustaining the springflow in the middle-Himalayan headwaters. J Earth Syst Sci 132: e124. https://doi.org/10.1007/s12040-023-02136-8

Tarafdar S, Bruijnzeel LA, Kumar B (2019) Improved understanding of spring and stream water responses in headwaters of the Indian lesser Himalaya using stable isotopes, conductivity and temperature as tracers. Hydrol Sci J. https://doi.org/10.1080/02626667.2019.1600698

Tague CJ, Band LE (2004) RHESSys: Regional Hydro Ecologic Simulation System - an object-oriented approach to spatially distributed modeling of carbon, water and nutrient cycling. Earth Interact 8:19. https://doi.org/10.1175/1087-3562(2004)81:RRHSSO2.0.CO;2

Tiwari PC (1988) Kumaun Himalaya: land use system and environmental degradation. Indian Geogr J 63:58–63

Tiwari PC, Joshi B (2012) Environmental changes and sustainable development of water resources in the Himalayan headwaters of India. Water Resour Manag 26:883–907. https://doi.org/10.1007/s11269-011-9825-y

Tiwari KR, Sitaula BK, Bajracharya RM, Borresen T (2009) Runoff and soil loss responses to rainfall, land use, terracing and management practices in the Middle Mountains of Nepal. Acta Agric Scand Sect B - Soil Plant Sci 59:197–207. https://doi.org/10.1080/09064710802006021

Tóbon C, Bouten IW, Dekker SC (2000) Forest floor water dynamics and root water uptake in four forest ecosystems in Northwest Amazonia. J Hydrol 237:169–183. https://doi.org/10.1016/S0022-1694(00)00302-4

Toky OP, Ramakrishnan PS (1981) Run-off and infiltration losses related to shifting agriculture (Jhum) in Northeastern India. Environ Conserv 8:313–321. https://doi.org/10.1017/s0376892900028125

Tomar JMS, Kar SK, Ahmed A et al (2022) Hydrological fluxes from Sal (*Shorea robusta*) forests under various disturbance gradients of Himalayan foothills. Ecohydrol Hydrobiol 22:541–552. https://doi.org/10.1016/j.ecohyd.2021.12.002

Tripathi P, Patel NR, Kushwaha SPS, Dadhwal VK (2014) Upscaling of leaf area index in Terai forest plantations using fine- and moderate-resolution satellite data. Int J Remote Sens 35:7749–7762. https://doi.org/10.1080/01431161.2014.976886

Tyagi JV, Qazi NQ, Rai SP, Singh MP (2013) Analysis of soil moisture variation by forest cover structure in lower western Himalayas, India. J For Res 24:317–324. https://doi.org/10.1007/s11676-013-0355-8

Upadhyay H, Vinke K, Bhardwaj S et al (2021) Locked houses, fallow lands: climate change and migration in Uttarakhand, India. Potsdam Institute for Climate Impact Research, Potsdam, Germany (PIK) and The Energy & Resources Institute, New Delhi (TERI), p 106. https://www.international-climate-initiative.com/en/iki-media/publication/locked_houses_fallow_lands_climate_change_and_migration_in_uttarakhand_india/

Upgupta S, Sharma J, Jayaraman M et al (2015) Climate change impact and vulnerability assessment of forests in the Indian Western Himalayan region: a case study of Himachal Pradesh, India. Clim Risk Manag 10:63–76. https://doi.org/10.1016/j.crm.2015.08.002

Valdiya KS, Bartarya SK (1989) Diminishing discharges of mountain springs in a part of Kumaun Himalaya. Curr Sci 58:417–426. https://www.researchgate.net/publication/277843040

Valdiya KS, Bhattarya SK (1991) Hydrological studies of springs in the catchment of the Gaula River, Kumaon lesser Himalaya, India. Mnt Res Dev 11:239–258. https://doi.org/10.2307/3673618

Van Dijk AIJM, Bruijnzeel LA (2001) Modelling rainfall interception by vegetation of variable density using an adapted analytical model. Part 1. Model description. J Hydrol 247:230–238. https://doi.org/10.1016/S0022-1694(01)00392-4

Van Meerveld HJ, Zhang J, Tripoli R, Bruijnzeel LA (2019) Effects of reforestation of a degraded *Imperata* grassland on dominant flow pathways and streamflow responses in Leyte, The Philippines. Water Resourc Res 55:WR023896. https://doi.org/10.1029/2018WR023896

Van Stan II JT, Coenders-Gerrits M, Dibble M et al (2017) Effects of phenology and meteorological disturbance on litter rainfall interception for a *Pinus elliottii* stand in the Southeastern United States. Hydrol Proc 31:3719–3728. https://doi.org/10.1002/hyp.11292

Verma AK, Garkoti SC (2019) Population structure, soil characteristics and carbon stock of the regenerating banj oak forests in Almora, Central Himalaya. For Sci Technol 15:117–127. https://doi.org/10.1080/21580103.2019.1620135

Vertessy RA, Watson FGR, O'Sullivan SK et al (1998) Predicting water yield from mountain ash forest catchments. CRCCH Industry Rep 98/4. Cooperative Research Centre for Catchment Hydrology, CSIRO, Canberra, p 38

Vertessy RA, Watson FGR, O'Sullivan SK (2001) Factors determining relations between stand age and catchment water balance in mountain ash forests. For Ecol Manag 143:13–26. https://doi.org/10.1016/S0378-1127(00)00501-6

Wallace JS, Young A, Ong CK (2005) The potential of agroforestry for sustainable land and water management. In: Bonell M, Bruijnzeel LA (eds) Forests, water and people in the humid tropics. Cambridge University Press, Cambridge, UK, pp 652–670

Wallace JS, Macfarlane C, McJannet D et al (2013) Evaluation of forest interception estimation in the continental scale Australian Water Resources Assessment – Landscape (AWRA-L) model. J Hydrol 499:210–223. https://doi.org/10.1016/j.jhydrol.2013.06.036

Walsh RPD, Voigt PJ (1977) Vegetation litter: an underestimated variable in hydrology and geomorphology. J Biogeogr 4:253–274. https://doi.org/10.2307/3038060

Wangda P, Ohsawa M (2006) Gradational forest change along the climatically dry valley slopes of Bhutan in the midst of humid eastern Himalaya. Plant Ecol 186:109–128. https://doi.org/10.1007/s11258-006-9116-5

Wangdi N (2016) Drought stress tolerance and climate change adaptation potentials of main forest types in Bhutan. Dissertation,. University of Natural Resources and Life Sciences (BOKU), Vienna, Austria, p 192

Wani MA, Manhas RK (2012) Rainfall interception in relation to the tree architecture of *Pinus wallichiana* A.B. Jackson. Curr Sci 103:821–827. https://www.currentscience.ac.in/Volumes/103/07/0821.pdf

Ward RC (1984) On the response to precipitation of headwater streams in humid areas. J Hydrol 74:171–189. https://doi.org/10.1016/0022-1694(84)90147-1

Waterloo MJ (1994) Water and nutrient dynamics of *Pinus caribaea* plantations on former grassland soils in Viti Levu, Fiji. Dissertation, Vrije Universiteit Amsterdam, The Netherlands, p 478. https://www.researchgate.net/publication/234115361

Webb EL, Gautam AP (2001) Effects of community forest management on the structure and diversity of a successional broadleaf forest in Nepal. Int For Rev 3:146–157. https://www.jstor.org/stable/42609370

Wen XH, Gonzalez-Hernandez JJ (1996) Upscaling hydraulic conductivities in heterogeneous media: an overview. J Hydrol 183:9–32

Wester HTJ (2013) Hydraulic survey to asses water scarcity in the Salyan district, middle Mountains, western Nepal. Masters thesis, VU University Amsterdam, the Netherlands 52

White T, Brantley S, Banwart S et al (2015) The role of critical zone observatories in critical zone science. Dev Earth Surf Proc 19:15–78. https://doi.org/10.1016/B978-0-444-63369-9.00002-1

Wijngaard RR, Lutz AF, Nepal S et al (2017) Future changes in hydro-climatic extremes in the Upper Indus, Ganges and Brahmaputra River basins. PLoS One 12:e0190224. https://doi.org/10.1371/journal.pone.0190224

Wilson AM, Jetz W (2016) Remotely sensed high-resolution global cloud dynamics for predicting ecosystem and biodiversity distributions. PLoS Biol 14(3):e1002415. https://doi.org/10.1371/journal.pbio.1002415

Wolf JHD (1993) Diversity patterns and biomass of epiphytic bryophytes and lichens along an altitudinal gradient in the Northern Andes. Ann Missouri Bot Garden 80:928–960. https://doi.org/10.2307/2399938

Woods RA, McDonnell JJ (2010) Hillslope threshold response to rainfall: A field based forensic approach. J Hydrol 393:65–76. https://doi.org/10.1016/j.jhydrol.2009.12.015

Woolhiser DA, Smith RE, Giraldez JV (1996) Effects of spatial variability of saturated hydraulic conductivity on Hortonian overland flow. Water Resour Res 32:671–678. https://doi.org/10.1029/95WR03108

Yadav YP, Vasisth HB (1989) Infiltration capacity of forest soils under *Cryptomeria japonica*. Indian For 115:435–441. https://www.i-scholar.in/index.php/indianforester/article/view/9067

Yang B, Wen X, Sun X (2015) Seasonal variations in depth of water uptake for a subtropical coniferous plantation subjected to drought in an East Asian monsoon region. Agric For Meteo 201:218–228. https://doi.org/10.1016/j.agrformet.2014.11.020

Zaw Y, Oue H (2024) Influences of tree character on throughfall and stemflow from rainfall and fog in Popa Mountain Park, Myanmar. J For Res 35:118. https://doi.org/10.1007/s11676-024-01760-4

Zhan L, Chen J, Zhang C et al (2020) Fog interception maintains a major waterfall landscape in Southwest China revealed by isotopic signatures. Water Resour Res 56:e2019WR025628. https://doi.org/10.1029/2019WR025628

Zhang K, Kimball JS, Running SW (2016) A review of remote sensing based actual evapotranspiration estimations. WIREs Water 3(6):834–853. https://doi.org/10.1002/wat2.1168

Zhang J, Bruijnzeel LA, Tripoli R, van Meerveld HJ (2018) Water budget and run-off response of a tropical multispecies "reforest" and effects of typhoon disturbance. Ecohydrology. https://doi.org/10.1002/eco.2055

Zhang J, Bruijnzeel LA, Quiñones MC et al (2019) Soil physical characteristics of a degraded tropical grassland and a 'reforest': implications for runoff generation. Geoderma 333:163–177. https://doi.org/10.1016/j.geoderma.2018.07.022

Zhao P, Tang X, Zhao P et al (2013) Identifying the water source for subsurface flow with deuterium and oxygen-18 isotopes of soil water collected from tension lysimeters and cores. J Hydrol 503:1–10. https://doi.org/10.1016/j.jhydrol.2013.08.033

Zhao P, Tang X, Zhao P et al (2018) Temporal partitioning of water between plants and hillslope flow in a subtropical climate. Catena 165:133–144. https://doi.org/10.1016/j.catena.2018.01.031

Zhao L, Meng P, Zhang J et al (2022) Effect of slopes on rainfall interception by leaf litter under simulated rainfall conditions. Hydrol Proc 36:e14659. https://doi.org/10.1002/hyp.14659

Zhu X, Liu WJ, Jiang XJ et al (2018) Effects of land-use changes on runoff and sediment yield: implications for soil conservation and forest management in Xishuangbanna, Southwest China. Land Degrad Dev 29:2962–2974. https://doi.org/10.1002/ldr.3068

Zhu X, Zhang W, Jiang XJ et al (2021) Conversion of primary tropical rainforest into rubber plantation degrades the hydrological functions of forest litter: insights from experimental study. Catena 200:105172. https://doi.org/10.1016/j.catena.105172

Zimmermann B, Zimmermann A, Schreckenbach HL et al (2013) Changes in rainfall interception along a secondary forest succession gradient in lowland Panama. Hydrol Earth Syst Sci 17:4659–4670. https://doi.org/10.5194/hess-17-4659-2013

Zobel DB, Garkoti SC, Singh SP (2001) Leaf conductance of primary and mature leaves of *Pinus roxburghii*: a comparison. J For 6:1–5. https://doi.org/10.1007/BF02762715

Zokaib S, Naser G (2011) Impacts of land uses on runoff and soil erosion A case study in Hilkot watershed Pakistan. Int J Sedim Res 36:343–352. https://doi.org/10.1016/S1001-6279(11)60098-X

Zwartendijk BW, van Meerveld HJ, Ghimire CP et al (2020) Soil water- and overland flow dynamics in a tropical catchment subject to long-term slash-and-burn agriculture. J Hydrol 582:124287. https://doi.org/10.1016/j.jhydrol.2019.124287

Zwartendijk BW, van Meerveld HJ, Teuling AJ et al (2023) Modelling the hydrological responses of a headwater catchment under shifting cultivation in upland eastern Madagascar. EGU General Assembly 2023, abstract no. EGU-23-8987

Zwartendijk BW, Bruijnzeel LA, Ghimire CP et al (2024) Effects of (assisted) natural regeneration on infiltrability and preferential flow pathways in the Khasi Hills (Meghalaya, NE India). Land Degrad Dev 35:e5445. https://doi.org/10.1002/ldr.5445

Tree Seed Ecology in the Himalayan Forest Ecosystem: How Much Do We Know? 13

Shyam S. Phartyal

Abstract

The Himalayan arc has wide elevational and precipitation gradients that lack parallels elsewhere. This wide gradient encompasses a variety of ecosystems from tropical to alpine, which support a high degree of floral/faunal diversity that makes the Himalayas home to two global biodiversity hotspots. Despite the high diversity of endemic trees, the Himalayas is one of the underexplored mountain forest ecosystems in terms of seed ecology that demands much more attention and offers an opportunity to study how seed traits modulate tree assembly along altitudinal and longitudinal gradients of the Himalayas. This chapter aimed to collate fruit/seed traits of Himalayan trees from published literature and to unravel how much seed ecological traits differ with wet-dry season and along large macroclimatic gradients of the Himalayas. The results revealed that 23% of the tree species fully and 57% partially synchronised seed dispersal and germination with monsoon. Desiccation-sensitive, non-dormant, and wet-season dispersed seeds germinate primarily in the monsoon, whereas desiccation-tolerant, dormant, and dry-season dispersed seeds germinate within or outside the monsoon season. The percentage of fleshy-fruited, zoochorous, desiccation-sensitive, non-dormant seeded species was higher in the eastern Himalayas, while the percentage of dry-fruited, anemochorous, desiccation-tolerant, dormant seeded species was higher in the western Himalayas; however, these differences were not significant. Contrary to the longitudinal trends, there were inconsistencies in the altitudinal trends in seed traits of the species from the lower to the higher Himalayas. Seed dormancy increases from monsoon-synchronised to -desynchronised species, and approximately 75% of species in intermediate/dry habitats produce dormant seeds. Deciduous and evergreen

S. S. Phartyal (✉)
Department of Forestry, Mizoram University, Aizawl, India

S. P. Singh, R. Thadani (eds.), *Himalayan Forests in a Changing World*,
https://doi.org/10.1007/978-981-95-0978-2_13

species with long-leaf lifespans produced more anemochorous, desiccation-tolerant, and dormant seeds than those evergreen species with short-leaf lifespans, which may drive their desynchronisation or synchronisation with monsoon, respectively. The study concluded a weak/non-significant longitudinal and altitudinal trend in tested seed ecological traits and that the microscale level variability of heterogenous landforms in the Himalayan arc may have masked seed trait variation prevalent at the macroscale level. However, even identifying weak trends in the seed ecological traits of tree species along the Himalayan gradient will help us better understand their regeneration ecology and vulnerability to climate. Finally, this chapter also proposed a few future research agendas requiring urgent attention to better understand the Himalayan forest ecosystem.

Keywords

Altitudinal gradient · Forest ecosystem · Functional traits · Himalayan arc · Latitudinal gradient · Monsoon synchronisation · Regeneration ecology · Seed desiccation behaviour · Seed dispersal syndrome · Seed dormancy and germination · Seed regeneration traits

Introduction

The Himalayan ecosystem is driven by heterogenous landforms with wide climatic, biogeographical, geological, and cultural gradients if one moves from east to west and from southern foothills to northern peaks. This wide gradient is dominated by forested landscapes of the Indomalayan and Palearctic biogeographical realms (Dinerstein et al. 2017). The Himalayan forest ecosystem differs significantly from temperate forest ecosystems in other parts of the world for its unique features like a high proportion of evergreen species with multiple leaf flush, varying leaf lifespan, high litter decomposition, and net primary productivity (Zobel and Singh 1997; Singh and Thadani 2024). The high variations in landforms, elevation, and precipitation make the Himalayan forest ecosystem high in endemism (Wani et al. 2024), with a unique and rich diversity of flora and fauna (Rawal et al. 2013). The change in precipitation pattern from wet to dry along the east-west arc of the Himalayas and the growing season from long to short along the south-north elevational gradients are known to change forest types from monodominant to mixed species stands, so their strategies to regenerate from the seeds (Singh et al. 2017a, b).

Although at the macroscale level, the variation in precipitation, temperature, and other climatic features from east to west and lower to higher Himalayas may be gradual, when we look at them at the microscale level, especially variation in edaphic or topographic features (like slope, aspects, rock strata, soil depth, soil moisture, soil temperature, and other soil physiochemistry), they may change abruptly within a short geographic distance. It is perceived that a warmer macroclimate is often strongly buffered by the presence of site-specific vegetation in geology, hydrology, and topography that results in a change in microclimate with altered temperature and moisture extremes (Graae et al. 2022). Furthermore, many organisms living on the soil surface or buried close to the soil surface experience only microscale soil temperatures and are less likely to relate themselves to free-air temperatures at the macroscale (Lembrechts et al. 2020, 2022). Recent studies suggest that in comparison to plants' adult life stage, microclimatic environmental filters (like soil temperature, moisture, pH, and light at the soil surface) play a more crucial role in their early life stage (seed dormancy, germination, seedling survival and growth) transitions (Grubb 1977; Walck et al. 2011; Bewley et al. 2012; Baskin and Baskin 2014; Fraaije et al. 2015; Jiménez-Alfaro et al. 2016). Thus, how gradual to abrupt changes in regional to local environment along the east-west longitudinal and south-north altitudinal gradients of the Himalayas modulate tree assembly requires a deeper understanding of their seed ecology.

Seeds play crucial roles in the early stages of plant life. They help them adapt to changing environments, maintain existing populations, disperse themselves in unexplored but suitable habitats via seed dispersal networks, and establish themselves as a new population (Poschlod et al. 2013; Saatkamp et al. 2019; Phartyal et al. 2020; Rosbakh et al. 2020, 2022, 2023; Fernández-Pascual et al. 2023). In general, the survival strategy of plants depends on when and how a species disperses its dispersules, the ability of dispersules to withstand desiccation and remain in a dormant or quiescence stage during harsh environments, and finally begins to germinate when the environment is suitable for seedling emergence and its subsequent survival (Baskin and Baskin 2014). Thus, both macroclimate and microclimate govern a species distribution and community assemblage through regional to local environmental filters by regulating the seasonal rhythms of seed maturation, dispersal, and dormancy, which eventually play a crucial role in synchronising the timing of seed germination with the optimal growing season suitable for the survival of newly emerged seedlings. For example, tree species of the outer Himalayas, dependent on monsoon precipitation, synchronise their seed dispersal and germination with the monsoon season. On the contrary, seeds of tree species of the inner Himalayas dependent on westerly precipitation (snowfall) disperse and germinate outside the monsoon season in autumn, winter, and spring, thus desynchronising from the monsoon pattern (Singh et al. 2017a, b).

In projected climate change scenarios, frequent heat waves (Schär et al. 2004), high-temperature extremes, and drought episodes are expected (Cubasch et al. 2001), which may be coupled with up to 20% increase in annual precipitation in many mountain regions, including the Hindu Kush and Himalayan region (Hock et al. 2019). Changes in precipitation with warming are predicted to influence seed production, dormancy, germination, seed persistence in the soil seed bank, and the proportion of germinated seeds that survive up to the seedling stage as a moist environment with cool/warm temperature is required for cold/warm stratification for seed dormancy break (Baskin and Baskin 2014; Walck and Hidayati 2022). Therefore, partitioning the effects of climate change variables, especially extreme events, from seed development to seedling establishment stages would allow us to understand better the underlying ecophysiological mechanisms of the early stages of plant life (Walck and Hidayati 2022). For example, high intensity/short duration rainfall events could wash a cohort of seeds out of their natural habitat or bury them deep in the soil, eventually affecting seedling emergence; warmer winter spells interrupt cold stratification, which leads to seed mortality; increased wildfires influence both seed mortality and seed germinability in the soil seed bank; and interactions between heat waves and precipitation affect warm stratification/after-ripening and mortality of desiccation-sensitive seeds (Walck and Hidayati 2022 and references therein). Therefore, plant regeneration from seeds results from a complex interaction of traits with pre- and post-germination environment conditions (Huang et al. 2018; Saatkamp et al. 2019), and we lack a clear understanding of how they will function and interact within a climate change framework (Saatkamp et al. 2019).

The amount/duration of snowfall in mountain ecosystems is projected to decline by 10–40% by 2031–2050 (Hock et al. 2019), and it plays a central role in temperate ecosystems by protecting seeds and seedlings from freezing air and fluctuating soil temperatures during harsh winters (Decker et al. 2003; Simons et al. 2010). Seed dormancy break and germination of some temperate species occur under snow cover, and those of other species coincide with snowmelt in spring, providing ample soil moisture for seedling establishment (Phartyal et al. 2009, 2012, 2014; Kondo et al. 2011, Baskin and Baskin 2022). Although seed traits play a crucial role in synchronising plant distribution and persistence along varied environmental gradients, our understanding of how climate change affects plant regeneration from seeds is limited and remains neglected (Poschlod et al. 2025), especially in Himalayan forest ecosystems. Maharjan et al. (2021) tried to find which plant traits best predict tree species distributions along the eastern Himalayan elevational gradient, and their study revealed that multiple trade-offs in plant size, hydraulics, and light competitiveness shape species distribution along the elevational gradient as these traits increase resource conservation and safety against harsh environment at higher elevations. However, as noticed in most of the other ecological studies and reported in Rosbakh et al. (2020, 2023) and Phartyal et al. (2020), the study by Maharjan et al. (2021) also suffered from ignorance of the regenerative (seed) traits that may have equally or better-predicted species distribution along environmental gradients than vegetative traits alone (Poschlod et al. 2025).

In a seasonal forest ecosystem such as the Himalayas, seed dispersal plays a fundamental role in colonising new habitats and maintaining population dynamics and community structure (Van der Pijl 1982; Howe and Smallwood 1982; Sautu et al. 2006). The specialised diaspore morphology in several plant species enables them to make efficient use of available biotic and abiotic dispersal vectors depending on their fruiting peak (Van der Pijl 1982; Wenny 2001; Poschlod et al. 2010, 2013) and the position of species within the canopy structure (Griz and Machado 2001; Butler et al. 2007; Jara-Guerrero et al. 2011). Furthermore, seed dispersal syndrome (see Box 13.1 for details) is known to vary with the season of peak fruiting. For example, anemochorous (wind-disperse) species disperse seeds towards the end of the dry season; autochorous (self-disperse) species disperse seeds mainly during the dry season, while zoochorous (animal-disperse) species disperse fruit throughout the year, but seed dispersal peaks in the rainy season (Van Schaik et al. 1993; Mendoza et al. 2017; Escobar et al. 2018; Miranda et al. 2023). Similarly, within the tree canopy structure, species with autochory and zoochory are widely frequent among under-canopy species, whereas differences in wind speed among top-canopy species lead to anemochory (Griz and Machado 2001; Butler et al. 2007). Furthermore, seed dispersal syndrome is also associated with species deciduousness (leaf lifespan) and habitat moisture conditions, as dryness favours leaf shedding (Singh and Thadani 2024) and wetness favours fleshy fruit ripening. Therefore, anemochory is supposed to be relatively common in dry and deciduous habitats, whereas zoochory dominates in wet and evergreen habitats (Frankie et al. 1974; Howe and Smallwood 1982; Willson et al. 1990; Griz and Machado 2001).

Box 13.1: Definitions of Technical Terms Used in the Text

Term	Definition
Seed dispersal syndrome	A set of morpho-physio-chemical traits of dispersules (fruit/seed unit as it dispersed) facilitates seed dispersal by a particular or a combination of dispersal vectors.
Anemochory	Seed dispersal by wind, especially if dispersules are small or have wings, pappus, or long hair.
Autochory	Seed dispersal through forceful ejection or explosion of fruits (capsules/pods), often due to osmotic pressure.
Polychory	Dispersules with polymorphic dispersal structures often disperse through two or more dispersal vectors that could operate sequentially.
Zoochory	Seed dispersal by frugivorous animals through the internal gut pass (endozoochory), external body attachment (epizoochory), and deliberately carried in the mouth/beak (synzoochory).
Desiccation-sensitive seeds	Seeds dispersed with high moisture content are susceptible to dehydration and low temperature and lose viability quickly (short-lived).
Desiccation-tolerant seeds	Seeds dispersed with low moisture content withstand dehydration and low temperatures without affecting viability (long-lived).
Seed longevity	The storability of a seed under ex-situ conditions, i.e. how long a seed retains its viability under a controlled dry environment (dry storage).
Seed lifespan	Seed timespan from dispersal to germination or loss of germinability due to seed ageing, decay, and predation. Both intrinsic (seed characteristics) and extrinsic (environmental factors) influenced seed lifespan in wet/dry soil (wet longevity).
Seed dormancy (D)	Fresh, mature, viable seed prevents germination under environmental conditions that are otherwise optimal for germination or germinate under a narrow set of environmental conditions.
Non-dormancy (ND)	Fresh, mature seed germinates over the broadest environmental conditions in a short period (usually in <4 weeks).
Physical dormancy (PY)	Fresh, mature seed fails to germinate because one or more layers of impermeable water palisade in the seed coat (or fruit) restrict the embryo from imbibing water. Dormancy break involves physical disruption in the seed coat (physical problem).

(continued)

Box 13.1 (continued)

Term	Definition
Morphological dormancy (MD)	Fresh, mature seed fails to germinate due to the presence of an underdeveloped embryo that must grow to a species-specific critical length to begin the emergence of the radicle (morphological problem).
Physiological dormancy (PD)	Fresh, mature seed fails to germinate due to the low growth potential of a fully developed embryo, but the growth potential increases as dormancy break occurs (physiological problem).
Morpho-physiological dormancy (MPD)	Fresh, mature seed fails to germinate first due to an underdeveloped embryo and then due to the low growth potential of a fully developed embryo (morpho-physiological problem).
Combinational dormancy (PY + PD)	Fresh, mature seeds do not germinate due to physical and physiological problems.
Soil seed bank	A reserve of viable seeds is buried in the soil, often in the dormant stage. They can provide a memory of past vegetation and serve as a repository of the structure of future vegetation.
Seedling bank	A cohort of seedlings persist in the stunted form under the canopy shades and wait for the creation of canopy gaps to likely thrive in the changed light environment.

Furthermore, the interaction of seasonality, seed dispersal syndrome, and seed dormancy mechanism (see Box 13.1 for detail) is known to play a crucial role in controlling seed germination time and seedling establishment (Garwood 1983; Sautu et al. 2007; Baskin and Baskin 2014; Singh et al. 2017a, b). In a seasonal tropical ecosystem, a species seed germination time in the early rainy season is considered to be controlled by its dispersal timing, in the late rainy season by seed dormancy, while in dry season, germination time of species is supposed to be controlled equally by seed dispersal timing and dormancy (Garwood 1983; Sautu et al. 2007). Often, those species that disperse seeds early in the rainy season are non-dormant and germinate quickly, and those that disperse at the end of the rainy or beginning of the dry season are dormant and germinate only during the following wet or rainy season (Garwood 1983; Blakesley et al. 2002; Sautu et al. 2007; Salazar et al. 2011; Silveira et al. 2012). However, we know little about how these ecological processes work in the Himalayan forest ecosystem; therefore, it is worth investigating them, as Himalayan forests have numerous unique characteristics that make them closer to tropical forests with respect to long growing seasons, nutrient turnover rate, high productivity, and a high proportion of evergreen trees (Zobal and Singh 1997; Singh and Thadani 2024).

In general, plant regeneration is based on the lifespan of seeds, which is influenced by both intrinsic seed characteristics (desiccation tolerance, dormancy, and longevity) and the extrinsic environment (Thompson et al. 1997; Leck 2012; Marques et al. 2018; Baskin and Baskin 2022). Seeds with long lifespans tend to be desiccation-tolerant and dormant, orchestrating an optimal strategy that is suitable not only for the initiation of the germination process but also for the survival of seedlings (Donohue et al. 2010; Marques et al. 2018). In a habitat with abundant and constant moisture regimes and highly predictable seasons, the probability of producing seeds with desiccation tolerance, dormant, and long lifespan is low because having these seed traits requires a high allocation of resources but does not serve any evolutionary ecological advantages. Therefore, such predictable seasonal habitats often dominate with specialist species and produce cost-effective, large, desiccation-sensitive, short-lived, non-dormant, moist seeds (Garwood 1989; Vázquez-Yanes and Orozco-Segovia 1993; Daws et al. 2005; de Souza et al. 2015; Marques et al. 2018; Singh et al. 2017a). Alternative ecological theory suggests that species that disperse desiccation-sensitive, non-dormant, fast-germination seeds take these traits as an ecological adaptation to avoid seed predation and pathogen infection (see Dalling et al. 2020 and references therein), but risk an entire cohort of newly emerged seedlings to predation (Daws et al. 2005). The association between desiccation sensitivity of large, moist seeds of shade-tolerant species with dormancy could be favoured by combining evolutionary biotic forces in the seed and seedling stages to minimise seedling predation risk. These seeds would not dry in the shaded understory, while a delay in germination would allow some seedlings to escape predation as part of the seedling bank. Therefore, large, moist, short-lived, dormant seeds are more frequent in the rainy season, especially in the late rainy season, with significant differences only for moisture content and kind of dormancy (Garwood 1983; Daws et al. 2005; Sautu et al. 2006, 2007). It is observed that the proportion of non-dormant species (Rosbakh et al. 2023) and desiccation-sensitive species decreases as the landscape/biome climate dries (Tweddle et al. 2003; Baskin and Baskin 2014).

Few studies have quantified seasonal variation in seed ecological traits at the community or ecosystem level (Ng 1978; Garwood 1983; Khurana and Singh 2001; Blakesley et al. 2002; Thapliyal and Phartyal 2005; Sautu et al. 2006, 2007; Salazar et al. 2011; Silveira et al. 2012; Yang et al. 2013; Singh et al. 2017a, b, Rosbakh et al. 2022), but hardly any studies exist at the broad landscape level such as Himalayan arc with a high degree of environmental heterogeneity. The lack of a large-scale study hinders our understanding of the regeneration ecology of the highly diverse Himalayan forest ecosystem. Therefore, this chapter aims to compile information on fruit/seed traits of Himalayan tree species from published literature and databases and to collate the available knowledge to unravel how the regeneration strategies (an overlooked aspect) of the Himalayan forest ecosystem vary spatially and temporally and cope with projected climate change. The heterogeneous landscapes of the Himalayas are expected to show a distinct set of seed ecological traits. Therefore, the study investigated and tested the following question and hypothesis: Do tree regeneration traits differ along the large macroclimatic gradients of the Himalayas? The main hypothesis is that a higher proportion of tree species bearing fleshy fruits, zoochorous dispersal, desiccation-sensitive, and non-dormant seeds would occur in the wet season and in the eastern and outer Himalayas, where precipitation is primarily derived from the summer monsoon. On the contrary, a higher proportion of species with dry fruits, anemochorous dispersal, desiccation-tolerant, and dormant

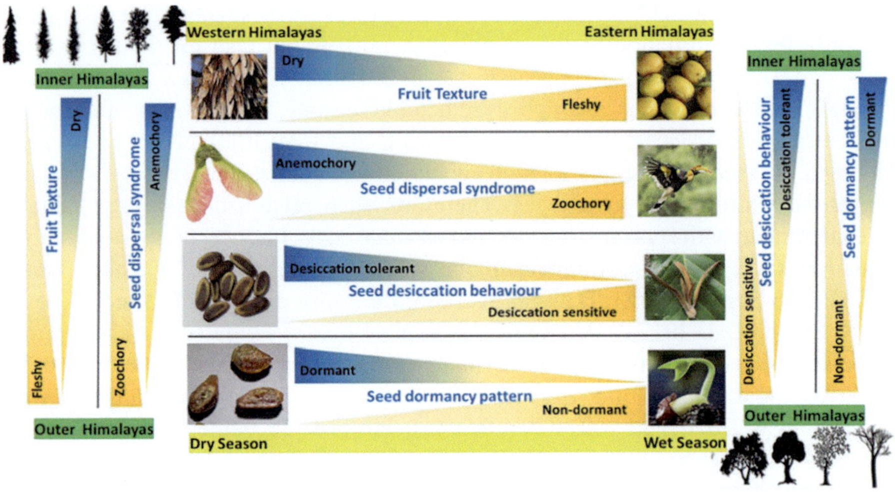

Fig. 13.1 Schematic profile of expected variations in seed ecological traits of Himalayan tree species along the wet-dry seasonal, east-west, and outer-inner Himalayan gradients. Triangles indicate a decreasing/increasing proportion of species from one gradient to another

seeds is expected in the drier western and inner Himalayas, which receive a substantial amount of precipitation from westerly winter rains or snowfall rather than from the summer monsoon (Fig. 13.1). To test the hypothesis, the variation in fruit texture, seed mass, dispersal syndrome, desiccation behaviour, and dormancy pattern for 201 Himalayan tree species was evaluated along (1) the dry to wet seasonal gradient, (2) the east-west (longitudinal) and south-north (altitudinal) gradient, (3) with habitat moisture conditions, and (4) with leaf phenology.

Compilation of Data on Species Distribution and Their Abiotic and Biotic Traits

The abiotic and biotic ecological traits, viz. altitudinal range, habitat moisture conditions, mature seed response to desiccation sensitivity/tolerance, seed mass, seed shedding and germination timing, and monsoon synchronisation strategy for 201 Himalayan tree species representing 50 families and 116 genera, were extracted from Singh et al. (2017a). For their study, they gathered information from Troup (1921), Champion and Seth (1968), Rao (1984), Ralhan et al. (1985), Singh and Singh (1987, 1992a, b), Khullar et al. (1991), Thapliyal et al. (2003), Thapliyal and Phartyal (2005), and the Royal Botanical Garden Kew Seed Information Database (now known as SER, INSR, RBGK, Seed Information Database 2017). Additional ecological traits such as the longitudinal distribution of species along the Himalayan arc, fruit texture, seed dispersal syndrome, seed dormancy pattern, leaf phenology, and leaf lifespan were collected and augmented in the existing database of Singh et al. (2017a). For the longitudinal distribution pattern, the species were initially

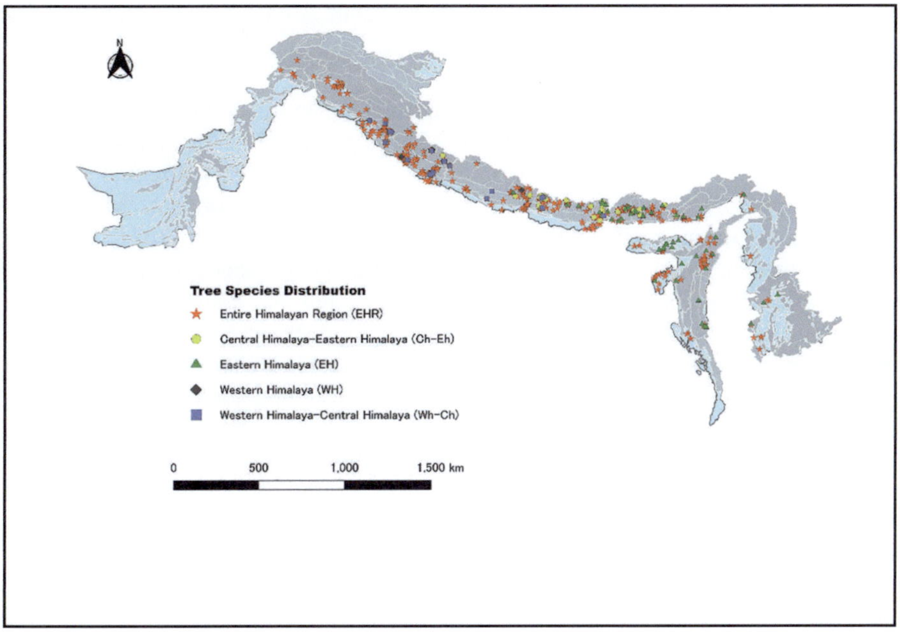

Fig. 13.2 Geographical distribution of the study tree species in the Himalayan arc

grouped into five categories according to their natural distribution: (1) along the entire Himalayan arc, (2) from the west to the central Himalayas, (3) from the centre to the eastern Himalayas, and only in (4) the eastern Himalayas and (5) the western Himalayas (Fig. 13.2). For analysis purposes, the study species were regrouped into two categories distributed in the eastern Himalayas (species with Indomalayan realm affinity plus more abundant from the central Himalayas eastward) and in the western Himalayas (species with Palearctic realm affinity plus more abundant from the central Himalayas to the westward) based on the individual species distribution map in the Global Biodiversity Information Facility (2023).

For each species, according to the botanical description in the taxonomical flora, the texture of the fruit was assigned as fleshy or dry depending on the presence or absence of conspicuous fleshy pericarps or appendages at maturity (Willson et al. 1990; Chen and Moles 2015; Chen et al. 2017). Seed dispersal syndrome was assigned as anemochory, autochory, polychory, and zoochory to each species according to the types of fruits and the morphological characteristics of the diaspores (Van der Pijl 1982; Jara-Guerrero et al. 2011; Escobar et al. 2018). The anemochorous species had a wing-like structure and were dispersed by wind; autochorous species were dispersed automatically through triggered pressure, explosion, or ballistic force. Some species have a polymorphic dispersal unit, and their seeds disperse through multiple dispersal vectors. Zoochorous species have fleshy fruit that provides rewards as food to dispersing animals or dispersules that have some hooks or awns that facilitate attachment with animal fur and feathers; any polychoric species with an animal as one of the seed vectors were grouped under zoochory for analysis purposes.

Information on seed dormancy was extracted from Baskin and Baskin (2014) and Rosbakh et al. (2020 and 2023). Each species was grouped into one of six classes of seed dormancy: non-dormant (ND), physiological dormant (PD), physical dormant (PY), morphological dormant (MD), morphophysiological dormant (MPD), and combinational dormant (PY + PD), following Baskin and Baskin (2021) (see Box 13.1 for explanation). For those species where seed dormancy information was unavailable, the probable dormancy class was inferred and assigned based on the most prevalent class of seed dormancy in their congeners or family's phylogenetic position, as per Baskin and Baskin (2014). Furthermore, if a species was known to produce both non-dormant and dormant seeds, it was upgraded to the respective dormancy class. Similarly, MD was upgraded to MPD and PY + PD with the PD class for analysis purposes. Based on leaf phenology, each species was assigned to one of the following classes: (1) deciduous, (2) evergreen with a short leaf lifespan of lt;450 days, and (3) evergreen with a long leaf lifespan of gt;450 days with a slight modification of the leaf-based classification of Himalayan tree species proposed by Singh and Thadani (2024).

Variation in Fruit/Seed Traits Along the Dry-Wet Seasonal Gradient

Overall, among the 201 species studied, 40 and 60% produce dry and fleshy fruits, 23 and 77% desiccation-sensitive and desiccation-tolerant seeds, 28 and 72% non-dormant and dormant seeds, and 41 and 54% anemochory and zoochory seeds, respectively (data not shown). Fruit/seed dispersal occurs throughout the year but peaks during the monsoon when 33% of species disperse their seeds, followed by 24% in post-monsoon and 21% each in winter and pre-monsoon summer months (Fig. 13.3a). As expected, fleshy fruits are proportionally more common during the wet season of monsoon (52%) and post-monsoon (43%) as compared to dry season of winter (17%) and pre-monsoon (34%). In Himalayan ecosystems, late post-monsoon to pre-monsoon seasons, coupled with low temperature and drought stress, are known to trigger leaf senescence both in deciduous and evergreen tree species (Singh and Thadani 2024); therefore, a high proportion of dry-fruited (up to 83%) and anemochory (up to 63%) species are found during these seasons (Fig. 13.3a, b). In contrast, a high proportion of fleshy-fruited (52%) and zoochory (70%) species were found in the monsoon season (Fig. 13.3a, b). The ecological explanation for these trends is that dry-fruited anemochory species are mainly desiccation-tolerant (Fig. 13.3c) and need a leafless forest canopy for high wind flow to facilitate long-distance seed dispersal (Griz and Machado 2001; Butler et al. 2007). Although fleshy fruited zoochory species are mostly desiccation-sensitive, they need to remain hydrated to maintain viability; therefore, they disperse in high proportion during monsoon and post-monsoon seasons, as reported in

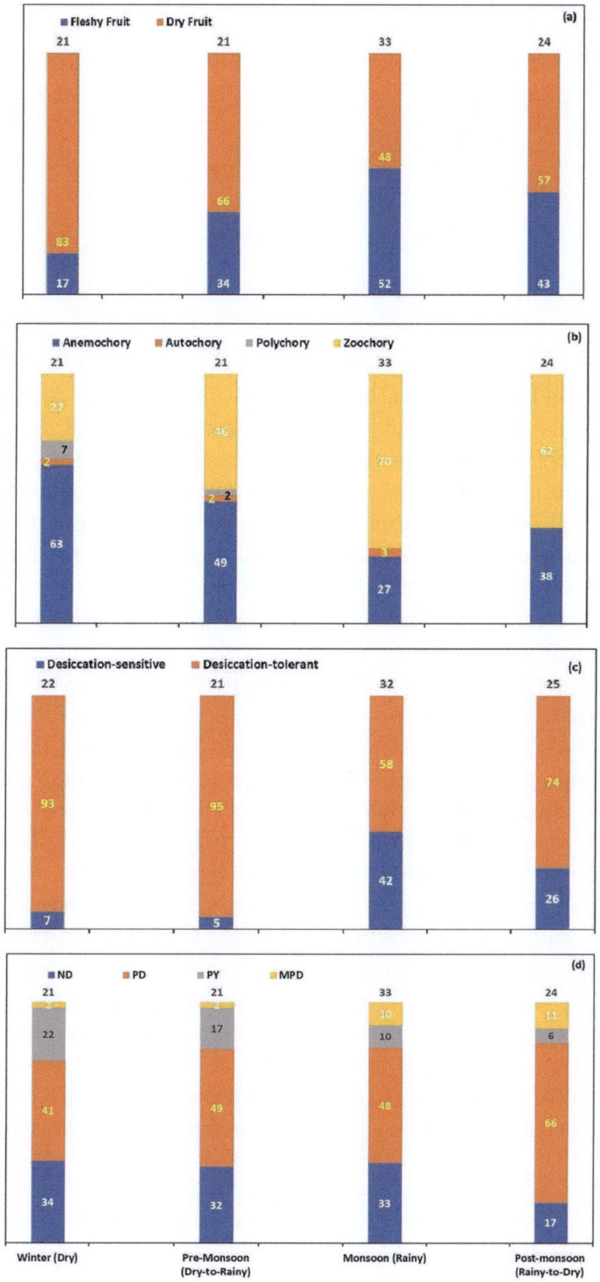

Fig. 13.3 The relative proportion of species (%) that fruit in different dispersal seasons according to their (**a**) fruiting texture, (**b**) dispersal syndromes, (**c**) seed desiccation, and (**d**) dormancy pattern (ND: Non-dormancy; PD: Physiological dormancy; PY: Physical dormancy; MPD: Morphophysiological dormancy). The digits inside and above the bar indicate the percentage of species in each category

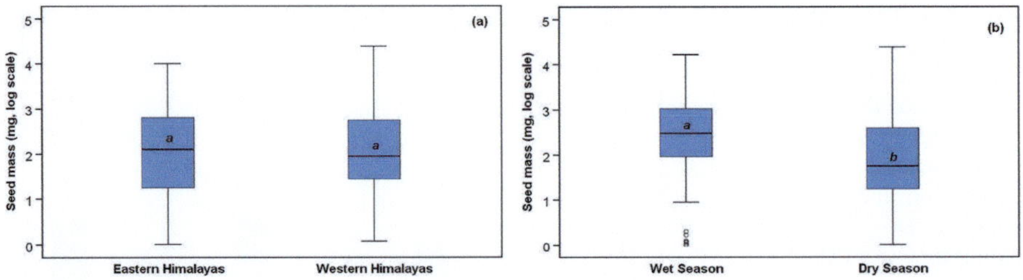

Fig. 13.4 Variation in the dry seed mass of the species in relation to the (**a**) east-west latitudinal gradient and (**b**) seasonal wet-dry gradient. Different letters indicate significance level ($P < 0.05$)

humid/moist tropical forest and savanna ecosystems of neotropics (Sautu et al. 2006, 2007; Salazar et al. 2011; Escobar et al. 2018; Miranda et al. 2023). Species dispersing seeds in the post-monsoon (autumn) to pre-monsoon (spring/summer) season are supposed to encounter an extended cool winter temperature and hot, dry spell of summer, thus producing a high proportion of dormant seeds compared to other seasons (Fig. 13.3d). Similar trends were reported in other parts of the world as described in the Introduction section.

Overall, dispersal of dry-fruited, desiccation-tolerant, dormant, and anemochory seeds is more prevalent during the dry season (from post-monsoon to pre-monsoon). However, the dispersal of fleshy, desiccation-sensitive, non-dormant, and zoochory seeds is more prevalent in the wet season (monsoon) with few exceptions like *Quercus leucotrichophora,* whose seeds dispersed during dry winters. These findings are consistent with a previous continental-scale study in which a dataset of about 4000 Australian plant species confirmed that wet, warm, and stable climates favoured fleshy-fruited species (Chen et al. 2017). To some extent, these observations support the proposed hypothesis (Fig. 13.1); however, seed ecological trait data from many Himalayan tree species are still needed to validate these seasonal patterns further. The mean seed mass of the species dispersed in the wet season (1337.4 mg) was approximately 1.8 times higher than that dispersed in the dry season (753.7 mg) (Fig. 13.4a) because most species dispersed in the wet season produce fleshy fruits with desiccation-sensitive and non-dormant seeds. Singh et al. (2017a) confirmed this trend as the mean seed mass of desiccation-sensitive and monsoon-synchronised species was higher than that of desiccation-tolerant seeds of partially monsoon-synchronised and monsoon-desynchronised species.

Variation in Fruit/Seed Traits Along the East-West Longitudinal Gradient

Contrary to our expectations, there was no prominent trend in the distribution of species in the eastern and western Himalayas that produce more fleshy vs. dry fruits or anemochory vs. zoochory seeds (Fig. 13.5a, b). We hypothesised that there are more fleshy fruits and zoochory species in the eastern Himalayas (Fig. 13.1), as it is wetter and more diverse in arboreal and avifauna, especially with frugivorous

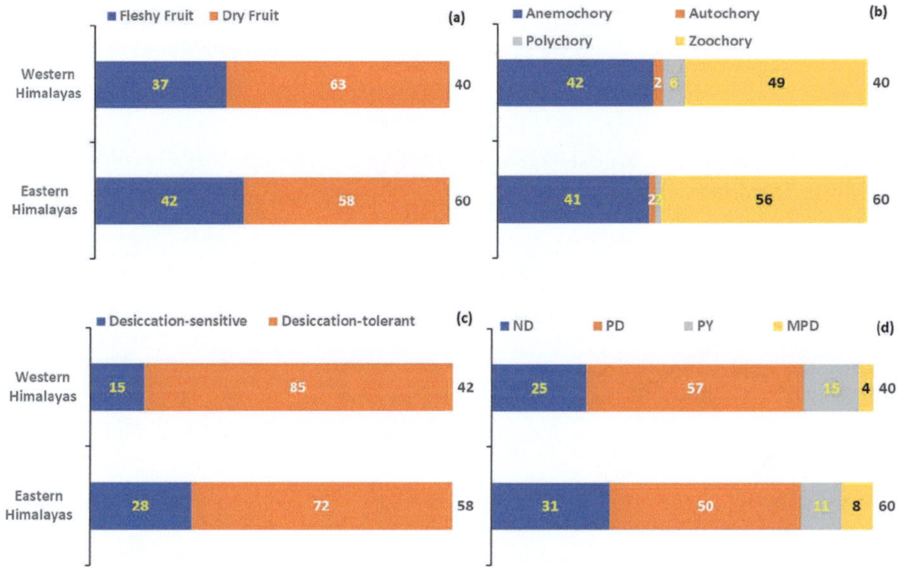

Fig. 13.5 The relative proportion of species (%) distributed in the eastern and western Himalayas according to their (**a**) texture of the fruit, (**b**) seed dispersal syndrome, (**c**) seed desiccation behaviour, and (**d**) dormancy pattern (ND Non-dormancy, PD Physiological dormancy, PY Physical dormancy, MPD Morphophysiological dormancy). The numbers beside the bar indicate the percentage of species in each category

animals, compared to the western Himalayas. In line with our expectations, the proportion of seeds sensitive to desiccation (28 vs. 15%) and non-dormant (31 vs. 25%) produced by tree species from the eastern Himalayas is comparatively higher than that from the western Himalayas (Fig. 13.5c, d). Approximately 2/3 of the species were desiccation-tolerant and dormant in the eastern and western Himalayas. However, it is worth mentioning that much of the western Himalayas' outer ranges are exposed to monsoon and covered by continuous monodominant forests of sal and oaks known to produce mostly desiccation-sensitive and non-dormant (even viviparous) seeds or by monodominant *Pinus roxburghii* but with desiccation-tolerant and non-dormant seeds (Box 13.2). We expected the mean seed mass of eastern Himalayan species to be higher than the western Himalayan species; however, contrary to our expectation, there was a non-significant ($p > 0.05$) reverse trend (807.5 mg in eastern Himalayas vs. 1167.6 mg in western Himalayas) (Fig. 13.4b). One of the possible reasons for the relatively low longitudinal variation in fruit/seed traits between the eastern and western Himalayas is that there is a distribution of exceptionally large-sized, desiccation-sensitive, non-dormant seeded species in both regions (e.g. *Quercus lamellosa* in eastern and *Aesculus indica* in western Himalayas). Furthermore, most of our study species (~80%) are distributed in the foothills and outer range of the Himalayan arc and experienced similar climatic conditions, except for a prominent variation in monsoonal precipitation from the eastern to the western Himalayan arc. Another possible reason may be that we have not captured the variability of fruit/seed traits of all Himalayan tree species, thus warranting more future studies on seed ecological traits of Himalayan trees covering the entire Himalayan arc.

Box 13.2: Seed Ecological Traits of Himalayan Tree Species Forming Monodominant Forest Canopy

Family	Species	Fruit texture type	Seed dispersal syndrome	Seed desiccation behaviour	Seed dormancy class	Monsoon synchronisation categories
Betulaceae	*Alnus nepalensis*	Dry	Polychory	DT	PD	PMS
Betulaceae	*Betula utilis*	Dry	Anemochory	DT	PD	MD
Cupressaceae	*Cupressus torulosa*	Dry	Anemochory	DT	PD	PMS
Dipterocarpaceae	*Dipterocarpus turbinatus*	Dry	Anemochory	DS	ND	MD
Dipterocarpaceae	*Shorea robusta*	Dry	Anemochory	DS	ND	MS
Ericaceae	*Rhododendron arboreum*	Dry	Anemochory	DT	PD	PMS
Fabaceae	*Acacia catechu*	Dry	Anemochory	DT	PY	PMS
Fabaceae	*Dalbergia sissoo*	Dry	Anemochory	DT	ND	PMS
Fagaceae	*Quercus lamellosa*	Dry	Zoochory	DS	ND	PMS
Fagaceae	*Quercus leucotrichophora*	Dry	Zoochory	DS	PD	PMS
Fagaceae	*Quercus semecarpifolia*	Dry	Zoochory	DS	ND	MS
Pinaceae	*Abies densa*	Dry	Anemochory	DT	PD	MD
Pinaceae	*Abies pindrow*	Dry	Anemochory	DT	PD	MD
Pinaceae	*Abies spectabilis*	Dry	Anemochory	DT	PD	MD
Pinaceae	*Cedrus deodara*	Dry	Anemochory	DT	PD	MD
Pinaceae	*Picea smithiana*	Dry	Anemochory	DT	PD	MD
Pinaceae	*Pinus kesiya*	Dry	Anemochory	DT	ND	PMS
Pinaceae	*Pinus roxburghii*	Dry	Anemochory	DT	ND	PMS
Pinaceae	*Pinus wallichiana*	Dry	Anemochory	DT	PD	PMS
Theaceae	*Schima wallichii*	Dry	Zoochory	DS	ND	PMS

DT Desiccation-tolerant, *DS* Desiccation-sensitive, *PD* Physiological dormancy, *PY* Physical dormancy, *ND* Non-dormancy, *MD* Monsoon desynchronisation, *PMS* Partial monsoon synchronisation, *MS* Monsoon synchronisation

Variation in Fruit/Seed Traits Along the South-North Altitudinal Gradient

Along the altitudinal belt, the proportion of fleshy fruited species decreased (45–26%) and dry fruited species increased (55–74%) from the foothill up to 3000 m (Fig. 13.6a). Subsequently, along timberlines and inner Himalayas (>3000 m), the proportion of fleshy fruited species increased (44%) due to the presence of *Juniperus* and *Sorbus* species and the availability of seed trait data on a limited number of species from inner Himalayas. Seeds of >50% of species from the lower/outer Himalayas dispersed through zoochory, while anemochory was the most prominent mode of seed dispersal in the higher/inner Himalayas (Fig. 13.6b). It is in the expected line, as most genera like *Abies, Acer, Betula, Cedrus, Fraxinus, Picea, Pinus, Rhododendron, Toona,* and *Ulmus*, known to produce samara and winged dispersules, are abundant in those altitudinal ranges. Except in the altitudinal range of the outer foothills (<1000m), the proportion of species with desiccation-sensitive and desiccation-tolerant seeds decreased (40–11%) and increased (60–89%) from the middle range (1001–2000 m) to the inner (>3000 m) Himalayan range (Fig. 13.6c). The possible reason for the considerably high proportion of species with desiccation-sensitive seeds in the middle Himalayas (40%) and foothills (20%) may be the abundance of tree species belonging to Anacardiaceae (Mango), Dipterocarpaceae (Sal), Fagaceae (Oaks), Lauraceae (*Litsea*), Lecythidaceae (*Careya*), Meliaceae (Neem), Moraceae (Mulberry, Jackfruit), and Rutaceae (Beal,

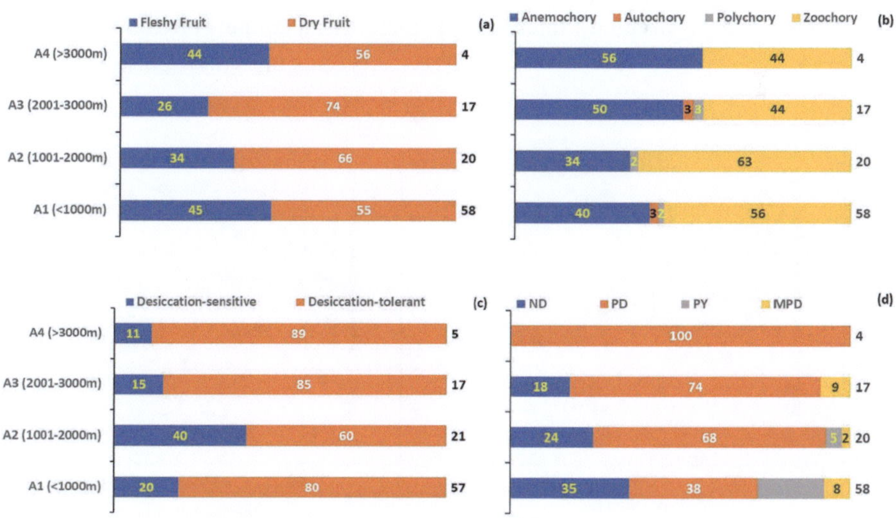

Fig. 13.6 Relative proportion of species (%) distributed along the south-north altitudinal gradient of the Himalayas according to (**a**) fruit texture, (**b**) seed dispersal syndrome, (**c**) desiccation behaviour, and (**d**) dormancy pattern (ND Non-dormancy, PD Physiological dormancy, PY Physical dormancy, MPD Morphophysiological dormancy). The numbers beside the bar indicate the percentage of species in each category

Citrus) families that taxonomically and phylogenetically are known to produce desiccation-sensitive seeds and most of them also have tropical affinities (Singh and Singh 1992a, b; Dickie and Pritchard 2002; Wyse and Dickie 2018; Singh et al. 2017a). In contrast, families with species known to produce seeds tolerant to desiccation are dominated in the higher/inner Himalayas such as Pinaceae (Pine, Abies), Cupressaceae (*Juniperus*), Taxaceae (*Taxus*), and Ulmaceae (Elm) and in foothills and outer Himalayas such as Fabaceae (*Acacia, Albizia, Bauhinia*), Malvaceae (*Grewia, Sterculia*), Lythraceae (*Lagerstroemia*), and Combretaceae (*Anogeissus, Terminalia*). Another reason for the dominance of species with desiccation-sensitive seeds in the outer and middle Himalayas is that much of these altitudinal ranges, directly exposed to monsoon thrust, receive up to 3000–4000 mm annual rainfall (Singh et al. 2017a, b), thus having a closer affinity with tropical moist forests compared to temperate forests (Singh 2014; Singh and Thadani 2013). Furthermore, the continuous close canopy nature of the gregarious monodominant forest stands of Dipterocarpaceae and Fagaceae makes the forest floor moist, helping desiccation-sensitive seeds retain adequate moisture content under canopy shading or below the moist litter layers (Joët et al. 2013, 2016).

Since high-altitude species produce more wind dispersal and desiccation-tolerant seeds, we expected some correlation of seed mass with altitude. However, there was no overall correlation of seed mass with altitude (Singh et al. 2017a), even if the species were separated according to the desiccation behaviour (Fig. 13.7a). On the contrary, there was a significantly positive correlation of seed mass with altitude according to two monsoon synchronisation strategies and a negative correlation with the monsoon desynchronisation strategy (Fig. 13.7b; see below for more information on monsoon synchronisation). Interestingly, there was a unique correlation between

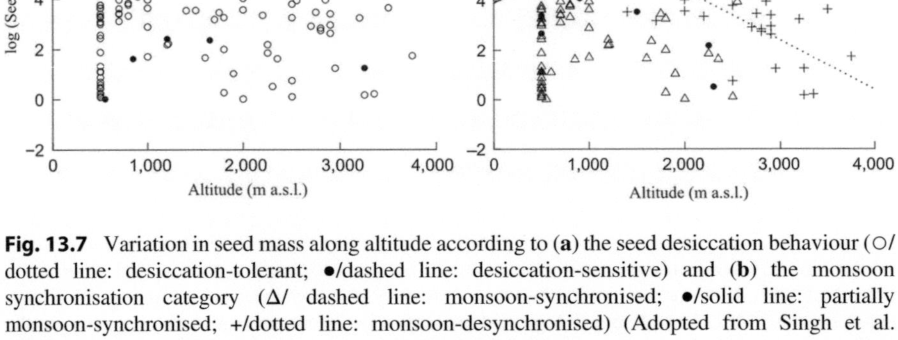

Fig. 13.7 Variation in seed mass along altitude according to (**a**) the seed desiccation behaviour (○/ dotted line: desiccation-tolerant; ●/dashed line: desiccation-sensitive) and (**b**) the monsoon synchronisation category (Δ/ dashed line: monsoon-synchronised; ●/solid line: partially monsoon-synchronised; +/dotted line: monsoon-desynchronised) (Adopted from Singh et al. (2017a); open access, Creative Commons Attribution License)

the pairs of cogeneric species, as higher-altitude species produce relatively heavier seeds than their lower-altitude counterparts. For example, low-altitude *Abies pindrow* (18.3 mg/seed) vs high-altitude *A. spectabilis* (37 mg/seed) and similar case is with *Betula alnoides* (0.11 mg/seed) vs *B. utilis* (0.19 mg/seed); *Pinus wallichiana* (50 mg/seed) vs *P. gerardiana* (349.4 mg/seed); *Q. floribunda* (3120 mg/seed) vs *Q. semecarpifolia* (7143 mg/seed)*; and Rhododerdron arboreum* (0.02 mg/seed) vs *R. hodgsonii* (0.24 mg/seed) (Singh et al. 2017a, see table 5 therein). They also reported that the mean seed mass of Himalayan tree species with desiccation-sensitive seeds (3095 mg/seed) was eight times higher than that of desiccation-tolerant seeds (388 mg/seed) following the global trend reported by other researchers (Dickie and Pritchard 2002; Daws et al. 2005 2006). The seed dormancy pattern shows a clear trend compared to the desiccation behaviour of the seeds. The proportion of species with non-dormancy (ND) and physical dormancy (PY) seeds declined from lower to higher altitudes. At the same time, the species with overall dormant seeds (primarily PD) increased (Fig. 13.6d). Studies on seed dormancy patterns at regional or global levels also reported that the proportion of species with dormant seeds increased from lower to higher elevational gradients or from benign aseasonal (tropical/subtropical) to highly seasonal (alpine/subalpine) ecosystem (Baskin and Baskin 2014; Fernandez-Pascual et al. 2021; Rosbakh et al. 2022, 2023; Zhang et al. 2022; Chen et al. 2023). The possible reason behind this is that species that face harsh environments (low temperature, drought, unpredictable frost events, short growing season) must synchronise their seed germination and seedling emergence with the growing season through a dormancy mechanism. The reason for a high proportion of species with PY at lower and PD at higher altitudes is due to the dominance of plant families such as Fabaceae (*Bauhinia, Cassia*) at lower altitudes and Fagaceae (Oaks) or Pinaceae (*Abies, Picea, Juniperus*) Sapindaceae (chestnut, maple) at higher altitude, respectively.

Interaction Between Seed Traits with the Monsoon Synchronisation Strategy and Habitat Moisture Conditions

As indicated in an earlier study by Singh et al. (2017a), approximately 80% of the study species synchronise their seed dispersal and germination timing with the monsoon either fully (MS) or partially (PMS) and in the remaining 20% of the species, there was a complete desynchronisation of seed dispersal and germination timing with the monsoon (MD) (Table 13.1). Among the MS species, about 42% produce short-lived desiccation-sensitive seeds, while among the PMS and MD species, only 18 and 8% had desiccation-sensitive seeds, respectively (Table 13.1). In general, species with desiccation-sensitive seeds are known to produce large, fleshy seeds; therefore, the seed mass of MS was 2.7 times greater than the PMS and MD species (Fig. 13.8).

In the Himalayan ranges, much of the forest vegetation is dominated by several species of Fagaceae and Dipertocarpaceae (Oaks and Sals) that fall into the MS category known to produce a high proportion of heavy-weighted, short-lived,

Table 13.1 Percentage of Himalayan tree species according to monsoon synchronisation categories and their desiccation behaviour ($n = 197$) (Adopted from Singh et al. (2017a); open access, Creative Commons Attribution License)

Monsoon synchronisation category	Total species (%)	Species as per seed desiccation behaviour	
		Desiccation-sensitive (%)	Desiccation-tolerant (%)
Monsoon-synchronised (MS)	23	42	58
Partially monsoon-synchronised (PMS)	57	18	82
Monsoon-desynchronised (MD)	20	8	92
		21	**80**

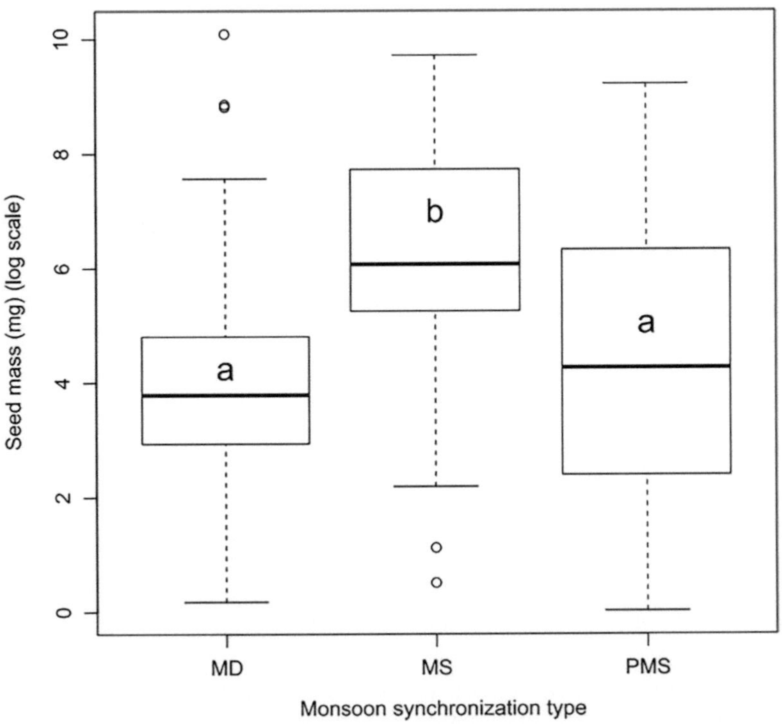

Fig. 13.8 The mean seed mass of Himalayan tree species ($n = 197$) according to their monsoon synchronisation strategy. Letters represent subsets with significant differences ($p < 0.001$) (Adopted from Singh et al. (2017a); open access, Creative Commons Attribution License)

desiccation-sensitive seeds compared to species in the PMS and MD category (Singh et al. 2017a). A similar trend was reported in dry tropical forests in northern Thailand (Blakesley et al. 2002) and Barro Colorado Island, Panama (Garwood, 1983), and in evergreen rain forests around the world (Tweddle et al. 2003). Since seeds of species

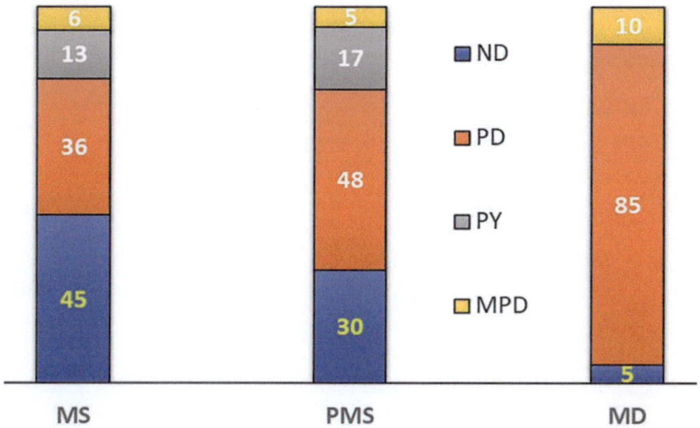

Fig. 13.9 Relative proportion of species (%) with seed dormancy pattern (ND Non-dormancy, PD Physiological dormancy, PY Physical dormancy, MPD Morphophysiological dormancy) vary according to monsoon synchronisation category (MS monsoon synchronised, PMS partial monsoon synchronised, MD monsoon desynchronised)

in the MS category have short viability due to their sensitivity to desiccation, a high proportion of their seeds fall into the non-dormant category. Non-dormancy decreases from MS (45%) to MD (5%). On the contrary, the proportion of dormancy increases from the MS (55%) to the MD (95%) category, specifically physiological dormancy from 36% in MS to 85% in MD (Fig. 13.9). It is worth highlighting that physical dormancy is more prominent in the PMS and MS categories, as some of the prominent species belong to families such as Fabaceae and Malvaceae fall under this category. Species of the MD category (*Abies* spp., *Acer* spp., *Aesculus* spp., *Cedrus deodara, Fraxinus* spp., *Picea* spp., *and Taxus baccata*) from the higher/inner Himalayas have a considerably higher proportion of physiological dormant seeds. Previous studies also reported that seed dormancy is prevalent in climatically unfavourable seasons and regions for seedling recruitment (Jurado and Flores 2005; Rubio de Casas et al. 2017; Wyse and Dickie 2018; Collette and Ooi 2021; Zhang et al. 2022; Rosbakh et al. 2022, 2023; Jayasuriya and Phartyal 2024). Thus, dormancy desynchronises germination and seedling emergence with the lowest probability of frost (temperate climates of higher Himalayas) or drought (inner dryer valley of the western Himalayas) events.

The pattern of dormancy, along with the moisture gradient of the habitat of the species, revealed that the species distributed in intermediate and dry habitats produce a high proportion (75%) of dormant seeds compared to moist habitats (65%) (Fig. 13.10). In the previous study on the same set of species, Singh et al. (2017a) found that the seed mass of species from the moist habitat was approximately 1.8 and 3.8 times heavier than that of the intermediate and dry habitats, respectively. In an expected line, 35% of the species of moist habitats with heavier seeds produce non-dormant seeds, as >90% are desiccation-sensitive and fall into the MS and PMS categories. It is interesting to note that the proportion of PY seeds doubled and tripled from moist (7%) to intermediate (14%) and dry (23%) habitats, supporting

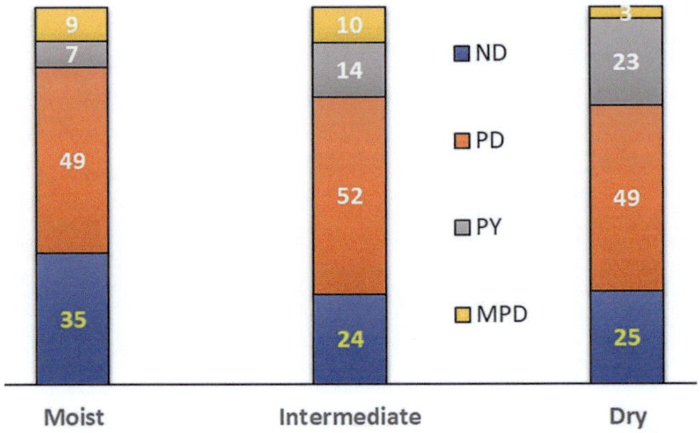

Fig. 13.10 Relative proportion of species (%) with seed dormancy pattern (ND Non-dormancy, PD Physiological dormancy, PY Physical dormancy, MPD Morphophysiological dormancy) varied according to habitat moisture conditions

earlier studies that PY was associated with dry habitats (Rosbakh et al. 2023; Jayasuriya and Phartyal 2024); as an impermeable seed coat with a closed water gap protects seeds from desiccation (Baskin and Baskin 2000, 2014) and helps prevent germination by maintaining a low seed moisture content (Pausas and Lamont 2022) during the unfavourable dry season. Since >50% of our study species fall into the PMS category and germinate during the monsoon season, 1/4 of them produce PY seeds. In tropical dry forests of northern Indian Gangetic plains, which share several common species with outer Himalayan forest ecosystems, seeds with PY dispersed in the dry season or late in the wet season and persist in the soil seed bank until the wet season of the following year to partially synchronise seed germination with monsoon (Singh and Singh 1992a, b; Khurana and Singh 2001). During the long dry season, the opening of the water gap or cracking (scarification) in the hard seed coat (i.e. breaking of PY) occurs when the seed coat experiences exceedingly high temperatures in the summer season (Vázquez-Yanes and Orozco-Segovia 1982; Baskin and Baskin 2014) or heat shock from forest fire (Ooi et al. 2014, 2022). Most Fabaceae and Malvaceae species are naturally distributed in dry habitats in our study ecosystem.

In Himalayan forest ecosystems, especially pine-dominant landscapes, frequent forest fires are a historical feature, and the major incidence of fires occurs at intervals of about 2–6 years from 1911 to 2016 (Dobriyal and Bijalwan 2017; Dogra et al. 2018). Climate change and anthropogenic disturbances are influencing the timing and duration of fire seasons in many ecosystems around the world and directly impact postfire recruitment from seeds as many regeneration traits are driven by fire cues, including resprouting, flowering, seed production, seed dispersal, dormancy break, and germination (Ooi et al. 2022). For example, fire-induced heat shock helps release canopy-stored (even grounded), non-dormant seeds from serotinous fruits/cones of several woody species and breaks PY. In contrast, fire-induced smoke,

smoke-water, and smoke-derived chemicals act as germination-stimulating agents, primarily on those species that had physiological components of dormancy like PD and MPD (Baskin and Baskin 2022 and Ooi et al. 2022 and references therein). So far, we have no empirical records in published literature indicating the presence of tree species with strict serotinous cone/fruit in the Himalayas. The absence of serotiny in the Himalayas indicates that crown fire has not been a natural phenomenon in the Himalayan forest ecosystem, possibly because of rapid litter decomposition or because the forest floor does not accumulate enough fuel load to result in a crown fire and act as seed dispersal of serotinous cone/fruit. How forest fires impact the regeneration (seed) ecology of Himalayan forest ecosystems has hardly been studied (but see Konsam et al. 2017, 2020); thus, a detailed investigation is needed, especially since recurring anthropogenic forest fires in Himalayan ecosystems (Konsam et al. 2017, 2020) are not natural disturbance agents but may act as a major ecological forces shaping the regeneration niche of plants (He et al. 2019; Ooi et al. 2022).

Throughout the outer monsoonal forest of the Himalayan ecosystems, Dipetrocarpace, Fagaceae, and Pinaceae species monodominate and form continuous forest landscapes with close canopy. In the Himalayas, Fagaceae has 63 species that are distributed naturally from foothills to timberlines, with a high diversity of *Quercus* (27 sp) in western and *Lithocarpus* (21 sp), *Castanopsis* (4 sp), and *Castanea* (1 sp) in the eastern Himalayas (Global Tree Search Database 2023; Yadav et al. 2024). Being economically and ecologically valuable species, most of the Fagaceae (especially *Quercus*) are also under heavy abiotic and biotic pressure (such as extreme weather, frequent drought and forest fire, extensive lopping for animal feed, frugivory, pest and disease attack, poor seed production) that affect their natural regeneration potential (Singh and Singh 1987; Thadani and Ashton 1995; Singh 2014; Shekhar et al. 2022). Recent literature synthesis on seed desiccation, dormancy, and germination behaviour of Fagaceae by Jaganathan and Phartyal (2025) shows that most species of Fagaceae rely exclusively on desiccation-sensitive acorns for their dispersal and regeneration. Seed dormancy and desiccation sensitivity, previously considered to coexist in a family rarely (Tweddle et al. 2003; Jaganathan 2025), tend to be widespread in Fagaceae, especially in the genus *Quercus*. This genus has desiccation-sensitive seeds either with physiological dormancy (PD) or epicotyl PD (ePD), and the ePD has been so far not appropriately studied and thus underestimated, as ePD is often ignored/overlooked or classified as non-dormancy (Jaganathan and Phartyal 2025; Jaganathan et al. 2025a, 2025b). PD is considered an adaptation to prevent seed germination before a harsh winter or until the onset of spring or wet season. In contrast, in ePD (Fig. 13.11), radicle emergence occurs within a few days after seed dispersal, but shoot emergence is delayed for a considerable period, usually 3–4 weeks or longer (Baskin and Baskin 2021; Jaganathan and Phartyal 2025; Jaganathan et al. 2025a, 2025b). The immediate emergence of radicles in ePD could be a bet hedge strategy to avoid drying-induced stress in desiccation-sensitive acorns or avoid frugivory, while, on the contrary, a delay in the emergence of the shoot in radicle-emerged acorns until the following growing season helped species maximise fitness, a possible best bet hedge strategy

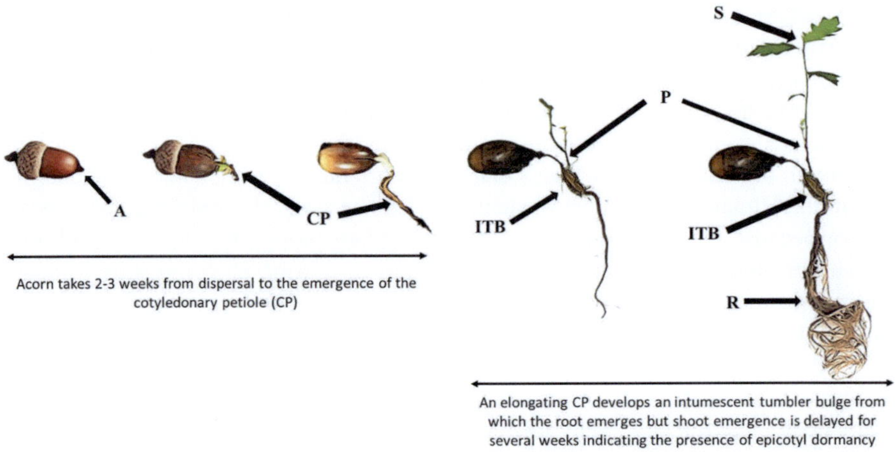

Fig. 13.11 A schematic diagram indicating the possible course of the germination process (epicotyl dormancy) in acorns of most Fagaceae species. A acorn apex, CP cotyledonary petiole, ITB intumescent tubular bulge, P plumule, R root, S shoot. (Modified after Jaganathan and Phartyal (2025)

(Jaganathan and Phartyal 2025; Jaganathan et al. 2025a, 2025b). Epicotyl PD has also been reported in desiccation-tolerant seeds of some of the Fabaceae species from tropical ecosystems (Jayasuriya et al. 2010, 2012) that were previously classified under the non-dormancy category due to a lack of detailed investigation (Jaganathan et al. 2025b). Therefore, there is a high possibility that several tree species in the Himalayan forest ecosystem, especially from Fagaceae and Fabaceae, classified as non-dormant, may also have ePD, and the mechanism that controls their germination is not yet clearly understood; thus, calls for more in-depth experimental studies.

Interaction Between Seed Traits and Leaf Phenology

Singh and Thadani (2024) advocated for a leaf-based classification for Himalayan forests because tree leaves, in terms of form and phenology, help explain some of the fine-scale variability in the ecological strategies of Himalayan forests. Leaf traits determine plant and stand characters (Reich et al. 1992), influence plant photosynthesis and defensive traits (Matsuki and Koike 2006), and vary in response to variations in resource availability (Zhang et al. 2020). On a similar line, plant regenerative traits also influence plant distribution and community assemblage (Grubb 1977). Therefore, we look at how seed dispersal syndrome, desiccation behaviour, and dormancy pattern interact with leaf phenology (leaf lifespan) in the Himalayan forest ecosystem. The study revealed that in evergreen species (short leaf lifespan), zoochory was the dominant dispersal syndrome (82%).

In comparison, between deciduous and evergreen species (long leaf lifespan), anemochory was the dominant dispersal syndrome (50–65%) (Fig. 13.12a). Both

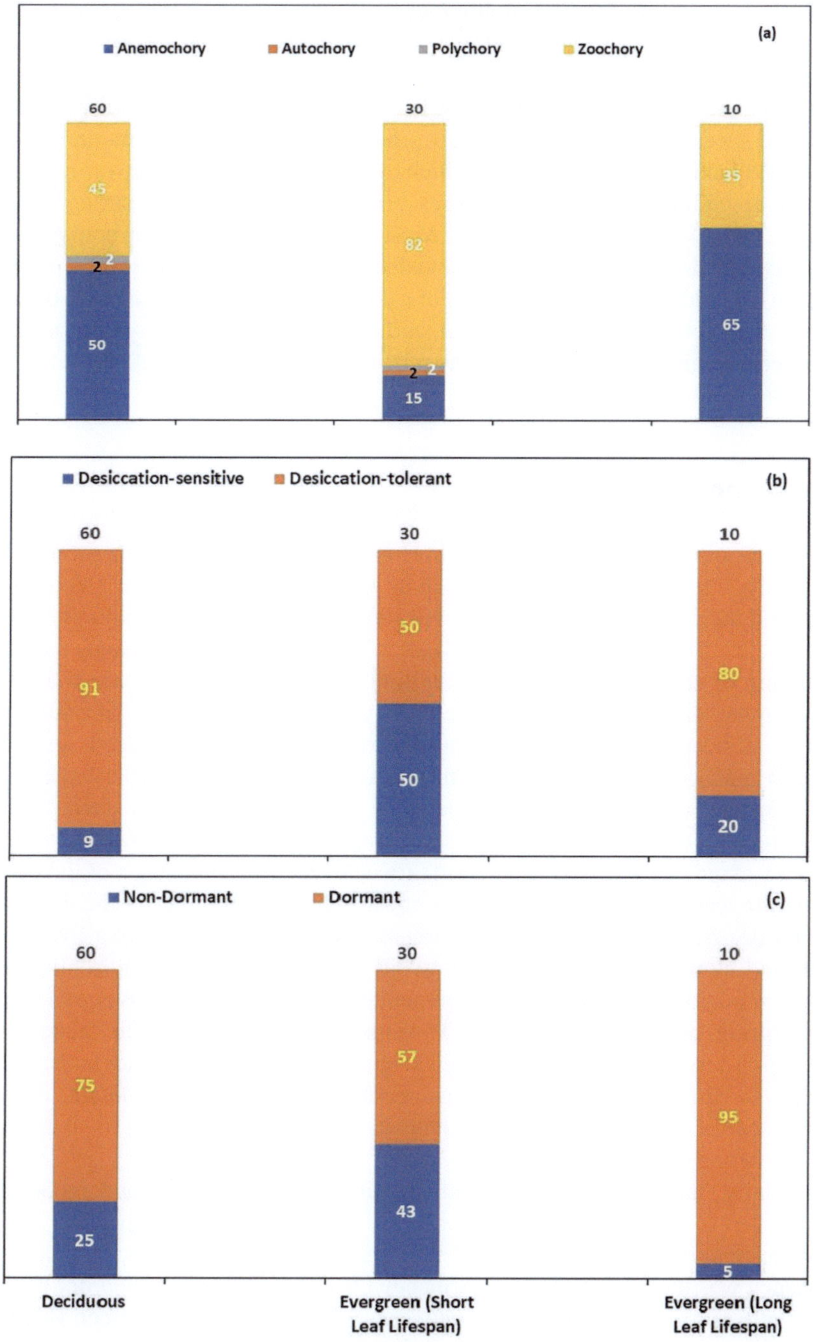

Fig. 13.12 Relative proportion of species (%) with (**a**) seed dispersal syndrome, (**b**) desiccation behaviour, and (**c**) absence/presence of dormancy varied according to leaf phenology and leaf lifespan

these later categories of leaf phenology were also dominated by desiccation-tolerant (>80%) as well as dormant (>75%) seed producing species (Fig. 13.12b, c). Globally in the temperate forest zone, most winter deciduous species (like maples), as well as evergreen species with long leaf lifespan (like conifers), disperse seeds (mainly desiccation-tolerant) primarily from late summer through autumn into early winter (Willson and Traveset 2000), which eventually exposed either warm summer/autumn or chilling winter temperatures prior to germinating in spring or early summer (Walck and Hidayati 2022). Most of these seeds have a component of physiological dormancy, which is alleviated through moist stratification at warm (autumn) or cold (winter) temperatures or a combination of both (Baskin and Baskin 2014). More detailed and experimental studies are needed on tree species in Himalayan ecosystems to validate further the relationship between leaf lifespan with seed desiccation and dormancy behaviour.

Conclusions

About 80% of the tree species studied in Himalayan forest ecosystems showed fully or partially synchronised seed germination with the monsoon (23% MS and 57% PMS) and their fruit/seed dispersal season (wet vs. dry), seed desiccation behaviour (sensitive vs. tolerant), and dormancy pattern (non-dormant vs. dormant) controlled the timing of seed germination and seedling establishment. Desiccation-sensitive, non-dormant, and wet-season dispersed seeds germinate mainly in monsoon (MS), whereas desiccation-tolerant, dormant, and dry-season dispersed seeds germinate either in monsoon (PMS) or outside the monsoon season (MD). To a large extent, seed dispersal during monsoon favours the evolution of fleshy-fruited, heavier, zoochorous, desiccation-sensitive (short-lived), non-dormant, fast-germinating seeds to avoid dry-stressed seed mortality and get full advantage of favourable season for seedling establishment. In contrast, seed dispersal during the dry season favours dry-fruited, lighter, anemochorous, desiccation-tolerant (long-lived), dormant, and slow-germinating seeds to avoid harsh/unfavourable environments for seed germination and seedling establishment until the onset of the favourable warm/moist spring or monsoon seasons. To some extent, fleshy-fruited, zoochorous, desiccation-sensitive, non-dormant seeded species were slightly higher in the eastern Himalayas. In contrast, dry-fruited, anemochorous, desiccation-tolerant, dormant seeded species were in the western Himalayas, but the difference in these traits was not strong enough as expected.

Similarly, the above traits (except dormancy) showed weak and inconsistent trends from lower to higher altitudes in the Himalayas. The probability of seed dormancy increases from species of the MS to MD category, and physiological dormancy was the most prevalent dormancy class. In intermediate and dry habitats, ~75% of tree species produce dormant seeds. Deciduous and evergreen (long leaf lifespan) species produced more anemochorous, desiccation-tolerant, and dormant seeds than evergreen (short leaf lifespan). Most tree species forming monodominant forests in higher (northern) Himalayas produce desiccation-tolerant and dormant

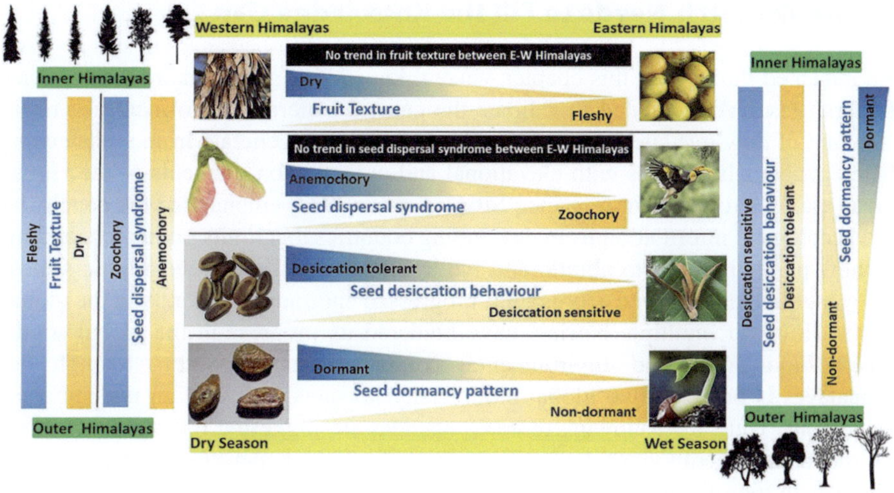

Fig. 13.13 Schematic profile of observed variations in seed ecological traits of Himalayan tree species along the wet-dry seasonal, east-west, and outer-inner Himalayan gradients. Triangles indicate decreasing/increasing, and rectangles indicate an equal proportion of species from one gradient to another. The black rectangles indicate no trend for the eastern-western Himalayan gradient for only those two seed traits

seeds that either desynchronise or partially synchronise with the monsoon. In contrast, in the lower (southern) Himalayas that receive the full force of the monsoon, they produce primarily desiccation-sensitive and non-dormant seeds that synchronise fully or partially with the monsoon, such as oaks and sal (Box 13.2).

In contrast to our hypothesis, there were no or weak longitudinal and altitudinal trends in the seed ecological traits of Himalayan forest ecosystems (Fig. 13.13). This could partly be because most of the analysis was based on species distributed in moist outer ranges receiving the direct monsoon thrust, which differ in moisture from west to east, but not to an extent to override the influence of seasonal similarity. The difference between monsoon and non-monsoon regions could be worth analysing as important species like oaks and chir pine do not penetrate the dry inner valleys of the Himalayas. The macroscale level heterogeneity of the Himalayan arc may have masked seed ecological trait variation prevalent at the microscale level, or in other words, microclimatic conditions might mask the effects of broad-scale climatic patterns. However, even identifying weak general patterns in the seed ecological traits of tree species along the Himalayan gradient will help us better understand their regeneration ecology and vulnerability to climate. It is also worth mentioning that the trait value of several tested species has been inferred or speculated from the trait value of their congeners; therefore, our results should be considered cautiously.

Future Research Needs to Fill the Knowledge Gap

As indicated throughout this chapter, several knowledge gaps remain; thus, further in-depth research is needed to validate the general patterns and to increase understanding of how tree species distribute, assemble, and regenerate from seeds across heterogeneous landforms with elevational and precipitation gradients of the Himalayan arc. In a recent opinion paper, Silveira et al. (2023) emphasised the paucity of seed ecological information from developing countries that limits our understanding of how plants respond to abiotic and biotic factors at different geographic scales. They highlighted that seed ecology has progressed enormously recently in the global north. However, multiple barriers have hindered its progress in the global south, where biodiversity and environmental complexity are highest. Although the Himalayas is home to two global biodiversity hotspots, it is still an understudied ecosystem. Thus, the following research agendas are suggested for urgent attention of the next generation of researchers to understand better the regeneration of Himalayan trees from seeds:

1. Document the timing of fruit/seed dispersal, seed germination, and seedling emergence using in situ studies to predict better species monsoon synchronisation strategies.
2. Measure fruit/seed morphological traits to know the seed dispersal mode, ecological interaction of species with abiotic and biotic vectors, and how seasonal variation in local to regional climate affects seed dispersal syndrome and species seed dispersal networks.
3. Investigate seed desiccation behaviour, seed storage physiology, seed bank longevity, and soil seed bank persistence to prioritise ex-situ and in-situ conservation of rare, vulnerable, and threatened tree species.
4. Determine the seed germination niche width (i.e. environmental minima, optima, and maxima required for seed germination) to understand better species distribution and community assemblage along micro- and macroclimatic gradients.
5. Screen species for the presence of dormancy; if so, investigate dormancy-breaking cues and class of dormancy (PD, MD, MPD, PY, PY + PD), including the probability of epicotyl dormancy, even if a species has non-dormant seeds, to understand species survival potential during the harsh environments, ability to persist in the soil, and bet-hedge strategies.
6. Investigate how forest fire, fire-induced heat, smoke, and smoke-water impact seed germination cues, dormancy, and soil seed bank persistence of species distributed in fire-prone zones.
7. Investigate how a delay in snowfall or early snow cover depletion impacts seed dormancy, germination, and seedling establishment of timberline tree species.
8. Investigate how monsoon-synchronised species' reproductive biology (seed production, maturation, dispersal, dormancy, germination, and seedling establishment) copes with annual variation in the arrival and departure of monsoon rain.

9. Determine how assisted (biotic) or unassisted (abiotic) seed dispersal of timberline species impacts their regeneration and establishment, especially when the requirements of cooler temperature optima for seed germination force upward migration but low soil moisture or drought conditions force downward migration.
10. Investigate how leaf phenology and leaf lifespan affect regeneration traits, precisely seed dispersal, desiccation, and germination traits, as leaf shading intensity and season may create canopy gaps with spatiotemporal variations that eventually decide the fate of freshly dispersed seeds accumulated on or below litter layers and may expose to different light and soil moisture conditions.
11. Analyse the relationship between species dominance and their seed ecological features, such as sal, oaks, birch, chir pine, deodar and spruce, and silver fir.

Acknowledgements Dr. Harsh Yadav, from the Graduate School of Environment and Information Sciences, Yokohama National University, Japan, is appreciated for his contribution to map preparation and data curation during the initial stage. Author also thanks Prof. S.P. Singh and Dr. Rajesh Thadani for their valuable comments on earlier versions of the chapter.

References

Baskin JM, Baskin CC (2000) Evolutionary considerations of claims for physical dormancy-break by microbial action and abrasion by soil particles. Seed Sci Res 10:409–413

Baskin CC, Baskin JM (2014) Seeds: ecology, biogeography, and evolution of dormancy and germination, 2nd edn. Academic/Elsevier, San Diego

Baskin JM, Baskin CC (2021) The great diversity in kinds of seed dormancy: a revision of the Nikolaeva–Baskin classification system for primary seed dormancy. Seed Sci Res 31:249–277

Baskin CC, Baskin JM (2022) Plant regeneration from seeds: a global warming perspective. Academic, London

Bewley JD, Bradford K, Hilhorst H (2012) Seeds: physiology of development, germination and dormancy. Springer Science & Business Media, Berlin

BGCI (2023) GlobalTreeSearch online database. Botanic Gardens Conservation International, Richmond, UK. Available at: https://tools.bgci.org/global_tree_search.php. Accessed 8 Sep 2023

Blakesley D, Elliott S, Kuarak C, Navakitbumrung P, Zangkum S, Anusarnsunthorn V (2002) Propagating framework tree species to restore seasonally dry tropical forest: implications of seasonal seed dispersal and dormancy. For Ecol Manag 164:31–38

Butler DW, Green RJ, Lamb D, McDonald WJF, Forster PI (2007) Biogeography of seed-dispersal syndromes, life-forms and seed sizes among woody rain-forest plants in Australia's subtropics. J Biogeogr 34:1736–1750

Champion HG, Seth SK (1968) General Silviculture for India. Govt. of India Publications Branch, Department of Printing and Stationery, Delhi, India

Chen SC, Moles AT (2015) A mammoth mouthful? A test of the idea that larger animals ingest larger seeds. Glob Ecol Biogeogr 24:1269–1280

Chen SC, Cornwell WK, Zhang HX, Moles AT (2017) Plants show more flesh in the tropics: variation in fruit type along latitudinal and climatic gradients. Ecography 40:531–538

Chen K, Chen ZH, Huang YY, Jiang ZH (2023) Elevation and phylogeny shape herbaceous seed dormancy in a biodiversity hotspot of Southwest China. Ecol Evol 13:e9986

Collette JC, Ooi MKJ (2021) Distribution of seed dormancy classes across a fire-prone continent: effects of rainfall seasonality and temperature. Ann Bot 127:613–620

Cubasch U, Meehl GA, Boer GJ, Stouffer RJ, Dix M, Noda A, Yap KS (2001) Projections of future climate change. In: Climate change 2001: the scientific basis. Contribution of WG1 to the Third Assessment Report of the IPCC (TAR). Cambridge University Press, Cambridge, pp 525–582

Dalling JW, Davis AS, Arnold AE, Sarmiento C, Zalamea P-C (2020) Extending plant defense theory to seeds. Annu Rev Ecol Evol Syst 51:123–141

Daws MI, Garwood NC, Pritchard HW (2005) Traits of recalcitrant seeds in a semi-deciduous tropical forest in Panamá: some ecological implications. Funct Ecol 19:874–885

Daws MI, Garwood NC, Pritchard HW (2006) Prediction of desiccation sensitivity in seeds of woody species: a probabilistic model based on two seed traits and 104 species. Ann Bot 97:667–674

de Souza TV, Torres IC, Steiner N, Paulilo MTS (2015) Seed dormancy in tree species of the tropical Brazilian Atlantic Forest and its relationships with seed traits and environmental conditions. Braz J Bot 38:243–264. https://doi.org/10.1007/s40415-014-0129-3

Decker KLM, Wang D, Waite C, Scherbatskoy T (2003) Snow removal and ambient air temperature effects on forest soil temperatures in northern Vermont. Soil Sci Soc Am J 67:1234–1242

Dickie JB, Pritchard HW (2002) Systematic and evolutionary aspects of desiccation tolerance in seeds. In: Black M, Pritchard HW (eds) Desiccation and survival in plants: drying without dying. CAB International, Wallingford, pp 239–259

Dinerstein E, Olson D, Joshi A, Vynne C, Burgess ND, Wikramanayake E, Saleem M (2017) An ecoregion-based approach to protecting half the terrestrial realm. Bioscience 67:534–545

Dobriyal M, Bijalwan A (2017) Forest fire in western Himalayas of India: a review. N Y Sci J 10: 39–46

Dogra P, Mitchell AM, Narain U, Sall C, Smith RW, Suresh S (2018) Forests and fire: strengthening prevention and management in India. The World Bank, Washington, DC

Donohue K, Rubio de Casas R, Burghardt L, Kovach K, Willis CG (2010) Germination, post-germination adaptation, and species ecological ranges. Annu Rev Ecol Evol Syst 41:293–319. https://doi.org/10.1146/annurev-ecolsys-102209-144715

Escobar DF, Silveira FA, Morellato LPC (2018) Timing of seed dispersal and seed dormancy in Brazilian savanna: two solutions to face seasonality. Ann Bot 121:1197–1209

Fernandez-Pascual E, Carta A, Mondoni A, Cavieres LA, Rosbakh S, Venn S, Satyanti A, Guja L, Briceno VF, Vandelook F et al (2021) The seed germination spectrum of alpine plants: a global meta-analysis. New Phytol 229:3573–3586

Fernández-Pascual E, Carta A, Rosbakh S, Guja L, Phartyal SS, Silveira FA, & Jiménez-Alfaro B (2023) SeedArc, a global archive of primary seed germination data. New Phytol 240(2): 466–470. https://doi.org/10.1111/nph.19143

Fraaije RG, ter Braak CJ, Verduyn B, Breeman LB, Verhoeven JT, Soons MB (2015) Early plant recruitment stages set the template for the development of vegetation patterns along a hydrological gradient. Funct Ecol 29:971–980

Frankie GW, Baker HG, Opler PA (1974) Comparative phenological studies of trees in tropical wet and dry forests in the lowlands of Costa Rica. J Ecol 62:881–919

Garwood NC (1983) Seed germination in a seasonal tropical forest in Panama: a community study. Ecol Monogr 53:159–181

Garwood NC (1989) Tropical soil seed banks: a review. In: Leek MA, Parker VT, Simpson RL (eds) Ecology of soil seed banks. Academic, San Diego, pp 149–209

Global Biodiversity Information Facility (2023) GBIF Occurrence Download. https://doi.org/10.15468/dl.bxmvnc. Accessed 28 Mar 2023

Graae BJ, Nystuen KO, Vandvik V, Eycott AE (2022) Effects of climate change on regeneration of plants from seeds in boreal, subarctic, and subalpine regions, pp 19–32

Griz LMS, Machado ICS (2001) Fruiting phenology and seed dispersal syndromes in caatinga, a tropical dry forest in the northeast of Brazil. J Trop Ecol 17:303–321

Grubb PJ (1977) The maintenance of species-richness in plant communities: the importance of the regeneration niche. Biol Rev 52:107–145

He T, Lamont BB, Pausas JG (2019) Fire as a key driver of Earth's biodiversity. Biol Rev 94:1983–2010

Hock R, Rasul G, Adler C, Cáceres B, Gruber S, Hirabayashi Y et al (2019) High mountain areas. In: Pörtner H-O, Roberts DC, Masson-Delmotte V, Zhai P, Tignor M, Poloczanska E, Mintenbeck K, Alegría A, Nicolai M, Okem A, Petzold J, Rama B, Weyer NM (eds) IPCC special report on the ocean and cryosphere in a changing climate. IPCC, Geneva, pp 131–202

Howe HF, Smallwood J (1982) Ecology of seed dispersal. Annu Rev Ecol Syst 13:201–228

Huang Z, Footitt S, Tang A, Finch-Savage WE (2018) Predicted global warming scenarios impact on the mother plant to alter seed dormancy and germination behaviour in Arabidopsis. Plant Cell Environ 41:187–197

Jaganathan GK (2021) Ecological insights into the coexistence of dormancy and desiccation-sensitivity in Arecaceae species. Ann For Sci 78:1–14

Jaganathan GK (2025) Mutual exclusivity in seed traits: implications for conservation and ecology. Bot Lett 1–10. https://doi.org/10.1080/23818107.2025.2521293

Jaganathan GK, Sánchez JA, Phartyal SS (2025a) Epicotyl physiological dormancy in desiccation-sensitive acorns of Quercus sagrana, a vulnerable endemic species from western Cuba. Seed Sci & Tech (in review)

Jaganathan GK, Phartyal SS, Jayasuriya KMGG (2025b) From oversight to insight: integrating epicotyl emergence to redefine germination and enhance the seed dormancy framework. Seed Sci Res 1–6. https://doi.org/10.1017/S0960258525100007

Jara-Guerrero A, De la Cruz M, Méndez M (2011) Seed dispersal spectrum of woody species in south Ecuadorian dry forests: environmental correlates and the effect of considering species abundance. Biotropica 43:722–730

Jayasuriya KG, Phartyal SS (2024) Dormancy, germination, and associated seed ecological traits of 25 Fabaceae species from northern India. Plant Biol 26:41–50

Jayasuriya KG, Wijetunga AS, Baskin JM, Baskin CC (2010) Recalcitrancy and a new kind of epicotyl dormancy in seeds of the understory tropical rainforest tree Humboldtia laurifolia (Fabaceae, Ceasalpinioideae). Am J Bot 97:15–26

Jayasuriya KG, Wijetunga AS, Baskin JM, Baskin CC (2012) Physiological epicotyl dormancy and recalcitrant storage behaviour in seeds of two tropical Fabaceae (subfamily Caesalpinioideae) species. AoB Plants 2012:pls044

Jiménez-Alfaro B, Silveira FA, Fidelis A, Poschlod P, Commander LE (2016) Seed germination traits can contribute better to plant community ecology. J Veg Sci 27:637–645

Joët T, Ourcival JM, Dussert S (2013) Ecological significance of seed desiccation sensitivity in Quercus ilex. Ann Bot 111:693–701

Joët T, Ourcival JM, Capelli M, Dussert S, Morin X (2016) Explanatory ecological factors for the persistence of desiccation-sensitive seeds in transient soil seed banks: Quercus ilex as a case study. Ann Bot 117:165–176

Jurado E, Flores J (2005) Is seed dormancy under environmental control or bound to plant traits? J Veg Sci 16:559–564. https://doi.org/10.1111/j.1654-1103.2005.tb02396.x

Khullar P, Thapliyal RC, Beniwal BS, Vakshasya RK, Sharma A (1991) Forest Seed. Indian Council of Forestry Research and Education, Dehradun, p 409

Khurana E, Singh JS (2001) Ecology of seed and seedling growth for conservation and restoration of tropical dry forest: a review. Environ Conserv 28:39–52

Kondo T, Mikubo M, Yamada K, Walck JL, Hidayati SN (2011) Seed dormancy in Trillium camschatcense (Melanthiaceae) and the possible roles of light and temperature requirements for seed germination in forests. Am J Bot 98:215–226

Konsam B, Phartyal SS, Kumar M, Todaria NP (2017) Life after fire for understory plant community in subtropical Chir pine forest of Garhwal Himalaya. Indian Forester 143:759–766

Konsam B, Phartyal SS, Todaria NP (2020) Impact of forest fire on soil seed bank composition in Himalayan Chir pine forest. J Plant Ecol 13:177–184

Leck MA (2012) Ecology of soil seed banks. Elsevier, New York
Lembrechts JJ, Aalto J, Ashcroft MB, De Frenne P, Kopecký M, Lenoir J et al (2020) SoilTemp: a global database of near-surface temperature. Glob Chang Biol 26:6616–6629
Lembrechts JJ, van den Hoogen J, Aalto J, Ashcroft MB, De Frenne P, Kemppinen J et al (2022) Global maps of soil temperature. Glob Chang Biol 28:3110–3144
Maharjan SK, Sterck FJ, Dhakal BP, Makri M, Poorter L (2021) Functional traits shape tree species distribution in the Himalayas. J Ecol 109:3818–3834
Marques A, Buijs G, Ligterink W, Hilhorst H (2018) Evolutionary ecophysiology of seed desiccation sensitivity. Funct Plant Biol 45:1083–1095
Matsuki S, Koike T (2006) Comparison of leaf life span, photosynthesis and defensive traits across seven species of deciduous broad-leaf tree seedlings. Ann Bot 97:813–817
Mendoza I, Peres CA, Morellato LPC (2017) Continental-scale patterns and climatic drivers of fruiting phenology: a quantitative Neotropical review. Global Planet Change 148:227–241
Miranda LD, Silva BT, Santos JP, Rodrigues MD, Trevisan SH, Menezes IS, Funch LS (2023) Seasonality, dispersal modes, and optimal germination times modulate the fruiting of tropical tree species. Rodriguésia 74:e00652022
Ng FSP (1978) Strategies of establishment in Malayan forest trees. In: Tomlinson PB, Zimmerman M (eds) Tropical trees as living systems, vol 1. Cambridge University Press, New York, pp 129–162
Ooi MK, Denham AJ, Santana VM, Auld TD (2014) Temperature thresholds of physically dormant seeds and plant functional response to fire: variation among species and relative impact of climate change. Ecol Evol 4:656–671
Ooi MK, Tangney R, Auld TD (2022) Fire and regeneration from seeds in a warming world, with emphasis on Australia. In: Baskin CC, Baskin JM (eds) Plant regeneration from seeds. Academic, London, pp 229–242
Pausas JG, Lamont BB (2022) Fire-released seed dormancy—a global synthesis. Biol Rev 97: 1612–1639
Phartyal SS, Kondo T, Baskin JM, Baskin CC (2009) Temperature requirements differ for the two stages of seed dormancy break in Aegopodium podagraria (Apiaceae), a species with deep complex morphophysiological dormancy. Am J Bot 96:1086–1095
Phartyal SS, Kondo T, Baskin CC, Baskin JM (2012) Seed dormancy and germination in the giant Himalayan lily (Cardiocrinum giganteum var. giganteum): an assessment of its potential for naturalisation in northern Japan. Ecol Res 27:677–690
Phartyal SS, Kondo T, Fuji A, Hidayati SN, Walck JL (2014) A comprehensive view of epicotyl dormancy in Viburnum furcatum: combining field studies with laboratory studies using temperature sequences. Seed Sci Res 24:281–292
Phartyal SS, Rosbakh S, Ritz C, Poschlod P (2020) Ready for change: seed traits contribute to the high adaptability of mudflat species to their unpredictable habitat. J Veg Sci 31:331–342
Poschlod P, Tackenberg O, Bonn S (2010) Plant dispersal potential and its relation to species frequency and coexistence. In: Maarel E (ed) Vegetation ecology. Blackwell Science, Nova Jersey, pp 147–171
Poschlod P, Mašková T, Chen SC, Phartyal S, Rosbakh S, Silveira FA, Vandelook F (2025) A handbook for standardised measurements of regenerative plant functional traits. EcoEvoRxiv. https://doi.org/10.32942/X27W7D
Poschlod P, Abedi M, Bartelheimer M, Drobnik J, Rosbakh S, Saatkamp A (2013) Seed ecology and assembly rules in plant communities. In: Evd M, Franklin J (eds) Vegetation ecology, 2nd edn. Wiley-Blackwell, Chichester
Ralhan PK, Khanna RK, Singh SP, Singh JS (1985) Phenological characteristics of the tree layer of Kumaun Himalayan forests. Vegetatio 60:91–101
Rao PB (1984) Regeneration of some trees of western Kumaun Himalaya. Ph.D. Thesis, Kumaun University, Nainital, India
Rawal RS, Bhatt ID, Sekar KC, Nandi SK (2013) The Himalayan biodiversity: richness, representativeness, uniqueness and life-support values. GB Pant Institute of Himalayan Environment and Development (GBPIHED), Kosi-Katarmal, Almora, Uttarakhand, India, p 84

Reich PB, Walters MB, Ellsworth DS (1992) Leaf lifespan in relation to leaf, plant, and stand characteristics among diverse ecosystems. Ecol Monogr 62:365–392
Rosbakh S, Phartyal SS, Poschlod P (2020) Seed germination traits shape community assembly along a hydroperiod gradient. Ann Bot 125(1):67–78
Rosbakh S, Chalmandrier L, Phartyal S, Poschlod P (2022) Inferring community assembly processes from functional seed trait variation along elevation gradient. J Ecol 110:2374–2387
Rosbakh S, Carta A, Fernández-Pascual E, Phartyal SS, Dayrell RL, Mattana E, Baskin C (2023) Global seed dormancy patterns are driven by macroclimate but not fire regime. New Phytol 240(2):555–564
Royal Botanic Gardens Kew (2017) Seed Information Database (SID). Version 7.1. Available at: http://data.kew.org/sid/. Accessed on 15 July 2017
Rubio de Casas R, Willis CG, Pearse WD, Baskin CC, Baskin JM, Cavender-Bares J (2017) Global biogeography of seed dormancy is determined by seasonality and seed size: a case study in the legumes. New Phytol 214(4):1527–1536
Saatkamp A, Cochrane A, Commander L, Guja LK, Jimenez-Alfaro B, Larson J, Walck JL (2019) A research agenda for seed-trait functional ecology. New Phytol 221(4):1764–1775
Salazar A, Goldstein G, Franco AC, Miralles-Wilhelm F (2011) Timing of seed dispersal and dormancy, rather than persistent soil seed-banks, control seedling recruitment of woody plants in Neotropical savannas. Seed Sci Res 21(2):103–116
Sautu A, Baskin JM, Baskin CC, Condit R (2006) Studies on the seed biology of 100 native species of trees in a seasonal moist tropical forest, Panama, Central America. For Ecol Manage 234(1–3):245–263
Sautu A, Baskin JM, Baskin CC, Deago J, Condit R (2007) Classification and ecological relationships of seed dormancy in a seasonal moist tropical forest, Panama, Central America. Seed Sci Res 17(2):127–140
Schär C, Vidale PL, Lüthi D, Frei C, Häberli C, Liniger MA, Appenzeller C (2004) The role of increasing temperature variability in European summer heatwaves. Nature 427(6972):332–336
Shekhar C, Ginwal HS, Meena RK, Shankhwar R, Martins-Ferreira MAC, Pandey S, Bhandari MS (2022) Spatio-temporal distribution of broad-leaved Quercus semecarpifolia indicates altitudinal shift in northwestern Himalayas. Plant Ecol 223(6):671–697
Silveira FA, Ribeiro RC, Oliveira DM, Fernandes GW, Lemos-Filho JP (2012) Evolution of physiological dormancy multiple times in Melastomataceae from Neotropical montane vegetation. Seed Sci Res 22(1):37–44
Silveira FA, Fuzessy L, Phartyal SS, Dayrell RL, Vandelook F, Vázquez-Ramírez J, Saatkamp A (2023) Overcoming major barriers in seed ecology research in developing countries. Seed Sci Res 33:172–181
Simons AM, Goulet JM, Bellehumeur KF (2010) The effect of snow depth on overwinter survival in Lobelia inflata. Oikos 119(10):1685–1689
Singh SP (2014) Attributes of Himalayan forest ecosystems: they are not temperate forests. Proc Indian Natl Sci Acad 80:221–233
Singh JS, Singh SP (1987) Forest vegetation of the Himalaya. Bot Rev 53:80–192
Singh JS, Singh SP (1992a) Forests of Himalayas, structure, functioning and impact of man. Gyanodaya Prakashan, Nainital, India
Singh JS, Singh VK (1992b) Phenology of seasonally dry tropical forest. Curr Sci 62:684–689
Singh SP, Thadani R (2013) Valuing ecosystem services flowing from Indian Himalayan states for incorporating into National accounting. In: Lowman M, Devy S, Ganesh T (eds) Tree Tops at Risk. Springer, New York, pp 423–432
Singh SP, Thadani R (2024) A leaf based classification for Himalayan forests. Int J Ecol Environ Sci 50(1):11–31
Singh SP, Phartyal SS, Rosbakh S (2017a) Tree seed traits' response to monsoon climate and altitude in Indian subcontinent with particular reference to the Himalayas. Ecol Evol 7(18): 7408–7419

Singh SP, Singh RD, Gumber S, Bhatt S (2017b) Two principal precipitation regimes in Himalayas and their influence on tree distribution. Trop Ecol 58(4):679–691

Thadani R, Ashton PMS (1995) Regeneration of banj oak (Quercus leucotrichophora A. Camus) in the central Himalaya. For Ecol Manage 78(1-3):217–224

Thapliyal RC, Phartyal SS (2005) Dispersal and germination syndromes of tree seeds in a monsoonal forest in northern India. Seed Sci Res 15:29–42

Thapliyal RC, Thapliyal M, Bahar N, Naithani KC, Bist JPS (2003) A handbook for testing Indian tree seeds. Forest Research Institute, Dehradun, p 40

Thompson K, Bakker JP, Bekker RM (1997) The soil seed banks of North West Europe: methodology, density and longevity. Cambridge University Press, Cambridge

Troup RS (1921) The silviculture of Indian trees. Published under the authority of His Majesty's Secretary of State for India in Council. Smithsonian Institution

Tweddle JC, Dickie JB, Baskin CC, Baskin JM (2003) Ecological aspects of seed desiccation sensitivity. J Ecol 91:294–304

Van der Pijl L (1982) Principles of dispersal in higher plants, 3rd edn. Springer, New York

Van Schaik CP, Terborgh JW, Wright SJ (1993) The phenology of tropical forests: adaptive significance and consequences for primary consumers. Annu Rev Ecol Syst 24:353–377

Vázquez-Yanes C, Orozco-Segovia A (1982) Seed germination of a tropical rain forest pioneer tree (Heliocarpus donnell-smithii) in response to diurnal fluctuation of temperature. Physiol Plant 56(3):295–298

Vázquez-Yanes C, Orozco-Segovia A (1993) Patterns of seed longevity and germination in the tropical rainforest. Annu Rev Ecol Syst 24:69–87

Walck JL, Hidayati SN (2022) Plant regeneration from seeds in the temperate deciduous forest zone under a changing climate. In: Baskin CC, Baskin JM (eds) Plant regeneration from seeds. Academic, London, pp 131–143

Walck JL, Hidayati SN, Dixon KW, Thompson K, Poschlod P (2011) Climate change and plant regeneration from seed. Glob Change Biol 17(6):2145–2161

Wani SA, Khuroo AA, Zaffar N, Rafiqi S, Farooq I, Afzal S, Rashid I (2024) Data synthesis for biodiversity science: a database on plant diversity of the Indian Himalayan Region. Biodivers Conserv 33(12):3377–3397. https://doi.org/10.1007/s10531-024-02784-2

Wenny DG (2001) Advantages of seed dispersal: a re-evaluation of directed dispersal. Evol Ecol Res 3(1):37–50

Willson MF, Traveset A (2000) The ecology of seed dispersal. In: Fenner M (ed) Seeds: the ecology of regeneration in plant communities. CABI, Wallingford, pp 85–110

Willson MF, Rice BL, Westoby M (1990) Seed dispersal spectra: a comparison of temperate plant communities. J Veg Sci 1(4):547–562

Wyse SV, Dickie JB (2018) Ecological correlates of seed dormancy differ among dormancy types: a case study in the legumes. New Phytol 217(2):477–479

Yadav H, Phartyal SS, Iwachido Y, Sasaki T (2024) Regional assessment of diversity and distribution of native tree species in South Asia. Biodivers Conserv 33:379–396. https://doi.org/10.1007/s10531-023-02756-y

Yang W, Liu F, Zhang S, An S (2013) Dispersal and germination syndromes of tree seeds in a seasonal evergreen monsoon rainforest on Hainan Island, China. Seed Sci Res 23(1):41–55

Zhang YL, Moser B, Li MH, Wohlgemuth T, Lei JP, Bachofen C (2020) Contrasting leaf trait responses of conifer and broadleaved seedlings to altered resource availability are linked to resource strategies. Plants 9(5):621

Zhang Y, Liu Y, Sun L, Baskin CC, Baskin JM, Cao M, Yang J (2022) Seed dormancy in space and time: global distribution, paleoclimatic and present climatic drivers, and evolutionary adaptations. New Phytol 234(5):1770–1781. https://doi.org/10.1111/nph.18099

Zobel DB, Singh SP (1997) Himalayan forests and ecological generalisations. Bioscience 47(11): 735–774

MIX
Papier aus verantwortungsvollen Quellen
Paper from responsible sources
FSC® C105338

If you have any concerns about our products,
you can contact us on
ProductSafety@springernature.com

In case Publisher is established outside the EU,
the EU authorized representative is:
Springer Nature Customer Service Center GmbH
Europaplatz 3, 69115 Heidelberg, Germany

Printed by Libri Plureos GmbH
in Hamburg, Germany